Statistics for Biology and Health

Series Editors

Mitchell Gail, Division of Cancer Epidemiology and Genetics, National Cancer Institute, Rockville, MD, USA

Jonathan M. Samet, Department of Environmental & Occupational Health, University of Colorado Denver - Anschutz Medical Campus, Aurora, CO, USA

Statistics for Biology and Health (SBH) includes monographs and advanced textbooks on statistical topics relating to biostatistics, epidemiology, biology, and ecology.

More information about this series at http://link.springer.com/series/2848

Matthew P. Fox • Richard F. MacLehose
Timothy L. Lash

Applying Quantitative Bias Analysis to Epidemiologic Data

Second Edition

 Springer

Matthew P. Fox
Department of Epidemiology
Boston University School of Public
Health
Boston, MA, USA

Timothy L. Lash
Department of Epidemiology
Rollins School of Public Health, Emory
University
Atlanta, GA, USA

Richard F. MacLehose
Division of Epidemiology and Community
Health
University of Minnesota School of Public
Health
Minneapolis, MN, USA

ISSN 1431-8776   ISSN 2197-5671   (electronic)
Statistics for Biology and Health
ISBN 978-3-030-82675-8   ISBN 978-3-030-82673-4   (eBook)
https://doi.org/10.1007/978-3-030-82673-4

This Springer imprint is published by the registered company Springer Nature Switzerland AG
The registered company address is: Gewerbestrasse 11, 6330 Cham, Switzerland

Preface

Bias analysis quantifies the influence of systematic error on an epidemiology study's estimate of occurrence or estimate of association. The fundamental methods of bias analysis in epidemiology have been well-described for decades, yet are seldom applied in published presentations of epidemiologic research. More complex approaches to bias analysis—such as probabilistic bias analysis, direct bias modeling, and Bayesian bias analysis—appear even more rarely. We suspect that there are both supply-side and demand-side explanations for the scarcity of bias analysis. On the demand side, journal reviewers and editors seldom request that authors address systematic error aside from listing them as limitations of their study. This listing is often accompanied by explanations for why the limitations should not pose much concern. On the supply side, methods for bias analysis receive little attention in most epidemiology curricula and are often scattered throughout textbooks or absent from them altogether, and they cannot be implemented easily using standard statistical computing software. Although there has been some increase in the use of these methods since the publication of the first edition of this text, and the accompanying spreadsheets and code have improved the accessibility of the methods, they are still strongly underutilized. Our objective in this text is to reduce these supply-side barriers, with the hope that demand for quantitative bias analysis will follow.

The first edition of this text was launched after writing a few bias analysis papers in epidemiology journals. The American College of Epidemiology invited a day-long workshop on the methods of bias analysis, which became an outline, which in turn became the first edition of this text. The text begins with a description of the need for bias analysis and continues with two chapters on the legwork that must be done at the outset of a research project to plan and implement a bias analysis. This introductory section is followed by three chapters that explain simple bias analysis methods to address each of the fundamental threats to the validity of an epidemiology study's estimate of association: selection bias, classification errors, and uncontrolled confounding. We then extend the bias analysis methods from these early chapters to probabilistic bias analysis, direct bias modeling, Bayesian bias analysis, and ultimately multiple bias analysis methods. The book concludes with two chapters on the presentation and interpretation of the results of a bias analysis.

This second edition contains substantial additions beyond what appeared in the first edition. We include a wider array of simple bias analysis methods to address each of the fundamental threats to validity. We have divided the chapters on probabilistic bias analysis into a presentation of methods applied to summary level data and a presentation of methods applied to record level data. Conceptual connections between the two approaches are woven into all chapters. We have added a chapter on direct bias modeling, with particular emphasis on regression methods to address mismeasurement of continuous exposures, covariates, and outcomes. We have also added a chapter on Bayesian bias analysis methods, which in many ways provide theoretical underpinnings for many of the other methods in this book.

Readers might use the text as an independent resource to address bias analysis as they conduct epidemiologic research, as a secondary text in a class on epidemiologic methods, or as the central text in an advanced class on data analysis in epidemiologic research that focuses on bias analysis. We hope that students will find, as we have, that once they have completed one bias analysis, it becomes hard to imagine analyzing epidemiologic data without it. We have aimed the text at readers who have familiarity with epidemiologic research and intermediate data analysis skills. For those without those skills, we suggest a comprehensive methods text, such as *Modern Epidemiology*, which can be used in conjunction with this text to provide a foundation in epidemiologic terminology, study design, and data analysis.

An important adjunct resource for this textbook is the suite of freely available spreadsheets and software available for download at https://sites.google.com/site/biasanalysis/. We have endeavored to provide spreadsheets, SAS, R code, and Stan and JAGS code to replicate the analyses presented herein. We encourage readers to download the software, follow the examples in the text, and then modify the code to implement their own bias analysis. We would be delighted to hear from anyone who improves the tools or detects an error, and we will post revised tools as they become available. Likewise, we welcome comments, criticisms, and errata regarding the text from readers and will maintain a log of this feedback at the same website.

Funding for the first edition, some of which now carries over to this second edition, was made possible by grant G13LM008530 from the National Library of Medicine, NIH, DHHS. The views expressed in any written publication, or other media, do not necessarily reflect the official policies of the Department of Health and Human Services, nor does mention by trade names, commercial practices, or organizations imply endorsement by the U.S. Government.

Boston, MA, USA   Matthew P. Fox
Minneapolis, MN, USA   Richard F. MacLehose
Atlanta, GA, USA   Timothy L. Lash

Acknowledgments

The authors gratefully acknowledge the important contribution of Aliza K. Fink to the first edition of this text. Although she did not participate in preparing this second edition, her work and ideas remain foundational to its content.

We also thank our friends and colleagues who contributed to the text directly or indirectly. We appreciate Charles Poole's suggestion to the American College of Epidemiology that a short course on bias analysis would be of value, and we appreciate John Acquavella's decision to accept that suggestion on behalf of the college. As noted in the Preface, this day-long seminar provided the outline for the first edition. Sander Greenland participated in the development and presentation of the American College of Epidemiology workshops and has been instrumental in improving the methods of bias analysis. We are grateful for his input and dedication to the topic. We also thank our colleagues who have a particular interest in bias analysis methods; they have challenged us to develop our ideas and to communicate them clearly. We cannot list them all, so we acknowledge especially Carl Phillips, George Maldonado, Anne Jurek, Ken Rothman, Rebecca Silliman, Soe Soe Thwin, Dan Brooks, Steve Cole, Tyler VanderWeele, Thomas Ahern, Lindsay Collin, Paul Gustafson, Lawrence McCandless, and Hailey Banack. We especially thank Paul Gustafson for reviewing parts of Chapters 8 and 11, and Ghassan Hamra and Haitao Chu for their review of Chapter 11.

These collaborations and reviews have been immensely helpful in preparing the text; the authors are responsible for any error or omission.

We further acknowledge that parts of some chapters borrow, with permission, from work we have previously published. In particular:

Parts of Chapter 1 include content that first appeared in Lash TL. Heuristic thinking and inference from observational epidemiology. *Epidemiology*. 2007;18:67–72.

Parts of Chapters 13 and 14 include content that first appeared in Lash TL, Fox MP, MacLehose RF, Maldonado G, McCandless LC, Greenland S. Good practices for quantitative bias analysis. *Int J Epidemiol*. 2014;43:1969–85.

Contents

Chapter 1
Introduction, Objectives, and an Alternative

Introduction: Biases in Health Research

Epidemiologic investigations have a direct impact on all aspects of health. Studies of social, environmental, behavioral, medical, and molecular factors associated with the incidence of disease or disease outcomes lead to interventions aimed at preventing the disease or improving its outcomes. However, biases are commonplace in epidemiologic research. Even the most carefully designed and conscientiously conducted study will be susceptible to some sources of systematic error. Study participants may not answer survey questions correctly (information bias), there may be confounders that were uncontrolled by study design or left unmeasured (uncontrolled confounding), or participants may have been enrolled differentially or incompletely followed (selection bias). Every epidemiologic study has some amount of information bias, selection bias and uncontrolled confounding.

Although randomized controlled trials (RCTs) are often thought of as the "gold standard" in health research, they are subject to these biases as well. If blood tests are not perfect or subjects incorrectly respond to survey questions, RCTs are subject to information bias. If subjects are lost to follow-up, RCTs are subject to selection bias. The most attractive aspect of an RCT is that it can, in some settings, effectively deal with unmeasured confounding. The process of randomizing assignment to the treatment groups will yield an expectation of equal distribution of all confounders (both measured and unmeasured) between the intervention and control arms of the study at the outset of follow-up. However, this advantage only holds in expectation; in small or even moderate RCTs, confounding can remain a concern. A second advantage is that the randomization provides a rationale for assigning probability distributions to each of the potential study results, and thereby lends meaning to the conventional statistical analyses [1].

In interpreting epidemiologic data, then, scientists are often confronted with the question of how to incorporate quantitative consideration of these biases into results and inferences. Unfortunately, as discussed below, a common approach is to assume

© Springer Nature Switzerland AG 2021
M. P. Fox et al., *Applying Quantitative Bias Analysis to Epidemiologic Data*,
Statistics for Biology and Health, https://doi.org/10.1007/978-3-030-82673-4_1

that these biases do not exist or to simply describe the possibility of these biases in the discussion section of the paper. Neither of these approaches allows analysts or readers of research to understand accurately the impact these biases might have had on their inferential findings.

The objective of this text is to reduce the barriers to routine implementation of quantitative bias analysis. We have collected existing methods of quantitative bias analysis, explained them, illustrated them with examples, and linked them to tools for implementation. We note that worked examples are provided only for the purpose of illustrating the methods. We make no claim to the validity or causal interpretation of examples before or after implementation of the bias analysis methods. The bias analysis methods are an aid towards causal inference goals, but such inference requires contextual knowledge and an understanding of the complete evidence base, and these are beyond the scope of this book, so not incorporated into the worked examples.

The second chapter provides a guide to choosing the method most appropriate for the problem at hand and for making inference from the method's results. Chapter 3 provides possible sources of data to support the values assigned to the parameters of the bias models used in bias analyses. Chapters 4, 5, and 6 present simple methods to bias-adjust aggregate data for selection bias, uncontrolled confounding and information bias. Chapters 7, 8, 9, 10, 11, and 12 present more sophisticated methods to adjust for biases at the summary or individual record level while also incorporating random error. Chapters 13 and 14 present general guidance on good practices for implementing bias analysis and for presenting and interpreting the results in research papers. For many of the bias analysis techniques presented in the book, we provide worked examples to aid the reader in following along and eventually implementing them. We also provide software tools and code on the text's website to automate the analysis in familiar software such as Excel, SAS and R and to provide output that reduces the resources required for implementation and presentation.

There are sources of bias that we do not address. For example, none of the methods described herein would be useful to detect or address fraudulent research practices, selective presentation of results, or their less nefarious and more prevalent cousins HARKing [2] and p-hacking [3, 4]. Properly specifying a model to link the independent variables with dependent variables using the proper functional form, while accounting for the assumed causal structure, is also critical to valid estimation. Model specification and the use of causal graphs to identify a sufficient set of variables for control have both been extensively addressed in other texts [5, 6], so are not addressed here. Publication bias and other meta forces that influence the accumulated evidence base are also not addressed here. Our focus is strictly on the three most common threats to internal validity—selection bias, uncontrolled confounding, and information bias—and quantitative methods to bias-adjust for them.

Statistical Inference in Public Health Research

To evaluate the importance of biases in epidemiologic research, and methods to adjust for those biases, we must first review some fundamental aspects of statistical inference. Statistical inference in epidemiologic research often focuses on using data collected in a study to estimate an effect of interest (such as a risk or rate difference, risk or rate ratio, hazard ratio or odds ratio) and a corresponding estimate of variance for that parameter to measure the random error. In this section, we will denote the true value of the parameter of interest (such as the risk difference), which we can never know, as θ. This unknown parameter θ that we wish to estimate is known as an estimand, and we will generally use Greek letters to denote estimands. The choice of which estimand to focus on should be guided by scientific and practical concerns, such as ease of interpretation, as well as statistical concerns, such as collapsibility in the absence of confounding [7]. The causal effect θ can never be directly seen and instead we measure it using an estimator, $\hat{\theta}$, where the hat (or circumflex) is a reminder that this is an estimate of the true parameter and not the estimand itself.

Statistical estimators have properties that are important to the discussion of quantitative bias analyses; among these are bias, variance, and mean squared error (MSE). If we use the estimator $\hat{\theta}$ for inferences about a true parameter θ, we define the estimator as unbiased if it equals the truth in expectation: $E(\hat{\theta}) = \theta$. The amount of bias is quantified as the difference between an estimator's expectation and the truth: $Bias(\hat{\theta}) = E(\hat{\theta}) - \theta$. For example, if we repeated a study in the same way many times (and index the studies using the subscript i) with no errors other than sampling error, the average of the $\hat{\theta}_i$ across the studies would be expected to equal the truth. Information bias, uncontrolled confounding, and selection bias are three processes that can bias an estimator, such that we would expect $E(\hat{\theta}) \neq \theta$. Epidemiologists refer to this type of error as systematic error. Much of this book is aimed at proposing alternative estimators that are expected to be less biased in the presence of information bias, uncontrolled confounding, or selection bias.

The variance of an estimator is a description of how far an estimator deviates from its mean, on average: $V(\hat{\theta}) = E([\hat{\theta} - E(\hat{\theta})]^2)$. Variance quantifies variability around the mean of an estimator: if a study were repeated many times, how far would individual estimates be from their overall mean, on average. In epidemiology, variance is viewed as arising from the random occurrence of an outcome or sampling error, even if it is not the result of a formal random sampling process such as random treatment assignment or random sampling of individuals. It is possible to reduce random error, and hence variance, of an estimator by increasing the sample size. In general, however, systematic error does not shrink with increasing sample size. It is important to note that variance is not a measure of the variability of the estimator around the truth. This latter quantity is referred to as the mean squared error: $MSE(\hat{\theta}) = E([\hat{\theta} - \theta]^2)$. There is a well-known relationship between variance, bias and MSE: $MSE(\hat{\theta}) = V(\hat{\theta}) + [bias(\hat{\theta})]^2$. MSE can be reduced by reducing the variance (e.g., by increasing the sample size) and by decreasing the bias (e.g., by controlling for confounding factors). While many estimators we use are chosen specifically because they are unbiased, this is not always the case. Some estimators are chosen to be biased by design; a slightly biased estimator may have smaller MSE if the bias is offset by a reduction in variance. Hierarchical models, for example, may produce estimates that are somewhat biased but have lower overall MSE because they are accompanied by a large reduction in variance [8].

Formal statistical inference is often taught as being divided into two schools of thought: frequentist and Bayesian statistics [9]. Frequentist statistics deals with expectations over long-run series of events and is far more commonly applied than Bayesian statistics as an inferential framework in applied epidemiologic research. Subjective Bayesian inference requires specification of personal prior beliefs before beginning an analysis. Both approaches to inference can be useful to epidemiologists, and we will discuss each in detail. First, however, we point out that while there has been much written about disagreements between frequentist and Bayesian statisticians, we view these disputes as largely misguided. From the point of view of practicing epidemiologists, both frequentist and Bayesian inference have a great deal to offer and each has its own limitations. Bayesian inference provides relatively straightforward techniques for generating new estimators, which is particularly useful in complicated settings like bias analysis; however, its reliance on prior information makes it susceptible to manipulation. Frequentist inference is particularly useful in judging how well those estimators work (e.g., through estimates of MSE, bias and variance); however, the long-run interpretation of probability makes interpreting frequentist statistics problematic in a single study [8, 10, 11].

Frequentist theory is the most common approach to statistical inference in epidemiologic research. From a frequentist perspective, probability statements relate to expectations over long-run series of events; if the probability of a coin flip resulting in heads is 50%, that means if a person flips an unbiased coin an infinite number of times half of the flips will be heads. This is a clear interpretation that is useful in some situations and has led to seemingly endless suffering in others. The frequentist interpretation will be most useful when we truly care about repeated sampling. One such instance has already been seen: when we are interested in evaluating an estimator, our attention is focused on how the estimator behaves in a long run series of events (such as the bias, variance and MSE of the estimator). However, statistical inference in epidemiology often focuses on the interpretation from one set of study results and not from a hypothetical infinite number of study results. In this circumstance, frequentist statistical interpretations can be confusing.

A common way to express the random error in an epidemiologic study is to re-express the variance in terms of a 95% confidence interval. A frequentist definition of a 95% confidence interval is that, if the underlying statistical model is correct and there is no bias, a confidence interval derived from a valid test will, over unlimited repetitions, contain the true parameter with a frequency no less than its confidence level (*e.g.*, 95%). This long-run interpretation of the confidence interval is useful when thinking about what might happen if every analyst investigating a hypothesis estimated the same effect with a design free of bias, but less so when interpreting results in one given study. A common *misinterpretation*, for example, is that a 95% confidence interval for a given study is the interval within which we are 95% certain that the causal effect resides. This type of interpretation, while appealing, is incorrect and closer in line with Bayesian interpretations of intervals.

A simple example illustrates the difference. Imagine a single experiment in which an *unbiased* coin is flipped 100 times and 40 heads are observed. What is the probability of observing heads when the coin is next flipped? The data in this experiment provide an estimate of 40% (40/100), with a Wald 95% frequentist confidence interval of (30.4%, 49.6%). One might be tempted to say that the probability of heads for the next flip is 40%, and that the probability that the true probability is 50% is quite small because the value 50% lies outside the 95% confidence interval. However, recognize that the experimenter has perfect prior knowledge; the coin was defined as unbiased, so the probability of heads for the flip is, by definition, 50%. The probability that the truth is in the 95% frequentist confidence interval is, in fact, 0%. Were this experiment repeated an infinite number of times, 95% of the frequentist confidence intervals generated by this design would cover the true value of 50%; this just happens to be one of the 5% of the experiments where the interval did not cover the truth. Unfortunately, there is nothing about the experiment that would allow the analyst to diagnose that it was one of the 5% of experiments where the interval failed to cover the truth, except perfect prior knowledge.

A correct interpretation of a frequentist 95% confidence interval for a single study would therefore be that the probability is either 0 or 1 that the true effect lies in the interval. This interpretation is uninteresting and, as a result, epidemiologists have sometimes relied on very informal interpretations of confidence intervals. Wide confidence intervals reflect imprecise effect estimates, so suggest that random error could have had substantial impact on the observed result. Narrow confidence intervals reflect precise effect estimates, so suggest that random error was less likely to have had a substantial impact on the observed result. Statements about what is precise or not should be made in the context of the existing evidence base on the topic, and in relation to the size of the effect estimate. If we conducted a study and estimated a risk ratio of 1.5 with 95% confidence interval 1.1–2.3, we might view this as relatively precise with a ratio of the upper to lower confidence limit of 2.3/ 1.1 = 2.1. However, if there was already such a substantial literature in the field that a previous analyst had conducted a meta-analysis and found a summary risk ratio of 1.4 with 95% confidence interval of 1.4–1.6, we might view our result as less precise. A risk ratio of 1.01 with 95% confidence interval 0.9–1.1 might be viewed as imprecise if one is interested in the departure of the estimate from the null (e.g., a study relating air pollution to cardiovascular mortality with a time series analysis), and might be viewed as precise if one is interpreting the estimate as null (e.g., a study relating air pollution to death in a case-crossover study of deaths in automobile accidents).

The discussion of confidence intervals as precise or not is therefore subjective. Greenland [1] and others [12] have suggested an alternative interpretation of a 95% confidence interval as the set of point estimates that are broadly compatible with the observed data, assuming a valid statistical model and that there is no other source of error, other than randomness. This interpretation has the advantage of forcing epidemiologists to carefully consider all points within the confidence interval as being compatible with the data, as opposed to focusing exclusively on the point estimate at the center of the confidence interval. It also forces epidemiologists to consider the validity of the underlying statistical model and, most importantly for this text, whether the assumption of no other sources of error is defensible.

A related frequentist concept is the p-value, a continuous measure of how well a test statistic conforms to a set of assumptions and a test hypothesis. If the p-value is incompatible with any of the assumptions or the test hypothesis, a small p-value will be expected. Particular focus is often placed on the null hypothesis (that the risk difference is 0 or the risk ratio is 1, for example) as the most important hypothesis and the unfortunate common practice is that all other assumptions are ignored. In this case, a small p-value is taken as evidence against the null hypothesis. However, as has been explained elsewhere [12, 13], the overall set of assumptions and hypotheses typically include assumptions of no selection bias, no information bias, no uncontrolled confounding, as well as valid modeling assumptions. A low p-value could result from any combination of the assumptions or hypotheses being false. The attention to the null hypothesis is symptomatic of a general unhealthy focus on null hypothesis significance testing in the field and, in the context of this book, it is important to remember that low p-values can be obtained simply by having bias in a study.

Subjective Bayesian inference is an alternative model for statistical inference (there are other schools of Bayesian inference, as well, such as objective Bayes, but those are less germane to this text). Rather than base probabilities on a long-run series of events as in frequentist statistics, Bayesian statistics define probabilities in terms of belief. If the probability of a coin flip returning heads is 50%, that equates to a personal belief that a coin flip returning heads is 50% [10, 14]. This interpretation can be formalized in terms of betting; a 50% probability is one in which a gambler would view even odds on either outcome occurring as a fair bet. Bayesian inference, not surprisingly, relies on Bayes theorem:

$$f(\theta|x) = \frac{f(x|\theta)f(\theta)}{\int f(x|\theta)f(\theta)d\theta}$$

where θ is the parameter of interest and x are the observed data. The posterior distribution, $f(\theta|x)$, represents belief in the parameter of interest after seeing data. It is proportional to the likelihood of the observed data given a particular parameter value, $f(x|\theta)$ times the prior distribution, $f(\theta)$. The integral in the denominator of Bayes theorem is often impossible to compute in closed form. Instead, computer intensive methods like Markov chain Monte Carlo techniques or approximations like data augmentation or nested Laplace approximations are used. The prior distribution in subjective Bayesian inference is a distribution that represents belief about the effect of the parameter of interest prior to conducting the study. The prior distribution encompasses a potential strength and weakness of the Bayesian approach. On the one hand, accurately specifying a prior distribution can result in a posterior estimator with lower MSE than a corresponding frequentist estimator [8, 15]. For example, in the coin flipping example, if one professed some belief that the coin was unbiased, the resulting posterior distribution would be centered on a value somewhere between the observed (frequentist) point estimate of 40% and the prior belief of 50%. The Bayes result would decrease the bias and decrease the MSE, in this case, because the prior distribution was correctly chosen to be centered on the truth. Indeed, Bayesian estimates can perform well even if the prior is not perfectly centered on the truth; being close may be good enough to improve the performance of the estimator in many situations. On the other hand, if a prior distribution is poorly chosen, results from the posterior distribution could be gravely misleading. In the coin flipping example, if one believed strongly that the coin was unfair and was designed to only come up heads 25% of the time when, truly, the coin was fair, then one would arrive at a posterior belief that was even further from the true value of 50%. The error in the posterior distribution derives entirely from an error in prior belief, not from errors or biases in the data. Finally, imagine that you were told at the outset that the coin was *unfair* and that its flipping would be computer simulated. The only value we can rule out as the true probability of heads is 50%; all other values for heads from >0% to <100% are equally likely. Now the prior information about the coin is almost completely uninformative. A reasonable estimate of the probability that the next flip will be heads, given the observed data, is 40% and the 95% frequentist interval (30.7%, 49.6%) may also be an accurate representation of our posterior Bayesian belief that the true probability of a coin coming up heads is in that interval with 95% probability. When the prior information is substantial (*i.e.*, that the coin is unbiased), the data-based estimate and interval contribute nothing to the estimate of the probability of heads for the next flip. When the prior information is non-informative, the data-based estimate and interval contribute almost all of the information to the estimate of the probability of heads for the next flip. We discuss Bayesian inference in more detail in Chapter 11.

A Bayesian approach can be very flexible; specifying a more complicated model, such as the bias models we describe in this book, does not always substantially increase the overall difficulty of a Bayesian estimation process. This potential flexibility of the Bayesian approach is attractive when conducting quantitative bias analysis. As will be discussed in subsequent chapters, adjusting for biases relies on assigning values to the bias parameters of a bias model (such as the sensitivity and specificity of misclassification). Using Bayesian methods, it is relatively straightforward to assign distributions of values to these bias parameters as prior distributions. While explicit Bayesian methods for bias analysis are not mentioned until Chapter 11 of this book, many of the methods introduced in earlier chapters have a direct relationship to Bayesian methods. In particular, there is good evidence that the majority of the methods to adjust for uncontrolled confounding, selection bias and information bias in this book are good approximations of Bayesian methods [16–18].

The Treatment of Uncertainty in Nonrandomized Research

The objective of epidemiologic research is to obtain a valid, precise, and generalizable estimate of the distribution of a health parameter in a population (descriptive epidemiologic research) or of the effect of an exposure on the occurrence of a health outcome (etiologic epidemiologic research). A valid estimate in epidemiology is one that is not biased by the systematic errors mentioned above: uncontrolled confounding, information bias, or selection bias. A precise estimate is one that has a relatively small standard error or a narrow confidence interval. A generalizable estimate is one that can be used to make inferences in a target population of interest, which may be different than the study population. To accomplish this, analysts have a two-fold obligation. First, they must design their investigations to enhance the precision and validity of the effect estimate that they obtain. Second, recognizing that no study is perfect, they must inform interested parties (collaborators, colleagues, and consumers of their research findings) how near the precision, validity, and generalizability objectives they believe their estimate of effect might be. Note that we presume throughout a well-defined research question of health relevance, a topic that itself has been widely discussed [19–21].

To enhance the precision of an effect estimate (*i.e.,* to reduce random error), epidemiologists design their studies to gather as much information as possible [22], apportion the information efficiently among the strata of variables that affect the outcome [23], and undertake precision-enhancing analyses such as pooling [24], regression [25], and scoring [25]. The methods to enhance a study's validity and the precision of its estimate of effect are well-described in epidemiologic textbooks such as *Modern Epidemiology* [6]. Even with an efficient design and analysis, epidemiologists customarily present a quantitative assessment of the remaining random error about an effect estimate. Although there has been considerable [26–29] and continuing [12, 30–34] debate about methods of describing random error, as discussed above, a common way is to use frequentist confidence intervals [35].

To enhance the validity of an effect estimate (*i.e.,* to reduce systematic error), epidemiologists design their studies to assure comparability of the exposed and unexposed groups (through approaches like randomization, matching and restriction) [36, 37], reduce differential selection forces [38–41], and reduce measurement error by obtaining accurate information [42–44]. When validity might be compromised by confounding after implementation of the design, epidemiologists employ analytic techniques such as stratification and standardization [45] or regression-based equivalents of both [46] to improve the validity of the effect estimate. Analytic adjustments for selection forces or measurement error or uncontrolled confounding are seldom seen, and quantitative assessments of the remaining systematic error about an estimate are even more rare [47–51].

To enhance the generalizability of an effect estimate, some epidemiologic studies are designed to enroll a representative sample of subjects from the target population [21]. This representative sample can be enrolled as a simple random sample, or by stratified sampling such that strata with rare combinations of traits are over-represented, which can then be adjusted for the sampling by reweighting

[52, 53]. Representativeness of the study sample, or the ability to reweight, is paramount for surveillance or other descriptive epidemiology. Sampling and reweighting can be effectively incorporated into the design of etiologic studies as well; the case-cohort design provides a straightforward example. Because the sampling fraction is known, information on risks in the whole cohort can be recovered by reweighting controls by the inverse of the sampling fraction. More generally, inference to a target population can be obtained by oversampling some subgroups and then standardizing or reweighting the study data to match the target population distribution.

The importance of sampling to achieve representativeness in etiologic research may depend on the maturity of the research topic [54]. In the earliest research phases, analysts may be willing to trade representative study samples for efficient study samples, such as study samples enhanced in the extremes of the exposure distribution or study samples at particularly high risk for the outcome [21]. Once the nature and at least the order of magnitude of an effect are established by studies designed to optimize validity, transportability to other, unstudied target populations becomes a matter of reweighting from the distribution of effect modifiers in the studies' samples to the distribution of these modifiers in the target populations, assuming that effects have been well-measured within strata of all relevant effect modifiers. The importance of this reweighting to account for differences in the prevalence of effect modifiers has been well understood for some time, yet quantitative treatments to make inference from research study populations to target populations seldom appear.

Thus, the quantitative assessment of the error about an effect estimate usually reflects only the precision of the estimate, with little quantitative attention to the estimate's validity or generalizability. As noted above, much has been written and many examples proffered about the abuses made of these quantitative assessments of random error. The near complete absence of quantitative assessments of residual systematic error and lack of generalizability in published epidemiologic research has received much less attention. Several reasons likely explain this inattention. First, existing custom does not expect a quantitative assessment of the systematic error about an effect estimate. For example, the uniform requirements for manuscripts submitted to biomedical journals instructs authors to "quantify findings and present them with appropriate indicators of measurement error or uncertainty (such as confidence intervals)." [55] Conventional frequentist confidence intervals measure only residual random error and typically assume no systematic error—including measurement error—after customary regression adjustments. With no habit to breed familiarity, few epidemiologists are comfortable with the implementation of existing bias analysis methods. Second, the established methods for quantitative bias analysis require more time and effort than standard methods. Analysts have more demands on their time than ever before and pressure to publish is often intense, making quantitative bias analysis less attractive. Finally, the statistical computing tools often used by epidemiologists provide quantitative assessments of residual random error about effect estimates, but seldom contain automated methods of assessing residual systematic error. Quantitative assessment of random error is an established custom and requires little expertise or effort to obtain an apparently rigorous measurement of residual random error from widely available statistical computing software. Quantitative bias analysis to address systematic errors and lack of generalizability enjoy none of these advantages. The first edition of this text was accompanied by freely available software in Microsoft Excel, and since its publication numerous packages in statistical software including SAS, Stata and R have been developed to allow for simple and probabilistic bias analyses. All of these should make routine implementation of such quantitative bias analysis easier and less daunting. We maintain a website where we post links to these tools and files that apply them to examples in this book.

When Bias Analysis Will Be Most Useful

Although most studies would benefit from a careful identification of the sources of bias that affect its results, and quantification of the impact of those biases would help to elucidate both the direction and magnitude of the bias, bias analysis will be more useful in studies with certain characteristics. Medium to large studies with sufficient sample size such that standard errors are small and conventional frequentist 95% confidence intervals are narrow, are well suited for bias analysis. With increasing sample size, the random error approaches 0, but systematic error generally remains unchanged in expectation. In such cases, systematic errors become the dominant source of study error and it becomes imperative to assess the impact of systematic error through quantitative bias analysis. Bias analysis is particularly important because overreliance on statistical significance and incorrect interpretations of 95% confidence intervals can lead one to draw conclusions about the results of a study as if the study had no systematic error or as if systematic error could be separated from random error in drawing study conclusions. When this happens, because large studies will produce narrow confidence intervals and low p-values, analysts may develop a high level of confidence in biased results. Quantitative bias analysis can help to prevent this overconfidence by showing the impact that the bias may have had on study results, adjusting the study effect estimate for that bias, and widening intervals to account for the uncertainty created by systematic error. In the era of "big data," when massive datasets can be analyzed to produce very precise results, bias analysis has become increasingly important. Meta-analyses are another area in which bias analysis has much to contribute, as such studies pool the results of a body of studies on a topic and the resulting summary estimates often have small standard errors. However, this pooling does not overcome the systematic error in the individual studies, making careful incorporation of the impact of these systematic errors into point estimates and error intervals essential for appropriate inferences [46, 47]. We provide a more comprehensive guide for when bias analysis is most useful in Chapter 13.

Judgments Under Uncertainty

As emphasized above, epidemiologic research is an exercise in measurement. Its objective is to obtain a valid, precise, and generalizable estimate of either the occurrence of disease in a population or the effect of an exposure on the occurrence of disease or other health condition. Conventionally, epidemiologists present their measurements in three parts: a point estimate (*e.g.,* a risk ratio), a frequentist statistical assessment of the uncertainty (*e.g.,* a confidence interval, but also sometimes a p-value), and a qualitative description of the threats to the study's validity. This description is often accompanied by a judgment about the importance of these threats, and especially whether they should be considered as an important barrier to viewing the measurement as a reliable basis for inference pertaining to the research question. Every incentive aligns towards viewing them as unimportant, and this incentive structure favoring inferential overconfidence is compounded by fundamental human tendencies towards overconfidence when reasoning under uncertainty. This section briefly reviews these human tendencies with the goal of further motiving the use of quantitative bias analysis to offset them.

Without randomization of study subjects to exposure groups, point estimates, confidence intervals, and p-values lack a basis for frequentist interpretations, as described above [1]. Randomization and an assumption about the expected allocation of outcomes–such as the null hypothesis allow one to assign probabilities to the possible study outcomes. One can then compare the observed test statistic (such as the risk ratio divided by its standard error), with the distribution of possible study outcomes to estimate the probability of the observed test statistic, or those more extreme, under the test hypothesis. This comparison provides an aid to causal inference [1] because it provides a probability that the outcome distribution is attributable to chance lack of comparability between the exposed and unexposed groups as opposed to the effects of exposure, if no association truly existed. The comparison is therefore at the root of frequentist statistical methods and inferences from them; it provides a link between randomized trials and the statistical methods that produce measures of the extent of random error, such as confidence intervals. However, since these methods are based on assumptions of randomness, interpretation in observational epidemiology is problematic. Some have advocated interpreting these measures of variance and confidence intervals as conservative estimates [56]. We suggest that the lack of randomization should lead epidemiologists to interpret standard statistical methods with great caution. Not only do standard methods only focus on one type of error that is often not seen in epidemiologic studies, but they also generally ignore systematic error that is present. Results from these methods conflate the two types of error and proper inference therefore requires an educated guess about the strength of the systematic errors compared with the strength of the exposure effects.

These educated guesses can be accomplished quantitatively by likelihood methods [57–61], Bayesian methods [17, 62–64], regression calibration [65–68], missing data methods [69, 70], or Monte Carlo simulation [71–73] [*see* Greenland

(2005) for a review and comparison of methods [16]]. These methods will be described in later chapters. The conventional approach, however, is to make the guess qualitatively by describing the study's limitations. An assessment of the strength of systematic errors, compared with the strength of exposure effects, therefore becomes an exercise in reasoning under uncertainty. Human ability to reason under uncertainty has been well-studied and shown to be susceptible to systematic bias resulting in predictable mistakes. A brief review of this literature, focused on situations analogous to epidemiologic inference, suggests that the qualitative approach will frequently fail to safeguard against tendencies to favor exposure effects over systematic errors as an explanation for observed associations. Few graduate science programs require training in cognition and its influence on individual inferences that rest on uncertain evidence [74], which is one reason why we include this introduction to the topic here.

The Dual-Process Model of Cognition

A substantial literature from the field of cognitive science has demonstrated that humans are frequently biased in their judgments about probabilities and at choosing between alternative explanations for observations [75–77], such as epidemiologic associations. Some cognitive scientists postulate that the mind uses dual processes to solve problems that require such evaluations or choices [78–80]. The first system, labeled the "Associative System," uses patterns to draw inferences. We can think of this system as intuition, although any pejorative connotation of that label should not be applied to the associative system. The second system, labeled the "Rule-Based System," applies a logical structure to a set of variables to draw inferences. We can think of this system as reason, although the label alone should not connote that this system is superior. The Associative System is not necessarily less capable than the Rule-Based System; in fact, skills can migrate from the Rule-Based System to the Associative System with experience. The Associative System is in constant action, while the Rule-Based System is constantly monitoring the Associative System to intervene when necessary. This paradigm ought to be familiar; we have all said at some time "Wait a minute – let me think," by which we do not mean that we have not yet thought, but that we are not satisfied with the solution our Associative System's thought has delivered. After the chance to implement the Rule-Based System, we might say "On second thought, I have changed my mind," by which we mean that the Rule-Based System has overwritten the solution initially delivered by the Associative System.

The process used by the Associative System to reach a solution relies on heuristics. A heuristic reduces the complex problem of assessing probabilities or predicting uncertain values to simpler judgmental operations [81]. An example of a heuristic often encountered in epidemiologic research is the notion that nondifferential misclassification biases an association toward the null. Heuristics often serve us well because their solutions are correlated with the truth, but they can sometimes lead to systematic and severe errors [81]. Nondifferential and indepen-dent misclassification of a dichotomous exposure leads to associations that will be biased toward the null *on average*, but individual studies may be closer or further from the truth [82]. For example, any particular association influenced by nondifferential misclassification may not be biased toward the null [83], dependent errors in classification can substantially bias an association away from the null–even if classification errors are nondifferential [84], nondifferential misclassification of disease may not lead to any bias in some circumstances [43], and a true association may not provide stronger evidence against the null hypothesis than the observed association based on the misclassified data–even if the observed association is biased toward the null [85]. Application of the misclassification heuristic without deliber-ation can lead to errors in interpreting an estimate of the strength and direction of the bias [86], as is true for more general cognitive heuristics [81].

Cognitive scientists have identified several classes of general heuristics and biases, three of which are described below because they may be most relevant to causal inference based on observational epidemiologic results. While studies that have elicited an understanding of these heuristics have most often been conducted in settings that are not very analogous to causal inference using epidemiologic data, one such study of epidemiologists has been conducted and its results corresponded to results elicited in the cognitive science setting [87]. In addition, these heuristics have been shown to affect evidence-based forecasts of medical doctors, meteorologists, attorneys, financiers, and sports prognosticators [88]. It seems unlikely that epide-miologists would be immune.

Anchoring and Adjustment

The first heuristic relevant to causal inference based on observational epidemiologic results is called "anchoring and adjustment" [81]. When asked to estimate an unknown but familiar quantity, respondents use a heuristic strategy to select (or receive) an anchor, and then adjust outward from that anchor in the direction of the expected true value. Adjustments are typically insufficient. For example, one might be asked to give the year in which George Washington was elected as the first president of the USA [89]. Most respondents choose a well-known date in the era as the anchor:1776, the year that the USA declared independence. Respondents adjust upward to later years, because they know the US Constitution was not ratified in the same year. The average response equals 1779, and the correct value equals 1788. Why, on average, is the upward adjustment insufficient? The predictably insufficient adjustment may arise because respondents adjust outward from the anchor until their adjusted estimate enters a range they deem plausible. The true value, more often, lies toward the center of the plausible range. When the anchor is below the true value, as in the year that Washington was first elected, the estimate is predictably lower than the true value. Conversely, when the anchor is above the true value, the estimate is predictably higher than the true value. For example, one might be asked to give the temperature at which vodka freezes [89]. Most respondents choose the anchor to be 32 °F (0 °C), the temperature at which water freezes. Respondents adjust downward to lower temperatures, because they know alcohol freezes at a lower temperature than water. The average response equals 1.75 °F (−16.8 °C), and the correct value equals −20 °F (−28.9 °C). Importantly, the anchoring and adjustment heuristic operates in the same manner regardless of whether the anchor is self-generated or provided by an external source, so long as the respondent is aware of the anchor and it is on the same scale as the target [90].

How might the anchoring and adjustment heuristic affect inference from observational epidemiologic results? Consider the point estimate associating an exposure with a disease, derived from a study's results, to be an anchor. Further consider that interested parties (the investigator, collaborators, readers, and policymakers) may be aware of the direction of an expected bias (*e.g.*, toward the null). Can the interested parties be expected to adjust the point estimate sufficiently to account for the bias? An understanding of the anchoring and adjustment heuristic suggests that the adjustment might be predictably insufficient. Interested parties should be expected to adjust the association to account for the bias only so far as is plausible, which adjustment will, on average, be insufficient.

Overconfidence

The second bias relevant to causal inference based on observational epidemiologic results is called "overconfidence." When asked to estimate an unknown but familiar quantity, respondents can be trained to provide a median estimate (the estimate about which they feel it is as likely that the true value is higher as it is that the true value is lower), as well as an interquartile range. The interquartile range is defined by the respondent's estimate of the 25th percentile (the estimate about which they feel it is 75% likely that the true value is higher and 25% likely that the true value is lower) and the respondent's estimate of the 75th percentile. For a well-calibrated respondent, it should be 50% likely that the true value would fall into the interquartile range. For example, one might be asked to give the average annual temperature in Boston, Massachusetts, USA. A respondent might provide a median estimate of 50 °F (10 °C), a 25th percentile estimate of 40 °F (4.4 °C), and a 75th percentile estimate of 60 °F (15.6 °C). The true average annual temperature in Boston equals 51.3 °F (10.7 °C). Were one scoring this respondent's answers, she would receive one point because her interquartile range contains the true value. A second respondent might provide a median estimate of 45 °F (7.2 °C), a 25th percentile estimate of 40 °F (4.4 °C), and a 75th percentile estimate of 50 °F (10 °C). Were one scoring this respondent's answers, he would receive no point because his interquartile range does not contain the true value. Note that the difference in respondents' scores derives more from the narrow width of the second respondent's interquartile range than from the distance of the median estimate from the truth. Were the second respondent's interquartile range as wide as the first respondent's (and still centered on the median estimate), then the second respondent would also have received a positive score. Setting too narrow an uncertainty interval is the hallmark of the overconfidence heuristic.

In one experiment, a cognitive scientist asked 100 students to answer ten questions like the above question about the average temperature in Boston [91]. For a well-calibrated student, one would expect the true value to lie in the interquartile range for five of the ten questions. Using the binomial distribution to set expectations, one would expect 5 or 6 of the 100 students to give answers such that 8, 9, or 10 of the true values fell into their interquartile ranges. None of the students had scores of 8, 9, or 10. One would also expect 5 or 6 of the 100 students to give answers such that 2, 1, or 0 of the true values fell into their interquartile ranges. Thirty-five of the students had scores of 2, 1, or 0. How would the distribution skew so strongly toward low scores? The skew toward low scores arises because respondents provide too narrow a range of uncertainty, so the true value lies outside the interquartile range much more often than it lies inside it. The overconfidence heuristic acts in the same way when respondents are asked to give extreme percentiles such as the first and 99th percentile [91], is most pronounced when tasks are most difficult [92], has been observed to act in many different populations and cultures [93], and does not depend strongly on the accuracy with which respondents estimate the median [91]. In fact, the discrepancy between correctness of response and overconfidence increases with the knowledge of the respondent. That is, when a response requires considerable reasoning or specialized knowledge, the accuracy of experts exceeds the accuracy of novices, but the extent of their overconfidence – compared with novices – increases faster than the extent of their accuracy [75].

How might the overconfidence heuristic affect inference from observational epidemiologic results? Consider the conventional frequentist confidence interval about a point estimate associating an exposure with a disease, derived from a study's results, to be an uncertainty range like the interquartile range described above. Further consider that interested parties may be aware that the interval fails to account for uncertainty beyond random error, so should be considered a minimum description of the true uncertainty [1, 34]. If asked to, could interested parties appropriately inflate the interval sufficiently to account for sources of uncertainty aside from random error? An understanding of the overconfidence heuristic suggests that the intuitive inflation will be predictably insufficient.

Failure to Account for the Base-Rate

The final bias relevant to causal inference based on observational epidemiologic results is called "failure to account for the base-rate." When asked to estimate the probability of a target event on the basis of the base-rate frequency of the event in a relevant reference population and specific evidence about the case at hand, respondents systematically focus on the specific evidence and ignore the base-rate information [94]. For example, 60 medical students were asked [95]:

If a test to detect a disease whose prevalence is 1/1000 has a false positive rate of 5%, what is the chance that a person found to have a positive result actually has the disease, assuming you know nothing about the person's symptoms or signs?

An implicit assumption is that all who are truly diseased will also test positive. Almost half of the respondents answered 95%, which takes account of only the specific evidence (the patient's positive test) and completely ignores the base-rate information (the prevalence of the disease in the population). The correct response of 2% was given by 11 students. The same inferential error is made when analysts misinterpret the frequentist p-value as the probability that the null hypothesis is true [12, 13]. Failure to account for the base-rate does not derive solely from innumeracy [96]. Rather, the specific evidence is ordinarily perceived to be concrete and emotionally interesting, thereby more readily inspiring the respondent to write a mental script to explain its relevance. Base-rate information is perceived to be abstract and emotionally uninteresting, so less likely to inspire the Associative System to write an explanatory script that contradicts the specific evidence.

How might failure to account for the base-rate affect inference from observational epidemiologic results? Consider a conventional epidemiologic result, composed of a point estimate associating an exposure with a disease and its frequentist confidence interval, to be specific evidence about a hypothesis that the exposure causes the disease. Further consider that interested parties have devoted considerable effort to generating and understanding the research results. Can the interested parties be expected to take account of the base-rate of "true" hypotheses among those studied by epidemiologists, which may not be very high [13, 97]? An understanding of this last heuristic suggests that interested parties are not likely to adequately account for the base-rate, despite exhortations to use base-rate information in epidemiologic inference [10, 35, 98–100].

Conclusion

Epidemiologists are not alone among scientists in their susceptibility to the systematic errors in inference engendered, in part, by influence of the heuristics described above. For example, a review of measurements of physical constants reported consistent underestimation of uncertainty [101]. Measurements of the speed of light overestimated the currently accepted value from 1876 to 1902 and then underestimated it from 1905 to 1950. This pattern prompted one analyst to hypothesize a linear trend in the speed of light as a function of time and a second analyst to hypothesize a sinusoidal relation. In reaction, Birge adjusted a set of measurements for systematic errors and produced adjusted values and intervals that correctly reflected uncertainty, and concluded that the speed of light was constant [102]. Henrion and Fischoff attribute the consistent underassessment of uncertainty in measurements of physical constants to analysts using the standard error as the full expression of the uncertainty regarding their measurements, to the impact on their inferences of heuristics such as those described above, and to "real-world" pressures that discourage a candid expression of total uncertainty [101]. These same three forces certainly affect inference from observational epidemiology studies as well.

They recommend three solutions [101]. First, those who measure physical constants should strive to account for systematic errors in their quantitative assessments of uncertainty. Second, with an awareness of the cognitive literature, those who measure physical constants should temper their inference by subjecting it to tests that counter the tendencies imposed by the heuristics. For example, overconfidence arises in part from a natural tendency to overweight confirming evidence and to underweight disconfirming evidence. Forcing oneself to write down hypotheses that counter the preferred (e.g., causal) hypothesis can reduce overconfidence in that hypothesis. Finally, students should be taught how to obtain better measurements, including how to better account for all sources of uncertainty and how to counter the role of heuristics and biases in reaching an inference. These same recommendations would well serve those who measure epidemiologic associations.

It may seem that an alternative would be to reduce one's enthusiasm about conclusions resting on research results. In cognitive sciences, this approach is called debiasing, and sorts into three categories [103]: resistance–"a mental operation that attempts to prevent a stimulus from having an adverse effect," remediation–"a mental operation that attempts to undo the damage done by the stimulus," and behavior control–"an attempt to prevent the stimulus from influencing behavior." These strategies have been shown to be ineffective solutions to tempering the impact of the heuristics described above [103].

Nonetheless, epidemiologists are taught to rely on debiasing when making inference. We are told to interpret our results carefully, and to claim causation only with trepidation. Consider, for example, the advice offered by three commentators on the apparent difference between randomized [104] and nonrandomized [105] studies of the association between hormone replacement therapy and cardiovascular disease. Each included in their commentary some advice tantamount to a warning to "be careful out there." One wrote that we should be "reasonably cautious in the interpretation of our observations" [106], the second wrote that "we must remain vigilant and recognize the limitations of research designs that do not control unobserved effects" [107], and the third wrote that "future challenges include continued rigorous attention to the pitfalls of confounding in observational studies" [108]. Similar warnings about the importance of careful interpretation are easy to find in classroom lecture notes or textbooks.

The reality is that such trepidation, even if implemented, is ineffective. Just as people overstate their certainty about uncertain events in the future, we also overstate the certainty with which we believe that uncertain events could have been predicted with the data that were available in advance, had it been more carefully examined. The tendency to overstate retrospectively our own predictive ability is colloquially known as 20–20 hindsight. Cognitive scientists, however, label the tendency "creeping determinism" [75].

Creeping determinism can impair one's ability to judge the past or to learn from it. It seems that when a result such as the trial of hormone therapy becomes available, we immediately seek to make sense of the result by integrating it into what we already know about the subject. In this example, the trial result made sense only with the conclusion that the nonrandomized studies must have been biased by unmeasured confounders, selection forces, and measurement errors, and that the previous consensus must have been held only because of poor vigilance against biases that act on nonrandomized studies. With this reinterpretation, the trial results seem an inevitable outcome of the reinterpreted situation. Making sense of the past consensus is so natural that we are unaware of the impact that the outcome knowledge (the trial result) has had on the reinterpretation. Therefore, merely warning people about the dangers apparent in hindsight, such as the recommendations for heightened vigilance quoted above, has little effect on future problems of the same sort [75]. A more effective strategy is to appreciate the uncertainty surrounding the reinterpreted situation in its original form. Better still is to appreciate the uncertainty in current problems, so that similar problems might be avoided in the future.

In fact, cataloging the uncertainty surrounding events is the one method that reliably reduces overconfidence. Simply weighing whether a hypothesis is true– equivalent to the "vigilance" recommended by editorialists on the putative cardioprotective effect of hormone replacement therapy–actually increases belief of the validity of the hypothesis because a person focuses more on explanations as to why it could be true than why it could be false [103]. We tend to seek and overweight information that is consistent with the favored hypothesis and tend to avoid and underweight information that is inconsistent with it [90]. To mitigate this confirmation bias, a person must consider the opposite, that is, imagine alternative outcomes or hypotheses and concretely evaluate their explanatory value. It is not enough to simply be aware of alternative evidence and explanations; in fact, awareness alone can adversely bolster the original belief. One must force the alternative beliefs and evidence out of the Associative System and into consideration by the Rule-Based System. Concrete action, such as implementation of quantitative bias analysis, is one way to force this consideration. This general recommendation has been more explicitly suggested to epidemiologists as a constructive and specific method to reveal uncertainties [109].

As presently practiced, epidemiology seems destined to repeat an endless cycle of overstating our certainty about an uncertain event until a watershed observation washes away the overconfidence, at which point epidemiologists will look back and say that the truth should have been easily realized had we only been more vigilant. Promises to be more vigilant, however, apply only in hindsight; they are inevitably swamped by the tendency to overstate certainty as we look forward. The best hope to break the cycle is to make quantifying random and systematic error common practice, which will mitigate the tendency toward overconfidence regarding the uncertain event. We recognize, however, that the design, implementation, and interpretation of the bias analysis may also be subject to these heuristics and biases [110]. We address this concern throughout, and generally recommend that all results include a phrase such as "conditional on the accuracy of the bias model" (a phrase that applies in equal measure to conventional results, but is almost never explicitly attached to them).

Unfortunately, conventional observational research treats nonrandomized data as if it were randomized in the analysis phase. In the inference phase, analysts usually offer a qualitative discussion of limitations, which are frequently also discounted as important threats to validity. Alternatives that quantify threats to validity do exist but have been infrequently adopted. Instead, students are taught about threats to validity, and occasionally to think about the likely direction of the bias, but not the methods by which their effect might be quantified, and are warned to make inference with trepidation because of the potential for bias. This paradigm corresponds quite well with circumstances known from the cognitive sciences literature to be ripe for the impact of the heuristics described above. In the presence of sparse evidence and a low base-rate of true hypotheses, those who assess the probability of an event–such as the truth of the hypothesis–are overconfident on average [88].

Epidemiologists could explicitly specify alternatives to the causal hypothesis and quantify the uncertainty about the causal association that is induced by each alternative hypothesis. This process removes the guesswork about sizes of biases, compared with the size of exposure effects, from the Associative System and places it under the purview of the Rule-Based System. The second paradigm prompts analysts to list explanations for results that counter their preferred hypothesis, and requires that they incorporate these hypotheses into their assessments of uncertainty. The result is a more complete description of total uncertainty and an effective counter to the impact of the heuristics described above.

References

1. Greenland S. Randomization, statistics, and causal inference. Epidemiology. 1990;1:421–9.
2. Kerr NL. HARKing: Hypothesizing after the results are known. Personality and Social Psychology Review. 1998;2:196–217.
3. Motulsky HJ. Common misconceptions about data analysis and statistics. British Journal of Pharmacology. 2015;172:2126–32.
4. Simmons JP, Nelson LD, Simonsohn U. False-positive psychology. Psychological Science. 2011;22:1359–66.
5. Hernan MA, Robins JM. Causal Inference: What if. Boca Raton: Chapman & Hall/CRC; 2020.
6. Lash T, VanderWeele TJ, Haneuse S, Rothman KJ, Haneuse S. Modern Epidemiology. fourth ed. Philadelphia: Wolters Kluwer; 2021.
7. Greenland S, Robins JM, Pearl J. Confounding and collapsibility in causal inference. Statistical Science. 1999;14:29–46.
8. Greenland S. Principles of multilevel modelling. International Journal of Epidemiology. 2000;29:158–67.
9. Wasserman L. All of statistics: a concise course in statistical inference. Springer Science & Business Media; 2013.
10. Greenland S. Bayesian perspectives for epidemiological research: I. Foundations and basic methods. Int J Epidemiol. 2006;35:765–75.
11. Robins JM, Hernán MA, Wasserman L. On Bayesian estimation of marginal structural models. Biometrics. 2015;71:296.

12. Greenland S, Senn SJ, Rothman KJ, Carlin JB, Poole C, Goodman SN, et al. Statistical tests, P values, confidence intervals, and power: A guide to misinterpretations. Eur J Epidemiol. 2016;31:337–50.
13. Lash TL. The harm done to reproducibility by the culture of null hypothesis significance testing. Am J Epidemiol. 2017;186:627–35.
14. Gelman A, Carlin JB, Stern HS, Dunson DB, Vehtari A, Rubin DB. Bayesian Data Analysis. CRC press; 2013.
15. Carlin BP, Louis TA. Bayesian Methods for Data Analysis. CRC Press; 2008.
16. Greenland S. Multiple-bias modeling for analysis of observational data. J R Stat Soc Ser A. 2005;168:267–308.
17. MacLehose RF, Gustafson P. Is probabilistic bias analysis approximately Bayesian? Epidemiology. 2012;23:151–8.
18. Steenland K, Greenland S. Monte Carlo sensitivity analysis and Bayesian analysis of smoking as an unmeasured confounder in a study of silica and lung cancer. Am J Epidemiol. 2004;160: 384–92.
19. Hernán MA. Does water kill? A call for less casual causal inferences. Ann Epidemiol. 2016;26:674–80.
20. Keyes KM, Galea S. The limits of risk factors revisited: Is it time for a causal architecture approach? Epidemiology. 2017;28:1–5.
21. Poole C. Some thoughts on consequential epidemiology and causal architecture. Epidemiology. 2017;28:6–11.
22. Rothman KJ, Greenland S. Planning study size based on precision rather than power. Epidemiology. 2018;29:599–603.
23. Haneuse S. Stratification and Standardization. In: Lash T, VanderWeele TJ, Haneuse S, Rothman KJ, editors. Modern Epidemiology. fourth ed. Philadelphia: Wolters Kluwer; 2021.
24. Rothman KJ, Lash TL. Precision and Study Size. In: Lash T, VanderWeele TJ, Haneuse S, Rothman KJ, editors. Modern Epidemiology. fourth ed. Philadelphia: Wolters Kluwer; 2021. p. 333–66.
25. Haneuse S. Regression Analysis Part 1: Model Specification. In: Lash T, VanderWeele TJ, Haneuse S, Rothman KJ, editors. Modern Epidemiology. fourth ed. Philadelphia: Wolters Kluwer; 2021. p. 473–503.
26. Thompson WD. Statistical criteria in the interpretation of epidemiologic data. Am J Public Health. 1987;77:191–4.
27. Thompson WD. On the comparison of effects. Am J Public Health. 1987;77:491–2.
28. Poole C. Beyond the confidence interval. Am J Public Health. 1987;77:195–9.
29. Poole C. Confidence intervals exclude nothing. Am J Public Health. 1987;77:492–3.
30. The Editors. The value of P. Epidemiology. 2001;12:286–286.
31. Weinberg CR. It's time to rehabilitate the P-value. Epidemiology. 2001;12:288–90.
32. Gigerenzer G. Mindless statistics. J Socioeconomics. 2004;33:587–606.
33. Wasserstein RL, Lazar NA. The ASA Statement on p-Values: Context, process, and purpose. The American Statistician. 2016;70:129–33.
34. Wasserstein RL, Schirm AL, Lazar NA. Moving to a World Beyond "p < 0.05." The American Statistician. 2019;73(sup1):1–19.
35. Poole C. Low P-values or narrow confidence intervals: which are more durable? Epidemiology. 2001;12:291–4.
36. Greenland S, Robins JM. Identifiability, exchangeability, and epidemiological confounding. Int J Epidemiol. 1986;15:413–9.
37. Greenland S, Robins JM. Identifiability, exchangeability and confounding revisited. Epidemiol Perspect Innov. 2009;6:1–9.
38. Miettinen OS. Theoretical epidemiology. New York: Wiley; 1985.
39. Wacholder S, McLaughlin JK, Silverman DT, Mandel JS. Selection of controls in case-control studies. I. Principles. Am J Epidemiol. 1992;135:1019–28.

40. Hernan MA, Hernandez-Diaz S, Robins JM. A Structural approach to selection bias. Epidemiology. 2004;15:615–25.
41. Hernán MA. Selection bias without colliders. Am J Epidemiol. 2017;185:1048–50.
42. Greenland S. The effect of misclassification in the presence of covariates. Am J Epidemiol. 1980;112:564–9.
43. Brenner H, Savitz DA. The effects of sensitivity and specificity of case selection on validity, sample size, precision, and power in hospital-based case-control studies. Am J Epidemiol. 1990;132:181–92.
44. Flegal KM, Brownie C, Haas JD. The effects of exposure misclassification on estimates of relative risk. Am J Epidemiol. 1986;123:736–51.
45. Greenland S, Rothman KJ. Introduction to stratified analysis. In: Rothman KJ, Greenland S, Lash TL, editors. Modern Epidemiology. third ed. Philadelphia: Lippincott Williams & Wilkins; 2008. p. 258–82.
46. Greenland S. Introduction to regression modeling. In: Rothman KJ, Greenland S, editors. Modern Epidemiology. second ed. Philadelphia, PA: Lippincott-Raven; 1998. p. 401–34.
47. Lash TL, Fox MP, MacLehose RF, Maldonado G, McCandless LC, Greenland S. Good practices for quantitative bias analysis. International Journal of Epidemiology. 2014;43:1969–85.
48. Lash TL, Fox MP, Cooney D, Lu Y, Forshee RA. Quantitative bias analysis in regulatory settings. American Journal of Public Health. 2016;106:1227–30.
49. Fox MP, Lash TL. On the need for quantitative bias analysis in the peer-review process. American Journal of Epidemiology. 2017;185:865–8.
50. Jurek A, Maldonado G, Church T, Greenland S. Exposure-measurement error is frequently ignored when interpreting epidemiologic study results. American Journal of Epidemiology. 2004;159:S72–S72.
51. Hunnicutt JN, Ulbricht CM, Chrysanthopoulou SA, Lapane KL. Probabilistic bias analysis in pharmacoepidemiology and comparative effectiveness research: a systematic review. Pharmacoepidemiol Drug Saf. 2016;25:1343–53.
52. Cole SR, Stuart EA. Generalizing evidence from randomized clinical trials to target populations: The ACTG 320 trial. Am J Epidemiol. 2010;172:107–15.
53. Westreich D, Edwards JK, Lesko CR, Stuart E, Cole SR. Transportability of trial results using inverse odds of sampling weights. Am J Epidemiol. 2017;186:1010–4.
54. Lesko CR, Buchanan AL, Westreich D, Edwards JK, Hudgens MG, Cole SR. Generalizing study results: a potential outcomes perspective. Epidemiology. 2017;28:553-561.
55. Bailar JC, Mosteller F. Guidelines for statistical reporting in articles for medical journals. Ann Intern Med. 1988;108:266–73.
56. Meier P. Damned liars and expert witnesses. Journal of the American Statistical Association. 1986;81:269–76.
57. Espeland MA, Hui SL. A general approach to analyzing epidemiologic data that contain misclassification errors. Biometrics. 1987;43:1001–12.
58. Lyles RH. A note on estimating crude odds ratios in case-control studies with differentially misclassified exposure. Biometrics. 2002;58:1034–6.
59. Lyles RH, Allen AS. Estimating crude or common odds ratios in case-control studies with informatively missing exposure data. American Journal of Epidemiology. 2002;155:274–81.
60. Lyles RH, Lin J. Sensitivity analysis for misclassification in logistic regression via likelihood methods and predictive value weighting. Statistics in Medicine. 2010;29:2297–309.
61. Lyles RH, Tang L, Superak HM, King CC, Celentano DD, Lo Y, et al. Validation data-based adjustments for outcome misclassification in logistic regression: an illustration. Epidemiology. 2011;22:589–97.
62. Gustafson P. Measurement Error and Misclassification in Statistics and Epidemiology: Impacts and Bayesian Adjustments. Boca Raton, Florida: Chapman & Hall/CRC; 2003.
63. Chu H, Wang Z, Cole SR, Greenland S. Sensitivity analysis of misclassification: a graphical and a Bayesian approach. Ann Epidemiol. 2006;16:834–41.

64. Chu R, Gustafson P, Le N. Bayesian adjustment for exposure misclassification in case–control studies. Statistics in Medicine. 2010;29:994–1003.
65. Sturmer T, Schneeweiss S, Avorn J, Glynn RJ. Adjusting effect estimates for unmeasured confounding with validation data using propensity score calibration. Am J Epidemiol. 2005;162:279–89.
66. Spiegelman D. Approaches to uncertainty in exposure assessment in environmental epidemiology. Annu Rev Public Health. 2010;31:149–63.
67. Spiegelman D, Rosner B, Logan R. Estimation and inference for logistic regression with covariate misclassification and measurement error in main study/validation study designs. J Am Stat Assoc. 2000;95:51–61.
68. Spiegelman D, Carroll RJ, Kipnis V. Efficient regression calibration for logistic regression in main study/internal validation study designs with an imperfect reference instrument. Statistics in Medicine. 2001;20:139–60.
69. Little RJA, Rubin DB. Statistical Analysis with Missing Data. second ed. New York: Wiley; 2002.
70. Robins JM, Rotnitzkey A, Zhao LP. Estimation of regression coefficients when some regressors are not always observed. J Am Stat Assoc. 1994;89:846–66.
71. Lash T, Fink AK. Semi-automated sensitivity analysis to assess systematic errors in observational data. Epidemiology. 2003;14:451–8.
72. Fox MP, Lash TL, Greenland S. A method to automate probabilistic sensitivity analyses of misclassified binary variables. Int J Epidemiol. 2005;34:1370–6.
73. Greenland S. Interval estimation by simulation as an alternative to and extension of confidence intervals. Int J Epidemiol. 2004;33:1389–97.
74. Greenland S. The need for cognitive science in methodology. Am J Epidemiol. 2017;186:639–45.
75. Piattelli-Palmarini M. Inevitable Illusions. New York: Wiley; 1994.
76. Kahneman D, Slovic P, Tversky A. Judgment Under Uncertainty: Heuristics and Biases. New York: Cambridge University Press; 1982.
77. Gilovich T, Griffin D, Kahneman D. Heuristics and Biases: The Psychology of Intuitive Judgment. New York: Cambridge University Press; 2002.
78. Kahneman D, Frederick S. Representativeness revisited: Attribute substitution in intuitive judgment. In: Heuristics and Biases: The Psychology of Intuitive Judgment. New York: Cambridge University Press; 2002. p. 49–81.
79. Sloman S. Two systems of reasoning. In: Heuristics and Biases: The Psychology of Intuitive Judgment. New York: Cambridge University Press; 2002. p. 379–96.
80. Kahneman D. Thinking, Fast and Slow, first ed. New York: Farrar, Straus and Giroux; 2011.
81. Tversky A, Kahneman D. Judgment Under Uncertainty: Heuristics and Biases. In: Kahneman D, Slovic P, Tversky A, editors. Judgment Under Uncertainty: Heuristics and Biases. New York: Cambridge University Press; 1982. p. 3–22.
82. van Smeden M, Lash TL, Groenwold RHH. Reflection on modern methods: five myths about measurement error in epidemiological research. Int J Epidemiol. 2020;49:338–47.
83. Jurek AM, Greenland S, Maldonado G, Church TR. Proper interpretation of non-differential misclassification effects: expectations vs observations. Int J Epidemiol. 2005;34:680–7.
84. Kristensen P. Bias from nondifferential but dependent misclassification of exposure and outcome. Epidemiology. 1992;3:210–5.
85. Gustafson P, Greenland S. Curious phenomena in Bayesian adjustment for exposure misclassification. Stat Med. 2006;25:87–103.
86. Lash TL, Fink AK. Re: "Neighborhood environment and loss of physical function in older adults: evidence from the Alameda County Study." Am J Epidemiol. 2003;157:472–3.
87. Holman CD, Arnold-Reed DE, de Klerk N, McComb C, English DR. A psychometric experiment in causal inference to estimate evidential weights used by epidemiologists. Epidemiology. 2001;12:246–55.

88. Koehler D, Brenner L, Griffin D. The calibration of expert judgment: Heuristics and biases beyond the laboratory. In: Gilovich T, Griffin D, Kahneman D, editors. Heuristics and Biases: The Psychology of Intuitive Judgment. New York: Cambridge University Press; 2002. p. 686–715.

89. Epley N, Gilovich T. Putting adjustment back in the anchoring and adjustment heuristic. In: Heuristics and Biases: The Psychology of Intuitive Judgment. New York: Cambridge University Press; 2002. p. 139–49.

90. Chapman G, Johnson E. Incorporating the irrelevant: Anchors in judgments of belief and value. In: Gilovich T, Griffin D, Kahneman D, editors. Heuristics and Biases: The Psychology of Intuitive Judgment. New York: Cambridge University Press; 2002. p. 120–38.

91. Alpert M, Raiffa H. A progress report on the training of probabilisty assessors. In: Kahneman D, Slovic P, Tversky A, editors. Judgment Under Uncertainty: Heuristics and Biases. New York: Cambridge University Press; 1982. p. 294–305.

92. Lichtenstein S, Fischoff B, Phillips L. Calibration of probabilities: The state of the art to 1980. In: Kahneman D, Slovic P, Tversky A, editors. Judgment Under Uncertainty: Heuristics and Biases. New York: Cambridge University Press; 1982. p. 306–34.

93. Yates J, Lee J, Sieck W, Choi I, Price P. Probability judgment across cultures. In: Gilovich T, Griffin D, Kahneman D, editors. Heuristics and Biases: The Psychology of Intuitive Judgment. New York: Cambridge University Press; 2002. p. 271–91.

94. Tversky A, Kahneman D. Evidential impact of base-rates. In: Kahneman D, Slovic P, Tversky A, editors. Judgment Under Uncertainty: Heuristics and Biases. New York: Cambridge University Press; 1982. p. 153–62.

95. Casscells W, Schoenberger A, Graboys TB. Interpretation by physicians of clinical laboratory results. N Engl J Med. 1978;299:999–1001.

96. Nisbett R, Borgida E, Crandall R, Reed H. Popular induction: Information is not necessarily informative. In: Kahneman D, Slovic P, Tversky A, editors. Judgment Under Uncertainty: Heuristics and Biases. New York: Cambridge University Press; 1982. p. 101–16.

97. Ioannidis JP. Why most published research findings are false. PLoS Med. 2005;2:e124.

98. Wacholder S, Chanock S, Garcia-Closas M, El GL, Rothman N. Assessing the probability that a positive report is false: an approach for molecular epidemiology studies. J Natl Cancer Inst. 2004;96:434–42.

99. Greenland S, Robins JM. Empirical-Bayes adjustments for multiple comparisons are sometimes useful. Epidemiology. 1991;2:244–51.

100. MacLehose RF, Hamra GB. Applications of Bayesian methods to epidemiologic research. Curr Epidemiol Rep. 2014;1:103–9.

101. Henrion M, Fischoff B. Assessing uncertainty in physical constants. In: Gilovich T, Griffin D, Kahneman D, editors. Heuristics and Biases: The Psychology of Intuitive Judgment. New York: Cambridge University Press; 2002. p. 666–77.

102. Birge R. The general physical constants: as of August 1941 with details on the velocity of light only. Reports on Press in Physics. 1941;8:90–134.

103. Wilson T, Centerbar D, Brekke N. Mental contamination and the debiasing problem. In: Gilovich T, Griffin D, Kahneman D, editors. Heuristics and Biases: The Psychology of Intuitive Judgment. New York: Cambridge University Press; 2002. p. 185–200.

104. Rossouw JE, Anderson GL, Prentice RL, LaCroix AZ, Kooperberg C, Stefanick ML, et al. Risks and benefits of estrogen plus progestin in healthy postmenopausal women: principal results From the Women's Health Initiative randomized controlled trial. JAMA. 2002;288: 321–33.

105. Stampfer MJ, Colditz GA. Estrogen replacement therapy and coronary heart disease: a quantitative assessment of the epidemiologic evidence. Prev Med. 1991;20:47–63.

106. Michels KB. Hormone replacement therapy in epidemiologic studies and randomized clinical trials - are we checkmate? Epidemiology. 2003;14:3–5.

107. Piantadosi S. Larger lessons from the Women's Health Initiative. Epidemiology. 2003;14:6–7.
108. Whittemore AS, McGuire V. Observational studies and randomized trials of hormone replace-
     ment therapy: what can we learn from them? Epidemiology. 2003;14:8–10.
109. Savitz DA, Wellenius GA. Interpreting Epidemiologic Evidence: Connecting Research to
     Applications. 2 edition. Oxford; New York: Oxford University Press; 2016. 240 p.
110. Lash TL, Ahern TP, Collin LJ, Fox MP, MacLehose RF. Bias Analysis Gone Bad. Am J
     Epidemiol. 2021;190:1604–12.

Chapter 2
A Guide to Implementing Quantitative Bias Analysis

Introduction

Estimates of association from nonrandomized epidemiologic studies are susceptible to two types of error: random error and systematic error. Random error (sometimes thought of as sampling error) is often called chance, and decreases toward zero as the sample size increases. The amount of random error in an estimate of association is measured by its precision, which is the inverse of the variance. Systematic error, often called bias, does not necessarily decrease toward zero as the sample size increases. That is, while random error can be dealt with by collecting more data, systematic error cannot be overcome in the same manner. Just as precision is the reduction of random error, we define validity as the reduction of systematic error. Reducing systematic error can be accomplished before the study has been conducted through design strategies or after the study has been conducted, through regression techniques or through quantitative bias analysis techniques.

Conventional frequentist confidence intervals are a measure of the precision, or random error, of an estimate of association and give no information about the amount of systematic error. The objective of quantitative bias analysis is to estimate quantitatively the systematic error that remains after implementing a study design and analysis and adjust the effect estimate for that systematic error. This text assumes that the reader has a thorough understanding of established principles for study design and analysis and recognizes that the systematic error remaining after implementing those principles merits quantification and presentation. For comprehensive guidance on study design and analysis, the reader should consult an epidemiology methods textbook, such as *Modern Epidemiology* [1].The next sections briefly review principles of study design and data analysis that have presumably been applied in a study under consideration for quantitative bias analysis. They are followed by sections on planning for quantitative bias analysis and a brief overview of the types of bias analyses described in this chapter.

Reducing Error

The objective of study design and statistical analysis in etiologic epidemiologic research is to obtain an accurate (precise and valid) estimate of the effect of an exposure on the occurrence of a disease or other outcome. Epidemiologic studies should be designed and analyzed with this objective in mind, but epidemiologists should realize that this objective can never be completely achieved. All study protocols require an exercise in the judicious use of limited resources and therefore every study has limited sample size, which means that every study contains some random error. Similarly, every study is susceptible to sources of systematic error. Even randomized studies, which are sometimes thought of as the gold standard research design, are susceptible to selection bias from losses to follow-up and to misclassification of analytic variables. Since the research objective can never be achieved perfectly, epidemiologists should instead strive to reduce the impact of all sources of study error as much as possible. These efforts are made in the design and analysis phases of the study.

Reducing Error by Design

To reduce random error in a study's design, epidemiologists can increase the size of the study or alter the distribution of the collected data, in specific ways, to produce favorable distributions of the exposure and outcome to thereby reduce the variance of the estimate of association. Increasing the size of the study requires enrolling more subjects and/or following the enrolled subjects for a longer period; this additional information will generally reduce the estimate's standard error. A second strategy to improve an estimate's precision is to improve the efficiency with which the data are distributed into categories of the analytic variables. To understand this second strategy, consider the standard error of the odds ratio, which equals the square root of the sum of inverses of the frequencies of the interior cells of a 2×2 table. As displayed in Table 2.1, the 2×2 table is the simplest contingency table relating exposure to disease.

With this data arrangement, the odds ratio (OR) equals $(a/c)/(b/d)$ and the standard error of the ln(OR) equals $\sqrt{(1/a + 1/b + 1/c + 1/d)}$. In a study with 100 subjects and each interior cell frequency equal to 25, the odds ratio equals its null value of 1.0 and the standard error of the ln(OR) equals $\sqrt{(1/25 + 1/25 + 1/25 + 1/25)} = 0.4$. The 95% confidence interval about the odds ratio equals 0.46 to 2.19. Alternatively, had 20, rather than 50, diseased individuals been available for

Table 2.1 The 2×2 contingency table relating exposure to disease.

	Exposed	Unexposed
Diseased	a	b
Undiseased	c	d

the study, but the sample size remained constant by changing the ratio of those with and without the disease to 1 to 4, rather than 1 to 1, the odds ratio would remain null. The odds ratio's standard error would then equal $\sqrt{(1/10 + 1/10 + 1/40 + 1/40)} = 0.5$ and the 95% confidence interval would range from 0.38 to 2.66. Although the sample size (100 subjects) did not change, the standard error and the width of the confidence interval (measured on the log scale) have both increased by 25% due only to the less efficient distribution of the subjects within the contingency table. This example illustrates how the distribution of data within the categories of analytic variables affects the study's precision, given a fixed sample size, even when no bias is present. It is important to note that the optimal distribution of data within even a simple table like this will depend on the measure of association that an analyst wishes to estimate.

Improving the efficiency with which the data are distributed requires an understanding of the distribution of the exposure and disease in the base population. If one is interested in studying the relation between sunlight exposure and melanoma incidence, then a population in the northern United States might not have an efficient distribution of the exposure compared with a population in the southern United States where sunlight exposure is more common. If one is interested in studying the relation between tanning bed exposure and melanoma incidence, then a population in the northern United States might have a more efficient distribution of the exposure than a population in the southern United States. Careful selection of the source population is one strategy that analysts can use to improve the efficiency of the distribution of subjects within the categories of the analytic variables.

Matching is a second strategy that can, in some situations, improve the efficiency of this distribution. In a case-control study, matching a predetermined number of controls to each case on potential confounders assures that the controls will appear in a constant ratio to cases within the categories of the confounder. For example, skin type (freckled vs unfreckled) might confound the relation between sunlight exposure and melanoma incidence. Cases of melanoma may be more likely to have freckled skin than the base population that gives rise to cases, and people with freckled skin might have different exposure to sunlight than people with unfreckled skin. Without matching, most of the cases will be in the category of the confounder denoting freckled skin, and most of the controls will be in the category of the confounder denoting unfreckled skin because it is more common in the source population. While this can be adjusted for in the analysis phase, it can sometimes yield an inefficient analysis, and therefore a wider confidence interval. Matching controls to cases ensures that controls appear most frequently in the stratum where cases appear most frequently (e.g., freckled skin). This provides for an efficient distribution within which the analyst can control for the confounding factor in the analysis and has the potential to yield an estimate with higher precision. Matching unexposed to exposed persons in cohort studies can sometimes achieve a gain in efficiency as well, though is often expensive to do in prospective studies that require enrolling people in a study and possibly turning away eligible subjects. It is important to note, however, that matching is not guaranteed to improve study efficiency, therefore careful consideration is required prior to implementing a matched design.

Systematic errors in epidemiology are typically divided into two types of validity: external validity and internal validity. External validity refers to how well a measure of effect that is estimated in one population can be used to estimate the effect that would have been estimated in a separate population. External validity is, generally, beyond the scope of this book; we do, however, discuss it briefly in the chapter on selection bias. We focus in this text on systematic errors that influence internal validity. These systematic errors influence how well a study's measure of association reflects the effect in the source population in which the study is conducted. There are three main sources of systematic error that typically concern epidemiologists: selection bias, uncontrolled confounding, and information bias. Bias caused by these three sources may be impossible to eliminate completely but can be mitigated, in many instances, by study design.

To reduce systematic error in a study's design, epidemiologists should focus on the fundamental criterion that must be satisfied to obtain a valid comparison of the measure of disease frequency in the exposed group with the measure of disease frequency in the unexposed group. That is, each group (exposed and unexposed) must have the measure of disease frequency that the other group (unexposed or exposed, respectively) would have had if they, counter to the fact, received the opposite exposure [2, 3], within the strata of measured confounders. This assumption is referred to as full exchangeability; however, other assumptions, such as partial exchangeability, may be sufficient to ensure lack of bias due to confounding in specific target populations. The ideal study would compare the disease occurrence in some target population (either the total cohort, exposed portion of the cohort or, perhaps, some external population) that would have occurred if the entire target population had been exposed to the disease occurrence that would have occurred if that same target population had been unexposed. Since this ideal comparison involves counterfactuals (in which we need to estimate the disease incidence among unexposed people had they been exposed and among exposed people had they been unexposed) and can never be realized, the disease incidence is measured in surrogate groups. The exposed portion of the study population is a surrogate group for the disease occurrence that would have occurred among the unexposed, had they been exposed; and, similarly, the unexposed portion of the study population is a surrogate for the disease occurrence that would have occurred among the exposed, had they been unexposed. The assumption of exchangeability, that those surrogate populations can be used to gain a valid measure of the required disease frequency, cannot be empirically verified. The analyst must strive for the desired expectation of exchangeability in the data, which is achievable asymptotically by randomization, or assume that it holds within strata of measured confounders [4].

With this criterion in mind, the design principles to enhance validity follow directly. The study population should be selected such that participation is not conditional on both exposure status and disease status. Many study designs enroll participants conditional on either their exposure (such as matched cohort studies) or disease (case-control studies). However, when both exposure status and disease status affect the probability that a member of the base population participates in the study, the estimate of association will be susceptible to selection bias. Enrolling subjects and/or documenting their exposure status before the disease occurs (i.e., prospectively) assure that disease status is less likely to be associated with initial participation.

Second, the study population should be selected such that the net effects of all other predictors of the outcome, aside from exposure itself, are in balance between the exposed and unexposed groups. This balance is commonly referred to as having no confounding or as exchangeability. Randomization achieves this objective within limits that are statistically quantifiable [4]. When exposure status cannot be assigned by randomization, which is usually the situation in studies of disease etiology, the analyst can limit confounding by restricting the study population to one level of the confounder or ensuring that data are collected on potential confounders so that their effects can be controlled in the analysis.

Finally, and unrelated to exchangeability assumptions, the data should be collected and converted to electronic form with as few measurement or classification errors as possible. Some errors in measurement or classification are, however, inevitable. Analysts often strive to assure that the rates of classification errors do not depend on the values of other variables (e.g., rates of exposure classification errors do not depend on disease status, which is called nondifferential exposure misclassification) or on the proper classification of other variables (e.g., errors in classification of exposure are as likely among those properly classified as diseased as among those improperly classified as diseased, which is called independent exposure misclassification). This second objective can be readily achieved by using different methods to collect information on disease status from those used to collect information on exposure status (as well as information on confounders). The data collection for disease status should be conducted so that the data collector is blinded to the information on exposure and confounder status. Nondifferential and independent errors in classification often yield the most easily predictable, and therefore most readily addressable, bias of the estimate of association. Nonetheless, one may choose to select a design expected to yield relatively small differential classification errors in preference to a design expected to yield relatively large nondifferential classification errors, since the former could yield less bias and uncertainty. Generalized advice always to balance information quality across compared categories (i.e., to strive for nondifferential classification errors) ignores the potential for this trade-off to favor small differential errors.

Reducing Error in the Analysis

Following implementation of a design that reduces random and systematic error to the extent practical, a well-designed analysis of the collected data can further reduce error. Data analysis should begin with a clearly specified definition of each of the analytic variables. The conversion algorithm and variable type should be defined for each analytic variable, after careful consideration of the variability in dose, duration, and induction period that will be characterized in the analysis.

After completing the definition and coding of analytic variables, the analysis proceeds to a descriptive characterization of the study population. The descriptive analysis shows the demographic characteristics of the people in the study. For example, it might show the proportion of the population enrolled at each of the study centers, the distribution of age, and the proportion belonging to each sex. Descriptive analyses should include the proportion of the study population with a missing value assigned to each analytic variable. The proportion with missing data helps to identify analytic variables with problems in data collection, definition, or format conversion. Analysis at this stage is very useful for identifying potential errors in the data (extreme or impossible values) as well as informing coding decisions (what cut points will be used for categorization) and to identify areas of the data that are sparse.

Examination of the bivariate relations between analytic variables is the third step in data analysis. Bivariate relations compare proportions, means or medians for one study variable within categories of a second. These comparisons inform the analyst's understanding of the data distributions and can also identify data errors that would prompt an inspection of the data collection, variable definitions, or format conversions. Analytic decisions regarding what confounders might be included in a regression model are often informed by the bivariate examination of the potential confounder and the outcome as well as the potential confounder and the exposure. The number of bivariate relations that must be examined grows exponentially as the number of analytic variables increases. If the number grows too large to be manageable, the analyst should restrict the examination to pairs that make sense a priori (e.g., variables that are thought to be potential confounders). However, whenever possible, all pairs ought to be examined because a surprising and important finding might easily arise from a pair that would be ignored a priori.

The comparisons of the proportions (or means) of exposed and unexposed with the disease of interest within the categories of the analytic variables are a special subset of bivariate comparisons. These proportions can be explicitly compared with one another by difference or division, yielding estimates of association such as the risk difference, risk ratio, or a difference in means. When estimates of association are calculated as a part of the bivariate comparison, the analysis is also called a stratified analysis and the resulting associations are referred to as crude or unadjusted estimates. Often one comparison is a focus of the stratified analysis, which is the comparison of the disease proportions in those exposed to the condition of interest with those unexposed to the condition of interest. This comparison relates directly to the original objective: a valid and precise estimate of the effect of an exposure on the occurrence of a disease. To continue the stratified analysis, the comparisons of disease proportions in exposed versus unexposed are expanded to comparisons within levels of other analytic variables. For example, the risk ratio comparing exposed with unexposed might be calculated within each of three age groups. An average risk ratio can be calculated by standardization or pooling. Comparison of this average or standardized risk ratio with the crude or collapsed risk ratio (including all ages in one stratum) can indicate whether age is an important confounder of the risk ratio [5]. If the adjusted risk ratio is substantially different from the crude risk ratio then the adjusted risk ratio may provide an estimate of association that removes some of the systematic bias due to confounding by age. The correspondence between noncollapsibility and confounding does not generally hold for the odds ratio, hazard ratio, rate ratio, and rate difference [5]. That is, an odds ratio may be unconfounded even if constant stratum specific effects do not equal the crude effects [6]. If the risk of disease is low (<10%) in every combination of the categories of exposure and the categories of controlled confounders then these measures are good approximations of corresponding risk ratios and risk differences, which are collapsible under the no confounding condition and can be used safely to evaluate the strength of confounding by the stratification variable. However, when the risk of disease is greater than 10% in at least one combination of the exposure and a stratification variable evaluated as a potential confounder, these estimates of association are not necessarily collapsible across levels of the stratification variable, even if that variable is not a confounder. In such cases, avoiding these estimates of effects as an etiologic measure of association is considered good practice.

The analysis can proceed by further stratification on a second variable (e.g., natal sex groups) and standardizing or pooling to simultaneously adjust for confounding by both age and natal sex. The number of strata increases geometrically as additional variables are analyzed. As the number of strata increase, it may be difficult to examine the data in each stratum. Of more practical concern, the data quickly become too sparse to allow estimation in some of the strata. A common solution to the problem engendered by this geometric progression is to use regression modeling rather than stratification. Regression models yield estimates of association that simultaneously adjust for multiple confounders. The interpretation of estimates from regression models is similar to stratified estimates: the association between the exposure and disease conditional on a fixed level of the potential confounders. Their advantage over stratification is that regression models do not become cumbersome or suffer from small numbers as easily as multiple stratification. Care should be used with regression models, as they resolve issues of sparse data by making modeling assumptions such as no interactions and linearity that should be carefully evaluated. Without these assumptions, estimates from regression models would be plagued with the same problems as stratified analysis approaches. Further, regression modeling does not show the data distribution, so should not be used without first conducting the bivariate analysis and stratification on the critical confounders.

This analytic plan describes the conventional epidemiologic approach to data analysis. It yields a quantitative assessment of random error by producing confidence intervals about the crude or adjusted estimates of association. It also adjusts the estimate of association for confounding variables included in the stratification or regression model. However, there is typically no adjustment for selection bias, information bias, confounding by uncontrolled confounders, or residual confounding by measured confounders that are poorly specified or poorly measured. Nor is there any quantification of uncertainty arising from these sources of bias. Quantitative bias analysis addresses these shortcomings in the conventional approach to epidemiologic data analysis.

Quantifying Error

The goal of study design and analysis is to reduce the amount of random and systematic error in an estimate of association to the extent possible. Because it is impossible to completely eliminate either source of error, it is important that analysts quantify how far they are from this goal. Quantifying the impact of random error using confidence intervals is common in epidemiology. Quantitative bias analysis quantifies the extent to which systematic bias may affect estimates of association. Conducting a study that will yield a measure of association with as little bias as is practical requires careful planning and choices in the design of data collection and analysis. Similarly, quantifying the amount of residual bias requires careful planning and choices in the design of data collection and analysis. Since conducting a high-quality bias analysis follows the same steps as conducting a high-quality epidemiologic study, plans for both should be integrated at each phase of the study, as depicted in Figure 2.1.

Evaluating the Potential Value of Quantitative Bias Analysis?

Before discussing the steps involved in planning and conducting a quantitative bias analysis, it is important to first consider when it makes the most sense to conduct a bias analysis. Quantitative bias analysis is most valuable when a measure of effect from a study is relatively precise [7], such that the conventional confidence interval will be narrow. A narrow interval reflects a small amount of residual random error, tempting interested parties to underestimate the true uncertainty and overstate their

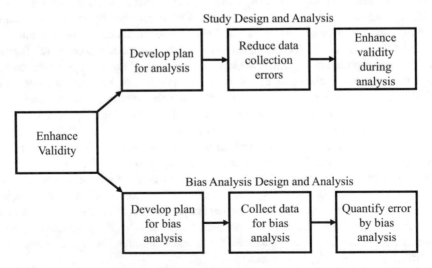

Figure 2.1 Integration of planning for bias analysis with conventional study design and analysis.

confidence that an association is truly causal and of the size estimated by the study. When a wider interval is obtained, inference from the study's results should be tenuous because of the substantial random error, regardless of whether systematic error has also been estimated quantitatively. Note that this formulation assumes that analysts use conventional confidence intervals as a measure of the extent to which random error could impact results. Analysts who simply note whether the interval includes the null, a surrogate for statistical significance testing, will often be misled by statistically significant, but substantially imprecise estimates of association [8].

Quantitative bias analysis is most valuable when a study is likely susceptible to one or more sources of bias about which there is reasonable information [7]. Studies susceptible to multiple sources of bias about which little is known may not be good candidates for quantitative bias analysis because the total error will be so large that meaningful inference will be difficult. These studies are similar to studies that yield wide conventional confidence intervals: the analyst or interested party should recognize that no inference may be reliable, so the effort of a quantitative bias analysis may not be productive. There are, of course, exceptions to this rule such as if a policy is going to be implemented on the basis of a study for which there may be substantial bias. Demonstrating overall uncertainty in this case may be useful. Quantitative bias analysis is likely to be most valuable when there is good evidence regarding the magnitude of the potential biases (such as validation studies). In the absence of evidence regarding the magnitude of bias, the quantitative bias analysis is likely to produce results with very large errors regarding the uncertainty of the estimate of association. Generally, studies with large conventional errors or that are susceptible to many large systematic errors might instead be useful for generating ideas for better-designed and larger subsequent studies. They should not, generally, provide a basis for inference or policy action, so the additional effort of quantitative bias analysis would seldom be an efficient use of resources.

Quantitative bias analysis is therefore most valuable when studies yield narrow conventional confidence intervals – so have little residual random error – and when these studies are susceptible to a limited number of systematic errors. Such studies often appear to be an adequate basis for inference or for policy action, even though only random error has been quantified by the conventional confidence interval. Quantification of the error due to the limited number of biases will safeguard against inference or policy action that takes account of only random error [7, 9]. Without a quantitative assessment of the second important source of error – systematic error – the inference or policy action would be premature.

Planning for Bias Analysis

Quantitative bias analysis is best accomplished with foresight and planning, just as with all aspects of epidemiologic research. The process of conducting a well-designed bias analysis goes beyond simply understanding the methods used for the analysis. It also includes a thorough planning phase to ensure that the information needed for quantification of bias is carefully collected and is itself valid. To facilitate this collection, analysts should consider the important threats to the validity of their research while designing their study [9]. This consideration should immediately suggest the quantitative analyses that will explore these threats and should inform the data collection that will be required to complete the quantitative analyses.

For example, an analyst may design a retrospective case-control study of the relation between leisure exposure to high vs low levels of sunlight and the occurrence of melanoma. Cases of melanoma and controls sampled from the source population will be interviewed by telephone or web-enabled survey regarding their exposures to sunlight and other risk factors for melanoma. The analyst should recognize the potential for selection bias to be an important threat to the study's validity: cases may be more likely than controls to agree to the interview, and those who spend substantial time in sunlight might be concerned about their risk of cancer and be more motivated to participate than those who do not spend much time in the sun. To quantitatively address the potential selection bias (Chapter 4), the analyst will need to know the participation proportions in cases and controls, within groups of high and low exposure to sunlight. Case and control status will be known by design, but to characterize each eligible subject's sunlight exposure, the analyst will need the participant to complete the interview. Sunlight exposure will not, therefore, be known for subjects who refuse to participate. However, in planning for a quantitative bias analysis, the analyst might ask even those who refuse to participate whether they would be willing to answer a single question regarding their sunlight exposure. If the proportion of refusals who did agree to answer this one question was high, this alone would allow the analyst to crudely compare sunlight exposure history among cases and controls who refuse to participate, and to bias-adjust the observed estimate of association for the selection bias.

To continue the example, the analysts might be concerned about the accuracy of subjects' self-report of history of leisure-time sunlight exposure. In particular, melanoma cases might recall or report their history of sunlight exposure with greater accuracy than controls sampled from the base population. This threat to validity would be an example of information bias (Chapter 6), which can also be addressed by quantitative bias analysis. To implement a bias analysis, the analysts would require estimates of the accuracy of self-reported sunlight exposure classification among melanoma cases and controls. Classification error rates might be obtained by an internal validation study (e.g., comparing self-report of sunlight exposure history with a diary of sunlight exposure kept by subsets of the cases and controls) or by external validation studies (e.g., comparing self-report of sunlight exposure history with a diary of sunlight exposure kept by melanoma cases and noncases in a similar second population).

Finally, imagine that the analyst was concerned that the relation between leisure time exposure to sunlight and risk of melanoma was confounded by exposure to tanning beds. Subjects who use tanning beds might have different leisure time exposure to sunlight than those who do not use tanning beds, and tanning bed use itself might be a risk factor for melanoma. If each subject's use of tanning beds was not queried in the interview, then tanning bed use would be an uncontrolled confounder (Chapter 5). While tanning bed use would ideally have been assessed during the interview, it is possible that its relation to melanoma risk was only understood after the study began. To plan for bias analysis, the analyst might turn to the published literature on similar populations to research the strength of association between tanning bed use and leisure time exposure to sunlight, the strength of association between tanning bed use and melanoma, and the prevalence of tanning bed use. In combination, these three factors would allow a quantitative bias analysis of the potential impact of the unmeasured confounder on the study's estimate of the association of leisure time exposure to sunlight on risk of melanoma.

In these examples, planning ahead for the quantitative bias analysis facilitates the actual analysis. Selection forces can be best quantified if the analyst plans to ask for sunlight information among those who refuse the full interview. Classification error can be best quantified if the analyst plans for an internal validation study or assures that the interview and population correspond well enough to the circumstances used for an external validation study. Unmeasured confounding can be best quantified if the analyst collects data from publications that studied similar populations to quantify the bias parameters. Table 2.2 outlines the topics to consider while planning for quantitative bias analysis. These topics are further explained in the sections that follow.

Table 2.2 Planning for quantitative bias analysis.

General tasks	Tasks for each type of bias		
	Information bias	Selection bias	Confounding
Determine likely threats to validity	Ask whether measurement error of any important analytic variable is a likely threat	Ask whether selection bias or loss-to-follow-up is a likely threat	Ask whether residual confounding or unmeasured confounding is likely
Determine data needed to conduct bias analysis	Internal validation study or external validation data	Collect information on selection proportions	Collect information on prevalence of confounder and its associations with exposure and disease
Consider the population from which to collect data	Study population or similar external population	Source population	Source population or similar external population
Allocate resources	Develop databases that allow for recording of data, allocate time for data collection and analysis of data, write protocols for substudies, collect and understand software for analysis		
Set order of bias-adjustments	Usually first	Usually second	Usually third
Consider data level	Record-level data corrections or summary data		
Consider inter-relations between biases	Important interrelations should be assessed with record-level data and multiple biases modeling		
Select a technique for bias analysis	Each method in Table 2.3 can be applied to each bias. Consider the number of biases to be analyzed, the interrelations between biases, the inferential question, the bias model fit to the data, and computational requirements		

Creating a Data Collection Plan for Bias Analysis

As described in the preceding examples, planning for quantitative bias analysis during the study design will produce the most effective analyses [10]. Analysts should consider the study design before data collection begins and ask, "What will likely be the major threats to validity once the data have been collected?" The answer to this question may prompt changes to the study protocol to reduce those biases or it may inform the plans for data collection necessary to conduct the quantitative bias analysis. If selection bias is a concern, then the analyst should collect the data required to calculate participation or follow-up proportions within strata defined by the exposure, disease status and important covariates. If measurement or classification errors are a concern, then the analyst should collect the data required to validate the study's measurements. If this is not feasible, information from other studies may help inform the extent of the misclassification. We provide further guidance on selecting and implementing an internal or external design in Chapter 3. If an important candidate confounder has not been, and cannot be, measured or controlled by design, then the analyst should plan to use internal and external data (sometimes in combination) to estimate the bias from the uncontrolled confounder.

If the analyst collects validation data, careful consideration of the target population from which this data will be collected is warranted. For example, confounding arises at the level of the source population, so data used to bias-adjust for an uncontrolled confounder should arise from the same source population (or a suitably similar population), but should not necessarily be limited to the population sampled for the primary study. Selection bias arises from disease and exposure-dependent participation into or drop out from the primary study. In assessing selection bias, the exposure and disease information are available for study participants, so the information required should be collected from nonparticipants. This implies the data should be collected in the same source population as the primary study sample. In contrast to this, information bias from classification or measurement error arises within the actual study population, so the data required for assessing information bias should be collected from a subset of participants (an internal validity study) or from a population similar to the participants (an external validity study). Careful consideration of the target population will lead to a more effective bias analysis.

Once the major threats to validity have been ascertained, and the population from which validation data will be collected has been identified, the analyst should devise a plan for collecting the validation data needed for the bias analysis. If the validation data will be external, then the analyst should conduct a systematic review of the published literature to find applicable validation studies. For example, if the analyst of the sunlight-melanoma relation is concerned about errors in reporting of sunlight exposure, then she should collect all of the relevant literature on the accuracy of self-report of sunlight exposure. Studies that separate the accuracy of exposure by melanoma cases and noncases will be most relevant. From each of these studies, she should abstract the sensitivities and specificities (or predictive values if that is all that is available) of self-report of sunlight exposure. Some estimates might be discarded if the population is not similar to the study population. For example, studies of the accuracy of self-report of sunlight exposure in teenagers would not provide good external validity information for a study of melanoma cases and controls, because there would be little overlap in the age range of the teenagers who participated in the validity study and the melanoma cases and controls who participated in the analyst's study.

Even after discarding the poorly applicable validity data, there will often be a range of values reported in the multiple articles, and the analyst should decide how to best use these ranges. An analyst may approach this like a meta-analysis, aggregating the parameters (in this case, sensitivities and specificities among the diseased and undiseased) from the relevant literature. For instance, if five studies report the sensitivity and specificity of self-reported sunlight exposure, those five sets of estimates can be meta-analyzed to produce one summary sensitivity and one summary specificity. Like a regular meta-analysis, study results should be examined for heterogeneity. In the presence of meaningful heterogeneity, results should not be pooled. Instead, values of individual studies can be used in a multidimensional bias modeling approach (described below). If the results are homogenous, an average value can be used with simple bias analysis, or the 95% confidence interval range can be incorporated into a multidimensional bias analysis, probabilistic bias analysis, or multiple biases model.

If the validation data will be internal, then the analyst should allocate study resources to conduct the data collection required for the quantitative bias analysis and this requires planning. If nonparticipants will be crudely characterized with regard to basic demographic information such as age and sex, so that they can be compared to participants, then the data collection system and electronic database should allow for designation of nonparticipant status and for the data items that will be sought for nonparticipants. If a validation substudy will be implemented to characterize the sensitivity and specificity of exposure classification, then resources should be allocated to accomplish the substudy. A protocol should be written to sample cases and controls to participate in the diary verification of self-reported sunlight exposure. The substudy protocol might require additional informed consent, additional recruitment materials, and will certainly require instructions for subjects on how to record sunlight exposure in the diary and a protocol for data entry.

These examples do not fully articulate the protocols required to plan and collect the data that will inform a quantitative bias analysis. The same principles for designing well-conducted epidemiologic studies apply to the design of well-conducted validation studies, although many epidemiologists are less familiar and confident about designing validation substudies [11]. The reader is again referred to texts on epidemiologic study design, such as *Modern Epidemiology* [1], for the details of valid study design. The larger point, though, is that the data collection effort for a validation substudy should not be underestimated. The analyst should plan such studies at the outset, should allocate sufficient study resources to the data collection effort, and should assure that the validation substudy is completed with the same rigor as applied to the principal study to ensure the results also reach validity and precision standards.

Creating an Analytic Plan for a Bias Analysis

Valid epidemiologic data analysis should begin with an analytic strategy that includes plans for quantitative bias analysis at the outset. The plan for quantitative bias analysis should make the best use of the validation data collected per the design described above.

Type of Data: Record-Level Versus Summary

Some of the techniques for quantitative bias analysis described herein assume that the analyst has access to record-level data. That is, they assume that the original data set with information on each subject in the study is available for analysis. Record-level data allows for a wider range of methods for quantitative bias analysis. With record-level data, bias-adjustments for any of the three systematic errors can be made at the level of the individual subjects, which preserves correlations between the study variables and allows the analyst to adjust for these biases while also controlling for known, measured confounders, as in a typical analysis, or to conduct more sophisticated mediation or interaction analyses.

Some of the techniques described herein apply to summary or aggregate data. That is, they apply to data displayed as frequencies in summary contingency tables or as estimates of association and their accompanying conventional confidence intervals. Analysts or interested parties with access to only summary data (e.g., a reader of a published epidemiology study) can use these techniques to conduct quantitative bias analysis. Of course, analysts with access to record-level data can generate these summary data and might also use these techniques. In general, use of record level data in bias analyses will provide the analyst with greater flexibility when implementing these approaches. Analysts with access to record-level data will often prefer to use the analyses designed for record-level data over analyses designed for summary data.

Type of Bias Analysis

Table 2.3 summarizes the analytic strategies available to accomplish a quantitative bias analysis. The first column provides the names of the bias analysis techniques. The second column explains how bias parameters are treated in the corresponding techniques. Bias parameters are the values that are required to complete the quantitative bias analysis and are typically the most important factors in determining the magnitude of the bias. For example, to analyze bias due to classification errors, the sensitivity and specificity of the classification method (or its predictive values) are required. The sensitivity and specificity of classification are therefore the bias

Table 2.3 Summary of quantitative bias analysis techniques.

Analytic technique	Treatment of bias parameters	Number of biases analyzed	Output	Combines random error?	Computationally difficult/ intensive?
Simple sensitivity analysis	One fixed value assigned to each bias parameter	One at a time	Single bias-adjusted estimate of association	Usually no	Low/low
Multidimensional analysis	More than one value assigned to each bias parameter	One at a time	Range of bias-adjusted estimates of association	Usually no	Low/low
Probabilistic analysis (summary level)	Probability distributions assigned to each bias parameter	One at a time	Frequency distribution of bias-adjusted estimates of association	Yes	Moderate/ moderate
Probabilistic analysis (record level)	Probability distributions assigned to each bias parameter	One at a time	Frequency distribution of bias-adjusted estimates of association	Yes	High/moderate
Direct bias modeling and missing data methods	Estimates and variance obtained from information internal or external to the dataset	One at a time	Distribution of bias-adjusted estimates of association	Yes	Moderate/ moderate
Bayesian bias analysis	Probability distributions assigned to each bias parameter	Multiple biases at once	Distribution of bias-adjusted estimates of association	Yes	High/high
Multiple bias modeling	Probability distributions assigned to each bias parameter	Multiple biases at once	Frequency distribution of bias-adjusted estimates of association	Yes	High/high

parameters of that bias analysis. The third column shows whether biases are analyzed individually or jointly for the corresponding techniques. The fourth column describes the output of the technique and the fifth column answers whether random error can be combined with the output to reflect the total error in the estimate of association. The last column depicts the computational difficulty of each technique, in terms of both the difficulty of writing the computing code and the computing time and resources required to run the analysis. Note that different analytic techniques refer to a class of methods used to adjust for biases, but do not refer to any particular bias. Each could be used to bias-adjust for selection bias, misclassification, or an unmeasured confounder.

Analysts designing or implementing a quantitative bias analysis should weigh three considerations as they choose the appropriate technique. When more than one bias will be analyzed, most of the analytic methods will treat them individually and/or independently. If more than one bias will be analyzed, analysts should consider using methods in the lower rows of the table. Second, analysts should consider the inferential goal, which relates most closely to the output in the foregoing table. The most common inferential goal is to adjust the observed estimate of association to take account of the bias. This goal can be accomplished with all of these analytic methods. Another common inferential goal is to adjust the observed confidence interval to reflect total error: the sum of the systematic error and the random error. This goal can be accomplished with only probabilistic bias analysis, direct bias modeling, or Bayesian bias analysis (when only one bias will be analyzed) or multiple bias modeling (when more than one bias will be analyzed). A last common inferential goal is to determine whether a non-null estimate of association can be completely attributed to the bias. This goal requires examination of the bias from different combinations of the bias parameters, along with a determination of whether the combinations that yield a null result are reasonable. Because each combination is individually examined, and multiple combinations are required, this inferential goal is best accomplished by multidimensional bias analysis.

The third consideration is the computational difficulty of the quantitative bias analysis. Many of the analytic techniques associated with the types of bias analysis listed in Table 2.3 can be accomplished using spreadsheets or computing code available on the text's website (including links to other developers' resources). However, the computational difficulty varies widely by bias analysis type. Some of the bias analysis techniques, such as Bayesian methods discussed in Chapter 11, can require special software and specific knowledge of intermediate statistical techniques. As the computational difficulty grows, the analyst should expect to devote more time and effort to completing the analysis, and more time and presentation space to explaining and interpreting the method. In general, analysts should choose the computationally simplest technique that satisfies their inferential goal given the number of biases to be examined and whether multiple biases can be appropriately treated as independent of one another. When only one bias is to be examined, and only its impact on the estimate of association is central to the inference, then computationally straightforward simple bias analysis is sufficient. When more than one bias is to be examined, the biases are not likely independent, and an assessment of total error is required to satisfy the inferential goal, then the computationally most difficult and resource-intensive bias modeling will be required.

Further, an analyst interested in performing a bias analysis should carefully consider the information available to specify the required bias parameters. Consider an example of bias-adjusting an effect estimate for misclassification of exposure. If a substantial literature exists and gives estimates of sensitivity and specificity of exposure classification, using probabilistic bias methods that require specification of the distribution of the bias parameter may be useful. However, if no information is available regarding the sensitivity and specificity of classification, conducting a multidimensional bias analysis examining the bias-adjusted estimate of effect at various combinations of sensitivity and specificity could be more scientifically useful as it allows visualization of many possible combinations of plausible sets of bias parameters.

Finally, and most importantly, the state of the scientific research and impact of the bias should help the analyst decide which bias analysis approach to use. Preliminary research that will not be immediately used for any policy decision may be most suitable to relatively simple bias analysis techniques (the top rows of Table 2.3). As an area of research is more fully developed, the requisite knowledge to specify distributions for bias parameters should naturally develop with it and, as such, more complicated bias analysis techniques, which incorporate the distribution of the bias parameter, should be considered. When estimates of effect will be used to inform policy, it is incumbent on the analyst to fully model all possible sources of bias [7]. In these cases, multiple bias modeling or other more computationally intense bias modeling should be seen as a necessary approach.

Order of Bias Analysis Adjustments

When multiple sources of systematic error are to be assessed in a single study, the order of adjustments in the analysis can influence results of the bias analysis and should be carefully considered. However, whether the order of adjustments matters depends on the technique used for the bias analysis. A formal statistical approach to bias analysis would require the analyst to write down a probability model for each random variable under consideration, which would include a model for the outcome and a model for the various sources of bias. This approach will be discussed in Chapter 11. The individual probability models would be combined to form a joint probability model and bias-adjusted estimates would be derived from that model. In this formal approach, there is no need to consider the order of adjustments as all bias adjustments are, loosely speaking, implemented simultaneously. That is, there is no need to consider whether to adjust for selection bias before misclassification because both are addressed by a joint bias model.

This formal approach is often difficult to implement and many of the bias analysis techniques presented in this book offer excellent approximations that are much easier to implement. With these approximate techniques, order of adjustments must be carefully considered. The way in which we ease implementation of the bias modeling is to have specific bias parameters that are used to adjust for each type of bias. Because these approximate techniques are conditional on values for specific bias parameters, the information needed to inform these bias parameters will typically dictate the order of adjustment. For instance, consider a situation in which there is both selection bias and outcome misclassification. As discussed in Chapter 6, one set of bias parameters necessary to adjust for outcome misclassification are sensitivity and specificity. As discussed in Chapter 4, the bias parameters necessary to adjust for selection bias are the probability of selection into the study, conditional on true exposure and true outcome status. In this case, logic dictates that, to adjust for selection bias, we must first adjust for outcome misclassification, so we can obtain an estimate of the true outcome status for people in our study. If we had, instead, adjusted for selection bias before misclassification, we would obtain a less accurate result because the selection bias-adjustment would assume the misclassified outcome was in fact the true outcome. Alternatively, it is possible that information for the selection probabilities was available based only on misclassified outcome status. In this case, adjustment for selection bias is only possible with the misclassified outcome data, reversing the order of bias-adjustments would be appropriate, and one would first adjust for selection bias and then adjust for misclassification.

As a rule of thumb, it is often the case that an analyst should adjust for biases in the reverse order in which the errors arose. Errors in classification arise in the study population, as an inherent part of the data collection and analysis, so should ordinarily be bias-adjusted first. Selection bias arises from differences between the study participants and the base population, so should ordinarily be bias-adjusted second. Confounding (or at least the relationship between variables that leads to confounding) exists at the level of the source population, so error arising from an unmeasured confounder should ordinarily be analyzed last. While this rule of thumb may often hold, exceptions will occur as described above. In short, one should follow the study design in reverse to determine the appropriate order of bias analysis. See Chapter 12 on multiple bias analysis for a more complete discussion of the order of bias-adjustments.

Bias Analysis Techniques

Simple Bias Analysis

With a simple bias analysis, the estimate of association obtained in the study is adjusted a single time to account for only one bias (uncontrolled confounding, selection bias or information bias. The output is a single bias-adjusted estimate of association. This estimate of association may or may not incorporate random error. For example, Marshall et al. (2003) investigated the association between little league injury claims and type of baseball used (safety baseball or traditional baseball) [12]. They observed that use of safety baseballs was associated with a reduced risk of ball-related injury (rate ratio = 0.77; 95% CI 0.64, 0.93). They were concerned that injuries might be less likely to be reported when safety baseballs were used than when traditional baseballs were used, which would create a biased estimate of a protective effect. To conduct a simple bias analysis, they estimated that no more than 30% of injuries were unreported and that the difference in reporting rates was no more than 10% (the bias parameters). Their inferential goal was to adjust the estimate of association to take account of this differential underreporting. With this single set of bias parameters, the estimate of association would equal a rate ratio of 0.88. They concluded that a protective effect of the safety ball persisted after adjusting for the potential for differential underreporting of injury, at least conditional on the accuracy of the values assigned to the bias parameters.

Cain et al. (2006) conducted a simple bias analysis with the inferential goal of evaluating whether their estimate of association could be completely attributed to bias [13]. Their study objective was to estimate the association between highly active antiretroviral therapy (HAART) and multiple acquired immunodeficiency syndrome (AIDS)-defining illnesses. Averaging over multiple AIDS-defining illnesses, the hazard of an AIDS-defining illness in the HAART calendar period was 0.34 (95% CI 0.25, 0.45) relative to the reference calendar period. The authors were concerned that differential loss-to-follow-up might account for the observed protective effect. They conducted a "worst-case" simple bias analysis by assuming that the 68 men lost-to-follow-up in the HAART calendar period had an AIDS-defining illness on the date of their last follow-up, and that the 16 men lost-to-follow-up in the calendar periods before HAART was introduced did not have an AIDS-defining illness by the end of follow-up. With these bounding assumptions, the estimated effect of HAART equaled a hazard ratio of 0.52. The inference is that differential loss-to-follow-up could not account for all of the observed protective effect of HAART against multiple AIDS-defining illnesses, presuming that this analysis did in fact reflect the worst-case influence of this bias.

Note that in both examples, the estimate of association was adjusted for only one source of error, that the adjustment was not reflected in an accompanying interval (only a point estimate was given), and that random error was not simultaneously incorporated to reflect total error. These are hallmarks of simple bias analysis.

Multidimensional Bias Analysis

Multidimensional bias analysis, or tabular bias analysis [14], is an extension of simple bias analysis. The analyst adjusts an effect measure for one type of bias but does so repeatedly at multiple values or combinations of values of the bias parameters, rather than a single value. Multiple bias analysis will often be preferred when relatively little information is available regarding the values to assign to the bias parameters. Rather than specifying a single value for the bias parameter, as in simple bias analysis, or an entire distribution for the bias parameter, as in probabilistic bias analysis, the analyst will repeat the bias analysis at various plausible bias parameter values. For example, Sundararajan et al. (2002) investigated the effectiveness of 5-fluorouracil adjuvant chemotherapy in treating elderly colorectal cancer patients [15]. Patients who received 5-fluorouracil therapy had a lower rate of colorectal cancer mortality than those who did not (hazard ratio = 0.66; 95% CI 0.60, 0.73). The analysts were concerned about bias from confounding by indication because the therapy assignment was not randomized. To assess the potential impact of this unmeasured confounder, they made assumptions about the range of (1) the prevalence of an unknown binary confounder, (2) the association between the confounder and colorectal mortality, and (3) the association between the confounder and receipt of 5-flourouracil therapy (these are the bias parameters). The inferential goal was to determine whether confounding by indication could completely explain the observed protective effect. Most combinations of the bias parameters also yielded a protective estimate of association; only extreme scenarios resulted in near-null estimates of association. The set of bias adjusted estimates of association, which do not incorporate random error, is the output of the multidimensional bias analysis. The authors wrote, "Confounding could have accounted for this association only if an unmeasured confounder were extremely unequally distributed between the treated and untreated groups or increased mortality by at least 50%." They therefore concluded that the entire protective effect could not be reasonably attributed to confounding by indication, which answered their inferential goal, at least conditional on the accuracy of the ranges assigned as values to the bias parameters. While multidimensional bias analysis provides more information than simple bias analysis in that it provides a set of bias-adjusted estimates, it does not yield a frequency distribution of adjusted estimates of association. Each adjusted estimate of association is evaluated individually, so the analyst or reader gains no sense of the most likely adjusted estimate of association (i.e., there is no central tendency) and no sense of the width of the distribution of the adjusted estimate of association (i.e., there is no frequency distribution of corrected estimates). Multidimensional bias analysis also addresses only one bias at a time, a disadvantage that it shares with simple bias analysis, and typically does not incorporate random error.

Probabilistic Bias Analysis

Probabilistic bias analysis is an extension of the previous techniques; the analyst assigns probability distributions to the bias parameters [16–18], rather than assigning them single fixed values (as with simple bias analysis) or a series of discrete values within a range (as with multidimensional bias analysis). The analyst repeatedly samples from the bias parameter distribution. For each set of sampled bias parameters, the analyst performs a simple bias analysis outlined above. The results of these repeated simple bias analyses, accounting for all modeled sources of uncertainty, give a frequency distribution of adjusted estimates of association, which can be presented and interpreted similarly to a conventional point estimate. Random error in the data and uncertainty in the bias parameters are both incorporated in a simulation interval (rather than a frequentist confidence interval). Like the earlier methods, only one bias at a time is examined. For example, in a study of the association between periconceptional vitamin use and preeclamptic risk, Bodnar et al. (2006) were concerned about confounding by fruit and vegetable intake, which had not been measured [19]. The odds ratio associating regular periconceptional use of multivitamins with preeclampsia equaled 0.55 (95% CI 0.32, 0.95). High intake of fruits and vegetables is more common among vitamin users than nonusers and also reduces the risk of preeclampsia. Bodnar et al. created a distribution of the potential relative risk due to confounding using external information about the strength of association between fruit and vegetable consumption and vitamin use, strength of association between fruit and vegetable consumption and preeclamptic risk, and prevalence of high intake of fruits and vegetables. They used Monte Carlo methods to integrate the conventional odds ratio, distribution of the relative risk due to confounding, and the random error to generate output that reflects both an adjusted point estimate and uncertainty intervals. As expected, this probabilistic bias analysis suggested that the conventional results were biased away from the null. The conventional OR (0.55) was attenuated to 0.63 (95% simulation interval: 0.56, 0.72 after accounting for only systematic error; 95% simulation interval: 0.36, 1.12 after accounting for both systematic and random error). Unlike with the previous bias analysis methods, probabilistic bias analysis provides a measure of the central tendency of the bias-adjusted estimate of association (0.63) and the amount of uncertainty in that estimate (as portrayed by the simulation intervals), and random error is integrated with systematic error. The bias analysis suggests that vitamin use is associated with a reduced risk of preeclampsia, even after taking account of the unmeasured confounding by fruit and vegetable intake and random error, assuming a valid bias model, which satisfies the inferential goal. The original analysis somewhat overestimated the protective effect and the conventional interval somewhat underestimated the total error, at least conditional on the accuracy of the assigned bias parameters.

Multiple Bias Modeling

Multiple bias modeling is also an extension of simple bias analysis in which the analyst assigns probability distributions to the bias parameters, rather than single values or ranges, and now the analyst examines the impact of more than one bias at a time [16–18]. For example, Lash and Fink (2004) conducted a case-control study of the effect of pregnancy termination (induced and spontaneous) on breast cancer risk among parous residents of Massachusetts ages 25–55 years at breast cancer diagnosis [20]. The study included all Massachusetts breast cancer cases reported to the Massachusetts cancer registry between 1988 and 2000 arising from the population of women who gave birth in Massachusetts between 1987 and 1999. The conditional adjusted odds ratio estimate of the risk ratio of breast cancer, comparing women who had any history of pregnancy termination with women who had no history of pregnancy termination, equaled 0.91 (95% CI 0.79, 1.0). Information on history of pregnancy termination and potential confounders was recorded on birth certificates before the breast cancer diagnosis, so errors in recall or reporting of this history should have been nondifferentially and independently related to breast cancer status. It may be that the observed near-null result derives from nondifferential, independent misclassification of history of termination, thereby masking a truly nonnull result. In addition, the study may have been subject to a selection bias if women who migrated from Massachusetts between the time they gave birth and the time they developed breast cancer differed from those who did not migrate with respect to pregnancy terminations. The inferential goal was to adjust the estimate of association and its interval to account for these two biases. We first implemented a probabilistic bias analysis with the following bias parameters: (1) a triangular distribution of sensitivity of termination classification ranging from 69% to 94% with a mode of 85%, (2) a triangular distribution of specificity of termination classification ranging from 95% to 100% with a mode of 99%, and (3) a prevalence of termination in the study base ranging from 20% to 30% with a mode of 25% [21–23]. To allow for small deviations from perfectly nondifferential misclassification, we allowed the sensitivity and specificity of termination classification in cases, versus controls, to vary independently of one another between 0.9-fold and 1.1-fold (e.g., if the sensitivity in cases was chosen to be 85%, then the sensitivity in the controls could be no less than 76.5% and no greater than 93.5%). These were the bias parameters used to address misclassification. The probabilistic bias analysis yielded a median odds ratio estimate of 0.90 (95% simulation interval 0.62, 1.2). Conditional on the accuracy of the distributions assigned to the bias parameters, this probabilistic bias analysis (which only accounts for one source of bias) supports the notion that the result is unlikely to arise from a bias toward the null induced by nondifferential nondependent misclassification of the dichotomous exposure variable, because the central tendency remained near null and the interval remained narrow.

   The expectation was that the estimates of association should be immune to initial selection bias because study eligibility did not require active participation. However, loss-to-follow-up by migration out of Massachusetts after giving birth may have been differentially related to history of pregnancy termination and breast cancer incidence. In a previous investigation of similar design, Tang et al. (2000) used state-specific migration data to estimate that the loss-to-follow-up may have led to a 5% underestimate of the relative effect [24]. To implement a multiple bias model, we combined the probabilistic misclassification bias analysis results above with a triangular bias parameter distribution ranging from 1 to 1.1 with mode 1.05 to account for the potential selection bias induced by migration. This multiple bias model, which accounts for the selection bias and the misclassification as well as random error, yielded a median estimate of 0.95 (95% simulation interval 0.65, 1.3; with random error incorporated). While the multiple bias analysis median estimate and the conventional point estimate are nearly identical, the width of the multiple bias model's simulation interval on the log scale is more than twice the width of the conventional 95% confidence interval, which conveys the additional uncertainty arising from the systematic errors. Taken together and conditional on the accuracy of the distributions assigned to the bias parameters, the bias analysis shows that the null result is unlikely to have arisen from misclassification of termination status or from selection bias arising from differential migration of subjects between the date of giving birth and the date of cancer diagnosis record, and that the total uncertainty is larger than reflected in the conventional confidence interval, but still not large enough to infer a nonnull result with any confidence.

Direct Bias Modeling and Missing Data Methods

Direct bias modeling is the term we use for a group of related methods that use study data directly to obtain a bias-adjusted estimate and its interval. For misclassification problems, these methods use estimates of sensitivities and specificities (or positive and negative predictive values) and their variances from validation data or external sources. These estimates are directly entered into equations to generate misclassification-adjusted effect measures and their confidence intervals. Like probabilistic bias analysis, these methods incorporate uncertainty in the bias parameters; however, they do so without generally needing to resort to simulations. We also include missing data methods and regression calibration (an approximate missing data method) in this group of methods. While regression calibration is used for measurement error in continuous variables, missing data methods more generally can be used to model any of the systematic biases: selection bias, uncontrolled confounding or information bias. Indeed, all of these systematic biases can be conceived as a special type of missing data: missing data on people not selected into the study; missing data on the confounder of interest; missing data on the gold standard exposure or outcome. Like probabilistic bias analysis, these methods also yield a bias-adjusted point estimate and an accompanying interval. The information to inform these bias-adjustments is typically internal to the original study (though direct methods can accommodate external validation data), for example using an internal validation study that compares a mismeasured variable with its gold-standard measurement to inform the bias-adjustment. Direct and missing data methods therefore incorporate the conventional sampling error and the sampling error of the values assigned to the bias parameters to yield an overall estimate of the variance of the bias-adjusted estimate.

Direct methods may have an advantage in that the methods rely on straightforward equations and do not require computer intensive simulations. Answers can be obtained relatively easy. Missing data methods require simulations but all modern statistical software programs include functions to address missing data. The accessibility of these methods has a downside as well. The direct methods are somewhat limited in the biases for which they can adjust and the effect measures they can estimate. Missing data methods, as commonly implemented, are constrained to situations in which internal validation data are the source of bias parameters. Incorporating external information is difficult, as is adjusting for multiple biases simultaneously.

As an example of direct methods for exposure misclassification, Greenland et al. (1988) estimated the association between self-reported use of antibiotics during pregnancy and the occurrence of Sudden Infant Death (SIDS) in the offspring [25]. The conventional odds ratio associating self-reported antibiotic use during pregnancy with SIDS was 1.42 (95% CI: 1.11, 1.83). The analysts recognized that the accuracy of self-reported antibiotic use during pregnancy was likely imperfect and may have been different between cases and controls given the retrospective design. They compared self-report of antibiotic use with medical records to validate the self-reported information among a subset of participants. Using the internal validation data and the bias-adjusted estimator of Marshall (1990) [26], one obtains a bias-adjusted estimate of the odds ratio of 1.21 with 95% confidence interval (0.79, 1.87). This estimate accounts for the errors in reporting that were observed in the internal validation study, as well as the variance of the estimates of these predictive values arising from the fact that they are proportions measured in a fixed sample, so with binomial error. The bias-adjusted estimate is nearer to the null than the conventional estimate and has an interval 1.4 times wider.

Bayesian Bias Analysis

Bayesian bias analysis methods use formal Bayesian statistical methods that incorporate systematic errors as data generating mechanisms. Many of the probabilistic bias analysis methods that we discuss in this text are approximately Bayesian methods [27, 28]. Inferential results obtained from a probabilistic bias analysis would typically be similar to inferential results obtained from a corresponding Bayesian analysis. In some settings, however, a formal Bayesian approach may yield quite different results and inferences than the corresponding probabilistic bias analysis. Although there has not been a formal theoretical examination of exactly when the two approaches differ to an appreciable extent, there is some literature we can use to identify when this might occur, which we review and illustrate with examples in Chapter 11.

By way of example, Bodnar et al. were interested in the relation between pre-pregnancy body mass index (BMI) and preterm delivery [29, 30]. Pre-pregnancy BMI categories were derived from self-reported pre-pregnancy body weight collected to complete the birth certificate. Because this information was collected after delivery, the accuracy of self-reported pre-pregnancy body weight may have depended on preterm versus term birth. The analysts validated pre-pregnancy BMI category on the birth certificate against pre-pregnancy BMI category derived from medical record review. Forty-eight groups were created for validation (pre-term versus term delivery; two categories of race/ethnicity, three categories of gestational weight gain, and four categories of pre-pregnancy BMI). The goal was to validate information from 30 birth certificates against medical record information for each of the 48 groups. Information in some groups was quite sparse because of the combinations of low prevalence conditions. For example, less than 20% of births were pre-term, less than 20% of mothers were African-American, and the combination of high gestational weight gain and under-weight pregnancy BMI category was quite rare, so the simultaneous combination of these four characteristics was rare in the population available for validation. As a result, there were fewer than 30 women to validate in some of the 48 categories. Rather than use the predictive values exactly as measured in each of the 48 categories, MacLehose et al. used a two-stage semi-Bayesian hierarchical model to borrow strength between group-specific bias parameters to adjust for exposure misclassification [31]. Bias parameters from this hierarchical Bayesian model were often more substantively reasonable and often had smaller variance. Model results supported evidence of an increased risk of early preterm birth among severely obese mothers, relative to normal weight mothers, and addressed reasonable concerns that the retrospective design may have biased results away from the null, conditional on the accuracy of the bias model.

Assigning Values and Distributions to Bias Parameters

In the interpretation of each of the preceding examples, the inference was always described as conditional on the accuracy of the values or distributions assigned to the bias parameters (or more generally, conditional on the accuracy of the bias model). It is, of course, impossible to know the accuracy of these assignments. Further, as described in future chapters, small inaccuracies in bias parameter specification can result in substantial inferential changes. In many cases, it will be very difficult to develop intuition for how much a slight change in a bias parameter will affect results. If an interested stakeholder supports a different set of bias parameter values, the bias analysis can and should be repeated with the alternate set of values to see whether the results of the bias analysis and the inference change substantially.

As will be described in Chapter 3, the assignment of values and distributions to bias parameters is a challenging process. In many instances, very little information may be available to inform assignments of values or distributions to bias parameters. Educated guesses may be required to implement a quantitative bias analysis. It is imperative, therefore, that in any bias analysis the values assigned to the bias parameters are explicitly given, the basis for the assignment explicitly provided, and any inference resting on the results of the bias analysis explicitly conditioned on the accuracy of the assignments.

Directed Acyclic Graphs

Directed Acyclic Graphs (DAGs), a form of causal diagrams introduced into the epidemiology literature in 1999 [32], have proven to be a useful structural approach to clarifying notions of confounding, selection bias and misclassification [33]. In this section, we give a brief introduction to DAGs and focus on aspects that will be most useful in quantitative bias analysis. Hernan and Robins [33] and VanderWeele and Rothman [34] are excellent introductions to DAGs. A more extensive and theoretical description of DAGs can be found in Pearl, 2017 [35].

DAGs are a graphical representation of an analyst's prior beliefs concerning the relations between factors in an epidemiologic study of interest. They describe the

data generating mechanism as the analyst understands it. A DAG depicts hypothesized causal relations between variables that are important to the study, whether
measured or not measured. The DAG should be informed from the existing literature
and as much prior knowledge as possible, including an understanding of the study's
design and planned analyses. Whether a DAG will be scientifically helpful will
depend on faithful specification of the DAG, so analysts are urged to be as thorough
as possible when depicting a DAG and to avoid simplifications.

Consider, again, the study described above in which we wish to estimate the
effect of sunlight exposure (high vs low levels) on melanoma in a case-control study.
Figure 2.2 depicts a DAG for this study. The variables of interest (which, importantly, are not limited to those we have measured) are depicted as nodes in the graph:
"sunlight," "melanoma," "age," "tanning beds," and "S". Suppose the analysts were
interested in age as a possible confounder and measured it in the case-control study.
Recall from the description above that "tanning bed" frequency was not ascertained
in the actual study. The node "S" is a selection node that depicts whether an
individual was selected into this case-control study. Nodes are connected to one
another in a DAG via unidirectional arrows that represent causal effects. In Figure 2.2, the arrow from the node "sunlight" to the node "melanoma" represents the
causal effect of interest. The arrow from "melanoma" to "S" represents the increased
probability of selection into the study if a person has melanoma (since they will be
invited to participate as a case with probability $=1$). The arrow from sunlight to "S"
represents the concern that people are more likely to consent to be in the study if they
have accumulated a large sunlight exposure and are concerned about their melanoma
risk. A node at the beginning of an arrow (e.g., "age") is referred to as a parent while
the node at the end of an arrow (e.g., "sunlight") is referred to as a child.

A path in a DAG is a series of adjacent arrows. Paths are often defined relative to
the exposure and outcome of interest, though they do not have to be. In Figure 2.2,
there are six paths from "sunlight" to "melanoma:"

1. "Sunlight" → "Melanoma"
2. "Sunlight" ← "Age" → "Melanoma"
3. "Sunlight" ← "Tanning Beds" → "Melanoma"
4. "Sunlight" ← "Age" → "Tanning Beds" → "Melanoma"
5. "Sunlight" ← "Tanning Beds" ← "Age" → "Melanoma"
6. "Sunlight" → "S" ← "Melanoma"

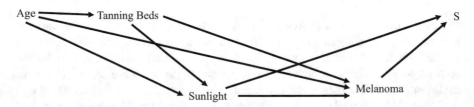

Figure 2.2 Example of a Directed Acyclic Graph encoding assumptions about the relation
between exposure to sunlight and the occurrence of melanoma.

A directed path is one in which all arrows in the path follow the same direction. Of the paths above, only path (1) is a directed path from the exposure to the outcome. To prevent a logical impossibility where a node causes itself, feedback loops are prohibited from occurring in DAGs; no directed path may lead back to the variable from which it begins. Any directed path from the exposure of interest to the outcome of interest represents a main effect (either a direct or indirect effect).

A path can be either blocked or unblocked. There are two common ways in which a path can be blocked: (1) by controlling for a variable along the path; or (2) by a collider (or child of a collider) that has not been controlled being present on the path. Controlling for a variable can occur via statistical analysis or by design. Common statistical analysis methods aimed at controlling for a variable are regression adjustment, standardization or stratification. A common design strategy in epidemiology that can block a path is by only enrolling people with a given value of that variable into the study, which is known as restriction (e.g., only enrolling people aged 20–25 into the study). Restriction is also sometimes employed in the analysis. In cohort studies, another design strategy that can block a path is matching on a variable in the path. The second way a path can be blocked is via a collider. A collider is a node along a path that has two parent nodes in the path such that the path enters through the head of one arrow and exits through the head of another. For instance, the "S" node is a collider of the path from sunlight to S to melanoma in Figure 2.2, since both "melanoma" and "sunlight" are parent nodes of S. A path that includes a collider is blocked because information must flow forward in time from one parent of the collider to the other. Colliders represent nodes where information "collides" and cannot continue to move forward in time. Thus path (6) is blocked by the collider "S." We note that variables are not colliders in and of themselves, but only in relation to a path, as a variable can be a collider for one path and a non-collider for another.

Backdoor paths are paths (not necessarily following the direction of the arrows) between the exposure and outcome that begin with an arrow pointing into the exposure. In this example, paths (2), (3), (4) and (5) are backdoor paths. Backdoor paths are crucial to a structural understanding of confounding. All four of these paths are unblocked backdoor paths and the presence of an unblocked backdoor path signals the possibility of confounding. Unblocked backdoor paths represent routes through which the exposure and disease can be associated, even if there is no effect of the exposure on the disease. This is how we typically represent confounders (variables associated with the exposure and the outcome and not on the causal pathway). For example, if path (1) did not exist there would be no causal effect of sunlight on melanoma. However, if path (2) exists but (1) does not, it is possible that a study would find an association between sunlight and melanoma because (for example) older study participants are more likely to get melanoma and less likely to be exposed to sunlight. This is an example of confounding bias, which in DAGs is often referred to as arising from common causes. We show how DAGs are useful in detecting the potential for confounding in the next paragraph.

In this example, we should be concerned that the crude effect of sunlight on melanoma is confounded because of unblocked backdoor paths (2), (3), (4), and (5). The confounding induced by these backdoor paths can be adjusted by controlling (for example via regression analysis) for variables along the unblocked backdoor paths. For instance, if age was measured in this study, we can adjust for age in a regression model with melanoma as the dependent variable and sunlight as the independent variable and this adjustment will block paths (2), (4) and (5) since they include age along the path. However, unless we also adjust for a variable in path (3), we will still have uncontrolled confounding along that path.

It is important to recognize the utility of DAGs for identifying confounders that should be controlled analytically. In some cases, those confounders are measured (as with "age") and confounding can be controlled through typical statistical procedures. In other cases, confounders will be unmeasured (as with "tanning beds") and failure to control for them potentially leads to biased estimates. The quantitative bias analysis methods presented in Chapter 5 allow users to adjust for unmeasured confounders by specifying values for bias parameters.

Although unblocked backdoor paths offer a structural understanding of confounding, colliders help offer a structural understanding of selection bias. When a collider is controlled (either through how participants are selected into the study or through statistical analysis), the parents of the collider will become associated, even if there is no association between them before controlling for the collider. In fact, not only will the parents of the collider be associated, but all ancestors of the collider will also be associated. This induced association will open up a pathway between the parents that can lead to new unblocked pathways from the exposure to the outcome that did not exist prior to control for the collider. In Figure 2.2, notice that our study design explicitly stratifies on the node "S," which happens to be a collider, since only those in the study (S = 1) will be analyzed. As a result, the parents of that node, "sunlight" and "melanoma" [path (6)] will be associated. Those who have melanoma are more likely to be enrolled in the study by virtue of this being a case-control study. Those with high sunlight exposure will selectively agree to enroll in the study because of concerns about their sunlight exposure. Even if there were no association between sunlight exposure and melanoma [i.e., no path (1)], we would observe an excess of melanoma cases who had been exposed to high levels of sunlight and estimate a spurious effect estimate. We refer to this bias as collider stratification bias and note that it is one mechanism that gives rise to selection bias, as will be discussed in more detail in Chapter 4, but not the only one [36, 37].

DAGs offer a visual and straightforward method to understand possible biases. Drawing a DAG will help analysts encode their assumptions about causal structures, allowing them to enumerate the possible sources of bias in their study. There is a temptation when drawing DAGs to only focus on variables that have been (or will be) measured in a given study and to ignore selection into the study. We strongly encourage analysts to draw a DAG that is faithful to their understanding of the causal process as possible to understand the possible sources of bias. Enumerating these biases will help direct quantitative bias analysis. Further, analysts should note that drawing a DAG is a subjective process and, while it should be based on literature in the field and their understanding of the causal processes and study design elements, it is possible that two scientists will draw different DAGs for the same study question. This difference is an excellent opportunity to determine why their prior beliefs may differ and, if no resolution is found, to analyze the data under both sets of DAGs. It can also guide additional data collection that would be necessary before the question of interest can be answered.

Using DAGs for bias analyses involving selection bias will be discussed in Chapter 4 and confounding will be discussed in Chapter 5. The extension of DAGs to misclassification will be discussed in Chapter 6.

Conclusion

Well-conducted epidemiologic research begins with a sound design, including valid methods of data collection and assurance that the collected data will yield a sufficiently precise and valid estimate of association. Nonetheless, even the estimate of association obtained from a well-designed study will inevitably be susceptible to residual error. Analysts conventionally calculate a confidence interval to quantify residual random error; quantitative bias analysis similarly quantifies residual systematic error. Just as analysts plan for their conventional analyses as they design their study, so too should they plan for quantitative bias analysis as they design their study. By incorporating quantitative bias analysis into all elements of study design and analysis, analysts will be best able to achieve the overarching objective of obtaining a valid and precise estimate of the effect of an exposure on the occurrence of disease.

References

1. Lash T, VanderWeele TJ, Haneuse S, Rothman KJ, Haneuse S. Modern Epidemiology. 4th ed. Philadelphia: Wolters Kluwer; 2021.
2. Greenland S, Robins JM. Identifiability, exchangeability, and epidemiological confounding. Int J Epidemiol. 1986;15:413–9.
3. Greenland S, Robins JM. Identifiability, exchangeability and confounding revisited. Epidemiol Perspect Innov. 2009;6:1–9.
4. Greenland S. Randomization, statistics, and causal inference. Epidemiology. 1990;1:421–9.
5. Greenland S, Robins JM, Pearl J. Confounding and collapsibility in causal inference. Statistical Science. 1999;14:29–46.
6. Cummings P. The relative merits of risk ratios and odds ratios. Archives of pediatrics & adolescent medicine. 2009;163:438–45.
7. Lash TL, Fox MP, MacLehose RF, Maldonado G, McCandless LC, Greenland S. Good practices for quantitative bias analysis. Int J Epidemiol. 2014;43:1969–85.
8. Poole C. Low P-values or narrow confidence intervals: which are more durable? Epidemiology. 2001;12:291–4.

9. Lash TL, Fox MP, Cooney D, Lu Y, Forshee RA. Quantitative bias analysis in regulatory settings. Am J Public Health. 2016;106:1227–30.
10. Fox MP, Lash TL. Quantitative bias analysis for study and grant planning. Annals of Epidemiology. 2020;43:32–6.
11. Fox MP, Lash TL, Bodnar LM. Common misconceptions about validation studies. International Journal of Epidemiology. 2020;49:1392–6.
12. Marshall SW, Mueller FO, Kirby DP, Yang J. Evaluation of safety balls and faceguards for prevention of injuries in youth baseball. JAMA. 2003;289:568–74.
13. Cain LE, Cole SR, Chmiel JS, Margolick JB, Rinaldo CR Jr, Detels R. Effect of highly active antiretroviral therapy on multiple AIDS-defining illnesses among male HIV seroconverters. Am J Epidemiol. 2006;163:310–5.
14. Gustafson P, McCandless LC. When Is a sensitivity parameter xxactly that? Statist Sci. 2018;33:86–95.
15. Sundararajan V, Mitra N, Jacobson JS, Grann VR, Heitjan DF, Neugut AI. Survival associated with 5-fluorouracil-based adjuvant chemotherapy among elderly patients with node-positive colon cancer. Ann Intern Med. 2002;136:349–57.
16. Lash T, Fink AK. Semi-automated sensitivity analysis to assess systematic errors in observational data. Epidemiology. 2003;14:451–8.
17. Phillips CV. Quantifying and reporting uncertainty from systematic errors. Epidemiology. 2003;14:459–66.
18. Greenland S. Multiple-bias modeling for analysis of observational data. J R Stat Soc Ser A. 2005;168:267–308.
19. Bodnar LM, Tang G, Ness RB, Harger G, Roberts JM. Periconceptional multivitamin use reduces the risk of preeclampsia. Am J Epidemiol. 2006;164:470–7.
20. Lash TL, Fink AK. Null association between pregnancy termination and breast cancer in a registry-based study of parous women. Int J Cancer. 2004;110:443–8.
21. Holt VL, Daling JR, Voigt LF, McKnight B, Stergachis A, Chu J, et al. Induced abortion and the risk of subsequent ectopic pregnancy. Am J Public Health. 1989;79:1234–8.
22. Werler MM, Pober BR, Nelson K, Holmes LB. Reporting accuracy among mothers of malformed and nonmalformed infants. Am J Epidemiol. 1989;129:415–21.
23. Wilcox AJ, Horney LF. Accuracy of spontaneous abortion recall. Am J Epidemiol. 1984;120: 727–33.
24. Tang MT, Weiss NS, Malone KE. Induced abortion in relation to breast cancer among parous women: a birth certificate registry study. Epidemiology. 2000;11:177–80.
25. Greenland S. Variance estimation for epidemiologic effect estimates under misclassification. Stat Med. 1988;7:745–57.
26. Marshall RJ. Validation study methods for estimating exposure proportions and odds ratios with misclassified data. J Clin Epidemiol. 1990;43:941–7.
27. Steenland K, Greenland S. Monte Carlo sensitivity analysis and Bayesian analysis of smoking as an unmeasured confounder in a study of silica and lung cancer. Am J Epidemiol. 2004;160: 384–92.
28. MacLehose RF, Gustafson P. Is probabilistic bias analysis approximately Bayesian? Epidemiology. 2012;23:151–8.
29. Bodnar LM, Abrams B, Bertolet M, Gernand AD, Parisi SM, Himes KP, et al. Validity of Birth Certificate-Derived Maternal Weight Data. Paediatric and Perinatal Epidemiology. 2014;28: 203–12.
30. Bodnar LM, Siega-Riz AM, Simhan HN, Diesel JC, Abrams B. The impact of exposure misclassification on associations between prepregnancy body mass index and adverse pregnancy outcomes. Obesity (Silver Spring). 2010;18:2184–90.
31. MacLehose RF, Bodnar LM, Meyer CS, Chu H, Lash TL. Hierarchical semi-Bayes methods for misclassification in perinatal epidemiology. Epidemiology. 2018 Mar; 29(2):183–90.
32. Greenland S, Pearl J, Robins JM. Causal diagrams for epidemiologic research. Epidemiology. 1999;10:37–48.

33. Hernan MA, Robins JM. Causal Inference: What if. Boca Raton: Chapman & Hall/CRC; 2020.
34. VanderWeele TJ, Rothman KJ. Formal causal models. In: Lash T, VanderWeele TJ, Haneuse S, Rothman KJ, editors. Modern Epidemiology. 4th ed. Philadelphia: Wolters Kluwer; 2021. p. 33–51.
35. Pearl J. Causality. New York: Cambridge University Press; 2000.
36. Greenland S. Response and follow-up bias in cohort studies. Am J Epidemiol. 1977;106:184–7.
37. Hernán MA. Invited Commentary: Selection bias without colliders. Am J Epidemiol. 2017;185: 1048–50.

Chapter 3
Data Sources for Bias Analysis

Bias Parameters

All bias analyses adjust a conventional estimate of effect to account for bias introduced by systematic error. These quantitative modifications combine the data used for the conventional estimate of association (*e.g.,* a risk difference or a rate ratio) with equations that adjust the conventional estimate for the expected impact of the systematic error. The equations comprise the bias model, which links the conventional estimate with the bias-adjusted estimate. These equations have parameters, called bias parameters, which ultimately determine the direction and magnitude of the bias-adjustment. For example:

- The proportions of all eligible subjects who participate in a study, simultaneously stratified into subgroups of persons with and without the outcome and within categories of at least the exposure variable of interest, are bias parameters. The values assigned to these parameters determine the direction and magnitude of selection bias.
- The sensitivity and specificity of exposure classification, within subgroups of persons with and without the disease outcome of interest, are bias parameters. The values assigned to the parameters determine the direction and magnitude of bias introduced by exposure misclassification.
- The strength of association between an unmeasured confounder and the exposure of interest and between the unmeasured confounder and the disease outcome of interest are bias parameters. The values assigned to these parameters affect the direction and magnitude of bias introduced by uncontrolled confounding.

Complete descriptions of the bias parameters required to bias-adjust estimates for selection bias, uncontrolled confounding, and information bias will be presented in their corresponding chapters, along with a comprehensive definition of each bias parameter. The current chapter pertains to the data sources that users of quantitative

© Springer Nature Switzerland AG 2021
M. P. Fox et al., *Applying Quantitative Bias Analysis to Epidemiologic Data*,
Statistics for Biology and Health, https://doi.org/10.1007/978-3-030-82673-4_3

bias analysis can examine to assign numeric values or probability distributions to each of the required bias parameters for the specific problem being interrogated.

Internal Data Sources

In many studies, the information to assign values or probability distributions to bias parameters can be obtained from a subgroup of the study population that yielded the conventional estimate of effect. When a substudy of those included in the larger study yields information about the values to assign to the bias parameters and satisfies necessary assumptions, then analysts may have greater confidence in the applicability of those bias parameters since they are derived from the study population of interest.

Selection Bias

When some eligible participants refuse to participate in a study, or some initial participants do not complete the study, the nonparticipation and/or loss to follow-up from the study could introduce selection bias. In particular, if selection is related to both the exposure and the outcome, such that the proportion of participants who are included in the study depends on both exposure and outcome status, then the conventional estimate of effect will be biased (other scenarios can also introduce selection bias, as described in Chapter 4). One can bias-adjust the conventional estimate of effect for the different participation proportions, but only if reasonable estimates of the probabilities can be assigned to all four combinations of disease and exposure (assuming both are operationalized as dichotomous variables). This assignment is challenging however, as one seldom knows both the exposure and outcome status of nonparticipants or those lost to follow-up; continued participation in the study is generally required to learn their exposure and outcome status as well as any other relevant information about them necessary to estimate the exposure-outcome relation. One way to get at this information is to ask nonparticipants to answer only a brief questionnaire that provides limited information on exposure and disease status. If all nonparticipants agree to answer the brief questionnaire, then the necessary selection proportions can be calculated from the combined information from participants and nonparticipants in a simple bias analysis. Such a circumstance would be very unusual, however, as it is highly likely that many nonparticipants would refuse to answer even the brief questionnaire. Those who agree to answer the brief questionnaire would likely differ from those who refuse or were lost-to-contact, and these differences may well relate to their exposure or outcome status. The values assigned to the bias parameters from such an internal data source would, therefore, themselves be measured with some systematic and random error. This additional uncertainty can be taken into account by multidimensional bias analysis or one of the advanced bias analysis methods described in Chapters 7, 8, 9, 10 and 11.

Uncontrolled Confounding

Ordinarily, a potential confounder that was not measured and not controlled for by design would be missing in all members of the study population, not just in a subgroup of the study population. However, it is possible that, by design or happenstance, an unmeasured confounder would be available for a subsample of the study population. For example, in a study of the relation between an occupational exposure and lung cancer occurrence, smoking history might be an unmeasured confounder. However, were the occupational cohort assembled from several different factory locations, it is possible that information from some factories would include smoking history and the information from other factories would not. Thus, smoking would, by happenstance, be unmeasured in only a portion of the study population. Alternatively, the analyst may not have the resources (time and money) to survey all members of the occupational cohort with respect to their smoking history, but might have the resources to survey a sample of the cohort. For example, in an occupational study of asbestos exposure on lung health in northern Minnesota, the primary source of data was occupational and morality records [1]. Smoking history was not measured for the whole cohort. However, an ancillary study administered a survey to a sample of this population that asked, among other things, about history of smoking. This ancillary study was used as a source of information for some of the bias parameters necessary to bias adjust for smoking in the main study [2]. Thus, smoking, by design, would be unmeasured in the members of the occupational cohort who are not sampled for the survey or who refused to participate in the survey. While it may seem counterproductive to gather information on only a portion of the study population, such a design strategy can yield very valuable information to inform an adjustment for the confounder by bias analysis methods, described in Chapter 5 and 10. It also has the advantage of rationalizing the use of limited study resources when it is more important to spend resources on getting good information on other study variables or on increasing the study size by enrolling more participants or extending the duration of follow-up.

In both examples, smoking information from the internal subgroup could be used to inform a quantitative bias analysis of the impact of the unmeasured confounder (smoking history) on the estimate of the effect of the occupation on lung cancer occurrence. The information necessary to bias-adjust is the strength of association between occupation and smoking categories, the association between smoking categories and lung cancer mortality, and the prevalence of smoking among those who did not participate in the occupation. It is common to express these three bias parameters in different ways, such as the prevalence of smoking among those who participated in the occupation, the prevalence of smoking among those who did not participate in the occupation, and the association between smoking categories and

lung cancer mortality. The same information is contained in these different parameterizations; they are simple transformations of one another. The values assigned to
these bias parameters determine the impact of the unmeasured confounder on the
estimate of effect between the exposure and the outcome derived from the entire
study population (without consideration of the possible correlation between smoking
history and other measured and controlled confounders). One should realize, however, that these bias parameters might themselves have been measured with error in
the subsample. The subcohorts in which smoking information was available might
not be representative of the entire cohort. It may be, for example, that factories where
smoking history was recorded and retained encouraged healthy habits in their
workforce, whereas factories where smoking history was not recorded did not or
did so to a lesser extent. Even when smoking history was collected by design, it may
be that the smoking history obtained from those who agreed to participate in the
survey would differ with respect to smoking history from those who refused to
participate in the survey. At the very least, the subsample is of finite size, so
sampling error must be taken into account for any bias-adjusted estimate of effect
that also reports an interval (*e.g.,* a confidence or simulation interval).

Information Bias

Often, a variable can be measured by many different methods, and each of these
methods would have its own accuracy. In some cases, the most accurate method may
be the most expensive, most time-consuming or most invasive, making it difficult to
use or at least difficult to use on everyone in the population. In this case, another less
expensive, less time-consuming or less invasive method could be used to collect the
information about the variable for all the participants. An internal validation study
would collect information about this same variable in a subsample of the study
population from both the less accurate method and the more accurate method. The
information about the variable would then be compared in the subsample, and this
comparison would yield values for the bias parameters (*e.g.,* sensitivity and specificity, or positive and negative predictive values) used to bias-adjust for classification errors in the complete study population.

   If the more accurate method is thought to be perfect, then it is often called the
"gold-standard" method for measuring the variable. For example, the gold-standard
method for measuring sobriety (dichotomized as above or below a threshold blood
alcohol concentration) might be a laboratory assay of blood alcohol content. A less
expensive, less time-consuming or less invasive assay would be a breathalyzer test,
which presumes a standard correlation between breath concentration of alcohol and
blood concentration of alcohol. To the extent that the correlation varies across
persons or within persons and across different breaths of air, the breathalyzer test
would be a less perfect measure of blood alcohol concentration. An even less
expensive test might be a field sobriety test, for which a subject is asked to perform
a set of tasks and poor performance correlates with lack of sobriety. However, to the

extent other factors can affect performance, and these factors vary across persons or across different attempts by the same person, the correlation with blood alcohol concentration is less than perfect, perhaps even worse than the breathalyzer. A study in which sobriety is a variable that needs to be measured might choose the breathalyzer or field sobriety test to assign sobriety values to its participants. A subsample of participants might also be assessed by blood alcohol concentration, and the sobriety classifications assigned by the breathalyzer or field sobriety test would be compared with the sobriety classifications assigned by the blood alcohol test in the same people within that subsample. One would then obtain estimates of the classification errors made by the breathalyzer or field sobriety test (*e.g.*, sensitivity and specificity), and these estimates of the bias parameters would be used to bias-adjust for misclassification in the entire study population. Derivation of these bias parameters in a subsample of the study population is called an internal validation study.

Design of Internal Validation Studies

Internal validation studies are often the most effective strategy to inform values for bias parameters. One can conceptualize internal validation studies as intentional reallocation of study resources (time and money) from efforts that would marginally reduce random error by enrolling more participants into the main study or following participants for longer duration. The reallocation of resources would be towards efforts that would reduce systematic error by measuring the values needed to assign to the parameters of a bias model. Unfortunately, many epidemiologists have little experience, education, or confidence in their ability to design or implement internal validation substudies [3]. In addition, the ongoing emphasis on achieving statistically significant results, regardless of their precision or validity, incentivizes allocation of all study resources towards reducing random error [4]. For these reasons, internal validation studies are seldom planned as an integral part of a study protocol, and when implemented, they often include a small or convenience sample that yields poor estimates of the values for the bias parameters.

By definition, internal validation studies include only a sample of the study or source population. Were data available to inform bias parameters for the entire study or source population, then the estimates of the values for the bias parameters would not be missing. If all members of the source population participated, then there would be no selection bias. If an uncontrolled confounder were measured in all participants, then the confounder could be controlled by conventional analysis methods. And if the gold standard measure of a mismeasured variable were available for all participants, then the mismeasured value could be discarded and replaced with the gold standard value.

The sampling strategy that yields participants in the internal validation study affects the precision and validity of the estimates of the values that will be assigned to the bias parameters. As described below, internal validation studies can themselves be subject to selection bias, which would affect the validity of the resulting

Table 3.1 Hypothetical cohort data with true data and expected data with nondifferential exposure misclassified (sensitivity=60% and specificity=90%).

	Truth		Expected	
	Exposed	Unexposed	Exposed	Unexposed
Cases	90	270	81	279
Non-cases	9910	89,730	14,919	84,721
Total	10,000	90,000	15,000	85,000
Risk	0.009	0.003	0.0054	0.0033
Risk difference	0.006		0.0021	
Risk ratio	3		1.65	

estimates. For internal validation studies meant to address classification errors, the design of the sampling strategy can affect the precision of the resulting estimates and the classification parameters that can be validly estimated. To illustrate, we begin with a synthetic example of a cohort study, as shown in Table 3.1. In the true study population, the estimated risk difference is 6 per 1000 and the true risk ratio is 3.0. If the dichotomous exposure is misclassified with 60% sensitivity and 90% specificity for both cases and non-cases, the expected risk difference would be 2.1 per 1000 and the expected risk ratio would be 1.65. These estimates, which are based on expectations, are biased towards the null by the non-differential and independent misclassification of a dichotomous exposure. We will define these terms and show how the expected table would be obtained in Chapter 6. Even though the sensitivity and specificity are the same for cases and controls, because the exposure is associated with the disease, the positive and negative predictive values of exposure classification will be different for the cases and controls. In this example, the positive predictive value is 66.7% among cases and 39.9% among non-cases. The negative predictive value is 87.1% among cases and 95.3% among non-cases.

A validation substudy would aim to estimate the correct values of the sensitivity and specificity and/or predictive values. A tempting strategy is to obtain a simple random sample of the expected data. Table 3.2, Panel A, shows the validation study data one would expect to obtain by validation of exposure status of a 10% simple random sample of the data in Table 3.1 One would obtain valid estimates of the sensitivity, specificity, positive predictive value, and negative predictive value. However, the estimates obtained in cases would be imprecise and, because of rounding error, inaccurate. There are 360 total cases, so a 10% sample would be expected to include 36 cases. Among those cases, only 8 (~10% of 81) would be classified as exposed by the mismeasured exposure classification. Of these 8, 5 (~60%) would be validated as exposed by the gold standard measure. The estimate of the positive predictive value that one would obtain would therefore be 5/8 = 0.625 with 95% confidence interval 0.28, 0.89. One cannot obtain the correct estimate of the positive predictive value (0.67) because of rounding error (participants must appear as integers in the validation data) and the sampling error yields a very wide confidence interval because the estimate of positive predictive value is based on such small numbers. Similar problems arise for the estimates of sensitivity, specificity,

Table 3.2 Estimates of sensitivity and specificity obtained from three validation subsamples of data shown in Table 3.1.

Panel A							
Cases, 10% sample				Non-cases, 10% sample			
	Truth				Truth		
Observed	E	U	Total	Observed	E	U	Total
E	5	3	8	E	595	897	1492
U	4	24	28	U	396	8076	8472
Total	9	27		Total	991	8973	
Sensitivity=0.556; 95% CI 0.24, 0.84				Sensitivity=0.600; 95% CI 0.57, 0.63			
Specificity=0.889; 95% CI 0.73, 0.97				Specificity=0.900; 95% CI 0.89, 0.91			
PPV=0.625; 95% CI 0.28, 0.89				PPV=0.399; 95% CI 0.37, 0.42			
NPV=0.857; 95% CI 0.69, 0.95				NPV=0.953; 95% CI 0.95, 0.96			
Panel B							
Cases, balanced, sampling observed				Non-cases, balanced, sampling observed			
	Truth				Truth		
Observed	E	U	Total	Observed	E	U	Total
E	54	27	81	E	32	49	81
U	10	71	81	U	4	77	81
Total	64	98	162	Total	36	126	162
Sensitivity=0.844; 95% CI 0.74, 0.92				Sensitivity=0.889; 95% CI 0.75, 0.96			
Specificity=0.724; 95% CI 0.63, 0.81				Specificity=0.611; 95% CI 0.52, 0.69			
PPV=0.667; 95% CI 0.56, 0.76				PPV=0.395; 95% CI 0.29, 0.50			
NPV=0.877; 95% CI 0.79, 0.94				NPV=0.951; 95% CI 0.89, 0.98			
Panel C							
Cases, balanced, sampling truth				Non-cases, balanced, sampling truth			
	Truth				Truth		
Observed	E	U	Total	Observed	E	U	Total
E	54	9	63	E	54	9	63
U	36	81	117	U	36	81	117
Total	90	90	180	Total	90	90	180
Sensitivity=0.600; 95% CI 0.50, 0.70				Sensitivity=0.600; 95% CI 0.50, 0.70			
Specificity=0.900; 95% CI 0.82, 0.95				Specificity=0.900; 95% CI 0.82, 0.95			
PPV=0.857; 95% CI 0.75, 0.93				PPV=0.857; 95% CI 0.75, 0.93			
NPV=0.692; 95% CI 0.60, 0.77				NPV=0.692; 95% CI 0.60, 0.77			

Red text indicates invalid estimates.
E exposed, *U* unexposed, *PPV* positive predictive value, *NPV* negative predictive value. Exact 95% confidence intervals using the mid-P method.

and negative predictive value in the cases. None of the estimates are accurate or precise. Among non-cases, all of the estimates of classification parameters are accurate and precise, because they are based on much larger numbers. The disease is rare, so a 10% sample of the expected data in Table 3.1 yields an abundance of non-cases (9964 of the 10,000 participants in the 10% sample are non-cases). The main problem with the simple random sample design is this discrepancy in accuracy of estimates among cases and non-cases, yielding estimates of classification parameter values for cases that are inaccurate and imprecise, despite designing a validation substudy that includes a large total number of participants.

To resolve this limitation of the simple random sample design, one must oversample the rare categories. A straightforward strategy is to use a balanced design [5]. In this design, one begins by identifying the cell of the 2 × 2 table with the lowest frequency of participants. To do so, we must work with the observed data, or in this case the "expected data" of Table 3.1, there are 81 exposed cases, and that is the lowest frequency. In the balanced design, all 81 of these exposed cases are included in the validation substudy. This frequency then sets the number of unexposed cases included in the validation substudy, and the number of exposed and unexposed non-cases included in the validation substudy. For each of the other three cells of the 2 × 2 sample, a simple random sample is drawn for inclusion in the validation substudy from the N > 81 participants available from each cell. The expected validation data are shown in Panel B of Table 3.2; note that the marginal total for the observed data is 81 for every combination of observed exposure status and disease status. The positive predictive value among cases estimated from this design equals the true value of 0.67 and has a much narrower 95% confidence interval (0.56, 0.76) than the simple random sample in Panel A. All predictive values are accurately and precisely estimated, and with a much smaller validation substudy (324 as opposed to 10,000 participants). This study design samples individuals based on their observed exposure status. Because of this, it can be used to estimate probabilities that also condition on the observed exposure status (positive and negative predictive values). However, it is very important to realize that the sensitivity and specificity of exposure classification cannot be directly estimated by this design because one marginal total has been manipulated by design. The sensitivity and specificity that are directly computed are invalid, so shown in red in Table 3.2, Panel B.

A third option is to implement a similar balanced design, but with sampling based on the true value of the exposure as opposed to the mismeasured value of the exposure. Of course, if one had the gold standard (true) value of the exposure on all participants, then the mismeasured value would be unimportant and a validation substudy would have no utility. It is possible, however, that the true value is available by design among a subset of the total study population. Imagine, for example, that the true data in Table 3.1 come from a single enrollment center, where the gold standard value of the exposure is available for all participants and the mismeasured value would have to be collected in a validation substudy. Then sampling on the basis of the true value would be an efficient design, and the results of this validation substudy would be useful to bias analyses for the other enrollment centers, where presumably only the mismeasured value of the exposure would be available. The design proceeds, as above, by identifying the cell of the 2 × 2 table with the lowest frequency of participants. In the true data of Table 3.1, there are 90 exposed cases, and that is the lowest frequency. All 90 of these exposed cases are included in the validation substudy. For each of the other three cells of the 2 × 2 sample, a simple random sample is drawn for inclusion in the validation substudy from the N > 90 participants available from each cell. The expected validation data are shown in Panel C of Table 3.2; note that the marginal total for the true data is 90 for every combination of true exposure status and disease status. The sensitivity among cases estimated from this design equals the correct value of 0.60 and has a narrow 95% confidence interval (0.50 0.70). All sensitivity and specificity values are accurately and precisely estimated, and with a small validation substudy (360 as opposed to 10,000 participants). It is important to realize that, because sampling was conducted conditional on true exposure, only probabilities conditional on the true exposure (sensitivity and specificity) can be directly estimated. The positive and negative predictive values of exposure classification cannot be validly estimated by this design because one marginal total has been manipulated by design. The positive and negative predictive values can be computed, but the computed estimates are invalid, so shown in red in Table 3.2, Panel C.

The main point of this example is that classification parameters can be accurately and precisely measured with relatively small subsamples of the study population, but this sample would seldom be a simple random sample. Strategic sampling allows this valid and precise estimation, but usually at the expense of valid estimation of one set of the classification parameters. These valid estimates might be able to be recovered by reweighting. More recently, adaptive validation designs have been proposed that allow sampling and re-estimation of classification parameters until some goal of the validation study has been achieved [6, 7]. These goals might be the precision of the estimate of the validation parameters, the precision of the estimate of a bias-adjusted estimate, or some other stopping rule. Adaptive designs allow the optimal allocation of study resources to achieve the objective of the validation substudy, with the potential for reallocation of remaining resources to other study objectives. Other options for cost-efficient validation designs have also been proposed [8–11].

Limitations of Internal Validation Studies

Estimates of bias parameters from internal validation studies are sometimes treated as if they are themselves measured without error. That treatment is, however, usually incorrect. At the least, these estimates are measured in a finite subsample, so the random error should be taken into account in any bias analysis that uses the bias parameters to adjust for the classification errors and yields an interval (*e.g.*, a confidence interval or simulation interval) for the adjusted estimate of association. More often, the bias parameters themselves are measured with uncertainty beyond the sampling error. For example, in the internal validation study of blood alcohol concentration, a large burden would be placed on those who agree to participate in the substudy. They would have to agree to a needlestick to donate a blood sample, whereas all other participants need only breathe into the breathalyzer or perform the field sobriety tasks. Therefore, study members who agreed to participate in the validation substudy would likely differ from those who refuse to participate in the validation substudy in ways that may relate to their true sobriety or to how their sobriety status was measured. Sober participants are likely to be more compliant, so would likely breathe into the breathalyzer exactly as instructed and would likely perform the field sobriety tasks to the best of their ability. Nonparticipants and those who are not sober are likely to be less compliant, so might not follow the breathalyzer instructions as carefully and might not be as motivated to perform the field sobriety tasks were they participants. In short, the internal validation substudy would itself be susceptible to selection and measurement biases. These biases should be taken into account by allowing the values assigned to the bias parameter to vary in ways that capture the expected direction of the biases, using multidimensional bias analysis or the more advanced methods described in Chapters 7, 8, 9, 10 and 11.

A second nuance to consider with all validation studies, including internal validation studies, is that the measurement of the variable may not perfectly equate with the concept that is being measured. In the blood alcohol example, sobriety is the concept that is being measured. Blood alcohol concentration over some threshold, which is an objectively measured laboratory value, has been set as a sharp line of demarcation to dichotomize study subjects into those who are sober enough to operate a motor vehicle, for example, and those who are not. However, given differences in experience with alcohol drinking and other factors, some participants with blood alcohol concentrations somewhat above the threshold may be sober enough to operate a motor vehicle and those below the threshold may not be sober enough to operate a motor vehicle. If the sobriety concept is task-oriented, the field sobriety test or some other task-oriented test may be a better method of measuring sobriety than blood alcohol content, albeit more difficult to enforce.

Finally, for many variables, there is no gold-standard measurement method. For example, human intelligence has been measured by many tests and methods, but there is no single test recognized as the gold standard by which intelligence could be optimally measured. When there is no gold-standard test, comparisons of measurement methods are reported as agreement (concordant and discordant classifications, or correlation of continuous measurements, by the two methods). These measures of agreement can be used to inform values assigned to the bias parameters, but should not be mistaken for direct estimates of the bias parameters.

External Data Sources

When no data are collected from a subsample of the study population to inform bias parameters, the values assigned to bias parameters must be informed by data collected from a second study population or by an educated guess about these values. These assignments rely on external data sources: the second study population or the experience relied upon to reach an educated guess. Some analysts resist these options; but such resistance is tantamount to assigning a value to the bias parameter such that no bias-adjustment would be made, effectively assuming that there is no bias. For example, refusing to assign the sensitivity and specificity of exposure classification observed in a second study population is equivalent to assigning perfect sensitivity and specificity to the exposure classification scheme, so that there is no need to conduct a quantitative bias-adjustment to account for exposed study participants who were misclassified as unexposed. Refusing to make an educated guess at the association between an unmeasured confounder and the outcome is equivalent to saying there is no confounding by the unmeasured confounder, so that there is no quantitative bias-adjustment to account for it. While values assigned to bias parameters from external data sources might well be imperfect, they are almost certainly more accurate than the assumption that there is no error, which is implicit in conventional analysis. Furthermore, the values assigned to bias parameters from internal data sources may be imperfect as well, as described in the preceding sections. Therefore, external data sources and educated guesses should be investigated as a valuable and valid option to inform the values assigned to bias parameters. Wide ranges of uncertainty about the values assigned to bias parameters and used in multidimensional bias analysis, or the more advanced methods described in Chapters 7, 8, 9, 10 and 11, will help to reflect the uncertainty about the true value of the bias parameter and about transporting information from one population to another.

Selection Bias

The bias parameters required to assess selection bias are often the proportions of eligible subjects included in the study, at baseline or over follow-up, within each combination of at least exposure and disease status. The subjects included in the study provide an initial internal estimate of the exposure prevalence and disease occurrence within each combination. To address selection bias, one can begin with the exposure prevalence and disease occurrence observed in those included in the study, and then adjust them both by making educated guesses about selection forces that would act in concert with the exposure prevalence and disease occurrence.

For example, in a case-control study, one ordinarily knows the proportion of cases and controls who agree to participate, so are included in the analysis. The exposure prevalence of participating cases and controls is known from the data gathered on participants. The exposure prevalence of cases and controls who did not participate is unknown and ordinarily requires an educated guess because there is rarely any internal validation data or data from another study population to inform estimates. In some cases, one might reason that exposure increases the probability that a case would participate in the study. Imagine, for example, that the exposure under study is an occupational hazard. Cases with a history of the occupation might be inclined to participate for reasons including an altruistic concern that the hazard should be identified or an interest in secondary gain (*e.g.,* worker's compensation). The prevalence of the occupational hazard among participating cases would therefore overestimate the true prevalence of the occupational hazard among all cases, so a reasonable educated guess of the prevalence of the occupational hazard in nonparticipating cases would be lower than its prevalence in participating cases. Controls, on the other hand, would not have the same secondary motivations as cases, so the exposure prevalence of the occupational hazard among nonparticipating controls might be about the same as the exposure prevalence of the occupational hazard among participating controls. In other situations, the reverse scenario could occur. For instance, if the exposure of interest was seen as socially undesirable, people with the exposure might be less likely to enroll in the study.

One might further inform these educated guesses at exposure prevalence or disease occurrence from research studies outside the study setting. For example, other studies might have investigated the occupational hazard and reported the exposure prevalence in a similar source population. This exposure prevalence might be used to inform the overall exposure prevalence expected in controls, and then one can solve for the exposure prevalence in nonparticipants by setting the second study's exposure prevalence equal to the average of the exposure prevalence in participating and nonparticipating controls, weighted by their proportions among all selected controls. When using exposure prevalence or disease occurrence information from other studies, one should assure that the second study's population is similar enough to the study population of interest to allow the information to be transported between the two populations. For example, if age and sex are related to disease occurrence, then the second study's distribution of age and sex ought to be similar to the distribution of age and sex in the study population of interest before the second study's estimate of disease occurrence is adopted to inform the selection bias parameters. When sufficient cross-tabulated data are available in both studies, standardization can be used to adjust for differences in the distributions. Multidimensional bias analysis or the more advanced methods described in Chapters 7, 8, 9, 10 and 11 can be implemented to incorporate uncertainty in the extrapolation or a range of educated guesses and information from more than one external source.

Unmeasured Confounder

A bias analysis to address an unmeasured confounder requires an estimate of the strength of association between the confounder and the outcome, the strength of association between the confounder and the exposure, and the prevalence of the confounder in the unexposed. Although alternative approaches rely on the prevalence of the confounder among both the unexposed and the exposed instead of the strength of association between the confounder and the exposure, this information ultimately leads to these same core concepts (one can calculate the strength of association between the confounder and the exposure from the prevalence of the confounder in the unexposed and exposed). Unlike with selection bias examples, this core information is often readily available from external data sources in the published literature or in publicly available surveillance data.

For example, imagine one was studying the association between diet and heart disease. The hypothesis might be that a "healthy diet," however defined (low fat

Mediterranean diet, vegetable rich, high in antioxidants, *etc.*), reduces the rate of first myocardial infarction compared with a "less healthy diet" (higher fat, vegetable poor, low in antioxidants). A reasonable concern would be that the association would be confounded by exercise habits, since persons with healthy diets might also be more likely to exercise than persons with less healthy diets, and exercise is known to reduce the risk of first myocardial infarction. Further imagine that information on exercise habits was not available in the study, so it was an unmeasured confounder. Nonetheless, the core information required to adjust for exercise should be readily available from external sources. Many studies have investigated the association between regular exercise and risk of myocardial infarction, so it could be easily identified with a literature search and review of relevant publications. These would provide estimates of the strength of association between the confounder and the outcome. These studies might also provide estimates of the association between exercise habits and diet habits and of the prevalence of different exercise habits (*e.g.*, this kind of information can often be found in the descriptive characterization of the study population, usually reported in Table 3.1 of a manuscript). Furthermore, many public data sets (*e.g.*, NHANES) report survey results on these types of behaviors, so can be used to estimate some of the requisite bias parameters (in this case, the association between diet and exercise habits, and the prevalence of exercisers in the study population).

Even if external estimates of the bias parameters are unavailable, one can assign educated guesses to their values. For example, given that risk-taking behaviors tend to correlate, one can reasonably assume that the prevalence of a risky behavior that is an unmeasured confounder will be higher in those with an exposure that is also a risky behavior than in the unexposed (*e.g.*, alcohol consumption is higher, on average, among cigarette smokers than nonsmokers). On the contrary, some behaviors preclude or reduce the prevalence of another behavior. For example, people gravitate to lifetime sports that may be mutually exclusive, on average, either because they are most common during the same seasons (*e.g.*, running or biking) or because they can only afford one choice (*e.g.*, golfing or downhill skiing). Of course, some runners also bike, and some skiers also golf, but the prevalence of bikers among runners and the prevalence of golfers among skiers may be lower than their respective prevalence among nonrunners and nonskiers, at least in some populations. Educated guesses at such relations should take best advantage of external data sources and the analyst's experience, and should also allow for uncertainty in the assigned values by using multidimensional bias analysis or the more advanced methods described in Chapters 7, 8, 9, 10 and 11.

Information Bias

External information about measurement error and classification rates is often available for potentially mismeasured or misclassified variables. Imagine, for example, a study of the association between receipt of vasectomy and subsequent risk of prostate cancer. In a cohort study of this association, one might ascertain receipt of vasectomy by baseline interview and occurrence of prostate cancer by linking the cohort to a mortality database that provides cause of death information. Vasectomy information is likely to be misclassified. Men who report they had a vasectomy very likely did, but some of the men who report they had not had vasectomy may well have had the procedure. They fail to report it either because they did not want to report the information or because they forgot or misunderstood the question. An external validation study of the association would presumably confirm this expectation. Such a study would compare self-report of vasectomy history with a comprehensive medical record in a random population of men similar to the men included in the original study. From such a study, one could estimate the sensitivity, specificity, positive predictive value, and negative predictive value of self-report of vasectomy. Were no such validation study available, one could reason as above to make an educated guess that the positive predictive value of self-report of vasectomy history is likely very high, and that the negative predictive value of self-report of no history of vasectomy may not be quite as high. Because the self-report of vasectomy was recorded at baseline, the sensitivity and specificity of self-report are unlikely to be affected by prostate cancer occurrence. Nonetheless, the positive and negative predictive values are functions of sensitivity, specificity, and the prevalence of the exposure. If vasectomy is associated with prostate cancer, which is the study hypothesis, then the prevalence of vasectomy would be different in cases and non-cases and we would expect positive and negative predictive values to differ for cases and non-cases. Estimates of the predictive values would therefore be required in cases and non-cases. Sensitivity and specificity do not depend on the prevalence of the exposure, so could be measured without conditioning on disease status. Had this been a study in which vasectomy information was collected retrospectively, such as by self-report after prostate cancer occurrence, we might make a different assumption and assume that sensitivity and specificity of self-report would be different among those without prostate cancer than among those with. Estimates of sensitivity, specificity, and predictive values would all be required within strata of cases and non-cases.

In this hypothetical study, information on prostate cancer occurrence is also likely recorded in error. Not all men who develop prostate cancer ultimately die of the disease, in fact most survive and ultimately die of some other cause. One could use external information that compares prostate cancer incidence with prostate cancer mortality to estimate the sensitivity and specificity of using prostate cancer mortality as a surrogate for prostate cancer incidence. We would expect that specificity of this disease classification method would be high. Among men who truly did not have incident prostate cancer, few would have been misclassified as having died of prostate cancer. On the other hand, the sensitivity of prostate cancer occurrence classified by this method is likely poor. Among men who truly did have incident prostate cancer, many would not be identified by only finding the men who died of the disease. A contingency table comparing incident cases with fatal cases over some time period would provide an estimate of the bias parameters. As will be discussed in Chapter 6, even if the sensitivity of disease classification is poor but does not depend on the exposure status, and there are few false-positive cases (very high or perfect specificity), risk ratio estimates of effect are not expected to be substantially biased by the disease classification errors [12]. This expectation will also hold approximately for the rate ratio, so long as these errors and the exposure do not substantially affect person-time, and for the odds ratio, so long as the outcome is rare.

In this example, one could reasonably expect to find external information to estimate the classification bias parameters. In addition, one could make reasonable educated guesses about which bias parameter was likely to be high and which to be low. These educated guesses are often sufficient to provide values for the bias parameters that allow an assessment of the direction and approximate magnitude of the bias and additional uncertainty arising from misclassification.

External information on these bias parameters should not be adopted for use in a study without consideration of its applicability. Consider, for example, the strategy of using visualization to assign natal sex categories (male or female) to study participants. In many adult populations, such a strategy would be cost-efficient and accurate. A validation study comparing the category assigned by visualization with the category assigned by karyotyping (a possible gold standard for natal sex) would often show high sensitivity and specificity. However, were visualization used as the method of assigning sex category in a newborn ward (presuming no visual clues like pink or blue blankets), it may not work so well. The sensitivity and specificity of visualization as a method for assigning natal sex category, as measured in a validation study in which the strategy was applied to adults, should obviously not be adopted to provide values for these bias parameters when visualization is used as a strategy to classify newborns as natal male or natal female. More generally, sensitivity and specificity are sometimes said to be characteristics of a test, so more readily applied to other populations in which the test is used than predictive values (as explained above and in more detail in Chapter 6, predictive values depend on the both test characteristics and the prevalence of the result in the study population). While this mantra is correct on its surface, the preceding example illustrates that sensitivity and specificity can also depend on the population in which the validation data were collected.

Expert Opinion

When no external data sources exist with which to parameterize a bias analysis, expert opinion is the only remaining option. Although experts bring their own biases and judgements to the process, this disadvantage can still be preferable to ignoring the opportunity to quantify the bias by assigning the implicit values to the bias parameters (such as assuming perfect classification by assigning values of 100% to the sensitivity and specificity) or to looking for combinations of bias parameters that would yield null values for the exposure-outcome association. When expert opinion is used, the biases brought by an individual expert may be mitigated by seeking input from multiple experts. One might consider using a Delphi approach [13], in which experts give their opinion without seeing what others say, then see what the group says and have a chance to revise their choices if they so desire. There are also methods to elicit prior information in Bayesian analyses [14, 15] Experts could be queried for their best estimate of the values for a particular bias parameter, or they could be asked for a range of values they consider likely to contain the true value (*e.g.*, they could be asked to provide their best guess and a 95% interval around that estimate). Results can then be summarized into a distribution that expresses the uncertainty in expert opinion, and these summary results could be fed into a probabilistic bias analysis or one of the other more advanced methods described in Chapters 7, 8, 9, 10 and 11.

Summary

Internal and external data sources can be used to inform the values assigned to bias parameters. In addition, expert opinion and educated guesses about the plausible range and distributional form of these values should be considered. Subsequent chapters will outline specific methods that make use of the bias parameters and the values assigned to them to estimate the direction and magnitude of the bias and uncertainty arising from systematic errors. In all cases, the best analysis makes optimal use of all three sources of information: internal validity data, external validity data, and the reason and judgment of the analyst and her colleagues.

References

1. Allen EM, Alexander BH, MacLehose RF, Ramachandran G, Mandel JH. Mortality experience among Minnesota taconite mining industry workers. Occup Environ Med. 2014;71:744–9.
2. Allen EM, Alexander BH, MacLehose RF, Nelson HH, Ramachandran G, Mandel JH. Cancer incidence among Minnesota taconite mining industry workers. Ann Epidemiol. 2015;25:811–5.
3. Fox MP, Lash TL, Bodnar LM. Common misconceptions about validation studies. Int J Epidemiol. 2020; 49:1392–6.
4. Lash TL. The harm done to reproducibility by the culture of null hypothesis significance testing. Am J Epidemiol. 2017;186:627–35.
5. Holcroft CA, Spiegelman D. Design of validation studies for estimating the odds ratio of exposure-disease relationships when exposure is misclassified. Biometrics. 1999; 55:1193–201.
6. Collin LJ, MacLehose RF, Ahern TP, Nash R, Getahun D, Roblin D, et al. Adaptive Validation Design. Epidemiol Camb Mass. 2020; 31:509–16.
7. Collin LJ, Riis AH, MacLehose RF, Ahern TP, Erichsen R, Thorlacius-Ussing O, et al. Application of the Adaptive Validation Substudy Design to Colorectal Cancer Recurrence. Clin Epidemiol. 2020; 12:113–21.
8. Spiegelman D, Gray R. Cost-Efficient Study Designs for Binary Response Data with Gaussian Covariate Measurement Error. Biometrics. 1991; 47:851–69.
9. Stram DO, Longnecker MP, Shames L, Kolonel LN, Wilkens LR, Pike MC, et al. Cost-efficient design of a diet validation study. Am J Epidemiol. 1995;142:353–62.
10. Spiegelman D. Cost-efficient study designs for relative risk modeling with covariate measurement error. J Stat Plan Inference. 1994;42:187–208.

11. Greenland S. Variance estimation for epidemiologic effect estimates under misclassification. Stat Med. 1988;7:745–57.
12. Brenner H, Savitz DA. The effects of sensitivity and specificity of case selection on validity, sample size, precision, and power in hospital-based case-control studies. Am J Epidemiol. 1990;132:181–92.
13. Hasson F, Keeney S, McKenna H. Research guidelines for the Delphi survey technique. J Adv Nurs. 2000;32:1008–15.
14. Johnson SR, Tomlinson GA, Hawker GA, Granton JT, Feldman BM. Methods to elicit beliefs for Bayesian priors: a systematic review. J Clin Epidemiol. 2010;63:355–69.
15. Johnson SR, Tomlinson GA, Hawker GA, Granton JT, Grosbein HA, Feldman BM. A valid and reliable belief elicitation method for Bayesian priors. J Clin Epidemiol. 2010;63:370–83.

Chapter 4
Selection Bias

Introduction

Selection bias arises when—in a *study population*—an estimate of disease occurrence, or an estimate of the effect of an exposure contrast on disease occurrence, differs from the estimate that would have been obtained in the study population's *source population* because of the way the study population was selected, either by design or analytic choice. The study population is the roster of persons, and their observed person-time, that are included in the analysis of the data that yielded the estimate. The source population is the roster of persons, and their observed person-time, eligible to be included in the analysis of the data that yielded the estimate. The difference between the source population and the study population—in the roster of included persons, in the observed person-time, or both—is what accounts for the selection bias. Selection bias is a systematic error and sometimes called a threat to internal validity. Selection bias can thus arise when participants enroll in a study, which is called "differential baseline participation," and it can also arise when participants withdraw from the study, which is called "differential loss-to-follow-up." Selection bias can occur because of study design choices or analysis choices. We use the word "differential" to denote that the baseline participation proportions or proportions lost to follow-up must have certain dependencies to induce a bias. These requisite dependencies will be described below. For now, it is important only to realize that some differential baseline selection forces or differential forces acting on loss-to-follow-up must exist. In the absence of other sources of bias, a study population comprised of a simple random sample of person-time from the source population would not be expected to yield a biased estimate of disease occurrence or a biased estimate of the effect of an exposure contrast on the occurrence of a disease.

A third population to consider is the target population, defined as the population to which information from the study will be applied. If the source population is a subsample of the target population, then we say that the results of the study will be "generalized" to the target population [1–3]. Note we are not saying the results are

© Springer Nature Switzerland AG 2021
M. P. Fox et al., *Applying Quantitative Bias Analysis to Epidemiologic Data*,
Statistics for Biology and Health, https://doi.org/10.1007/978-3-030-82673-4_4

necessarily "generalizable," only that we wish to generalize from our study population to the target population that encompasses our study population. If the source population is not a subsample of the target population, then we say that the results of the study will be "transported" to the target population [4]. The estimate of disease occurrence, or the estimate of the effect of an exposure contrast on the occurrence of a disease, may be different in the source population than in the target population, in which case we say the estimate is poorly generalizable or poorly transportable. This lack of generalizability or lack of transportability usually arises because characteristics of the target population are different from characteristics of the source population, and these differences affect the estimate of disease occurrence, or affect the estimate of the effect of the exposure contrast on the occurrence of the disease. Lack of generalizability or lack of transportability is sometimes called a threat to external validity. For example, a target population may be 50% natal males and 50% natal females. If natal males have a higher prevalence of the disease of interest, and the source population has a higher proportion of natal males than natal females, then the overall prevalence of the disease in the source population will be higher than one would observe in the target population because of the higher prevalence of males. If the target population overlaps the source population, we say the overall prevalence is not generalizable. If the target population does not completely overlap the source population, we say the overall prevalence is not transportable to the target population. This lack of generalizability or transportability will hold, even if the study population is a complete census of the source population, implying no selection bias according to the definition given above. The lack of generalizability and transportability does, however, arise from selection forces, the forces that caused natal males to be over-represented in the source population compared with the target population. Lack of generalizability and transportability therefore have similar structural underpinnings as selection bias and can be bias-adjusted using similar analytic methods. We will return to these topics towards the end of the chapter. Until then, the focus will be on selection bias arising from differences between the source population and the study population.

The difference in baseline or ongoing participation by members of the study population, compared with what would have been possible from the complete source population, is therefore the fundamental cause of selection bias. In descriptive epidemiologic research, the estimate of disease occurrence must be measured among participants, so is conditioned on participation. In etiologic epidemiologic research, the estimate of the effect of the exposure contrast on disease occurrence must be measured among participants, so is also conditioned on participation. Conditioning the measurement of disease occurrence, or the measurement of the effect of the exposure contrast on disease occurrence, on participation can bias the measured estimates [5], although it is not the only mechanism by which selection bias may be induced [6, 7].

Figure 4.1 illustrates an example of collider-induced selection bias in a study of the relation between exposure to a local hazardous waste site and occurrence of leukemia. If the study were conducted by retrospective case-control design, then one could imagine that leukemia cases might have been more likely to participate than

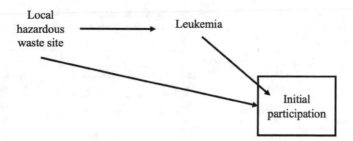

Figure 4.1 Causal graph showing selection bias affecting the relation between exposure to a hazardous waste site and occurrence of leukemia.

controls selected from the source population, since cases may have had greater interest in a study of their disease's etiology than controls. Similarly, persons who lived near the hazardous waste site (both cases and controls) may have perceived it to be a threat to their health, so may have been more motivated to participate in a study than persons who did not live near the hazardous waste site. In this example, both the exposure (residence near a local hazardous waste site) and the outcome (leukemia diagnosis) affected the probability of initial participation in the study. The association between residence near the hazardous waste site and leukemia occurrence could only be measured among the participants, so is conditioned on initial consent to participate. The proportion of exposed cases among participants (*i.e.,* exposed cases in the study population) would be expected to exceed the proportion of exposed cases among all those eligible to participate (*i.e.,* exposed cases in the source population). Conversely, the proportion of unexposed controls among participants (*i.e.,* unexposed controls in the study population) would be expected to be less than the proportion of unexposed controls among all eligible participants (*i.e.,* unexposed controls in the source population). That is, self-selection of participants into the study was likely to have induced an association between residence near the hazardous waste site and occurrence of leukemia, even if there was no association between them in the source population. We will show how to formally manipulate a DAG to demonstrate this later in this chapter. We note that it is technically possible (even if not probable) for the selection proportions to cancel one another's effect and result in no bias, which will be explained further below. When the association between variables measured in the study population is different from the association that would have been measured in the source population, the measured estimate of effect is biased by selection.

We further illustrate this source of selection bias with hypothetical data in Table 4.1. These data illustrate a mechanism by which such a bias may arise. Among all those eligible to participate (the source population), the incidence proportion among both those exposed and unexposed to hazardous waste sites equals 10% (100/1000). However, the study must be conducted among those who participated (the study population). Participation is affected by both the exposure to hazardous waste sites and the outcome, leukemia. The odds ratio associating participation with exposure equals $[(80 + 400)/(20 + 500)]/[(60 + 200)/(40 + 700)] = 2.63$,

Table 4.1 Hypothetical data illustrating that conditioning an estimate of association on participation can induce an association between exposure and disease, when no association exists among all those eligible to participate.

	Leukemia		No Leukemia	
	Hazardous waste	No hazardous waste	Hazardous waste	No hazardous waste
Participants	80	60	400	200
Nonparticipants	20	40	500	700
All eligible	100	100	900	900

reflecting the fact that a higher proportion of those exposed to hazardous waste sites (48%) participated than the those who were not exposed to hazardous waste sites (26%). The odds ratio associating participation with leukemia equals $[(80 + 60)/(400 + 200)]/[(20 + 40)/(500 + 700)] = 4.67$, reflecting the fact that a higher proportion of the people with leukemia participated (70%) than those without leukemia (33%). The net result is to induce an association between hazardous waste site exposure and leukemia in the study population (odds ratio = $[80/400]/[60/200] = 0.67$), when no association exists in the source population. This type of selection bias is due to differential participation by disease and exposure. Both exposure and disease cause participation in the study. Conditioning on that participation opens an unblocked backdoor path between exposure and disease, causing us to find an association even though one did not exist in truth.

Definitions and Terms

Below are definitions and terms that will be used to explain selection bias and quantitative bias analysis to bias-adjust for selection bias. The first section addresses concepts relevant to selection bias and the second section explains the motivation for bias analysis.

Conceptual

As noted above, selection bias arises in etiologic studies when differential selection forces affect study participation, meaning either initial participation or continued participation. From here forward we will focus on etiologic studies, though many of the same concepts readily apply to descriptive epidemiology. The design of an epidemiologic study requires the analyst to specify criteria for membership in the study population. These criteria list the inclusion criteria and exclusion criteria. To become a member of the source population, all the inclusion criteria must be met and none of the exclusion criteria may be met. Inclusion criteria specify the characteristics of the source population with respect to personal characteristics (*e.g.,* sex, age range, geographic location, and calendar period) and exposure or behavioral characteristics (*e.g.,* tobacco use, alcohol use, exercise regimen, diet, and occupation). Exclusion criteria specify the characteristics that restrict a subset of the persons who met the inclusion criteria from becoming members of the source population. For example, exclusion criteria may limit the population with respect to history of the disease under study (outcomes are often limited to first occurrence), language (*e.g.,* persons who are not fluent in the interview's language and those with poor hearing may not be able to complete an interview), or an administrative requirement (*e.g.,* those with a telephone or driver license, if these sources will be used to access participants).

The people who satisfy all of the inclusion criteria and none of the exclusion criteria constitute the source population for the study. Note that there may be persons who satisfy the inclusion criteria and none of the exclusion criteria but who are not identified. Their absence from the study does not necessarily induce a selection bias, although it may affect the study's generalizability to the target population. If any of the people in the source population do not participate in the study, the potential for selection bias arising at the stage of initial enrollment must be considered. The ability to identify participants is therefore an inherent component of the inclusion criteria. Note also that persons may not participate for many possible reasons. They may be invited to participate and refuse, for example by refusing to sign an informed consent form required by an ethical oversight committee. Refusing to sign an informed consent form is a type of active refusal. They may be identified but attempts to contact them with an invitation to participate are never answered. They may never answer telephone calls, or never reply to postal mail or electronic mail invitations. Failure to contact eligible participants is a type of passive refusal. There is no record of refusal to participate, but also no record of consent to participate. These persons must ordinarily be considered non-participants. A third possibility is that some members of the source population are excluded by design [8]. For example, case-control studies often endeavor to include all cases in the source population, but only a sample of the population giving rise to the cases. Sometimes, even cases are sampled from the source population. This third mechanism should not inherently generate selection bias because the sampling of controls (and possibly cases) should not depend on exposure status.

Most epidemiologic studies follow the participants who are enrolled for some period. If population members become lost to follow-up during this period, a second opportunity for selection bias arises. Differential loss-to-follow-up occurs when differential selection forces affect the probability of loss-to-follow-up. Because the longitudinal exposure–disease association can only be measured among those who are followed, the differential loss-to-follow-up causes a difference between the true and the measured association. Bias from differential continued participation is sometimes called attrition bias [9]. As with selection bias arising from differential baseline participation, selection bias from differential loss-to-follow-up can arise by several different mechanisms. First, ethical oversight committees require that persons who agreed to participate can withdraw their consent at any time and without accruing any adverse consequence. This type of loss-to-follow-up is active, analogous to active refusal to consent at baseline enrollment. Second, persons who were initially identified, contacted, and consented may subsequently be lost to follow-up. They may move to a different address or change their telephone number or email address, all without leaving information that allows study staff to recontact them. This type of loss-to-follow-up is passive, analogous to failure to contact eligible participants with an invitation for baseline participation. Study staff often invest substantial resources to avoid active and passive losses to follow-up. For example, study staff may regularly contact participants to maintain goodwill, even if the contacts have nothing to do with data collection for the study. These contacts might include health-related information or summaries of study findings, and sometimes even include some information about the study staff, with the hopes that this information will build a feeling of community and thereby discourage active withdrawal from the study. To avoid passive losses to follow-up, study staff often ask participants for multiple avenues by which they might be contacted, including information about a close friend or relative who would be able to reach the participant if study staff cannot. Study staff might also access public documents, such as credit reports and change of address databases, to learn whether and where participants have relocated [10]. There are also mechanisms of loss-to-follow-up analogous to baseline selection by design that involve dynamic sampling of person-time by design [11].

If all the study population participates initially and throughout follow-up, then the exposure–disease association will not be conditioned on participation (there is no stratum of nonparticipants). This point illustrates the connection between selection bias and missing data methods (see Chapter 10). For selection bias, the connection is quite simple: persons who refuse to participate do not contribute data to the analysis, so their data are completely missing. Persons who agree to participate but are subsequently lost to follow-up do not contribute complete data to the analysis, so some of their data are missing. In either case, if the missingness is not at random, for example is predicted by exposure and disease status, then analyses conducted among only the collected data will yield associations different from the association that would be observed, were the complete data from all subjects available.

Depicting Selection Bias Using Causal Graphs

As noted above, selection bias arises from the effects of differential selection forces. These differential selection forces can often be depicted using causal graphs [5, 12]. We will depict differential selection forces under two scenarios: under the null (exposure does not cause disease, so there is no direct path from exposure to disease) and off the null (exposure does cause disease, so there is a direct path from exposure to disease).

Figure 4.2 depicts selection bias under the null. An analyst is interested in the causal effect of exposure on disease, but we assume the null hypothesis is true. If

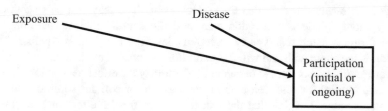

Figure 4.2 Causal graph depicting the general concept of selection bias under the null induced by conditioning on a collider requiring participation.

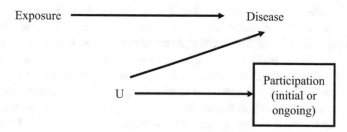

Figure 4.3 Causal graph depicting the concept of selection bias off the null induced by conditioning on a collider requiring participation. (Adapted from Hernán, 2017 [7]).

both the exposure and disease affect initial or ongoing participation in the study, and the conventional data analyses are completed using only the subjects with complete data, then the conventional data analyses are conditioned on participation. In the DAG in Figure 4.2, we illustrate that we are conditioning on P by drawing a box around it. Conditioning on this collider on the path between exposure (E) and disease (D), opens an unblocked backdoor path between E and D, which could cause bias. Sometimes, this induced correlation between E and D is illustrated with a double headed arrow on the DAG. We would expect a non-null association between E and D, even if the null hypothesis were true. Collider bias of this structure $(E \rightarrow \boxed{P} \leftarrow D)$ generally induces selection bias. In this DAG, where we assume the null is true and there is no causal effect of E on D, any observed E-D association would be due to this collider-stratification bias. If there were a causal effect of E on D, collider bias of this structure will also generally induce a bias, meaning part of the observed E-D association would be due to the selection bias and part of the observed E-D association would be due to the causal effect of the exposure on the disease. These parts would not be identifiable without additional information and analyses, such as the quantitative bias analyses described below and in Chapters 7–11.

Although all selection bias that occurs assuming the null is true arises from collider stratification bias, not all selection bias off the null arises from collider bias. Figure 4.3 depicts the causal graph structure used by Hernán (2017) to illustrate this point [7]. In this causal graph, there is a causal association between exposure (E) and disease (D). An unknown variable (U) affects both participation (P) and the disease (D). There is only one arrow entering the participation node so there is no collider stratification bias in this DAG. However, there may still be selection bias, even without collider stratification bias, and the causal graph is therefore not as useful for diagnosing this type of selection bias. The question is whether the E-D relation is the same with and without conditioning on P. This is a question of effect measure modification: does U (a parent of P) modify the effect of E on D? If so, there is an opportunity for selection bias in the absence of collider stratification. As an extreme example, imagine that the risk difference for the effect of E on D is 0.2 among those with U = 1 and − 0.2 among those with U = 0. Further, assume that the source population is comprised of 50% people who have U = 1, but in our study 75% are U = 1 and 25% are U = 0. In the source population, the true causal effect is a risk difference of 0; however, we would discover a risk difference of 0.1 in our study population. Had we measured U, we would be able to estimate the correct U-stratum specific effects. We could also standardize over U to obtain the correct average causal effect. However, in the absence of measured U, there is little we can do. Further, because effect modification is not well described by directed acyclic graphs, this type of selection bias is somewhat more difficult to detect from graphs. We can say that selection bias is possible, but we cannot say that it is expected. For selection bias to occur in this setting, the association between participation P and disease D must vary within levels of the exposure E on the scale (ratio or difference) used to estimate associations [6, 7]. Equivalently, the association between E and D must vary within levels of P. We note that this type of selection bias cannot occur if the sharp null is true because then the effect of E on D must be zero in every stratum of P. Selection bias may exist on the ratio scale but not the difference scale, on the difference scale but not the ratio scale, on both scales, or on neither scale. This dependence is analogous to the dependence on ratio or difference scale of measurements and inferences regarding interaction and effect measure modification. Causal graphs do not reliably encode biases that depend on the parameterization of the effect [7], which is why they have somewhat less utility for selection bias structures such as this that do not entail collider stratification.

Design Considerations

Protection against selection bias is sometime used as the basis for a distinction between case-control and cohort studies. Case-control studies select cases (usually all cases) from the source population and select controls as a sample of the source population. Because of this, there is already an association between disease and participation. As long as there is no association between exposure and participation, there is no collider stratification-style selection bias. However, if a different proportion of exposed and unexposed participate, then the opportunity arises for selection bias, because the D → P association exists by design. Cohort studies usually attempt to enroll a census of the source population and follow it forward in time to observe incident cases. Because the entire source population is meant to be included, and the population at risk excludes prevalent cases, this collider stratification bias is not so prominent for cohort studies at baseline.

This apparent distinction between susceptibility of case-control and cohort designs to selection bias provides one basis for a common misperception that case-control studies are inherently less valid than cohort studies. The distinction is not, however, as clear as the simple comparison in the preceding paragraph makes it seem. First, for selection bias from differences in initial participation to occur, both disease status and exposure status must affect initial participation. While case status may affect participation in a case-control study, exposure status may not, in which case no selection bias would be expected. Second, case-control studies that rely on registries in which both exposure and disease status are recorded before conception of the study hypothesis will be immune from selection bias. In these registry-based case-control studies, the complete study population participates, and selection of cases and controls is a matter of efficiency. There is no selection bias because all cases participate and controls are sampled without regard to exposure status. Similarly, case-control studies can be nested in prospective cohort studies. In these designs, all cases are sampled and controls are selected from the cohort that gave rise to the cases because some information required for the study would be too expensive to gather on the whole cohort. For example, case-control studies of gene–environment interaction were frequently nested in prospective cohort studies because genotyping was too expensive to complete for the entire cohort. Again, the design is dictated by cost efficiency. Third, cohort studies can be conducted retrospectively, so disease and exposure status may affect participation. For example, in a retrospective cohort study of the relation between mastic asphalt work and lung cancer mortality [13], two of the sources of information used to identify asphalt workers (*i.e.,* exposed subjects) were a registry of men enrolled in a benefit society and a list of benefit recipients [14]. The latter ought to have been a subset of the former; however, the fact that some benefit recipients were not listed as benefit society members suggested that the membership roster included only survivors at the time it was disclosed to the study. That is, members of the benefit society were deleted from its roster when they died. No similar force acted on the unexposed reference group (comparable Danish men). Thus, both disease status (mortality) and exposure status (asphalt work) were related to initial participation, giving rise to a selection bias in this cohort study. Finally, case-control studies and cohort studies are both susceptible to differential loss-to-follow-up, which is a selection bias.

To summarize, both case-control studies and cohort studies can be designed to prevent selection bias, and both can be conducted such that selection bias occurs. Each study requires examination of the design to assess its susceptibility to selection bias. The simple dichotomization of designs that suggests case-control studies are susceptible to selection bias, whereas cohort studies are not, will inevitably lead to avoidable errors. Each study must be examined on its own merits to ascertain its susceptibility to selection bias [15].

Bias Analysis

Motivation for Bias Analysis

The motivation for bias analysis to address selection bias follows directly from its conceptualization. One wishes to adjust an estimate of disease frequency or an estimate of association, measured only among initial or ongoing participants, to account for the bias introduced by conditioning on participation, when participation is affected by selection forces. The adjustment can be difficult since it often requires an assessment of the participation proportion among each of the four combinations (when exposure and outcome are dichotomous) of exposure and disease. Often the exposure status and disease status of the nonparticipants will be unknown—their participation is required to ascertain this information. In this case, one might approach the bias analysis from a different perspective. That is, one might ask whether reasonable estimates of the participation proportions could account for all of an observed association. A slight variation of this approach is to ask whether the participation proportions required to counter an observed association, so that the estimate adjusted for postulated selection bias equals the null, are reasonable. These approaches will be discussed in the methods for simple bias analysis below.

Sources of Data

A bias analysis to address selection bias should be informed by evidence that is ordinarily provided by data. The data may derive from a validation substudy, in which the participants in the substudy are also participants in the larger source population. Such a substudy would provide internal validation data. The data may also derive from a data source whose members do not overlap with the membership of the source population, in which case the study provides external validation data. The distinction between internal validation data and external validation data has two important consequences. First, internal data ordinarily address the problem of selection bias more credibly than external data, since the source of information also provides the data for the estimate of association from the study. Second, with internal validation data, only the subset of participants who did not participate in the validation substudy has missing data. The subset that did participate in the validation substudy has complete data (both the data to estimate the effect and the data to address the selection bias). With external validation data, no participants in the validation study provide data to estimate the effect. This nuance affects the analytic technique used to address the selection bias, as explained further below.

The last source of information about selection bias does not derive from validation data *per se*, but rather from the experience of the analyst. That is, the information used to inform the bias analysis is a series of educated guesses postulated by the analyst for the purpose of completing the bias analysis. Each educated guess derives from the analyst's experience and familiarity with the problem at hand. While the notion of making an educated guess to address selection bias may engender some discomfort, since the bias analysis seems to be entirely fabricated, the alternative is to ignore (at least quantitatively) the potential for selection bias to affect the study's results. This alternative is also a fabrication, and often runs contrary to evidence from the study, such as an observed difference in participation rates between cases and controls.

Simple Bias-Adjustment for Differential Initial Participation

Example

The example used to illustrate methods of simple bias analysis to address selection bias for differential initial participation derives from a case-control study of the relation between mobile telephone use and uveal melanoma [16]. Realize that the study's exposure period predates the era of widespread use of mobile telephones. Melanoma of the uveal tract is a rare cancer of the iris, ciliary body, or choroid. It is the most common primary intraocular malignancy in adults, with an age-adjusted incidence of approximately 4.3 new cases/million population (http://www.cancer.gov/cancertopics/pdq/treatment/intraocularmelanoma/healthprofessional, 2006). Cases are definitively diagnosed and treated at tertiary centers, where the necessary clinical resources can be concentrated at sufficient density to provide adequate care. A cohort study with self-reported use of mobile telephone use would be very unlikely to yield enough cases to provide a precise estimate of the association. Given the rarity of the disease, and the complexity of treating it, a case-control design was the only reasonable option.

Figure 4.4 illustrates the flow of participant enrollment into the study. Stang et al. identified 486 incident cases of uveal melanoma between September 2002 and

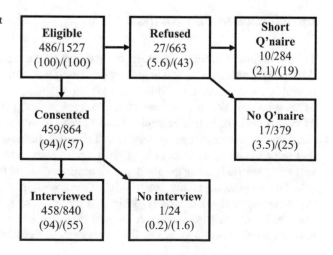

Figure 4.4 Flow of subject enrollment into the Stang et al., 2009 study of the association between mobile telephone use and the occurrence of uveal melanoma [16]. Participation frequencies are depicted as number of cases/number controls. Participation percentages, relative to the whole, are depicted as (% cases)/(% controls).

Eligible	Refused	Short Q'naire
486/1527	27/663	10/284
(100)/(100)	(5.6)/(43)	(2.1)/(19)

Consented		No Q'naire
459/864		17/379
(94)/(57)		(3.5)/(25)

Interviewed	No interview
458/840	1/24
(94)/(55)	(0.2)/(1.6)

September 2004 at a tertiary care facility in Essen, Germany that receives cases from all of Europe [16, 17]. Of these 486 cases, 458 (94%) agreed to participate in the case-control study and completed the interview. One hundred thirty-six (30%) of the interviewed cases reported regular mobile phone use and 107 (23%) reported no mobile phone use; the remainder used mobile phones irregularly. Three control groups were constructed; this example will use only the population-based controls who were matched to cases on age, sex, and region of residence. There were 1527 eligible population-based controls, of which 840 (55%) agreed to participate and were interviewed. Two hundred and ninety-seven (35%) of the 840 interviewed controls reported regular mobile phone use and 165 (20%) reported no mobile phone use. The odds ratio associating regular mobile phone use, compared with no mobile phone use, with uveal melanoma incidence equaled 0.71 (95% CI 0.51, 0.97). The substantial difference in participation rates between cases and controls (94% vs 55%, respectively) motivates a concern for the impact of selection bias on this estimate of association.

As shown in Figure 4.4, Stang et al. (2009) asked those who refused to participate whether they would answer a short-questionnaire to estimate the prevalence of mobile phone use among nonparticipants. Of the 27 nonparticipating cases, 10 completed the short questionnaire, and 3 of the 10 (30%) reported regular mobile phone use. Of the 663 nonparticipating controls, 284 completed the short questionnaire and 72 (25%) reported regular mobile phone use. Only two categories were available on the short questionnaire, so those who did not report regular mobile phone use were categorized as nonusers. Note that the prevalence of mobile phone use among participating cases (136/458 = 30%) was the same as among non-participating cases who answered the short questionnaire (3/10 = 30%), whereas the prevalence of mobile phone use among participating controls (297/840 = 35%) was higher than among non-participating controls who answered the short questionnaire (72/284 = 25%).

Introduction to Bias Analysis

To bias-adjust a relative estimate of association for selection bias induced by differential participation, one begins by depicting the participation in a contingency table. In most cases, one cannot allocate non-participants into exposure groups (if case-control design) or into outcome groups and possibly exposure groups (if cohort design). In this example, a subfraction of nonparticipants can be categorized into exposure groups, as shown in Table 4.2, because Stang et al. recognized that selection bias may affect their retrospective case-control design, so designed a validation substudy using the short questionnaire to inform the values assigned to the selection parameters of the bias model. This foresight must be recognized as an important strength of the study design. Without these estimates of the exposure prevalence in non-participating cases and controls, educated guesses would have been required to assign values to the parameters of the selection bias model.

Table 4.2 Depiction of participation and mobile phone use in a study of the relation between mobile phone use and the occurrence of uveal melanoma. Observed data from Stang et al., 2009 [16].

	Participants		Nonparticipants short questionnaire		Nonparticipants no questionnaire
	Regular use	No use	Regular use	No use	Cannot categorize
Cases	136	107	3	7	17
Controls	297	165	72	212	379

Using the data from participants and nonparticipants who completed the short questionnaire, one can see that the odds of participation depend on disease status [OR = (243/10)/(462/284) = 14.9] and on exposure status [which is examined only in the controls, OR = (297/72)/(165/212) = 5.3].

The crude odds ratio associating regular mobile phone use with uveal melanoma occurrence equals:

$$OR_{crude,participants} = \frac{136/297}{107/165} = 0.71$$

which approximately equals the matched odds ratio reported in the study. The matching will therefore be ignored for the purpose of illustrating the selection bias-adjustment.

Among nonparticipants who answered the short questionnaire, the crude odds ratio equals:

$$OR_{crude,nonparticipants} = \frac{3/72}{7/212} = 1.26$$

which is in the opposite direction from the crude odds ratio observed among participants. This difference illustrates the potential impact of selection bias.

Bias Analysis by Projecting the Exposed Proportion Among Nonparticipants

To bias-adjust for the selection bias, one could collapse the participant and nonparticipant data from persons who completed the short questionnaire, but that would leave out the nonparticipants who did not complete the short questionnaire. A simple solution would be to assume this second group of nonparticipants (*i.e.*, those who also did not participate in the short questionnaire) had the same exposure

prevalence, conditional on case/control status, as those who did agree to participate
in the short questionnaire. To accomplish this solution, divide those in the second
group of nonparticipants into exposure groups in proportion to the exposure prevalence observed among nonparticipating cases and controls who did complete the
short questionnaire. For example, multiply the 17 nonparticipant cases by 3/10 to
obtain the proportion expected to be regular users. The results are added to the
number of observed exposed cases (136) and the number of exposed cases who
answered the short questionnaire (3) to obtain an estimate of the number of total
exposed cases among those eligible for the study. Similar algebra is applied for the
unexposed cases and the exposed and unexposed controls, as shown in Equation 4.1:

$$OR_{bias-adjusted} = \frac{136 + 3 + \frac{3}{10}17}{297 + 72 + \frac{72}{284}379} \bigg/ \frac{107 + 7 + \frac{7}{10}17}{165 + 212 + \frac{212}{284}379} = 1.62 \qquad (4.1)$$

The bias-adjusted odds ratio is 1.62, which is substantially different than the
crude odds ratio among full participants (0.71), suggesting a substantial bias due to
differential selection in the study. As noted above, this bias-adjustment assumes that
the nonparticipants who did not answer the short questionnaire have the same
prevalence of mobile phone use, within strata of cases and controls, as the nonparticipants who did answer the short questionnaire. Although this assumption
cannot be tested empirically with the available data, one can explore alternative
bias analysis methods that explore the impact of violations of the assumption using
methods in Chapters 7–11. Table 4.3 foreshadows these methods by showing bias-
adjusted odds ratios obtained with different assumptions about the exposure prevalence in cases and controls who did not participate in either the full questionnaire or
the short questionnaire. If we assume that they have the same exposure prevalence as
those who participated in the short questionnaire, then the bias-adjusted odds ratio is
1.62, as shown in Equation 4.1. If we assume that they have the same exposure
prevalence as those who participated in the full questionnaire, then the bias-adjusted
odds ratio is 1.02. If we assume that none of them regularly used mobile phones, then
the bias-adjusted odds ratio is 2.17. It is only if we assume that all these
non-participants regularly used mobile phones that we obtained a bias-adjusted
odds ratio (0.69) similar to what was observed in the crude data (0.71). This scenario
seems implausible, given that only 25% to 35% of questionnaire respondents said
they regularly used mobile phones in answer to the full or short questionnaire. This
sensitivity analysis of the bias analysis strengthens the view that the crude estimate
(0.71) was substantially influenced by selection bias. Probabilistic bias analysis
(Chapters 8 and 9) will allow us to more completely take account of this information
as we complete sensitivity analyses of the bias analysis.

Table 4.3 Selection bias odds ratio and adjusted odds ratio in a study of the relation between the mobile phone use and the occurrence of uveal melanoma [16], assuming different exposure prevalences for nonparticipating cases and controls who did not answer the short questionnaire. Original data from Stang et al., 2009 [16].

Scenario	Exposed cases			Exposed controls			Unexposed cases			Unexposed Controls			
	Full Q'naire	Short Q'Naire	No Q'Naire	Full Q'naire	Short Q'Naire	No Q'Naire	Full Q'naire	Short Q'Naire	No Q'Naire	Full Q'Naire	Short Q'Naire	No Q'Naire	Odds ratio
Full participants only	136			297			107			165			0.71
Short questionnaire participants only		3			72			7			212		1.26
No questionnaire as short questionnaire	136	3	5.1	297	72	96.1	107	7	11.9	165	212	282.9	1.62
No questionnaire as full participants	136	3	9.5	297	72	243.6	107	7	7.5	165	212	135.4	1.02
No questionnaire all unexposed	136	3	0	297	72	0.0	107	7	17	165	212	379.0	2.17
No questionnaire all exposed	136	3	17	297	72	379.0	107	7	0	165	212	0.0	0.69

Table 4.4 Bias parameters required for simple bias analysis to address selection bias from differential initial participation. S is the probability of selection among cases and controls who are exposed (1) or unexposed (0).

	Exposure = 1	Exposure = 0
Cases	$S_{case,1}$	$S_{case,0}$
Noncases	$S_{control,1}$	$S_{control,0}$

Table 4.5 Selection proportions in a study of the relation between the mobile phone use and the occurrence of uveal melanoma. Original data from Stang et al., 2009 [16].

Selection probabilities	Regular users	Non users
Cases	$S_{case,1} = 136/(136 + 3 + (3/10) \cdot 17)$ $= 0.94$	$S_{case,0} = 107/(107 + 7 + (7/10) \cdot 17) =$ 0.85
Noncases	$S_{control,1} = 297/(297 + 72 + (72/284)$ $\cdot 379) = 0.64$	$S_{control,0} = 165/(165 + 212 + (212/284)$ $\cdot 379) = 0.25$

Bias Analysis Using Selection Proportions

The preceding example is unusual in that data on the exposure prevalence among nonparticipants are available. Ordinarily, only the number of nonparticipating cases and controls would be available, and perhaps the participation rate among eligible cases and controls, but no information about the exposure prevalence among non-participants would be known. In this circumstance, one must postulate selection proportions, guided by the participation rates in cases and controls, as shown in Table 4.4.

With these bias parameters, one can define a selection bias odds ratio as: $sOR = \frac{S_{case,1} S_{control,0}}{S_{case,0} S_{control,1}}$. This ratio can be used to adjust the observed odds ratio by multiplying the observed odds ratio by the inverse of the selection bias odds ratio to account for differential initial participation using Equation 4.2, where "S" represents a selection proportion, indexed by case/control status and exposure status (1 = regular mobile phone user, 0 = never mobile phone user):

$$\hat{OR}_{bias\ adjusted} = \frac{\hat{OR}_{crude}}{sOR} = \hat{OR}_{crude} \cdot \frac{S_{case,0} S_{control,1}}{S_{case,1} S_{control,0}} \tag{4.2}$$

Continuing with the example, we can calculate the selection proportions, again assuming that the exposure prevalence among the nonparticipants who answered the questionnaire equals the exposure prevalence among the nonparticipants who did not answer the short questionnaire. With this assumption, the selection probabilities are as shown in Table 4.5.

When these are multiplied together they give a selection bias odds ratio of 0.43. Using these numbers in Equation 4.2, the bias-adjusted odds ratio is estimated as

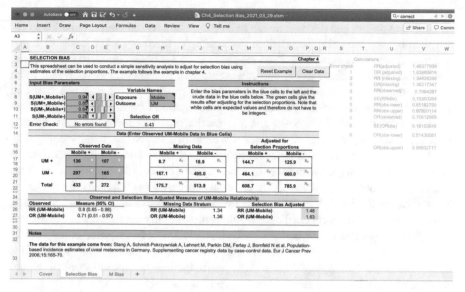

Figure 4.5 Screenshot of the solution to the selection bias problem using the selection proportion approach and data from Stang et al., 2009 study of the association between mobile telephone use and the occurrence of uveal melanoma [16].

$$\hat{OR}_{bias\,adjusted} = 0.71 \cdot \frac{0.85 \cdot 0.64}{0.94 \cdot 0.25} = 1.62$$

Note that the bias-adjusted odds ratio calculated by this method equals the same value as calculated by the first method earlier (and as is good practice, we do not round any of the components of the computation before obtaining the final answer). The two approaches are mathematically identical. Figure 4.5 shows a screen shot of the solution to this selection bias problem using the Excel® spreadsheet available on the text's website. The values assigned to the bias parameters and crude data are entered in the blue cells. The white cells show the missing data and the combination of the missing data and the observed data. The results, displayed in the green cells at the bottom, show the association between mobile phone use and uveal melanoma bias-adjusted for the selection bias.

As in the previous section, a reasonable second step in this selection bias example would be to calculate the limits of selection bias by presuming that all the non-participants who did not answer the short questionnaire were either regular mobile phone users or nonusers, and to subject the analysis to additional sensitivity analyses of the bias analysis.

Bias Analysis Using Inverse Probability of Participation Weighting

A third mathematically equivalent approach to selection bias adjustment in this example would be to reweight participants so that they represent themselves in the analysis, as well as other members of the source population who are like them (with respect to exposure and disease, in this example), but who did not participate, so are not members of the study population. For example, there were 136 cases of uveal melanoma who fully participated and regularly used a mobile phone. In the analysis, we would like to include them, and reweight them to represent anyone like them from among the non-participants. The selection proportion for exposed cases in Table 4.5 is 0.94. The inverse probability of participation for exposed cases therefore equals $1/0.94 = 1.06$. To account for non-participants like them, the participating exposed cases should therefore be given a weight of 1.06. Notice that by doing this, the 136 participating exposed cases represent $136 \cdot 1.06 = 144.2$, which is the same (with rounding error) as the number of total exposed cases $(136 + 3 + (3/10) \cdot 17)$ we estimated in row 3 of Table 4.3. If all exposed cases participated, the selection proportion would be 1.0, and then the inverse probability of participation would equal $1/1 = 1$, and each exposed case would represent only themselves in the inverse probability of participation weighted analysis. Only 25% of unexposed controls fully participated (see Table 4.5), so the inverse probability of participation weights for them would be $1/0.25 = 4$. Each fully participating unexposed control would represent themselves, and three other unexposed controls who did not fully participate. We can demonstrate this reweighting approach to bias analysis using the frequencies observed for exposed and unexposed cases and controls and the selection proportions from Table 4.5:

$$OR_{biasadjusted} = \frac{136 \frac{1}{S_{case,1}} / 297 \frac{1}{S_{control,1}}}{107 \frac{1}{S_{case,0}} / 165 \frac{1}{S_{control,0}}} = \frac{136 \frac{1}{0.94} / 297 \frac{1}{0.64}}{107 \frac{1}{0.85} / 165 \frac{1}{0.25}} = 1.62$$

As above, a reasonable next step would be to conduct a sensitivity analysis of the bias analysis by varying the values assigned to the selection proportions, which results in different weights for the inverse probability of participation weighting.

There are a number of important points to keep in mind regarding the previous three approaches. First, they are all mathematically equivalent and the answer an analyst obtains will not depend on the method. Second, choice of method will generally be based on information available. If information is lacking on individual selection probabilities and even a best guess is difficult to specify, it may be marginally easier to directly use the selection odds ratio since it only requires specifying one parameter. However, if information is lacking to this extent, we encourage analysts to be very cautious with their interpretation of bias adjustment results. We describe methods below that are extensions of the direct application of weights to bias adjustment. In these cases, it is easiest to directly weight the cell counts as in the third approach. The selection odds ratio method is convenient but only works if the analyst is estimating an odds ratio as the measure of effect for the study. This approach will not work for other measures of effect and selection weights are more versatile. Third, these approaches are as valid as the assumptions that are used to generate the results. In particular, we have assumed that selection depends only on exposure and disease and that we are able to perfectly specify selection probabilities. As we have noted, this may be hopeful at best. In the next section we describe more complicated settings in which we generate selection proportions based on all available information in the study. However, in general, for weights to adjust for selection-bias, the weights must be conditional on exposure and all covariates that are associated with both selection into the study and the outcome of interest.

Simple Bias-Adjustment for Differential Loss-to-Follow-up

As noted above, differential loss-to-follow-up should be viewed as a source of possible selection bias because it arises from differences in continued study participation that are related to both the exposure and the disease. If the information on loss-to-follow-up is missing at random, the bias can be addressed using missing data methods (Chapter 10). Below we describe an extension to the weighting approach, where weights are predicated on the knowledge available about the participants before their loss-to-follow-up.

Example

The example used to illustrate methods of simple bias analysis to address differential loss-to-follow-up derives from a cohort study of the relation between receipt of guideline breast cancer therapy and breast cancer mortality [18]. Although effective primary therapy for early-stage breast cancer has been well characterized and enjoys a broad consensus, this standard has not fully penetrated medical practice. The example study examined the effect of less than definitive guideline therapy on breast cancer mortality over 12 years of follow-up after diagnosis of local or regional breast cancer.

The study population comprised 449 women diagnosed with local or regional breast cancer at eight Rhode Island hospitals between July 1984 and February 1986 [19]. Patients' identifying variables were expunged after the enrollment project was completed to comply with the human subjects oversight committee. Subjects were reidentified for the follow-up study by matching patient characteristics to the Cancer Registry of the Hospital Association of Rhode Island. The Hospital Association of Rhode Island reidentified 390 of the original 449 patients (87%), and the remaining 59 patients (13%) were therefore lost to follow-up. The probability of reidentification depended most strongly on the hospital of diagnosis, because two hospitals participated in the Hospital Association of Rhode Island cancer registry for only part of the study's enrollment period.

The vital status of the 390 reidentified patients was ascertained by matching their identifying variables to the National Death Index. The outcome (breast cancer mortality) was assigned to subjects with a death certificate that listed breast cancer as the underlying cause or a contributing cause of death. The date of last follow-up was assigned as the date of death recorded on the death certificate for decedents. For

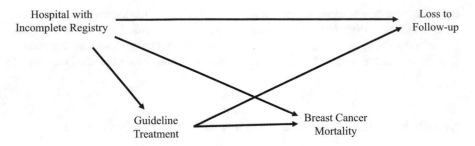

Figure 4.6 Directed acyclic graph of the study estimating the effect of treatment guidelines on breast cancer mortality at eight Rhode Island hospitals.

subjects with no National Death Index match, 31 December 1996 was assigned as the date of last follow-up.

The 59 patients lost to follow-up were not reidentified by the Hospital Association of Rhode Island, so their identifying information could not be matched to the National Death Index. It is not known, therefore, whether they were dead or alive by the end of follow-up and, if dead, whether they died of breast cancer. The hospital where these women were treated was known, and whether they received guideline therapy was known, since this information was ascertained before the identifying variables were deleted from the original data set. Given this information, one can conduct a simple bias analysis to assess the selection bias potentially introduced by differential loss-to-follow-up. A directed acyclic graph for this study is shown in Figure 4.6.

Bias Analysis by Modeling Outcomes

To bias-adjust the rate ratio for selection bias induced by differential loss-to-follow-up, one first depicts the follow-up in a contingency table as shown in Table 4.6. In this example, the rate of breast cancer mortality cannot be depicted among women who were lost to follow-up.

The crude rate difference associating less than guideline therapy (vs guideline) with breast cancer mortality equals:

$$IRD_{crude,participants} = \frac{40}{687} - \frac{65}{2560} = \frac{3.3}{100\,PY}$$

All that is known about those lost to follow-up is whether they initially received guideline therapy and other baseline characteristics, including their hospitals of diagnosis. To bias-adjust for their missing information, first estimate the number of missing person-years by multiplying the average follow-up duration among those with and without guideline therapy by the number of persons missing in each category.

Table 4.6 Depiction of breast cancer mortality over 12 years of follow-up among those receiving guideline therapy and those receiving less than guideline therapy, and the number lost to follow-up by receipt of guideline therapy or less than guideline therapy. Original data from Silliman et al., 1989 and Lash et al., 2000 [18, 19].

	Completed follow-up (N = 390)		Lost to follow-up (N = 59)	
	Less than Guideline	Guideline	Less than Guideline	Guideline
Breast cancer deaths	40	65		
Persons	104	286	13	46
Person-years	687	2560		
Crude rate	5.8/100 PY	2.5/100 PY		
Crude rate difference	3.3/100 PY	0.		
Crude rate ratio	2.3	1.		

$$PY_{<guideline} = \frac{687 \cdot 13}{104} = 85.9PY$$
$$\text{and}$$
$$PY_{guideline} = \frac{2560 \cdot 46}{286} = 411.7PY$$

(4.3)

To bias-adjust for potential differential loss-to-follow-up, multiply the estimated missing person-time by an educated guess of the breast cancer mortality rate that would have been observed in the missing person-time. Of course, guessing that it equals the rates in the observed person-time yields a bias-adjusted result that is the same as the crude result, suggesting that there would be no bias from the loss-to-follow-up. Instead, we use the mortality rates observed among those diagnosed at the two hospitals where the tumor registry operated for only part of the study. Recall that most of the women lost to follow-up were diagnosed at these two hospitals and note that the rate difference and rate ratio observed among the breast cancer patients diagnosed at these two hospitals who had complete data were much nearer the null than the associations observed in the whole population, which suggests the possibility of a differential loss-to-follow-up.

Table 4.7 shows the person-time and number of breast cancer cases imputed by this method for the women lost to follow-up. In the less than guideline care group, we estimated 85.9 missing person-years of follow-up (Equation 4.3) and observed a breast cancer mortality rate of 4.9/100 PY in the women who were successfully reidentified and who were diagnosed at the two hospitals where most of the losses to follow-up occurred. We estimate, therefore, that there were 85.9 PY•4.9/100 PY = 4.2 breast cancer deaths missed in the lost follow-up of the women who received less than guideline care. In the guideline care group, we estimated 411.7 missing person-years of follow-up (Equation 4.3) and observed a breast cancer mortality rate of 4.5/100 PY in the women who were successfully reidentified and who were diagnosed at the two hospitals where most of the losses to follow-up occurred. We estimate, therefore, that there were 411.7 PY•4.5/100 PY = 18.7 breast cancer deaths missed in the lost follow-up of the women who received less than guideline care. In both computations, we round only at the last step, so for example 411.7•4.5/100 PY equals 18.52, so would round to 18.5. However, both 411.7 and

Table 4.7 Depiction of breast cancer mortality over 12 years of follow-up among those receiving guideline therapy and those receiving less than guideline therapy, imputed for those lost to follow-up on the basis of breast cancer mortality rates observed for women diagnosed at two hospitals where tumor registries operated during only part of the study. Original data from Silliman et al., 1989 and Lash et al., 2000 [18, 19].

	Completed follow-up at two hospitals		Imputed loss-to-follow-up information	
	< Guideline	Guideline	< Guideline	Guideline
Breast cancer deaths	3	5	4.2	18.7
Person-years	60.8	110.2	85.9	411.7
Crude rate	4.9/100 PY	4.5/100 PY		
Crude rate difference	0.4/100 PY	0.		
Crude rate ratio	1.1	1.		

4.5/100 PY are themselves rounded. If we use the original numbers (2560•46/286)• (5/110.2) we obtain 18.68, which rounds to 18.7. As noted above, it is always good practice to round only at the end of such a computation.

To complete the simple bias analysis, sum the observed breast cancer cases and person-time within therapy groups to calculate the rate that would have been observed, had those lost to follow-up had a breast cancer mortality experience similar to that observed among the women who were followed after diagnosis at the two hospitals.

$$IR_{<guideline} = \frac{40 + 4.2}{687\,PY + 85.9\,PY} = 5.7/100PY$$

$$IR_{guideline} = \frac{65 + 18.7}{2560PY + 411.7PY} = 2.8/100PY.$$

Using these imputed rates, the incidence rate difference equals 2.9/100 PY and the incidence rate ratio equals 2.0. These associations are nearer the null than the associations observed among those with complete data, but not so near the null as the associations observed in just the two hospitals where most women lost to follow-up had been diagnosed. The bias analysis provides some assurance that the observed association between receipt of guideline therapy and breast cancer mortality was not entirely attributable to differential loss-to-follow-up, conditional on the accuracy of the values assigned to the bias parameters. To further buttress that inference, one could subject the analysis to the most extreme case, in which all of those lost to follow-up in the guideline therapy group died of breast cancer and none of those lost to follow-up in the less than guideline therapy group died of breast cancer:

$$IR_{<guideline} = \frac{40 + 0}{687\,PY + 85.9\,PY} = 5.2/100PY$$

$$IR_{guideline} = \frac{65 + 46}{2560PY + 411.7PY} = 3.7/100PY$$

Even with this extreme assumption, the rate difference (1.4/100 PY) remains above the null. This bounding simple bias analysis shows that the entire association between receipt of guideline therapy and breast cancer mortality could not be attributable to differential loss-to-follow-up, conditional on the assumptions about the amount of missing person-time.

Bias Analysis by Inverse Probability of Attrition Weighting

Bias analysis for loss-to-follow-up can also be addressed by inverse probability weighting [20]. To generate weights, we must recall the advice at the end of the previous section: weights must depend on the exposure of interest as well as all variables that predict retention in the study and the outcome [5]. In this case, we can examine the DAG in Figure 4.6 and see that the weights must depend on whether treatment guidelines were met and whether a person attended one of the two hospitals with an incomplete registry. Notice that the outcome does not impact participation in the study directly and selection weights do not need to depend on outcome status. This is in contrast with previous examples of selection bias in which selection weights had to depend on the outcome. For this reason, adjusting for selection bias due to loss-to-follow-up is often somewhat more tractable than selection bias in which the outcome directly influences participation.

Table 4.8 applies this weighting method to the breast cancer mortality example. The 449 women are stratified by exposure status (receipt of less than guideline therapy or guideline therapy) and hospital of diagnosis (diagnosed at one of the two hospitals where a tumor registry operated during only part of the study enrollment period or at one of the six hospitals where a tumor registry operated for the entire study enrollment period). Recall that these characteristics were known for all women, regardless of whether they were reidentified. This information allows a calculation of the probability of reidentification (or loss-to-follow-up, attrition) within each combination of exposure and hospital categories. In the subset of reidentified women, both the outcome of breast cancer mortality and the person-time of follow-up were observed. Inverse probability of attrition weighting (IPAW) reweights these observed rates to account for losses to follow-up. For example, in the women who received less than guideline care and were diagnosed at one of the six hospitals with complete registration, 96 of 99 women (97%) were reidentified. The inverse probability of reidentification equals 1/0.97 or 1.03. Each reidentified woman in this group counts for herself, and is slightly upweighted to account for the unobserved experience of the three women who were not reidentified. In the women who received guideline therapy and were diagnosed at one of the two hospitals with incomplete registration, only 13 of 48 women (27%) were reidentified. The inverse probability of reidentification in these women is 1/0.27 or 3.69. The mortality experience of these 13 women is substantially reweighted to account for the 35 women who were not reidentified.

Table 4.8 Use of inverse probability of attrition weighting to bias-adjust for loss-to-follow-up for breast cancer mortality over 12 years of follow-up among those receiving guideline therapy and those receiving less than guideline therapy, stratified by whether women were diagnosed at two hospitals where tumor registries operated during only part of the study.

	Diagnosed at other than two hospitals		Diagnosed at two hospitals	
	< guideline	guideline	< guideline	guideline
Cases in reidentified	37	60	3	5
Reidentified N	96	273	8	13
Reidentified PY	626.2	2449.8	60.8	110.2
Not reidentified N	3	11	10	35
Total N	99	284	18	48
Reidentification proportion	0.97	0.96	0.44	0.27
IPAW	1.03	1.04	2.25	3.69
Crude rate (/100 PY)	5.9	2.4	4.9	4.5
Stratum specific IRD (/100 PY)	3.5		0.4	
Stratum specific IRR	2.4		1.1	
Crude IRD (/100 PY)	3.3			
Crude IRR	2.3			
IPAW IRD (/100 PY)	3.0			
IPAW IRR	2.10			

Original data from Silliman et al., 1989 and Lash et al., 2000 [18, 19].
PY person years, *IPAW* inverse probability of attrition weights, *IRD* incidence rate difference, *IRR* incidence rate ratio.

As before, the crude incidence rate difference (3.3/100PY) and crude incidence rate ratio (2.29) suggest an association between receipt of less than guideline therapy, compared with receipt of guideline therapy, and the rate of breast cancer mortality. We obtained IPAW estimates by multiplying the observed data by their corresponding weights. For example, in the less than guideline therapy group, the IPAW rate is

$$IR_{IPAW} = \frac{A_1 wt_1 + A_2 wt_2}{PY_1 wt_1 + PY_2 wt_2}$$

$$IR_{<guideline, IPAW} = \frac{37 \cdot 1.03 + 3 \cdot 2.25}{626.2 \cdot 1.03 + 60.8 \cdot 2.25} = \frac{5.7}{100 \, PY} \qquad (4.4)$$

where A_i, PY_i, and wt_i are the number of cases, person years and the IPAW weights in the ith stratum. This rate is slightly lower than the crude rate (5.8/100PY). We can think of the crude rate as being calculated using weights of 1 (*i.e.*, for the crude rate, 1.03 and 2.25 are both replaced by 1 in Equation 4.4). IPAW weights the rate in the stratum of women diagnosed at the two hospitals with incomplete registration more highly than the rate among women diagnosed at the hospitals with complete registration (2.25 versus 1.03), because the former women were more likely to be lost to follow-up due to failure to be reidentified. The rate in the women who received less than guideline therapy, were diagnosed at the two hospitals with incomplete registration, and were reidentified (4.9/100PY) is lower than the rate in the women who received less than guideline therapy, were diagnosed at a hospital with complete registration, and were reidentified (5.9/100PY). Both the crude rate and the IPAW rate are a function of the weights applied to the two strata. The crude rate assigns an implicit weight of 1 to each stratum, whereas IPAW overweights the stratum with higher loss-to-follow-up (and a lower stratum-specific rate), resulting in a lower IPAW rate than crude rate.

Repeating the calculation in Equation 4.4 for those who met treatment guidelines gives

$$IR_{guideline,IPAW} = \frac{60 \cdot 1.04 + 5 \cdot 3.69}{2449.8 \cdot 1.04 + 110.2 \cdot 3.69} = \frac{2.7}{100\,PY}$$

The IPAW bias-adjusted estimates of association (IPAW IRD = 3.0/100PY; IPAW IRR = 2.10) are both nearer the null than the conventional estimates. Nonetheless, they suggest a substantial increased rate of breast cancer mortality associated with receipt of less than guideline breast cancer therapy, compared with receipt of guideline breast cancer therapy, conditional on the accuracy of the bias model. In this case, the key assumption of the bias model is that the women who were not reidentified had a breast cancer mortality experience similar to the women who were reidentified, within categories of guideline therapy and hospitals of diagnosis. Note also that the IPAW bias-adjusted estimates are slightly different than the bias-analysis estimates obtained by modeling outcomes (bias-adjusted IRD 3.0/100 PY versus 2.9/100 PY and bias-adjusted IRR 2.1 versus 2.0, respectively). This difference arises from the treatment of the women who were not reidentified. The bias-adjustment that modeled outcomes assumed that all women who were not reidentified have mortality rates like the women who were diagnosed at the two hospitals with incomplete registration. In fact, only 45 of 59 women who were not reidentified were diagnosed at these two hospitals. The IPAW bias-adjustment used the rates for these two hospitals for these 45 women, and used rates for the six hospitals with complete registration for the 14 women diagnosed at hospitals with complete registration. This slight difference in the bias model accounts for the small difference in the bias-adjusted results.

Multidimensional Bias Analysis for Selection Bias

Simple bias analysis implies that the analyst has one and only one estimate to assign to each of the values for the error model's bias parameters. In many situations, that is not the case. There are many bias parameters for which validation data do not exist, so the values assigned to the bias parameter are educated guesses. In this situation, the analyst is better served by making more than one educated guess for each value and then combining values in different sets. In other situations, multiple different estimates of the values for the bias parameter may exist, and there may be no basis for the analyst to select just one as the best estimate of the truth from among those available.

Multidimensional bias analysis is a direct extension of simple bias analysis whereby the methods for simple bias analysis are repeated with a range of values for the bias parameter(s). The methodologies and equations used for multidimensional bias analysis for selection bias are the same as described above. In a multidimensional bias analysis, the analytic procedures are simply repeated with multiple combinations of values assigned to the bias parameters. This method provides the analyst with some information regarding the range of estimates of bias-adjusted associations that are possible, given different assumptions regarding the values of the bias parameters. For example, if there are no data regarding the bias parameters, then multidimensional bias analysis could be used to determine the minimum amount of bias that would convert a positive association to a null association. The analyst could then assess the plausibility of the values that must be assigned to the bias parameters to accomplish the conversion.

Example

It is often difficult to obtain values to assign to bias parameters to estimate the impact of selection bias, since by definition the complete data on nonparticipants or those lost to follow-up is not available in the study's dataset. Therefore, analysts may not have any measured values to assign to the bias parameters used in the error model. One solution is to examine a range of plausible values assigned to the bias parameters to investigate whether selection bias could have spuriously produced an association or masked a true association.

For example, consider a study investigating the association between receipt of the intranasal influenza vaccine and the risk of Bell's palsy [21]. Bell's palsy is a condition that causes weakness or paralysis of the facial muscles because of trauma to the seventh cranial nerve. This case–control study reported a crude odds ratio equal to 33 (95% CI: 16, 71), indicating a potentially strong causal relation. However, of the 773 cases of Bell's palsy that met the study's eligibility criteria, 361 could not be enrolled because the cases' physicians refused to participate. A further 162 of the cases did not have an available matched control, so were also excluded (highlighting a downside with substantial consequences for matched case-control studies). Therefore, only 250 (32%) of potential cases were included in the study. There are at least two potential causes of selection bias. First, the exposure prevalence in the controls was about 1%, and therefore it is possible that even a slightly higher prevalence among the controls that were not recruited for the 162 cases could impact the results. Second, of the 250 included cases, 27% had the vaccine during the relevant time period, whereas of the 162 cases without available controls, 14% received the vaccine. In addition, the exposure prevalence was unknown for the 361 cases not recruited into the study. Because of the concern

Table 4.9 Multidimensional bias analysis to examine selection bias from not recruiting controls for 162 cases in a case–control study of the association between receipt of the intranasal influenza vaccine and the risk of Bell's palsy.

Control exposure prevalence	Vaccinated case/ control	Unvaccinated case/ control	Bias-adjusted Odds ratio
0.01	68/8.0	182/714.0	33.3
0.01	23/5.4	139/480.6	14.8
0.02	91/17.7	321/1190.3	19.0
0.03	91/22.6	321/1185.4	14.9
0.04	91/27.4	321/1180.6	12.2
0.05	91/32.3	321/1175.7	10.3

Adapted from Mutsch et al., (2004) [21].

over the potential selection bias, multidimensional bias analysis methods were used to assess these selection bias issues.

To evaluate the potential impact of incomplete control participation, the exposure prevalence in the controls matched to the 162 cases was varied. Even if the exposure prevalence in the controls was 5%, rather than the observed 1%, the odds ratio between receipt of the intranasal influenza vaccine and Bell's palsy would have equaled 10. Although the strength of the association would have been weaker, the observed effect cannot be explained completely by selection bias if the only source of selection bias was an underestimation of the exposure prevalence among controls and given the assumed values assigned to the bias parameters in Table 4.9. The first row of the table represents the observed data on cases who had matched controls. The second row of the table shows data for cases with no matched controls. The case data in this row gives the observed prevalence. Control data is created assuming 3 controls per case and an exposure prevalence equal to that of the observed control series (approximately 1%). If this had occurred, the OR would have been 14.8. The subsequent rows assume different prevalences of exposure for the hypothetical controls and combine the actual data from the first row of the table with the actual unmatched cases and their hypothetical controls (assuming exposure prevalences from 2% to 5%).

Next, the potential impact of selection bias from the nonparticipating cases was examined. It is unknown whether the exposure prevalence of the 361 cases would be the 27% observed in the included cases, the 14% observed among the cases without available controls, or some other value. To evaluate the impact from excluding these cases, the odds ratio was calculated with a range of exposure prevalence from 27% to 1% (Table 4.10). If the exposure prevalence in the excluded cases was 14%, as with the other cases not included in the case–control analysis because they had no matched controls, then the odds ratio would be 21. If the exposure prevalence in these cases equaled the 1% prevalence of the observed controls, then the overall odds ratio would have been 13. These calculations indicate that the observed effect cannot be attributed completely to selection bias, at least given this bias model for the selection bias and the values assigned to the model's bias parameters. Figure 4.7 shows a screenshot of an Excel spreadsheet to perform a general multidimensional bias analysis for selection bias using selection odds ratios.

Table 4.10 Multidimensional bias analysis to examine selection bias from excluding 361 cases in a case–control study of the association between receipt of the intranasal influenza vaccine and the risk of Bell's palsy.

Case exposure prevalence	Vaccinated case/ control	Unvaccinated case/ control	Bias-adjusted Odds Ratio
0.272	189.2/18.8	583.8/1786.2	30.7
0.26	184.9/18.8	588.1/1786.2	29.8
0.25	181.3/18.8	591.8/1786.2	29.1
0.2	163.2/18.8	609.8/1786.2	25.4
0.15	145.2/18.8	627.9/1786.2	21.9
0.14	141.5/18.8	631.5/1786.2	21.3
0.1	127.1/18.8	645.9/1786.2	18.7
0.05	109.1/18.8	664.0/1786.2	15.6
0.01	94.6/18.8	678.4/1786.2	13.2

Adapted from Mutsch et al., (2004) [21].
Assumes constant control exposure prevalence of 0.01.

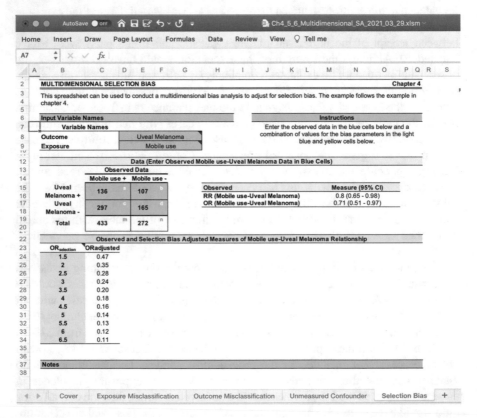

Figure 4.7 Screenshot of Excel spreadsheet to perform multidimensional bias analysis for selection bias.

References

1. Lesko CR, Buchanan AL, Westreich D, Edwards JK, Hudgens MG, Cole SR. Generalizing study results: A potential outcomes perspective. Epidemiology. 2017;28:553-561.
2. Pearl J. Generalizing experimental findings. J Causal Inference. 2015;3:259–66.
3. Balzer LB. All generalizations are dangerous, even this one.-Alexandre Dumas. Epidemiology. 2017;28:562-566.
4. Pearl J, Bareinboim E. External validity: From Do-Calculus to transportability across populations. Stat Sci. 2014;29:579–95.
5. Hernán MA, Hernández-Díaz S, Robins JM. A structural approach to selection bias. Epidemiology. 2004;15:615-25.
6. Greenland S. Response and follow-up bias in cohort studies. Am J Epidemiol. 1977;106:184–7.
7. Hernán MA. Invited commentary: Selection bias without colliders. Am J Epidemiol. 2017;185:1048-1050.
8. Wacholder S. The case-control study as data missing by design: estimating risk differences. Epidemiology. 1996;7:144–50.
9. Clough-Gorr KM, Fink AK, Silliman RA. Challenges associated with longitudinal survivorship research: attrition and a novel approach of reenrollment in a 6-year follow-up study of older breast cancer survivors. J Cancer Surviv. 2008;2:95-103.
10. Woolpert KM, Ward KC, England CV, Lash TL. Validation of LexisNexis Accurint in the Georgia Cancer Registry's Cancer Recurrence and Information Surveillance Program. Epidemiology. 2021;32:434-438.
11. Schildcrout JS, Schisterman EF, Mercaldo ND, Rathouz PJ, Heagerty PJ. Extending the case-control design to longitudinal data: Stratified sampling based on repeated binary outcomes. Epidemiology. 2018;29:67-75.
12. Greenland S, Pearl J, Robins JM. Causal diagrams for epidemiologic research. Epidemiology. 1999;10:37–48.
13. Hansen ES. Mortality of mastic asphalt workers. Scand J Work Env Health. 1991;17:20–4.
14. Cole P, Green LC, Lash TL. Lifestyle determinants of cancer among Danish mastic asphalt workers. Regul Toxicol Pharmacol. 1999;30:1–8.
15. Rothman KJ. Six Persistent Research Misconceptions. J Gen Intern Med. 2014;29:1060–4.
16. Stang A, Schmidt-Pokrzywniak A, Lash TL, Lommatzsch PK, Taubert G, Bornfeld N, et al. Mobile phone use and risk of uveal melanoma: results of the risk factors for uveal melanoma case-control study. J Natl Cancer Inst. 2009;101:120–3.
17. Schmidt-Pokrzywniak A, Jöckel KH, Bornfeld N, Stang A. Case-control study on uveal melanoma (RIFA): rational and design. BMC Ophthalmol. 2004;4:11.
18. Lash TL, Silliman RA, Guadagnoli E, Mor V. The effect of less than definitive care on breast carcinoma recurrence and mortality. Cancer. 2000;89:1739–47.
19. Silliman RA, Guadagnoli E, Weitberg AB, Mor V. Age as a predictor of diagnostic and initial treatment intensity in newly diagnosed breast cancer patients. J Gerontol. 1989;44:M46–50.
20. Weuve J, Tchetgen Tchetgen EJ, Glymour MM, Beck TL, Aggarwal NT, Wilson RS, et al. Accounting for bias due to selective attrition: the example of smoking and cognitive decline. Epidemiology. 2012;23:119-28.
21. Mutsch M, Zhou W, Rhodes P, Bopp M, Chen RT, Linder T, et al. Use of the inactivated intranasal influenza vaccine and the risk of Bell's palsy in Switzerland. N Engl J Med. 2004;350:896–903.

Chapter 5
Uncontrolled Confounders

Introduction

Confounding occurs when the disease experience in a reference population is not exchangeable with the counterfactual disease experience for which it is intended to substitute [1, 2]. For example, if one is interested in the effect of a dichotomous exposure on the disease experience in the exposed group, then an unbiased estimate of that effect requires that the reference (unexposed) population's factual disease experience must be exchangeable with the disease experience the exposed group would have had, had they been unexposed (counter to the fact). A sufficient set of confounders explains this lack of exchangeability; the effect of the exposure of interest mixes with the effects of these confounding variables, which must also be ancestors of the exposure [3]. According to causal graph theory, confounding occurs when the exposure and disease share common causal ancestors [4]. Understanding and controlling confounding in epidemiologic research, either in the design or analysis of a study, is central to obtaining an unbiased estimate of the effect of an exposure on the risk of an outcome. As students in introductory epidemiologic methods courses are taught, it is imperative to control confounding to assess causation from observational data because it can make an association appear greater or smaller than the true underlying effect and can even reverse its apparent direction. Confounding can also make a null effect (*i.e.*, no causal relation between the exposure and the disease) appear as either causal or preventive.

Confounding variables are those that occur along unblocked backdoor paths and are not affected by the exposure, the control of which would block the confounding path. In epidemiologic studies, confounding can be controlled through study design, such as with randomization, restriction, certain kinds of matched designs like cross-over studies and difference in difference designs. Some of these design-based approaches to confounding, such as randomization and matching-based approaches, can help control for confounders that were never measured. Alternatively, if there is confounding in a study and an analyst identifies a variable that would block an

© Springer Nature Switzerland AG 2021
M. P. Fox et al., *Applying Quantitative Bias Analysis to Epidemiologic Data*,
Statistics for Biology and Health, https://doi.org/10.1007/978-3-030-82673-4_5

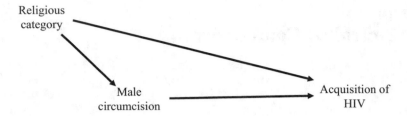

Figure 5.1 Causal graph showing confounding by religious category in the relation between male circumcision and male acquisition of HIV.

unblocked backdoor path, control is possible analytically through an ever-increasing number of methods, such as stratification, regression, standardization or g-methods to name a few. These analytic approaches to control for confounding typically rely on measured confounders to produce adjusted estimates.

As an example of confounding, a non-randomized study of the effect of male circumcision and male acquisition of human immunodeficiency virus (HIV) could be confounded by religious category, a marker for sexual behavior. This potential confounding is depicted in the causal graph in Figure 5.1, which encodes the presumed relations between the variables of interest. In the target population, being Muslim is a strong predictor of being circumcised compared with being a member of another religion. This tendency creates a relation between the covariate (religious category) and the exposure (circumcision). In addition, Muslim men are often at decreased risk of acquisition of HIV, even among those who are uncircumcised, which establishes the relation between the covariate and the disease among the unexposed or lowest risk exposure group. If this were the only confounder of the association and the DAG was faithful, to obtain a valid estimate of the association between male circumcision and male acquisition of HIV, religious category must be controlled in the design (*e.g.*, by restricting the study to either Muslims or non-Muslims) or in the analysis (*e.g.*, by adjusting for the effect of being Muslim on HIV acquisition in a stratified analysis or in a regression model).

A valid analysis of the causal effect of an exposure on the occurrence of disease must account for confounding of the crude association. When data have not been collected on a confounder during a study (whether it was because the confounder was known but unmeasured or because it was unknown to the analysts), the analyst cannot control for the effect of its confounding on the study results through standard methods like stratification, regression or g methods. In such cases, interested parties may question what impact the uncontrolled confounding may have had on the results, both in terms of the direction of the uncontrolled confounding and in terms of its expected magnitude (*i.e.*, the difference between the expected true causal effect and the observed estimate). In all cases, quantitative bias analysis can be a useful tool to evaluate the impact of an unknown or unmeasured confounder.

Key Concepts

Definitions

There are two types of confounding problems for which bias analysis is particularly useful. First, an analyst may be faced with a situation in which an important confounder was not measured during data collection and cannot be controlled for in the analysis. This situation can occur when the analyst is working with secondary datasets in which they had no control over data collection or in cases when it was prohibitively expensive or otherwise impractical to acquire accurate information about the confounder. In the case of a known but unmeasured confounder, the analyst will likely have, or be able to acquire, some knowledge about the confounder and its distribution in the study population. Ordinarily, this knowledge is the same information that gives rise to the notion that an unmeasured confounder has been left uncontrolled. The potential confounding that could be removed had the confounder been accurately measured is therefore known to the analyst and the scientific community, and adjustment would be considered important to estimate the true causal effect of the exposure on the outcome. Accordingly, when the confounder is known but unmeasured, it is reasonable to assume that the results of the study would likely be different, had the analysts been able to control for confounding by the unmeasured confounder through stratification or regression. We refer to this problem as an *unmeasured confounder*. For example, in an analysis of the relation between regular periconceptual multivitamin use and preeclampsia, Bodnar et al. (2006) were concerned that the association could be confounded by fruit and vegetable intake, a variable for which they had no data, because external evidence suggested that multivitamin users were more likely to eat fruits and vegetables and that fruit and vegetable intake reduces the risk of preeclampsia [5].

The second problem occurs when an observed association between an exposure and a disease is hypothesized to differ from the causal effect because it is confounded by a variable about which the analyst is unaware. This situation would be more likely to occur early in the study of the potential effect of an exposure on disease when less is known about the data generating mechanisms (*i.e.*, the underlying causal structures) that led to the observed data. For example, early studies on the relation between moderate alcohol intake and reduced risk of coronary heart disease were the subject of debate since the protective relation that had been observed could have been confounded by some as yet unknown confounder [6]. In such cases, the potential confounder is unknown to the analyst and as such it may be difficult to make reasonable assumptions about the relations between the confounder and the disease or its distribution in the study population. We are therefore often more interested in understanding the sensitivity of the observed results to the bias from this hypothetical confounder. This will be referred to as an *unknown confounder problem*. Unknown confounding problems are ideally dealt with through randomization, matching or, when this is not possible, though a method that can control for unknown confounding such as an instrumental variable design. When such approaches are not feasible, quantitative bias analysis can contribute to our understanding of the exposure disease relation. As the bias analysis process is the same whether the confounder is unknown or unmeasured, these two problems can be synthesized with the term *uncontrolled confounders*. We note that even measured confounders can leave some uncontrolled confounding when the confounder has been mismeasured (*i.e.,* residual confounding). We discuss mismeasurement problems in Chapters 6 and 10.

Motivation for Bias Analysis

To carry out a bias analysis to assess the impact of an uncontrolled confounder, the analyst must start with some information about the unmeasured confounder or theorize that information about an unknown confounder. In each case, the information will determine the strategy that the analyst employs to investigate the extent of the potential bias. A bias analysis informed by the most plausible bias parameters will be of most use to the analyst and the scientific community and will be more likely to provide convincing evidence.

The most common motivation for a simple bias analysis for an uncontrolled confounder is to understand what effect the confounder might have had on the study results. The goal is to bias-adjust the estimate of association between the exposure and the outcome to equal what it would have been, had the uncontrolled confounder been accounted for in the analysis. Typically, this bias-adjustment is made when the confounder can be identified (*i.e.,* it is unmeasured but not unknown) and when there is some information about its likely distribution in the study population in relation to the exposure and the outcome and possibly other covariates.

Alternatively, a bias analysis of an uncontrolled confounder could be conducted to determine combinations of bias parameters that could wholly explain the observed association if no effect of the exposure on the outcome truly existed (*i.e.,* the uncontrolled confounder made a null association appear causal or preventive). Such an approach is sometimes referred to as "nullification analysis." Such an analysis would be motivated by a desire to understand whether an unknown confounder could explain why an association was observed if no causal effect existed. The analyst can then draw inferences as to whether the required values for the bias parameters are plausible characteristics of an unknown confounder. While informative, we urge caution with this approach for several reasons. First, such an approach has parallels to null hypothesis significance testing in which the hypothesis of no effect of the exposure on the outcome is privileged above all others. The goal of etiologic epidemiologic research is to obtain a valid and precise estimate of the effect of the exposure on an outcome. Dichotomizing between whether there is or is not an effect is inconsistent with the estimation objective. Second, a nullification-based approach is also problematic in that many combinations of parameters could lead to a null effect, without any way for an analyst to determine if any of them are plausible. Finally, to conduct nullification analysis, one must pre-specify what will be considered a null effect, much as one does in an equivalency or non-inferiority trial. The exact null (ratio measure of 1.0 or difference measure of 0) may not be the only result that would be considered equivalent to no important effect. Still, such an approach may be the only option in cases where an unknown confounder is the problem. These and other considerations pertaining to nullification analyses will be discussed below in the section on E-values.

Another reason that a bias analysis of an uncontrolled confounder might be undertaken would be to assess what combination of bias parameters would be necessary to reverse the direction of a true effect from what was observed (*i.e.,* make a causal effect appear as a preventive association or vice versa). As with so called nullification analysis, caution is needed as many sets of parameters could lead to a reversal of the association and evaluating which are plausible can be fraught with indefensible judgments.

While the general approach described below can be used for a bias analysis motivated by any of these reasons, the reasons motivating the bias analysis should be considered carefully before choosing an approach. If the motivation is to evaluate what the observed data would have looked like, had the analyst been able to stratify by the uncontrolled confounder, then the simple bias analysis technique detailed below is appropriate. On the other hand, if the analyst wishes to understand what combinations of parameters might explain the observed results, then multidimensional techniques would be more appropriate. The reader should have a thorough knowledge of how to use the simpler methods before proceeding to multidimensional methods, as the logic underlying the analyses is the same for both.

Data Sources

As with any bias analysis (see Chapter 3), estimates of the values to assign to the bias parameters can come either from the subjects who were in the study that is being analyzed (typically data collected in a subset of the study population), from studies conducted in similar populations (typically acquired from searching the literature) or, failing that, a survey of experts' best educated guesses based on prior knowledge. We note that the survey of experts approach has its own important limitations that should be considered, the largest being that experts have their own personal biases that will inform the values assigned to the bias parameters. The problem of unmeasured confounding is no different. If the analyst is aware of a potential confounder for which they cannot collect data because it is too expensive, or otherwise impractical to measure in the entire study sample, then a substudy in which the confounder is measured in a sample of the population can be used to inform the bias analysis.

When no internal data have been collected, estimates from the literature are typically used. This method for bias analysis has been referred to as "indirect adjustment" because the values assigned to the bias parameters to complete the bias analysis are extracted from the literature and do not come from an internal substudy [7]. In these cases, the analyst should strive to find populations like the one under investigation. However, we note that if the uncontrolled confounding is due to a known confounder, it is typically because the confounder and its relation to the outcome have previously been scrutinized in a study in which the confounder was the primary exposure. In such cases, prior literature often provides enough information to assign values to bias parameters for an initial simple bias analysis.

The problem is more difficult when dealing with an unknown confounder because the analyst will not be able to acquire any data to inform the values to assign to the bias parameters used for the bias analysis. In this case, the confounder is hypothetical, and neither a substudy nor a search of the literature will provide insight into the distribution of the confounder or its effect on the outcome. The bias analysis will instead explore the impact of adjustment for unknown confounders with different size associations and distributions, which will be informed by educated guesses.

Introduction to Simple Bias Analysis

Approach

In the following discussion, unless otherwise stated, the simple bias analysis we will consider is the case in which an observed association between a dichotomous exposure (E) and a dichotomous outcome (D) is bias-adjusted for an uncontrolled dichotomous confounder (C). Each variable is coded such that 0 means subjects are in the reference, unexposed, or undiseased category of the variable and 1 denotes that subjects are in the index, exposed, or diseased category of the variable. We will make the simplifying assumption that the uncontrolled confounder of interest is independent of all other variables being controlled in the analysis.

To begin a bias analysis for an uncontrolled confounder, the analyst must first specify the bias model and its bias parameters. Given estimates of the values to assign to the bias parameters, one can bias-adjust the conventional estimate of association by estimating the data that would have been observed, had (1) we collected data on the uncontrolled confounder and (2) presuming the bias model, including the values assigned to the bias parameters, are valid. This second assumption is likely unrealistic in most cases (even in the case of internal validation data where random error in the measurement of the bias parameters still exists) and therefore violations of these assumptions should be tested. The methods explained below have a long history [8–13], although to date, use of them has been limited.

To begin, one must assign values to the bias parameters (informed by a substudy, external literature or data sources, or by guesses based on a survey of expert opinion). The mathematical equations representing the bias model can then be solved to postulate what results would have been observed, had data on the uncontrolled confounder been collected and the confounding been controlled in the analysis.

Introduction to the Example

To make the discussion more concrete, the example of the association between the male circumcision and the risk of male acquisition of HIV through heterosexual sexual contact, which might be confounded by religious category, will be continued below. Before randomized trials showed a protective effect of male circumcision [14–16], all data about this relation came from non-randomized studies, each of which adjusted for a different set of confounders [see Weiss et al. (2000) [17] and Siegfried et al. (2005) [18] for reviews]. Many of these studies reported a protective association between circumcision and risk of acquiring HIV, while some reported a null association, and—in a limited number of studies—harmful associations were reported.

A Cochrane Library Systematic Review published in 2003, so preceding trial data, concluded: "It seems unlikely that potential confounding factors were completely accounted for in any of the included studies" [19]. Although important confounders were likely unmeasured in some of the studies, this conclusion leaves no impression about which studies were more likely to have been confounded and to what degree the associations observed were confounded. A simple bias analysis can be used to bias-adjust the observed data for the unmeasured confounders and can be used when a study is already complete or when only summary data are available. We note further that bias analysis, while appropriate for most studies, is most warranted when random error is low such that systematic error is the main source of study error [20]. This would be the case when study results are combined to increase overall precision. At such points, bias analysis becomes more important to inform inference [[20], see also Chapter 13].

The example will use data from a cross-sectional study conducted among men with genital ulcer disease in Kenya. While cross-sectional studies have other problems like selection bias and temporality (difficulty establishing the temporal relationship between the exposure and the outcome), we will focus here on problems of uncontrolled confounding. Crude data from the study showed that men who were circumcised had about one-third the prevalence of HIV infection as men who were uncircumcised (Prevalence ratio, PR= 0.35; 95% CI 0.28, 0.50) [21]. Adjustment for measured potential confounders had little impact on the results, so the example will use the crude data.

The authors made no adjustment for religious category, a potentially important confounder. A study conducted in Rakai, Uganda noted that the protective effects of circumcision on HIV could be overestimated if there was no adjustment for religious category [22]. Being Muslim is often associated with a lower risk of male acquisition of HIV, and being Muslim strongly predicts male circumcision, so if male circumcision were truly protective as appears to be the case based on the randomized trials specifically designed to address this question [14–16], then studies that did not adjust for religious category would likely overestimate the protective effect of circumcision on HIV acquisition (that is, they would find an effect further from the null than the true protective effect). The purpose of the bias analysis is to address the question:

"What estimate of association between male circumcision and male acquisition of HIV would have been observed in this study, had the authors collected data on religious category and adjusted for it in the analysis?"

Bias Parameters

To bias-adjust for an uncontrolled confounder, the analyst must have knowledge of, or speculate about, three general parameters:

1. the association between the confounder and the outcome within strata of the exposure,
2. the association between the confounder and the exposure in the source population, and
3. the prevalence of the confounder among the unexposed in the source population.

Note that we make no statement about how the associations are measured, whether on the relative or additive scales nor whether the study design measures risks, odds or rates. All of these can be used, as will be discussed below. Further, as will be used below, the last two parameters listed can also be expressed in various ways that retain the same information but with different parameterizations. Many of the equations we introduce below replace parameters 2 and 3 with alternative parameters: the prevalence of the confounder among those without the exposure of interest (a specific case of parameter 3) and the prevalence of the confounder among those with the exposure of interest (which could be estimated by multiplying parameter 2 by parameter 3, for ratio measures). Throughout this chapter we will make clear precisely what specific bias parameters are needed as it will differ slightly depending on the effect measure of interest, whether there is an interaction between the exposure and confounder and, sometimes, the target population of interest.

To inform the bias analysis, one makes educated guesses (preferably informed by internal or external data) about the association between the unmeasured confounder and the outcome. Ideally this would be the direct effect; however, often only an estimate of the total effect might be available. This association is often expressed as an odds ratio or risk ratio but can also be expressed by other parameters (*e.g.,* rate ratio, rate difference, risk difference, *etc.*). In the study of the association between male circumcision and acquisition of HIV, if no data had been collected on religious category, and no substudy had been conducted, one could turn to the literature to estimate the effect of being Muslim on acquisition of HIV. Gray et al. reported that Muslim men in Uganda had 0.63-times (95% CI 0.21, 2.07) the rate of incident HIV relative to non-Muslim men [22] (Gray et al. 2000). Because they did not report the risk ratio (which is needed for this bias analysis), the example will use the incidence rate ratio from this Kenyan study as an estimate of the risk ratio. We note that because these data come from sources external to our study and because the analysis from which it was extracted may have also suffered from some bias, the estimate itself should be used with some caution and additional values should be tried to account for such uncertainties.

The second and third required bias parameters are the distribution of the confounder within strata of the exposure. The distribution is typically expressed as the proportion of subjects with the confounder ($C = 1$) among the exposed ($E = 1$) and the proportion of subjects with the confounder among the unexposed ($E = 0$) populations. In the Kenyan study, the exact distribution of circumcision in the population is unknown. To conduct a bias analysis, one might assume that 80% of circumcised and 5% of uncircumcised men were Muslim based on the prevalence reported in the Ugandan cohort [22]. Again, because the population may differ from the population under study, these estimates should not be assumed to be known with certainty and other values should be used as a sensitivity analysis of the bias analysis.

Assignment of these values to the bias parameters now permits a bias analysis to assess whether controlling for religious category through stratification and adjustment might explain part of the observed protective association between circumcision and acquisition of HIV.

Implementation of Simple Bias Analysis

Ratio Measures

The framework for analyzing the bias due to an unmeasured confounder was derived by Schlesselman (1978) [10], building on the work of Bross (1966) [8]. They developed a simple method to relate an observed risk ratio to the risk ratio bias-adjusted for an unmeasured confounder. We begin by assuming no interaction between the confounder and exposure in relation to the outcome on the risk ratio scale and relax that assumption in a section below. Given assumptions about the distribution of the confounder in the population and the effect of the confounder on the outcome in the absence of the exposure, the bias-adjusted risk ratio can be expressed as:

$$RR_{adj} = RR_{obs} \frac{RR_{CD} \cdot p_0 + (1 - p_0)}{RR_{CD} \cdot p_1 + (1 - p_1)} \tag{5.1}$$

where RR_{adj} is the risk ratio associating the exposure with the disease bias-adjusted for the uncontrolled confounder, RR_{obs} is the conventional observed risk ratio associating the exposure with the disease without adjustment for the uncontrolled confounder, RR_{CD} is the risk ratio associating the confounder with the disease (assuming no effect measure modification of the relative risk by the exposure), and p_1 and p_0 are the proportions of subjects with the confounder in the exposed and unexposed groups respectively. See Schneeweiss et al. (2005) [23] and Arah et al (2008) [24] for alternative ways of expressing this equation.

With credible estimates of the values to assign to these three bias parameters (RR_{CD}, p_1, and p_0), one can calculate the association between the exposure and the

Table 5.1 Data on the association between exposure E and disease D stratified by a dichotomous unknown or unmeasured confounder C.

	Total		C_1		C_0	
	E_1	E_0	E_1	E_0	E_1	E_0
D_1	a	b	A_1	B_1	A_0	B_0
D_0	c	d	C_1	D_1	C_0	D_0
	m	n	M_1	N_1	M_0	N_0

disease expected after bias-adjustment for the uncontrolled confounder directly using Equation 5.1. One can also project what the expected stratified data would have looked like had data on the confounder been collected, conditional on the validity of the bias model. Projecting the stratified data, given the bias parameters, is useful for understanding the distribution of the data within the strata of the confounder and can then be used to calculate a bias-adjusted relative measure such as a standardized risk ratio or Mantel-Haenszel risk ratio. We note that assuming no relative effect measure modification by the confounder of the effect of the exposure on the outcome, any summary measure across strata should be the same as the stratum-specific estimates, as will be seen below.

If the joint distribution of the exposure (E), disease (D), and confounder (C) were known, the data could be stratified as in Table 5.1. Small letters denote actual observed data and capital letters denote bias-adjusted projected data, had the confounder been measured given the assumptions about the bias parameters. The observed data collected in the study are in the far left of the table. To adjust for the uncontrolled confounder, one needs to complete the middle and far right stratified tables.

Given assumptions about the bias parameters, the stratified data (cells in the middle and right tables indexed by 1 and 0, respectively) can be projected from the collapsed, observed data (left table) using a few simple computations. First, complete the margins of the stratified tables using the observed data margins (m and n), the prevalence of the confounder among the exposed (p_1), and the prevalence of the confounder among the unexposed (p_0).

$$M_1 = m \cdot p_1 \text{ and } M_0 = m - M_1$$

and

$$N_1 = n \cdot p_0 \text{ and } N_0 = n - N_1$$

Having completed the margins of the table, use the estimate of the risk ratio for the confounder and the disease (RR$_{CD}$) to calculate the expected values in the A and B cells by recalling the assumption of no effect measure modification and noting that:

$$RR_{CD} = \frac{B_1/N_1}{B_0/N_0} \tag{5.2}$$

and then substituting $(b - B_1)$ for B_0 and $(n - N_1)$ for N_0:

$$RR_{CD} = \frac{B_1/N_1}{(b-B_1)/(n-N_1)} \qquad (5.3)$$

Rearrange this equation to solve for B_1

$$B_1 = \frac{RR_{CD}N_1 b}{RR_{CD}N_1 + n - N_1}$$

Because the b, n, and N_1 cells are already known or have already been calculated, and because of the assumption about the relation between the confounder and the disease, one can solve for B_1 and then for B_0 using $B_0 = (b - B_1)$.

Assuming no effect measure modification on the ratio scale, the same logic can be applied to the A and M cells.

$$RR_{CD} = \frac{A_1/M_1}{(a-A_1)/(m-M_1)}$$

which can be rearranged to:

$$A_1 = \frac{RR_{CD}M_1 a}{RR_{CD}M_1 + m - M_1}$$

We can then solve for A_0 as it is equal to $(a - A_1)$. For each column in the stratified tables, the number of noncases (e.g., C_1) equals the total number (e.g., M_1) less the number of cases (e.g., A_1). After solving for each of the cells in the stratified table, use the completed table to bias-adjust the estimate of effect for the confounder using a standardized morbidity ratio (SMR) or a Mantel-Haenszel (MH) estimate of the summary risk ratio. Throughout this book, when we refer to an SMR we specifically mean a measure of effect (in this case a risk ratio) that is standardized to the distribution of the confounder in the exposed population.

If, instead of risk data summarized by the risk ratio, one wanted to bias-adjust case-control data summarized by the odds ratio, one could substitute c and C for m and M, respectively, as well as d and D for n and N, respectively. Then solve for C_1 and D_1 as:

$$C_1 = c \cdot p_1 \text{ and } C_0 = c - C_1$$

and

$$D_1 = d \cdot p_0 \text{ and } D_0 = d - D_1$$

After solving for the C and D cells, one could solve for the A and B cells using an estimate of the odds ratio between the confounder and the disease as below.

$$A_1 = \frac{OR_{CD}C_1 a}{OR_{CD}C_1 + c - C_1}$$

and

$$B_1 = \frac{OR_{CD}D_1 b}{OR_{CD}D_1 + d - D_1}$$

Using the projected table, one can now summarize the data using the appropriate measure of effect for the study-design, such as stratum specific ORs, an SMR (OR standardized to the distribution of the confounder in the exposed population) or a Mantel-Haenszel OR estimator just as for the risk ratio. This same adjusted result can be accomplished directly using:

$$OR_{adj} = OR_{obs} \frac{OR_{CD} \cdot p_0 + (1 - p_0)}{OR_{CD} \cdot p_1 + (1 - p_1)}$$

where OR_{adj} is the odds ratio associating the exposure with the disease adjusted for the uncontrolled confounder, OR_{obs} is the observed odds ratio associating the exposure with the disease without adjustment for the uncontrolled confounder, OR_{CD} is the odds ratio associating the confounder with the disease (assuming no relative effect measure modification by the exposure), and p_1 and p_0 are the proportions of subjects with the confounder in the exposed controls and unexposed controls respectively.

The approach explained here assumes that the effect of the uncontrolled confounder on the outcome is constant across strata of the exposure (*i.e.*, no effect measure modification on the scale of the parameter being used in the bias-adjustment, the RR or OR). If effect measure modification is suspected, then multiple estimates of the effect of the confounder on the outcome within strata of the exposure need to be specified in addition to the prevalence of the confounder within levels of the exposure. The bias analysis method to address an unmeasured confounder in the presence of effect measure modification is described later in this chapter.

Example

The method described above, and the information on religious category and its association with HIV, is now used to bias-adjust the conventional estimate of male circumcision on HIV acquisition. The values assigned to the bias parameters are summarized in Table 5.2.

Table 5.2 Values assigned to bias parameters for a simple bias analysis of the association between male circumcision (E) and HIV (D) stratified by an unmeasured dichotomous confounder ($C_1=$ Muslim and $C_0=$ any other religion).

Bias parameter	Description	Values assigned to the bias parameter
RR_{CD}	Association between being Muslim and HIV	0.63
p_1	Prevalence of being Muslim among the circumcised	0.80
p_0	Prevalence of being Muslim among the uncircumcised	0.05

Table 5.3 Data on the association between male circumcision (E) and incident HIV (D) stratified by an unmeasured confounder (C) representing religious category, with observed crude data from Tyndall et al., 1996 [21].

	Total		C_1		C_0	
	E_1	E_0	E_1	E_0	E_1	E_0
D_1	105	85	A_1	B_1	A_0	B_0
D_0	527	93	C_1	D_1	C_0	D_0
Total	632	178	M_1	N_1	M_0	N_0

Table 5.4 Data on the association between male circumcision (E) and incident HIV (D) stratified by an unmeasured confounder (C) representing religious category, with observed crude data from Tyndall et al., 1996 [21], and with marginal totals in strata of C computed by Equation 5.4.

	Total		C_1		C_0	
	E_1	E_0	E_1	E_0	E_1	E_0
D_1	105	85	A_1	B_1	A_0	B_0
D_0	527	93	C_1	D_1	C_0	D_0
Total	632	178	505.6	8.9	126.4	169.1

The left side of Table 5.3 shows the observed data associating male circumcision (E) with incident HIV (D). The crude data and the assumptions about the bias parameters, in conjunction with the equations derived earlier, provide a solution to allow stratification by religious category. First solve for the M_1 and N_1 cells using the prevalence of the confounder among the exposed and the unexposed, and then for M_0 and N_0:

$$M_1 = 632 \cdot 0.80 = 505.6 \text{ and } M_0 = 632 - 505.6 = 126.4$$

and

$$N_1 = 178 \cdot 0.05 = 8.9 \text{ and } N_0 = 178 - 8.9 = 169.1 \qquad (5.4)$$

Table 5.3 can now be updated with the marginal totals, as in Table 5.4. The relation between being Muslim and HIV infection (0.63) can be used to solve for the interior cells of the stratified table (see Table 5.5):

Table 5.5 Data on the association between male circumcision (E) and incident HIV (D) stratified by an unmeasured confounder (C) representing religious category, with observed crude data from Tyndall *et al.*, 1996 [21], and with interior cells in strata of C computed by Equation 5.5.

	Total		C_1		C_0	
	E_1	E_0	E_1	E_0	E_1	E_0
D_1	105	85	75.2	2.7	29.8	82.3
D_0	527	93	430.4	6.2	96.6	86.8
Total	632	178	505.6	8.9	126.4	169.1
RR	0.35		0.49		0.49	

$$A_1 = \frac{RR_{CD}M_1 a}{RR_{CD}M_1 + m - M_1} = \frac{0.63 \cdot 505.6 \cdot 105}{0.63 \cdot 505.6 + 632 - 505.6} = 75.2$$

and

$$B_1 = \frac{RR_{CD}N_1 b}{RR_{CD}N_1 + n - N_1} = \frac{0.63 \cdot 8.9 \cdot 85}{0.63 \cdot 8.9 + 178 - 8.9} = 2.7 \qquad (5.5)$$

The remaining interior cells of Table 5.5 can be computed by differences. The crude risk ratio relating circumcision to male acquisition of HIV in this study was 0.35. The bias-adjusted stratified data allow bias-adjustment of the association between circumcision and incident HIV for confounding by religious category. After bias-adjusting for religious category, the SMR would be:

$$SMR = \frac{\frac{75.2+29.8}{505.6+126.4}}{\frac{505.6}{505.6+126.4} \cdot \frac{2.7}{8.9} + \frac{126.4}{505.6+126.4} \cdot \frac{82.3}{169.1}} = 0.49$$

Because we assumed no effect measure modification of the confounder-outcome relation, the exposure outcome relations within strata of C will each also be equal to 0.49, as shown in Table 5.5.

If the assumptions about the values assigned to the bias parameters and the bias model are valid, then had the authors collected data on religious category, they would have observed a protective effect of male circumcision on acquisition of HIV similar to the effect observed in one of the three key randomized trials on the topic [RR=0.40; 95% CI 0.24, 0.68; [14]]. Thus, this bias analysis provides support for the hypothesis that unmeasured confounding by religious category could explain some of the overestimate in the protective effect seen in the observational study, conditional on the validity of the bias model. However, even after adjusting for religious category, if the bias parameters were correct, circumcision would still have a substantial effect on reducing risk of HIV infection.

The results of a bias analysis provide an estimate of the impact of the bias caused by unmeasured confounding, assuming the bias parameters are correct. Accordingly, the results are only as valid as the estimates of the values assigned to the bias parameters used to conduct the analysis. The values used for this bias analysis came from a similar population, but not the same population, as the original study population. In addition, the analysis does not account for any other confounders already controlled in the main study analysis. While this could be accounted for by further stratifying data by confounders already adjusted, this solution is only possible when one has access to the raw data (or published summary data stratified by multiple confounders) and when the sample size is large enough to allow stratification by multiple confounders simultaneously. Therefore, while the method described in this section can be used to explore the impact of any unmeasured confounder, the method is limited by the accuracy of the values assigned to the bias parameters and the limitations of the data. Extensions of this approach, detailed in Chapters 7, 8, 9 and 10 show how to account for uncertainty about the values assigned to the bias parameters by making multiple estimates of them.

Other approaches to dealing with measured confounding when unmeasured confounding is also present include using results such as Equation 5.1 where the adjusted effect is substituted for the crude one or fitting a propensity score model to summarize the measured confounding and then conducting the analysis within levels of, say, quintiles of the propensity score or matching on the propensity score before conducting the bias analysis. A third approach would be to reweight the data using inverse probability weighting to account for measured confounders before correcting for the unmeasured confounder. Each of these approaches has benefits and drawbacks and requires different assumptions for the results to be valid, including assuming the unmeasured confounder is independent of the measured confounder. To account for this, one could attempt to specify bias parameters within levels of the measured confounders that already account for the correlations between measured and unmeasured confounders (*e.g.*, specifying the effect of the unmeasured confounder on the outcome independent of all measured confounders) but this can only be done if reasonable estimates of the covariance are available.

This simple approach to accounting for an uncontrolled confounder is an improvement over intuitive estimates of the impact of unmeasured confounding because the assumptions made are explicit and the impacts given those assumptions are quantified. Presentation of the results of a simple bias analysis allows others to dispute the values assigned to the bias parameters and to recalculate bias-adjusted estimates, given alternative values for the parameters. For example, one might argue that the effect of being Muslim on HIV infection in this population (RR_{CD}) has been overstated, particularly given adjustment for other variables. In that case, one could change the risk ratio associating being Muslim with incident HIV infection to 0.8; this change would yield a bias-adjusted estimate of 0.41 instead of the 0.49 presented above. In order to improve the quality of scientific debate on the impact of the unmeasured confounder, we strongly encourage analysts who use these methods to make the statistical computing code or software used for the analysis available to the research community to allow others to change the values assigned to the bias parameters should they see fit [20].

This example is also implemented in the Excel spreadsheet available on the text's website, which can be adapted to other data sets. Figure 5.2 shows a screenshot of the spreadsheet. The values assigned to the bias parameters and crude data are entered into the cells in the top right. Below that is a presentation of the bias-adjusted data within strata of the unmeasured confounder calculated using the method described above. The results displayed at the bottom of the spreadsheet

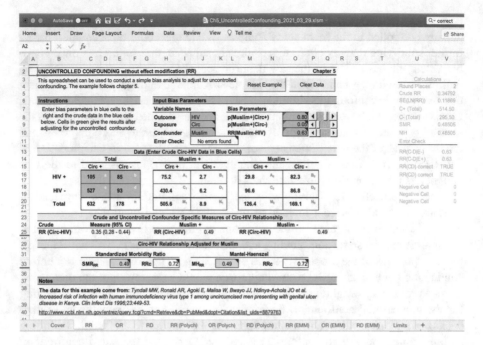

Figure 5.2 Excel spreadsheet used to bias-adjust for an unmeasured binary confounder when no effect measure modification is present.

show a bias-adjusted risk ratio of 0.49 after adjusting for religious category, which is the unmeasured confounder. This bias analysis corresponds to the results shown also in Table 5.5.

Note that the analyst could also have arrived at the same bias-adjusted estimate of association using Equation 5.1 and values for bias parameters in Table 5.2, without calculating the bias-adjusted cell counts:

$$RR_{adj} = 0.35 \frac{0.63 \cdot 0.05 + (1 - 0.05)}{0.63 \cdot 0.8 + (1 - 0.8)} = 0.49$$

The spreadsheet allows visual inspection of the interior cells of the stratified data as well as calculation of the adjusted SMR. Other free spreadsheets for external adjustment of unmeasured confounder using three-dimensional plots of the bias parameters can be found on the internet (Pharmacoepidemiology and Pharmacoeconomics Software Downloads. https://www.drugepi.org/dope/software#Sensitivity-analysis, Accessed 31 October 2021). Further we note that the Stata program Episensi (https://ideas.repec.org/c/boc/bocode/s456792.html, Accessed 2 January 2021) and the R package Episensi.R (https://rdrr.io/cran/episensr/, Accessed 31 October 2021) can also be used to conduct simple bias-adjustments for uncontrolled confounding.

Although analysts may have reasonable estimates of the values to assign to the bias parameters from either internal or external validation data, it is important to remember that these values are measured with their own sources of error, including random error. Therefore, it is unlikely that the values assigned to the parameters are the exact correct ones that should have been used. Because of this, we encourage analysts, even those using only simple bias analysis methods, to test the influence of small deviations from their assumptions. This sensitivity analysis of the bias analysis will evaluate whether small deviations from assumptions are likely to have much impact on the inferences drawn from the bias analysis. Some bias analyses are very sensitive to the assumptions made while others are very robust. Knowing this allows analysts to increase confidence in their inferences or to know where they should be cautious in interpreting the results of bias analyses.

Difference Measures

The above methods can be revised to complete the stratified table with bias parameters expressed as difference measures [12]. An estimate of the prevalence of the confounder in both the exposed and unexposed groups is still required, but instead of an estimate of the risk ratio associating the confounder with the outcome, one needs an estimate of the risk difference associating the confounder with the outcome (RD_{CD}). The M and N cells can be completed as above for ratio measures, but to determine the values in the A and B cells, one uses RD_{CD} as follows:

$$RD_{CD} = \frac{B_1}{N_1} - \frac{B_0}{N_0}$$

Substituting $(b - B_1)$ for B_0 and $(n - N_1)$ for N_0:

$$RD_{CD} = \frac{B_1}{N_1} - \frac{b - B_1}{n - N_1}$$

Rearranging this equation to solve for B_1

$$B_1 = \frac{RD_{CD}N_1(n - N_1) + b \cdot N_1}{n}$$

and similarly solving for A_1

$$A_1 = \frac{RD_{CD}M_1(m - M_1) + a \cdot M_1}{m}$$

Figure 5.3 shows a screenshot from an Excel spreadsheet to bias-adjust for unmeasured confounding when a risk difference is the desired measure of association. In this case, assuming a risk difference of -0.37 associating being Muslim to HIV and the same information as above on the proportion Muslim in the exposed and unexposed groups, the crude risk difference of -0.31 becomes -0.03 after bias-adjustment for religious category, assuming a valid bias model.

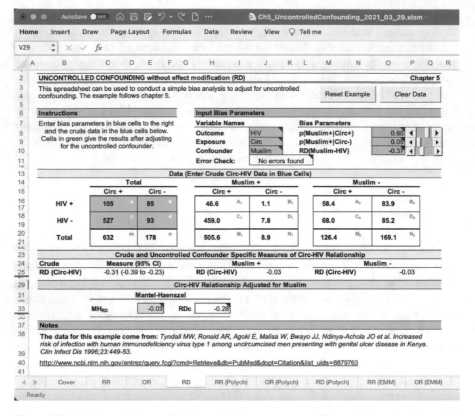

Figure 5.3 Excel spreadsheet used to bias-adjust for an unmeasured binary confounder when a risk difference is the measure of effect.

Alternatively, a numerically identical answer can be obtained using a equation similar to the Schlesselman equation in 5.1. Arah et al [24] present confounding adjustment equations for RR, RD, and OR for the exposed, unexposed and total target populations. Since we have been presenting SMRs that are standardized to the exposed distribution's confounder prevalence, we present the equation to bias-adjust a RD, standardized to the distribution of the confounder among the exposed. At the moment, however, choice of target population is irrelevant since we are still assuming homogeneity of effects on the risk difference scale.

$$RD_{adj} = RD_{obs} - RD_{CD} \cdot (p_1 - p_0)$$

Inserting the same values that were used in the spreadsheet, we find

$$RD_{adj} = -0.31 - (-0.37)(0.8 - 0.05) = -0.03$$

as we found when imputing the categorical data in the spreadsheet. These approaches are mathematically equivalent.

Person-time Designs

When the study design uses person-time instead of persons for denominators, and thus calculates rates instead of risks, some simplifying assumptions need to be made to be able to analytically account for the confounding. Confounders are factors that are associated with the individual units within the dataset, yet simple bias analysis methods focus on the expectation in the bias-adjusted estimates when using summary level data. Studies that use person-time rather than people as the denominator typically collect different amounts of follow-up time per person. Accordingly, individual level bias-adjustments as described in Chapter 9 will more appropriately assign the confounder to individuals and their corresponding person time. In the absence of that, if one is willing to assume that person time can be distributed among those with and without the confounder, slight modifications of the equations above—substituting person-time for persons—can be used to bias-adjust for an uncontrolled confounder. Such an assumption may be reasonable if exposure is expected to have limited impact on person time.

Unmeasured Confounder in the Presence of Effect Measure Modification

The preceding approaches assume that the analyst has estimates of the ratio or difference measure associating the confounder with the disease and of the distribution of the confounder within levels of the exposure. To this point, the methods have assumed that there is no effect measure modification on the scale in which the analyst is working. For example, we have assumed, when working on the ratio scale, that the effect of religious category on incident HIV is not modified by circumcision category. If the association between the confounder and outcome is different in strata of the exposure (*e.g.,* if the risk ratio associating Muslim, versus another religion, with male acquisition of HIV was different for males who were circumcised compared with those who were not), then one might wish to report the stratum-specific estimates rather than summarized or adjusted estimates of association. However, except in rare cases, some summarization is typically performed, and therefore, it is worthwhile to understand how to conduct the bias analysis for an unmeasured confounder in the presence of effect measure modification. This bias analysis method is more likely to apply to an unmeasured confounder than to an unknown confounder, because one would seldom have motivation to postulate or parameterize effect measure modification by an unknown variable.

If the risk ratio associating Muslim and incident HIV was $RR_{CD \mid E\,=\,1} = 0.4$ among circumcised men and $RR_{CD \mid E\,=\,0} = 0.7$ among uncircumcised men, then one substitutes these different values for RR_{CD} in Equation 5.5. Thus to calculate the interior cells of Table 5.6:

Table 5.6 Data on the association between male circumcision (E) and incident HIV (D) stratified by an unmeasured confounder (C) representing religious category, with observed crude data from Tyndall et al., 1996 [21], and with interior cells in strata of C computed by Equation 5.6, with the association between religious category and incident HIV modified by circumcision status.

	Total		C_1		C_0	
	E_1	E_0	E_1	E_0	E_1	E_0
D_1	105	85	64.6	3.0	40.4	82.0
D_0	527	93	441.0	5.9	86.0	87.1
Total	632	178	505.6	8.9	126.4	169.1
RR	0.35		0.38		0.66	

$$A_1 = \frac{RR_{CD|E=1}M_1 \cdot a}{RR_{CD|E=1}M_1 + m - M_1} = \frac{0.4 \cdot 505.6 \cdot 105}{0.4 \cdot 505.6 + 632 - 505.6} = 64.6$$

and

$$B_1 = \frac{RR_{CD|E=0}N_1 \cdot b}{RR_{CD|E=0}N_1 + n - N_1} = \frac{0.7 \cdot 8.9 \cdot 85}{0.7 \cdot 8.9 + 178 - 8.9} = 3.0 \qquad (5.6)$$

Table 5.4 can now be completed as Table 5.6 by differences. Use the stratified data to calculate the bias-adjusted risk ratio as:

$$SMR = \frac{\frac{64.6+40.4}{505.6+126.4}}{\frac{505.6}{505.6+126.4} \cdot \frac{3.0}{8.9} + \frac{126.4}{505.6+126.4} \cdot \frac{82.0}{169.1}} = 0.45 \qquad (5.7)$$

Alternatively, one can modify Equation (5.1) to make the bias-adjustment directly, again using Equation from Arah et al. [24] to directly adjust to the distribution in the exposed population:

$$RR_{adj} = RR_{obs} \frac{RR_{CD|E=0} \cdot p_0 + (1 - p_0)}{RR_{CD|E=0} \cdot p_1 + (1 - p_1)} \qquad (5.8)$$

It may seem curious, at first glance, that Equation 5.8 depends on $RR_{CD \mid E = 0}$ but not $RR_{CD \mid E = 1}$. Intuition can be gained by remembering that we are standardizing to the exposed population to match the SMR above. Notice that in the equations in Expression 5.6, only the first equation used $RR_{CD \mid E = 1}$. The purpose of this first equation is to redistributed the observed exposed cases between those with and without the confounder. Now, notice in the SMR Equation 5.7, that the numerator is simply the number of observed exposed cases (105) divided by the number of observed exposed (632). Regardless of how we redistribute exposed cases using the first Equation in 5.6, we will end up with 105 exposed cases and 632 exposed in the numerator of the SMR, precisely because we are standardizing to the distribution of the confounder in the exposed.

Now using the data on HIV and circumcision, the bias-adjusted risk ratio would be:

$$RR_{adj} = 0.35 \frac{0.7 \cdot 0.05 + (1 - 0.05)}{0.7 \cdot 0.8 + (1 - 0.8)} = 0.45$$

Alternative equations are available in Arah et al [24] for other effect measures (RR, OR, and RD) and other target populations (exposed, unexposed and total). Note that the advantage of calculating the interior cells over using Equation 5.7 is that the data can be used to calculate the risk ratios associating the exposure (circumcision) with the outcome (HIV acquisition) within strata of the confounder (religious category) and one does not need to derive separate equations for each target population and effect measure. Furthermore, if there is effect measure modification by the exposure of the association between the confounder and the outcome, there will also be effect measure modification by the confounder of the association between the exposure and the outcome, such that it is often more appropriate to present bias-adjusted stratified results rather than the overall bias-adjusted result. In this case, the risk ratio relating circumcision to male HIV acquisition was 0.38 among Muslim men and 0.66 among men of other religions (see Table 5.6). Figure 5.4 shows a screenshot from an Excel spreadsheet bias-adjusting for this example unmeasured confounder in the presence of effect modification.

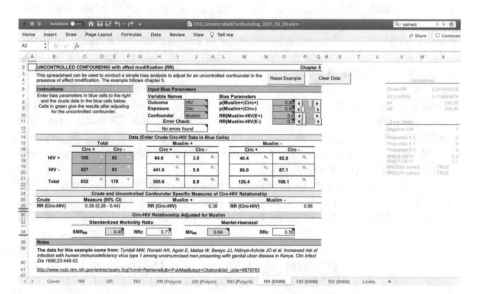

Figure 5.4 Excel spreadsheet used to bias-adjust for an unmeasured binary confounder when effect measure modification is present.

Table 5.7 Representation of data on the relation between male circumcision and incident HIV stratified by three religious categories (an unmeasured confounder), with observed crude data from Tyndall et al., (1996) [21].

	Total		Muslim		Christian		Other	
	E_1	E_0	E_1	E_0	E_1	E_0	E_1	E_0
D_1	105	85	A_2	B_2	A_1	B_1	A_0	B_0
D_0	527	93	C_2	D_2	C_1	D_1	C_0	D_0
	632	178	N_2	M_2	N_1	M_1	N_0	M_0

Polytomous Confounders

To this point, the bias analysis methods have focused on an uncontrolled confounder with two levels, (*e.g.*, a person either is Muslim or is another religion). In some cases, it will be more accurate to represent the uncontrolled confounder with more than two levels. For example, one might be interested in bias-adjusting for three levels of religious category: Muslim, Christian, and all other religions. Those who are Muslim might have 0.4 times the risk of incident HIV as those who are categorized as other, whereas those who are Christian might have only 0.8 times the risk as those categorized as other. In this case, one would want to incorporate into the bias analysis model the fact that the potential confounder has multiple levels and that the effect of each level of the confounder on the outcome is assumed to be different. One can expand the bias analysis framework to account for multiple level confounders by stratifying the observed data into three groups (Muslim, Christian, and other) as in Table 5.7. However, the number of bias parameters increases substantially, as now the associations between Muslim (vs other) and incident HIV (RR_{CD2}) and between Christian (vs other) and incident HIV (RR_{CD1}) must be estimated to calculate the expected cell counts, as well as the proportion of subjects among those circumcised who are Muslim (p_{21}) and Christian (p_{11}), and the proportion of subjects among those uncircumcised who are Muslim (p_{20}) and Christian (p_{10}). We assume no effect measure modification of the RR. When religious category was dichotomized, we assigned values to only three bias parameters; now values must be assigned to six bias parameters.

Although it now becomes more complicated to complete the interior cells (see the spreadsheet for interior cell calculations), one can use the same general bias analysis approach of Equation 5.1 to relate the bias parameters to the bias-adjusted data [7, 25]:

$$RR_{adj} = RR_{obs} \frac{RR_{CD2} \cdot p_{20} + RR_{CD1} \cdot p_{10} + (1 - p_{20} - p_{10})}{RR_{CD2} \cdot p_{21} + RR_{CD1} \cdot p_{11} + (1 - p_{21} - p_{11})}$$

For this example, assume that among the exposed population (circumcised), 60% are Muslim and a further 20% are Christian, while among the unexposed population (uncircumcised), 5% are Muslim and a further 20% are Christian. In this case, the bias-adjusted risk ratio would be:

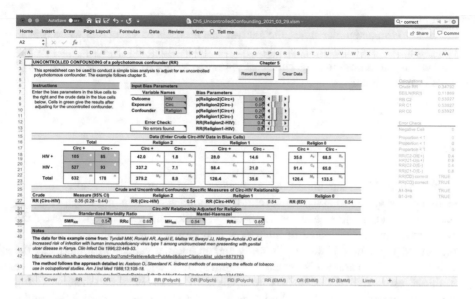

Figure 5.5 Excel spreadsheet used to bias-adjust for an unmeasured binary confounder when the confounder is polytomous.

$$RR_{adj} = 0.35 \frac{0.4 \cdot 0.05 + 0.8 \cdot 0.2 + 0.75}{0.4 \cdot 0.6 + 0.8 \cdot 0.2 + 0.2} = 0.54$$

Thus, the estimate of effect bias-adjusted for the multilevel confounder would be 0.54, assuming a valid bias model.

This same general framework could be adapted to confounders with four or more levels and could be adapted to scenarios in which there was also effect measure modification between the different levels of the confounder and the outcome with the exposure. While possible, caution should be used in such cases, as it would be unusual for an analyst to have sufficient data to parameterize such an analysis with confidence, yet still not have collected data on the unmeasured confounder from the outset. Figure 5.5 shows a screenshot from an Excel spreadsheet bias-adjusting for an unmeasured confounder when the confounder is polytomous using the example just described.

Multidimensional Bias Analysis for Unmeasured Confounding

Just as with selection bias, the bias parameters for an unmeasured confounder are not known with certainty and we often want to explore the impact of different sets of values. The bias parameters are rarely, if ever, known with certainty and as such, looking at the variation in results when using different values can be helpful. For example, when there are multiple external estimates from populations slightly different to the one under study, the analyst has no basis to select one value for the bias parameter over another. Multidimensional quantitative bias analysis can provide insight into the impact that the choice of values to assign to the bias parameters can have in these situations.

Example

A bias analysis of the impact of an uncontrolled confounder requires data or estimates for values to assign to several bias parameters, including the association between the confounder and the exposure variable, the association between the confounder and the outcome of interest, and the prevalence of the confounder in the source population. Unless one's study had an internal validation component, the values for these parameters would be unmeasured, and one may instead have multiple estimates of the values to assign to these bias parameters. Using multidimensional bias analysis allows the analyst to examine the association under different combinations of the values assigned to these bias parameters.

In Chapter 2, we gave an example regarding a study of 5-fluorouracil's chemotherapeutic impact on mortality from colon cancer [26]. The study reported a hazard ratio of 0.66 for colorectal cancer mortality among patients who received 5-fluorouracil compared with patients who did not. However, because this was a nonrandomized study of a therapy, there was concern about confounding by indication. The authors used multidimensional bias analysis to understand what the components of this unknown binary confounder would need to be to account for the observed protective effect. This analysis was accomplished by varying the values assigned to the bias parameters, which were the prevalence of the unknown confounder in the untreated population, the prevalence of the unknown confounder in the treated population (these first two establish the strength of association between the confounder and the exposure), and the strength of association between the confounder and colorectal cancer mortality. As depicted in Table 5.8 the multidimensional bias analysis indicated that the unknown confounder could only explain the entire observed beneficial effect if there was a substantial difference in its distribution between the treated and untreated groups and if the confounder was a strong predictor of colorectal cancer mortality. The authors argued that such a circumstance was unlikely. Figure 5.6 shows a screenshot of the Excel spreadsheet to bias-adjust for an unmeasured binary confounder using multidimensional bias analysis and the data from the circumcision and HIV example.

Table 5.8 Results of multidimensional bias analysis to assess the impact of confounding by indication by an unknown binary confounder (UBC) on a study of 5-fluorouracil (5-FU) and colorectal cancer mortality. Original data and bias adjusted estimates from Sundararajan et al., 2002 [26].

Prevalence of UBC in those without 5-FU	Prevalence of UBC in those with 5-FU	UBC Hazard ratio	Hazard ratio for 5-FU, adjusted for UBC
0.5	0.1	3.00	1.11
0.6	0.1	2.00	0.96
0.5	0.1	2.00	0.90
0.9	0.5	2.00	0.84
0.9	0.1	1.75	1.03
0.9	0.5	1.75	0.80
0.5	0.1	1.75	0.84
0.9	0.1	1.5	0.91
0.9	0.5	1.5	0.77
0.5	0.1	1.5	0.79
0.9	0.1	1.25	0.79
0.9	0.5	1.25	0.72
0.5	0.1	1.25	0.72
0.9	0.1	1.1	0.71
0.9	0.5	1.1	0.69
0.5	0.1	1 1	0.69

Bounding the Bias Limits of an Unmeasured Confounding

The method described above is useful when values can be assigned to all the bias parameters. This complete specification, however, may not always be possible. When only a subset of the bias parameters can be assigned values, bounds can be placed on the potential strength of confounding instead. For example, Flanders and Khoury (1990) derived equations that allow an estimate of the bias-adjustment bounds from an uncontrolled confounder when there is not enough information to assign values to all the bias parameters [27].

Analytic Approach

To assess the impact of an unmeasured confounder for which not all bias parameters can be assigned values, Flanders and Khoury computed bounds on the "relative risk due to confounding" (hereafter referred to as RR_{conf}). RR_{conf} is the ratio of the crude risk ratio (RR_{crude}) and the standardized risk ratio using the exposed group as the standard (RR_{adj}) [28].

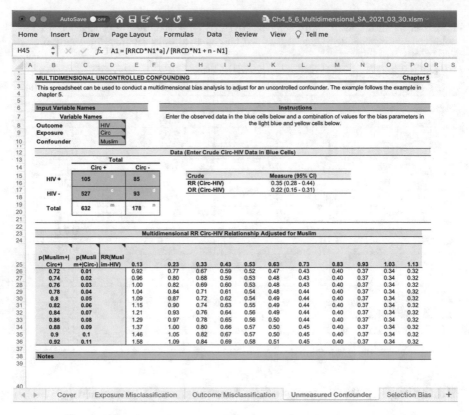

Figure 5.6 Excel spreadsheet used to bias-adjust for an unmeasured binary confounder using multidimensional bias analysis.

$$RR_{conf} = \frac{RR_{crude}}{RR_{adj}}, \text{ so } RR_{adj} = \frac{RR_{crude}}{RR_{conf}} \qquad (5.9)$$

Assuming the odds ratios associating the exposure and the confounder, OR_{CE}, and the risk ratio associating the confounder and the outcome, RR_{CD}, are greater than or equal to 1, that there is some prevalence of confounder among the unexposed, p_0, then the RR_{conf} must be greater than or equal to 1 and must be less than or equal to the minimum of the bounding factors listed in Table 5.9 [27].

These bounds are useful when only some of the information about the unmeasured confounder is available. Even if some of the bias parameters cannot be assigned values, knowing bias parameters with sufficient confidence to assign values will inform an estimate of the upper bound on the potential strength of confounding. In particular, note that since $RR_{adj} = [RR_{crude}/RR_{conf}]$, the minimum value of the factors in Table 5.9 (which we will refer to as $Parm_{min}$) can be substituted for RR_{conf}, since it is the largest possible value for the relative risk due to confounding, assuming a valid bias model and given the bias parameters that are

Table 5.9 Factors that bound the relative risk due to confounding when the associations between the confounder and the outcome and the confounder and the exposure are both greater than 1.

Bounding factors
OR_{CE}
RR_{CD}
$1/p_0$
$RR_{CD}/[(1 - p_0) + p_0 \cdot RR_{CD}]$
$OR_{CE}/[(1 - p_0) + p_0 \cdot OR_{CE}]$

assigned values. This substitution allows us to bound the bias-adjustment to the crude risk ratio as:

$$RR_{adj} = \frac{RR_{crude}}{Parm_{min}}$$

Given that RR_{CD} and OR_{CE} must be greater than 1, each of the values in Table 5.9 must be greater than 1; accordingly, $Parm_{min}$ must also be greater than 1. Thus, the bias-adjusted risk ratio must be between $RR_{crude}/Parm_{min}$ and RR_{crude}. This approach is most useful when little is known about the unmeasured confounder or when existing data on some of the factors (*e.g.*, prevalence of the confounder in the exposed and unexposed groups) is believed not to apply to the current study population. When either RR_{CD} or OR_{CE} is less than 1, one can consider recoding the variables so that both parameters are greater than 1 and specifying p_1 and p_0 for these newly coded relationships.

The E-Value and G-Value

The approach developed by Flanders and Khoury above is useful when some of the parameters are known and an analyst wishes to bound the bias-adjustment for an unmeasured confounder. A slightly different bounding approach can be obtained by considering combinations of the three bias parameters—RR_{EC}, RR_{CD}, and p_1—that would result in a null observed effect after adjustment for confounding. Because there are three variables, a contour plot is a convenient way to display the relationships with curves on the plot (isopleths) representing combinations of bias parameter values that yield the observed relative risk if in fact the adjusted relative risk was null (or, equivalently, that the strength of bias – the relative risk due to confounding –was equal to the observed association) [29]. One then evaluates the plausibility of these values to reach a judgment about whether uncontrolled confounding is likely to fully explain the observed association.

Winkelstein used this approach to evaluate whether uncontrolled confounding by an unknown factor might explain the association between smoking and cervical cancer, with an observed effect of $RR_{obs} = 2$ (a few years before human papilloma virus was found to be causally associated with cervical cancer) [30]. We approximate Winkelstein et al.'s approach by finding values of the three bias parameters that would render the observed association null had we adjusted for a confounder with these attributes. The three bias parameters are: the strength of the association between smoking and the unknown factor, RR_{EC}, the strength of the association between the unknown factor and cervical cancer, RR_{CD}, and the prevalence of the unknown factor among those who smoked, p_1.

To demonstrate this approach, we begin with Equation 5.1 from above, which we reproduce here as Equation 5.10, that expresses the relationship between the observed risk ratio, the bias parameters, and the adjusted risk ratio

$$RR_{adj} = RR_{obs} \frac{RR_{CD}p_0 + (1 - p_0)}{RR_{CD}p_1 + (1 - p_1)} \tag{5.10}$$

We are only interested in the situation in which $RR_{adj} = 1$ (there is a null effect after adjustment). Further, we wish to parameterize the model in terms of the bias parameter $RR_{EC} = \frac{p_1}{p_0}$. We substitute $RR_{adj} = 1$ and $p_0 = \frac{p_1}{RR_{EC}}$ in Equation 5.10 to obtain, after some rearranging:

$$RR_{CD} = \frac{RR_{EC} - RR_{EC}p_1 - RR_{EC}RR_{obs} + RR_{obs}p_1}{p_1(RR_{obs} - RR_{EC})} \tag{5.11}$$

At chosen prevalences, we generate curves by inserting values of RR_{EC} from 1 to 10 into Equation 5.11. These curves show the values of RR_{EC} and RR_{CD} that would create bias equal to the observed effect of smoking on cervical cancer ($RR_{obs} = 2$), thus rendering the effect null, had the confounder been controlled. We show these isopleth curves in Figure 5.7 at values of $p_1 = 0.25$, 0.5, and 1.0.

Figure 5.7 can be used to informally judge the plausibility of an unknown factor being completely responsible for the observed RR. For instance, if it is widely believed that there are no factors that are associated with smoking or cervical cancer with RR's greater than 2, it would appear to be unlikely that confounding could be entirely responsible for the observed effect, regardless of the prevalence of the factor

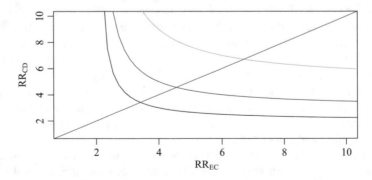

Figure 5.7 Contour plot showing values of RR_{EC} and RR_{CD} that are required to create a bias that would reduce $RR_{obs} = 2$ to the null at three values of p_1. The green line shows values of RR_{EC} and RR_{CD} at $p_1 = 0.25$; the red line shows values at $p_1 = 0.5$; the black line shows values at $p_1 = 1.0$. The blue diagonal line is the special case in which $RR_{EC} = RR_{CD}$.

among smokers. On the other hand, if it was believed that the prevalence of the factor among smokers was no more than 25%, and neither RR_{EC} nor RR_{CD} are thought to plausibly be larger than 6, we would deem it unlikely that this factor could completely explain the observed RR.

We pause to note a first limitation of this approach, which will carry over to the use of E-values. We are focusing our attention on parameter values necessary to completely explain an observed effect and thereby obtain $RR_{adj} = 1.0$, which may not be scientifically meaningful. A combination of bias parameter values deemed a-priori plausible may not lead to a bias that would have reduced the observed RR to the null; however, it is possible that they could reduce the observed RR to an adjusted RR near enough to the null to be substantively identical to the null (e.g., $RR_{adj} = 1.05$).

There are an infinite series of isopleth curves that could be drawn. Notice that if one were uncertain about the true value of p_1, one could attempt to be conservative by only paying attention to the isopleth where $p_1 = 1$ [29]. This focus is conservative in the sense that for a given value of, say, RR_{EC}, the value of RR_{CD} that would result in a bias sufficient to produce the observed RR will be smaller for $p_1 = 1$ than for any other value of p_1. We can insert $p_1 = 1$ in Equation 5.11 to produce an equation that yields only this isopleth curve:

$$RR_{CD} = \frac{RR_{obs} - RR_{EC}RR_{obs}}{RR_{obs} - RR_{EC}} \qquad (5.12)$$

This equation produces the black curve in Figure 5.7. Along this curve are pairs of RR_{EC} and RR_{CD} that create a bias exactly equal to the observed effect, if $p_1 = 1$.

For any RR_{EC}, RR_{CD} pair, other values of p_1 would create less bias.

We pause again in our description of the development of the E-value to note that we have made a very extreme assumption; namely, that $p_1 = 1$. This implies that the prevalence of the unknown factor among smokers is 100%; everyone who smokes also has this unknown factor. Deterministic relationships of this sort are extremely rare in epidemiology. The assumption that the prevalence is 100% results in a conservative set of estimates (under our narrow definition of conservative above). However, it is easy for this to be misleading as well as conservative. Imagine, continuing with our smoking and cervical cancer example, that there was a factor that had not been previously measured but was thought to be strongly associated with smoking ($RR_{EC} = 6.0$) and cervical cancer ($RR_{CD} = 4.0$). From Figure 5.7, we can see that if we use the approach in Equation 5.12 where we assume $p_1 = 1$, we would not rule out that this factor could have caused $RR_{obs} = 2.0$ when $RR_{adj} = 1.0$. However, if only 25% of those who smoked had this factor, even these extreme bias parameters could not have caused enough bias to produce $RR_{obs} = 2.0$ when $RR_{adj} = 1.0$.

Ding and VanderWeele [31] developed the E-value by making one additional simplification: they assumed $RR_{EC} = RR_{CD}$. That is, the effect of the exposure on the confounder is precisely the same as the effect of the confounder on the disease. We will refer to this common effect estimate as $RR_{EC} = RR_{CD} = \zeta$. Instead of an infinite set of RR_{EC} and RR_{CD} along the black curve in Figure 5.7, this simplification leads to one point along that line, where the blue diagonal line intersects with the black curve. Inserting ζ into Equation 5.12, we have

$$\zeta = \frac{RR_{obs} - \zeta RR_{obs}}{RR_{obs} - \zeta}$$

By solving for ζ using the quadratic equation, we obtain the value of $RR_{EC} = RR_{CD} > 1$ that would be necessary for the amount of confounding bias to completely explain away the observed effect, if the prevalence of the factor among the exposed was 100%. This value is called the E-value:

$$\mathrm{E-value} = RR_{obs} + \sqrt{RR_{obs}^2 - RR_{obs}} \tag{5.13}$$

In this example, we observed $RR_{obs} = 2$, which would result in an E-value = 3.41. If a confounder existed such that it had $RR_{EC} = RR_{CD} = 3.41$ and $p_1 = 1$, then such a confounder could, indeed, cause confounding such that $RR_{obs} = 2$ would be reduced to $RR_{adj} = 1$ had the confounder been controlled.

We pause again to point out a third extreme assumption that was made to derive the E-value formula: that the effect of the confounder on the exposure and the effect of the confounder on the outcome are precisely the same. This assumption is extremely unlikely to be true in practice, further limiting the use of E-values. If one of the two constituent confounder RRs is smaller than the E-value, the second RR must be larger in order for confounding to explain away RR_{obs}; however, this will not always hold and, even when it does, gaining intuition about how much larger the second RR must be is difficult as the function is non-linear. We have focused on three key assumptions in the derivation of the E-value: (1) a principal focus on $RR_{adj} = 1.0$; (2) assuming $p_1 = 1$; and (3) assuming $RR_{EC} = RR_{CD}$. These, and other, assumptions of the E-value have been criticized by multiple analysts, in part because the simplicity of Equation 5.13 allows for such easy implementation of the method even when users may not have a full understanding of the limitations [32–39].

Because the E-value is derived under the conservative, yet unrealistic, assumption that $p_1 = 1$, it can be thought of as a bound. Bounds can be tricky to interpret and require careful consideration. The E-value gives the value of $RR_{EC} = RR_{CD} > 1$ that would induce confounding bias that would "explain away" the observed effect if $p_1 = 1$. Examining Figure 5.7, we see that if p_1 were less than 1, those values of $RR_{EC} = RR_{CD}$ would not "explain away" the entirety of the observed effect (only part of it). In that sense, the E-value is a bound: it is the smallest value of $RR_{EC} = RR_{CD} > 1$ that could entirely explain away the observed effect due to confounding. Entirely explaining away the observed effect only occurs when $p_1 = 1$ and at any other value of p_1, larger values of $RR_{EC} = RR_{CD}$ would be required.

In fact, it is possible to relax the extreme assumption that $p_1 = 1$ and derive an alternative formula that depends on the prevalence of the confounder among the exposed. We denote this alternative formula the G-value(p_1) [40]:

$$G - \text{value}(p_1) = \frac{RR_{obs} + p_1 + RR_{obs}p_1 - 1 + \sqrt{(1 - p_1 - RR_{obs} - RR_{obs}p_1)^2 - 4RR_{obs}p_1^2}}{2p_1}$$

The G-value(p_1) gives the shared value of $RR_{CD} = RR_{EC} > 1$ that would be required to reduce an observed relative risk to the null for a given confounder prevalence among the exposed. We note that unlike the E-value, the G-value(p_1) is not a bound. It gives the exact values of the bias parameters that are required to reduce an observed relative risk to the null. The E-value is a special case that occurs when G-value($p_1 = 1.0$). When $p_1 < 1$, the result will be greater than the E-value since, as we mentioned above (and as can be seen in Figure 5.7), if the prevalence of the confounder is lower, then the effect of $RR_{EC} = RR_{CD}$ must be larger. The G-value(p_1) can be found in Figure 5.7 where the isopleths for $p_1 = 0.25$ and $p_1 = 0.5$ cross the diagonal line (where $RR_{EC} = RR_{CD}$): G-value($p_1 = 0.5$) = 4.6; G-value ($p_1 = 0.25$) = 6.7. We interpret these G-values as: if the prevalence of the unknown factor among smokers is 50%, $RR_{CD} = RR_{EC} = 4.6$ would induce a bias that would make a null effect appear to be a $RR_{obs} = 2$. Further, if the prevalence of the unknown factor were only 25%, a shared $RR_{CD} = RR_{EC} = 6.7$ would be required to reduce the observed RR to the null.

Uncontrolled confounding is an important concern in observational research. E-values attempt to limit the number of assumptions that are made when quantifying the impact of uncontrolled confounding. Although this may appear appealing, assumptions about bias factors are a necessary, unavoidable aspect of quantitative bias analysis. Specifying values for bias parameters can lead to healthy disagree-ments in the scientific community that can help to improve the literature. Often times, a substantial amount of information is known about bias parameters that will allow well informed bias analyses, which we detail throughout this book. Cases in which there is a concern that the entirety of an effect is due to an uncontrolled confounder and that nothing is known about that confounder are likely to be rare. For this reason, we suggest that analysts will often be better served by formal quantita-tive bias analyses laid out in this book or, perhaps, by the methods of Flanders and Khoury or the G-value(p_1), as described above.

Signed Directed Acyclic Graphs to Estimate the Direction of Bias

As noted in Chapter 2, DAGs have proven to be a useful approach to clarifying notions of confounding, selection bias and misclassification [4, 41–44]. DAGS are a graphical representation of an analyst's prior beliefs concerning the relations between factors in an epidemiologic study, describing the data generating mecha-nism as the analyst understands it. They can be used to identify a sufficient set of variables for control to estimate causal effects [4, 43, 44]—including variables that were unmeasured but are important to estimate causal effects—assuming that the graph validly depicts the true underlying causal relations.

While DAGs are typically presented without any reference to the magnitude or direction of the effect represented by an arrow from one variable to another, VanderWeele and colleagues have developed signed directed acyclic graphs that, under certain assumptions, allow identification of the direction of confounding bias from the common cause of an exposure and outcome [45, 46]. We here provide a description of the specific case when only a single unmeasured confounder is present within the analysis of an exposure disease relation and the exposure is a dichotomous variable. In this case, signs can be added to the arrows within a DAG if assumptions of monotonicity can be made. The monotonicity assumptions relate to the direction of causal effect between the confounder and exposure as well as the confounder and outcome. Assuming a positive sign for the monotonicity of the confounder and exposure implies, for example, that intervening on the confounder would increase the average value of the exposure. If an assumption is made that the sign is negative, intervening on the confounder is assumed to decrease the average value of the exposure variable in the population. When these assumptions can be made and positive or negative signs can be added to directed paths within the DAG, they can be used to determine the expected direction of bias due to uncontrolled confounding.

The rules that determine the direction of bias in a simple signed DAG are relatively simple and shown in Table 5.10. The confounding bias rules for signed DAGs follow the rules of multiplication: two positives or two negatives yield positive bias; signs in different directions yield negative bias. A positive direction of bias means that the crude effect estimate is larger than the truth. A negative direction of bias means the crude effect estimate is smaller than the truth.

For example, in Figure 5.1, prior knowledge would suggest that the node "Religious category" (the unmeasured confounder, which represents Muslim religion versus other religious groups) increases the prevalence of "Male Circumcision" (the exposure, which represents circumcised versus not circumcised) and decreases the incidence of "Acquisition of HIV" (the outcome, a dichotomous measure of the risk of incident HIV infection). We can assign a plus sign to the arrow from "Religious category" to "Male Circumcision" and a minus sign to the arrow from "Religious category" to "Acquisition of HIV." The sign of a path is defined to be the product of the signs of the edges that constitute that path as in Table 5.10. The backdoor path Male circumcision←Religious category→ Acquisition of HIV would

Table 5.10 Direction of confounding bias resulting from failure to adjust for confounder C that has either a positive or negative relationship with exposure and disease.

	E − C Association	C − D Association	Direction of Bias
Scenario 1	+	+	+
Scenario 2	+	−	−
Scenario 3	−	+	−
Scenario 4	−	−	+

therefore be assigned a minus sign. The observed crude effect of male circumcision on acquisition of HIV was protective and the backdoor path is assigned a negative sign, which suggests that the crude estimate was biased away from the null in the protective direction and that the true effect is larger (closer to the null) than the observed, as was seen in the quantitative bias analysis shown above. Had the product of the paths been positive, then one would know that the uncontrolled confounding would not account for the protective effect, and that the bias-adjusted estimate would have been even more protective.

References

1. Greenland S, Robins JM. Identifiability, exchangeability, and epidemiological confounding. Int J Epidemiol. 1986;15:413–9.
2. Greenland S, Robins JM. Identifiability, exchangeability and confounding revisited. Epidemiol Perspect Innov. 2009;6:1–9.
3. Kleinbaum DG, Kupper LL, Morgenstern H. Epidemiologic research: principles and quantitative methods. Belmont, California: Lifetime Learning Publications; 1982.
4. Greenland S, Pearl J, Robins JM. Causal diagrams for epidemiologic research. Epidemiology. 1999;10:37–48.
5. Bodnar LM, Tang G, Ness RB, Harger G, Roberts JM. Periconceptional multivitamin use reduces the risk of preeclampsia. Am J Epidemiol. 2006;164:470–7.
6. Poikolainen K, Vahtera J, Virtanen M, Linna A, Kivimaki M. Alcohol and coronary heart disease risk--is there an unknown confounder? Addiction. 2005;100:1150–7.
7. Axelson O, Steenland K. Indirect methods of assessing the effects of tobacco use in occupational studies. Am J Ind Med. 1988;13:105–18.
8. Bross ID. Spurious effects from an extraneous variable. J Chronic Dis. 1966;19:637–47.
9. Bross ID. Pertinency of an extraneous variable. J Chronic Dis. 1967;20:487–95.
10. Schlesselman JJ. Assessing effects of confounding variables. Am J Epidemiol. 1978;108:3–8.
11. Yanagawa T. Case-control studies: Assessing the effect of a confouding factor. Biometrika. 1984;71:191–4.
12. Gail MH, Wacholder S, Lubin JH. Indirect corrections for confounding under multiplicative and additive risk models. Am J Ind Med. 1988;13:119–30.
13. Greenland S. Basic methods for sensitivity analysis of biases. Int J Epidemiol. 1996;25:1107–16.
14. Auvert B, Taljaard D, Lagarde E, Sobngwi-Tambekou J, Sitta R, Puren A. Randomized, controlled intervention trial of male circumcision for reduction of HIV infection risk: the ANRS 1265 Trial. PLoS Med. 2005;2:e298.
15. Gray RH, Kigozi G, Serwadda D, Makumbi F, Watya S, Nalugoda F, et al. Male circumcision for HIV prevention in men in Rakai, Uganda: a randomised trial. Lancet. 2007;369:657–66.
16. Bailey RC, Moses S, Parker CB, Agot K, Maclean I, Krieger JN, et al. Male circumcision for HIV prevention in young men in Kisumu, Kenya: a randomised controlled trial. Lancet. 2007;369:643–56.
17. Weiss HA, Quigley MA, Hayes RJ. Male circumcision and risk of HIV infection in sub-Saharan Africa: a systematic review and meta-analysis. AIDS. 2000;14:2361–70.
18. Siegfried N, Muller M, Deeks J, Volmink J, Egger M, Low N, et al. HIV and male circumcision--a systematic review with assessment of the quality of studies. Lancet Infect Dis. 2005;5: 165–73.
19. Siegfried N, Muller M, Volmink J, Deeks J, Egger M, Low N, et al. Male circumcision for prevention of heterosexual acquisition of HIV in men. Cochrane Database Syst Rev. 2003;(2): CD003362.

20. Lash TL, Fox MP, MacLehose RF, Maldonado G, McCandless LC, Greenland S. Good practices for quantitative bias analysis. Int J Epidemiol. 2014;43:1969–85.
21. Tyndall MW, Ronald AR, Agoki E, Malisa W, Bwayo JJ, Ndinya-Achola JO, et al. Increased risk of infection with human immunodeficiency virus type 1 among uncircumcised men presenting with genital ulcer disease in Kenya. Clin Infect Dis. 1996;23:449–53.
22. Gray RH, Kiwanuka N, Quinn TC, Sewankambo NK, Serwadda D, Mangen FW, et al. Male circumcision and HIV acquisition and transmission: cohort studies in Rakai, Uganda. Rakai Project Team. AIDS. 2000;14:2371–81.
23. Schneeweiss S, Glynn RJ, Tsai EH, Avorn J, Solomon DH. Adjusting for unmeasured confounders in pharmacoepidemiologic claims data using external information: the example of COX2 inhibitors and myocardial infarction. Epidemiology. 2005;16:17–24.
24. Arah OA, Chiba Y, Greenland S. Bias Formulas for External Adjustment and Sensitivity Analysis of Unmeasured Confounders. Ann Epidemiol. 2008;18:637–46.
25. Greenland S. Quantitative methods in the review of epidemiologic literature. Epidemiol Rev. 1987;9:1–30.
26. Sundararajan V, Mitra N, Jacobson JS, Grann VR, Heitjan DF, Neugut AI. Survival associated with 5-fluorouracil-based adjuvant chemotherapy among elderly patients with node-positive colon cancer. Ann Intern Med. 2002;136:349–57.
27. Flanders WD, Khoury MJ. Indirect assessment of confounding: graphic description and limits on effect of adjusting for covariates. Epidemiology. 1990;1:239–46.
28. Miettinen OS. Components of the crude risk ratio. Am J Epidemiol. 1972;96:168–72.
29. Ding P, VanderWeele TJ. Sensitivity analysis without assumptions. Epidemiology. 2016;27:368–77.
30. Winkelstein W, Shillitoe EJ, Brand R, Johnson KK. Further comments on cancer of the uterine cervix, smoking, and herpesvirus infection. Am J Epidemiol. 1984;119:1–8.
31. VanderWeele TJ, Ding P. Sensitivity analysis in observational research: Introducing the E-Value. Ann Intern Med. 2017;167:268–74.
32. Ioannidis JP, Tan YJ, Blum MR. Limitations and misinterpretations of E-values for sensitivity analyses of observational studies. Ann Intern Med. 2019;170:108–11.
33. Blum MR, Tan YJ, Ioannidis JPA. Use of E-values for addressing confounding in observational studies—an empirical assessment of the literature. Int J Epidemiol. 2020;49:1482–94.
34. Greenland S. Commentary: An argument against E-values for assessing the plausibility that an association could be explained away by residual confounding. Int J Epidemiol. 2020;49:1501–3.
35. Poole C. Commentary: Continuing the E-Value's post-publication peer review. Int J Epidemiol. 2020;49:1497–500.
36. Kaufman JS. Commentary: Cynical epidemiology. Int J Epidemiol. 2020;49:1507–8.
37. VanderWeele TJ, Mathur MB, Ding P. Correcting misinterpretations of the E-Value. Ann Intern Med. 2019;170:131–2.
38. VanderWeele TJ, Ding P, Mathur M. Technical considerations in the use of the E-Value. J Causal Inference. 2019;7:20180007.
39. VanderWeele TJ, Mathur MB. Commentary: Developing best-practice guidelines for the reporting of E-values. Int J Epidemiol. 2020;49:1495–7.
40. MacLehose RF, Ahern TP, Lash TL, Poole C, Greenland S. The importance of making assumptions in bias analysis. Epidemiology. 2021;32:617-624.
41. Greenland S, Pearl J. Causal diagrams in Wiley StatsRef: Statistics Reference Online. John Wiley & Sons. 2017
42. Robins JM. Association, causation, and marginal structural models. Synthese. 1999;121:151–79.

43. Hernán MA, Robins JM. Causal inference: What if. Boca Raton: Chapman & Hall/CRC; 2020.
44. Glymour MM, Greenland S. Causal diagrams. In: Rothman KJ, Greenland S, Lash TL, editors. Modern Epidemiology, 3rd edition. Philadelphia: Lippincott Williams & Wilkins; 2008. p. 183–212.
45. VanderWeele TJ, Hernán MA, Robins JM. Causal directed acyclic graphs and the direction of unmeasured confounding bias. Epidemiology. 2008;19:720–8.
46. VanderWeele TJ, Robins JM. Signed directed acyclic graphs for causal inference. J R Stat Soc Ser B Stat Methodol. 2010;72:111–27.

Chapter 6
Misclassification

Introduction

The accurate measurement of exposure, disease occurrence, and relevant covariates is generally necessary to estimate causal relations between exposures and outcomes. However, in all epidemiologic research, there exists the opportunity for measurement errors. When the variables being measured are categorical, these errors are referred to as *misclassification*. We will consider two ways that misclassification can arise. The first is when information is not correctly recorded in the study database; these errors could be due to faulty instruments (e.g., the scale used to weigh people is incorrectly calibrated), respondents not answering truthfully to sensitive questions (e.g., their use of illegal substances), imperfect tests (e.g., blood tests that do not perfectly categorize patients with respect to a disease) or mistakes entering the data in medical records (e.g., a wrong ICD code was entered in a healthcare claim). The second source of misclassification is conceptual and occurs when there is discordance between the study's definition of the variable and the true definition. For example, this discordance can occur when the definition of an exposure or confounder is incorrectly operationalized with regard to dose, duration, or induction period. Studies of smoking and lung cancer may use smoking status at diagnosis, but this definition will cause some subjects to be incorrectly classified with respect to their etiologically relevant smoking status, which may have occurred decades before the time at which the study survey was administered. Former smokers might have had their lung cancer risk affected by their smoking history, and current smokers might have initiated smoking so near their diagnosis that it would have had no impact on their lung cancer risk. Regardless of the source of misclassification, when analyses divide study participants into categories pertaining to that covariate (*e.g,* exposed vs unexposed, diseased vs undiseased), some respondents will be classified in the wrong category, which can bias results.

Misclassification of measures of exposure, disease, and covariates (e.g., confounders, mediators and effect measure modifiers) are widespread in epidemiologic

M. P. Fox et al., *Applying Quantitative Bias Analysis to Epidemiologic Data*,
Statistics for Biology and Health, https://doi.org/10.1007/978-3-030-82673-4_6

research. Nutritional epidemiology, for example, often relies on food frequency questionnaires to assess nutrient and caloric intake of study participants. These surveys often ask participants to recall the frequency with which they consumed particular foods, on average, during a specified time period (e.g., the past year) and the nutrient or caloric variables derived from these measures may be subject to measurement error [1]. It is worth noting that a single variable can suffer from multiple sources of misclassification within the same study, as in the example just provided, there may be misclassification from incorrect nutrient values being assigned as well as errors in recall of foods and amount of foods consumed. In addition, data entry errors could lead to that information being incorrectly recorded.

Measurement errors in epidemiologic studies are common. A recent study using National Health and Nutrition Examination Survey found that 24% of women who self-reported overweight were, in fact, not overweight (two-thirds of them were actually obese, one-third were normal weight) [2]. Case-control studies may be more prone to exposure misclassification that varies based on outcome status, if the exposure is ascertained only after the outcome is known (i.e., retrospectively), such as in case-control studies of congenital anomalies and maternal exposures during pregnancy [3]. The impact of misclassification errors in epidemiologic studies is rarely explicitly addressed. Often, analysts have relied on beliefs that certain types of misclassification can bias results in known directions. We address the shortcomings of these beliefs later in this chapter and first present an introduction to misclassification as well as simple methods to adjust aggregate data for specified amounts of misclassification. When an analysis is susceptible to more than one set of classification errors, such as misclassification of both exposure and disease, the bias-adjustments below can be performed sequentially, so long as the classification errors are independent. When the classification errors are dependent, matrix methods described at the end of this Chapter or other bias modeling methods, such as Bayesian methods in Chapter 11, are required.

Definitions and Terms

Throughout this discussion of misclassification, we will use terms that describe the type of misclassification and measure the extent of the misclassification. We will begin by providing a conceptual definition and then below, where applicable, we will provide equations for calculating these different measures. These terms have comparable meanings whether the misclassified variable is the exposure, outcome, or a covariate. In this section, we will consider the misclassified variable as a dichotomous exposure variable. We will refine results below, when necessary, for covariate values with more than two levels.

Whenever data cannot be collected perfectly, some subjects' recorded values will differ from their, potentially unknown, true value. If we knew each subject's true categorization with respect to some exposure E, we could assess the probability that a classification scheme correctly separated subjects into exposed and unexposed groups. Typically, four measures are used to summarize the performance of a classification scheme: sensitivity, specificity, positive predictive value, and negative predictive value. Other measures, such as kappa, exist but are rarely used in quantitative bias analysis, so we omit discussion of them here.

The probability that a subject who was truly exposed was correctly classified as exposed is the classification scheme's *sensitivity* (SE). If the misclassified value of E is denoted by E^*, then the sensitivity can be expressed as $Pr(E^* = 1|E = 1)$. The proportion of those who were truly exposed who were incorrectly classified as unexposed is the *false-negative proportion* (FN), or $1 - SE = Pr(E^* = 0|E = 1)$. The probability that a subject who was truly unexposed was correctly classified as unexposed is the classification scheme's *specificity* (SP), or $Pr(E^* = 0|E = 0)$. The proportion of those who were truly unexposed who were incorrectly classified as exposed is the *false-positive proportion* (FP), or $1 - SP = Pr(E^* = 1|E = 0)$. Because these measures are proportions, their values range from 0 to 1. We present them throughout this chapter as percents, although they can also be presented as decimals or fractions.

A classification scheme's sensitivity and specificity measure how well the scheme sorts subjects into apparently exposed and unexposed groups. They condition on the people who truly have (for sensitivity) or do not have (for specificity) the exposure of interest, so they use the total number who truly do and do not have the exposure as their denominators. Sensitivity and specificity are measures of the quality of the classification scheme. It is commonly assumed that sensitivity and specificity remain constant across populations, that is, that the measurement scheme operates identically in different populations. If this assumption holds, sensitivity and specificity are assumed to be transportable across populations. There is little evidence that transportability is generally valid, although analysts frequently rely on it. In practice, analysts should evaluate whether their study population is similar enough to the source population in which the sensitivity and specificity were estimated such that the classification scheme can be expected to have reasonably similar performance in both populations.

One might also want to know, given that a subject was classified as exposed (or unexposed), the probability that the assigned classification was correct. These values are the *positive predictive value* (PPV), $Pr(E = 1|E^* = 1)$, and *negative predictive value* (NPV), $Pr(E = 0|E^* = 0)$, respectively. These measures condition on the people who are classified as being exposed or unexposed and use the number of subjects who are classified as exposed or unexposed in their denominators, respectively. The positive predictive value calculates the probability of truly being exposed if classified as exposed while the negative predictive value calculates the probability of truly being unexposed if classified as unexposed. Because the predictive values can be estimated by designing a validation study in which the true exposure is not yet known and validating it among some subjects who were and were not classified as exposed (i.e., the true value does not need to be known before the validation study begins), they are often simpler to estimate and therefore are more commonly estimated.

Predictive values are related to sensitivity and specificity via Bayes theorem, which we give below. It is important to recognize that the relation between the sensitivity/specificity and the predictive values depends on the prevalence of the true exposure status in the population. Whether sensitivities and specificities or positive and negative predictive values that were estimated in one population can be used in a different population is a question of transportability: how well can these values be transported from one population to another? Because of their dependence on prevalence, predictive values are less readily transported between populations than sensitivity and specificity. Thus, while predictive values may be easier to estimate in an internal validation study for reasons described above, it is not recommended that they be transported across populations where the prevalence of the covariate is likely to differ. Sensitivity and specificities, however, are typically thought to be more transportable, though evidence of this is hard to come by. To assess the transportability of sensitivity and specificity, an analyst could estimate them within strata of important predictors (perhaps sex or age), which would give insight into whether sensitivity and specificity parameters could be transported to a population with different distributions of these predictors.

In many cases, there is concern that the ability of the classification scheme to classify study participants with respect to some variable (e.g., exposure) may depend on the value of another variable (e.g., outcome). This concern about the classification errors can be described in terms of two important aspects: differentiality (where the classification scheme can be defined as either *differential* or *nondifferential*) and dependence (where the classification scheme can be defined as either *dependent* or *independent*). It is important to realize that these terms apply to the mechanism of classification and its susceptibility to interdependencies. While these terms are sometimes applied to a result, such as a contingency table, to say that the result was misclassified in some respect (e.g., nondifferentially and independently) requires comparison of the result with the classification that would have been obtained without error, which is seldom available.

Differential vs. Nondifferential Misclassification

Nondifferential misclassification is defined as the special case in which both the sensitivity and specificity of classification are independent of another key variable. When talking about exposure misclassification, we are typically, though not necessarily, referring to outcome status. To be more precise, if D is the true outcome status, nondifferential misclassification of exposure occurs when both of the following are true:

$$\Pr(E^* = 1|E = 1, D = 1) = \Pr(E^* = 1|E = 1, D = 0)$$

and

$$\Pr(E^* = 0|E = 0, D = 1) = \Pr(E^* = 0|E = 0, D = 0)$$

That is, the sensitivity among the diseased is the same as the sensitivity among the undiseased (the first expression above) and the specificity among the diseased is the same as the specificity among the undiseased (the second expression above). If either or both of these equalities does not hold, we say we have differential misclassification. Statistically speaking, nondifferential misclassification implies that E^* is independent of D, given E. However, the terms dependent and independent misclassification are reserved, in epidemiology, for a specific type of misclassification described in the next section. Instead, epidemiologists refer to dependence/independence of E^* and D using the terms differential/nondifferential. A mechanism by which nondifferential misclassification might occur is when exposure status is recorded before the onset of the disease or when the method of assigning exposure status is otherwise blind to the outcome status (so that errors in the classification of the exposure are not different within levels of disease status). For example, women might be asked about their smoking habits shortly after becoming aware of their pregnancy. In this case, in a study of the association between maternal smoking during pregnancy and orofacial clefts in the offspring, self-reported smoking might be misclassified, but it would likely be nondifferential because the exposure was recorded before the outcome diagnosis. Nondifferential exposure misclassification might also be expected when exposure is ascertained after the occurrence of the disease, but the ascertainment scheme relies on information that is independent of disease status. For example, in a study examining the association between maternal smoking during pregnancy and orofacial clefts, a post-partum analysis of hair samples for presence of cotinine or nicotine would likely result in nondifferential misclassification, even though the exposure was collected after the birth of the child when the outcome is already known. We assume that the cotinine deposition in hair created a record of tobacco smoke before the outcome diagnosis. An expectation of nondifferential exposure classification errors can also be created by blinding those collecting exposure data to participants' disease status and by blinding those collecting disease data to participants' exposure status.

Differential exposure misclassification is likely to occur when exposure is assessed after the disease has occurred (e.g., by subject interview); this expectation arises because having the disease may trigger cases to recall or report its perceived causes differently than noncases (i.e., *recall bias*). Continuing the previous example of maternal smoking and orofacial clefts, ascertaining smoking status via post-partum maternal survey could induce differential misclassification by multiple mechanisms. Mothers of infants with orofacial clefts may feel regret about smoking and report their smoking status with lower sensitivity than mothers of infants without orofacial clefts. Alternatively, mothers of infants with orofacial clefts may recall all aspects of their pregnancy with greater clarity and report smoking with a higher sensitivity than mothers of infants without orofacial clefts. Differential disease misclassification might also occur if exposed individuals were more likely to have their disease found than unexposed individuals (i.e., *detection bias*). For example, if study participants with diabetes are more likely to be screened for glaucoma than participants without diabetes, then false-negative cases of glaucoma may be more likely among nondiabetics.

It is important to realize that the definitions of nondifferential and differential misclassification pertain to the expectations of how misclassification mechanisms will operate given the study design. One can never verify that the realization of these mechanisms in the study data was nondifferential or differential. Verifying that the sensitivities and specificities of exposure classification were the same in cases and controls, so exposure misclassification was nondifferential, would require a measurement of the gold-standard (true) exposure classification in all study participants. Comparisons of measurement in samples would have to allow for the role of chance, so one could not readily discern between chance differences arising from nondifferential misclassification and differential misclassification. However, if one had the gold-standard measure in every participant, then the mismeasured classification would be obviated and the misclassification problem would be irrelevant. Nondifferential versus differential classification errors, and independent versus dependent classification errors (see below), are therefore expectations of how misclassification mechanisms will operate given the study design.

Misclassification may occur through multiple mechanisms for the same variable in the same study. As such, a mix of differential and nondifferential misclassification mechanisms can occur. In such cases, it can be difficult to anticipate the expected direction of the bias, making it important to use quantitative bias analysis informed by measures of the bias parameters that account for all of these mechanisms.

Dependent vs. Independent Misclassification

Dependent misclassification occurs when the probability of being misclassified with respect to two variables (typically exposure and outcome, though not exclusively) is greater (or less) than expected based on the probability of being misclassified with respect to either of the variables alone. For example, if the sensitivity of exposure misclassification is 90% and the sensitivity of outcome misclassification is also 90%, then we would expect, by chance, that $(1 - 0.9) \cdot (1 - 0.9) = 1\%$ of those who are truly exposed and have the outcome will be misclassified with respect to both variables (i.e., will be classified as unexposed and not having the outcome). If misclassification mechanisms generate this expectation, we say the errors are independent. If these misclassification mechanisms generate expectations of correlated errors, so that misclassification of a participant for one variable changes the probability that they will be misclassified on a second variable, then we say the errors are dependent. Alternatively, we can define dependent misclassification as a departure from any of the following equalities:

$$\Pr(E^* = 1 | E = 1, D = 1, D^* = 1) = \Pr(E^* = 1 | E = 1, D = 1, D^* = 0)$$

$$\Pr(E^* = 1 | E = 1, D = 0, D^* = 1) = \Pr(E^* = 1 | E = 1, D = 0, D^* = 0)$$

$$\Pr(E^* = 0 | E = 0, D = 1, D^* = 1) = \Pr(E^* = 0 | E = 0, D = 1, D^* = 0)$$

$$\Pr(E^* = 0 | E = 0, D = 0, D^* = 1) = \Pr(E^* = 0 | E = 0, D = 0, D^* = 0)$$

A similar set of four equations holds for the sensitivity and specificity of disease misclassification. For independent misclassification to hold, the classification of exposure (disease) must be independent of the reported disease (exposure) value. Note that it is possible to have dependent misclassification even if the misclassification is nondifferential. That is, the misclassification mechanism of exposure could depend on reported disease status but not the actual disease status. Dependent errors are also sometimes referred to as common source bias or correlated errors. When dependent misclassification of the exposure and outcome occurs, it can cause a strong bias away from the null even when the overall misclassification mechanism for both exposure and disease are nondifferential and even when the overall error rates appear low [4]. When exposures and confounders have dependencies in their error structures, confounder adjustment can bias results in unpredictable directions. We will discuss dependent misclassification in greater detail later in this chapter.

Misclassification should ideally be described along both axes at the same time. For example, misclassification of exposure may be independent and differential or independent and nondifferential, meaning that while the misclassification of exposure may be the same (nondifferential) or different (differential) within levels of the outcome, the rate of misclassification of the exposure is not dependent on the rate of misclassification of the outcome. Misclassification of exposure and outcome could also be dependent and nondifferential in which case the sensitivities and specificities of exposure and outcome classification would be the same within levels of the other variable, but the rate of double misclassification with respect to exposure and outcome is greater than expected. When dependent and differential misclassification occurs, the rate of double misclassification is different than expected, given their individual rates of misclassification, and at least one of the rates of misclassification of one variable within levels of the other must vary. Many characterizations of classification rates describe only whether the mechanisms that generated classification errors were nondifferential or differential, without specifying whether they were likely dependent or independent of other classification error rates. This is an important omission of the complete characterization of the mechanisms generating the misclassified data. A complete characterization of misclassification should consist of four parts: (1) the variable that is misclassified, (2) whether the design creates an expectation of dependent or independent misclassification, (3) whether the design creates an expectation of differential or non-differential misclassification, and (4) the variable with respect to which it is differentially/non-differentially, or dependently/independently, misclassified. For example, the exposure is expected by design to be independently and non-differentially misclassified with respect to the outcome.

Directed Acyclic Graphs and Misclassification

Whether misclassification mechanisms were differential or dependent can be depicted in graphical form using Directed Acyclic Graphs (DAGs), which were introduced in Chapter 2. Consider a true exposure E and true disease D that were measured with error resulting in E^* and D^*, respectively. In the DAG, following Hernán and Cole (2009) [5], U_E and U_D are all the unknown (latent) factors that directly contribute to the imperfect classification of E^* and D^*, respectively.

   To illustrate how to decide whether there is differential misclassification, we turn to Figure 6.1, in which an arrow exists from D to U_E. In turn, U_E affects the measured exposure status E^* (as does true exposure status, E), implying that true disease status affects the rate of exposure misclassification. For example, in a retrospective case-control study, cases (D) may recall or report exposure status (E^*) differently than controls. U_E represents the mechanisms that influence the accuracy of recall and reporting, such as rumination or social desirability. E represents the true exposure status, which also influences the recorded exposure status E^*. In this DAG, differential misclassification of exposure is possible and the sensitivity and specificity of exposure classification can depend on true disease status. Using the rules of DAGs, we can see that E^* and D are not d-separated because there exists an unblocked path from E^* to D. It is possible to also read probability expressions for misclassification from the DAG. To do this, we choose a node on the DAG, in this case E^*, and see that the probability of observing a given value of E^* is a function of all nodes that have open paths connecting them to E^*. For example, E^* is determined by both E and U_E and we can write the probability of observing $E^* = 1$ as $Pr(E^* = 1 | E = e, U_E = u_e)$; however, since U_E is not observed, it is generally more useful to replace it with the next node along the path from U_E: $Pr(E^* = 1 | E = e, D = d)$. Because this probability could be different for $D = 1$ and $D = 0$, we clearly have the potential for differential misclassification of exposure status by true disease status. Note that, in this example, disease misclassification remains nondifferential with respect to E since there is no open path between D^* and E. There is also no unblocked path between U_D and E^*, implying independent misclassification. We can again read the probabilities from the DAG as $Pr(D^* = 1 | D = d, U_D = u_d)$; however, since U_D is not

Figure 6.1 DAG representing independent, differential misclassification of exposure with respect to disease and independent, nondifferential misclassification of disease with respect to exposure.

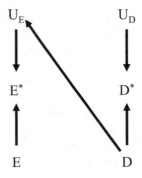

Figure 6.2 DAG representing dependent, nondifferential misclassification of disease with respect to exposure and dependent, differential misclassification of exposure with respect to disease.

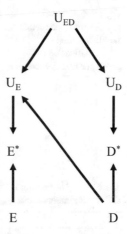

observed, we would typically write this as $Pr(D^* = 1 | D = d)$ and immediately see that the misclassification of D is nondifferential with respect to E since the misclassification probabilities do not depend on E. Therefore, this figure represents an example in which the exposure is differentially misclassified with respect to disease but disease is nondifferentially misclassified with respect to exposure.

Dependent misclassification is illustrated in Figure 6.2. Here, an unmeasured factor U_{ED} directly influences misclassification of both exposure and disease, as represented by the arrows from U_{ED} to U_E and U_{ED} to U_D. The unblocked path, $E^* \leftarrow U_E \leftarrow U_{ED} \rightarrow U_D \rightarrow D^*$, from E^* to D^* implies that the (mis)classification of E^* will be associated with the (mis)classification of D^*, independently of the true exposure or true disease status. This structure represents dependent misclassification and U_{ED} in this case might represent a latent propensity to give correct information about the exposure and disease on a survey. As above, we can read probability statements from the DAG. The probability of misclassification of E^* depends on E and U_E, but since U_E is not measured, we can write the probability of the next nodes that represent measured variables and are connected to U_D: D^* and D. That is, $Pr(E^* = 1 | E = e, D^* = d^*, D = d)$. That E^* is a function of D^* and D implies that misclassification of E is both dependent (conditional on D^*) and differential (conditional on D). Misclassification of D can also be read from the graph: it is dependent (conditional on E^*) and remains nondifferential (not conditional on E).

A common use of DAGs in epidemiology is to examine whether there is potential for bias when estimating an association. This use requires some additional care when misclassified variables are included. Typically, it is important to assess the potential for bias under the null hypothesis of no exposure effect on the outcome. For instance, Figure 6.3 shows a DAG in which the misclassification of exposure and outcome is nondifferential and independent. We might wish to estimate the effect of E on D, which we can see from the lack of an arrow, is null. Lacking data on these variables, we would estimate the association between E^* and D^*. As there is no path (either frontdoor or backdoor) from E^* to D^*, we would expect to estimate a null effect. This result would be unbiased, which is consistent with the expectation that

Figure 6.3 DAG
representing independent,
nondifferential
misclassification of
exposure and disease.

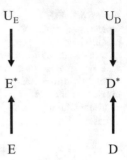

Table 6.1 Nomenclature for Equations 6.1, 6.2, 6.3, 6.4, 6.5 and 6.6 for calculating measures of classification accuracy.

	Truly exposed (E = 1)	Truly unexposed (E = 0)
Classified as exposed ($E^* = 1$)	A	B
Classified as unexposed ($E^* = 0$)	C	D

nondifferential misclassification produces no bias when the true effect is null. However, if there were an effect of exposure on the outcome and we drew an arrow from E to D, we would anticipate that we would find an association between E^* and D^*, but it is unclear from the DAG whether that association would be biased or not. In summary, using DAGs for the detection of bias of the main effect in misclassified data should generally be examined under the null hypothesis.

Calculating Classification Bias Parameters from Validation Data

We now turn to the equations used to calculate the measures of classification. Again, we will use dichotomous exposure classification as an example for this section, but the equations also apply to dichotomous disease and covariates. Assume that we had collected data on some exposure by self-report, which in this example is subject to error, and also had a way to know with certainty each subject's true exposure status. In that case, the data could be laid out as the 2×2 contingency table shown in Table 6.1, also called a validation table. Along the interior columns, subjects are classified according to their true exposure status (E), while along the interior rows subjects are classified according to their potentially mismeasured status (E^*). The equations assume that the participants in the validation study were representative of the study population (e.g., a simple random sample). Equations 6.1, 6.2, 6.3, 6.4, 6.5 and 6.6 show calculations for six measures of classification using the notation in Table 6.1.

$$\text{Sensitivity(SE)} = \frac{A}{A+C} \tag{6.1}$$

$$\text{Specificity(SP)} = \frac{D}{B+D} \tag{6.2}$$

$$\text{False negative proportion (FN)} = \frac{C}{A+C} \tag{6.3}$$

$$\text{False positive proportion (FP)} = \frac{B}{B+D} \tag{6.4}$$

$$\text{Positive predictive value (PPV)} = \frac{A}{A+B} \tag{6.5}$$

$$\text{Negative predictive value (NPV)} = \frac{D}{C+D} \tag{6.6}$$

These six measures of classification are proportions, so variance estimates and corresponding 95% confidence intervals can be estimated using standard methods. One simple approach that avoids impossible values for a proportion (i.e., outside the range [0–1]) is to use the $logit(x) = \ln(x/(1-x))$ transformation and its corresponding inverse $expit(x) = \exp(x)/(1+\exp(x))$. For example, 95% confidence intervals for sensitivity can be calculated as:

$$95\%\text{CI } SE = \text{expit}\left(\text{logit}(SE) \pm 1.96\sqrt{1/A + 1/C}\right)$$

Confidence intervals can be computed for each of the measures in similar fashion, replacing sensitivity (SE) with the measure of interest and replacing A and C with the cell counts in the denominator of Equations 6.1, 6.2, 6.3, 6.4, 6.5 and 6.6 for that measure. These confidence intervals are based on large sample asymptotic theory and alternative confidence interval computations (such as those by Clopper-Pearson, Wilson, or Agresti-Coull) may perform better with small samples sizes [6, 7].

To illustrate how these measures are calculated, we will use an example from a study comparing data provided by mothers on their child's birth certificate about their smoking status during pregnancy (E*) with data on smoking status recorded in their medical record (E) [8], as shown in Table 6.2. For this study, we will consider the medical record as the gold standard (true value) and the self-reported information on the birth certificate to be the imperfect classification scheme. We recognize that the medical record information on smoking status might also be subject to error,

Table 6.2 Validation of smoking status on the birth certificate by comparison with medical records from Piper, 1993 [8].

		Medical record		
		Smoker	Nonsmoker	Total
Birth certificate	Smoker	128	5	133
	Nonsmoker	36	465	501
	Total	164	470	634

and will discuss solutions to incorporate this additional source of uncertainty later in the chapter when we discuss multidimensional bias analysis and in Chapters 7, 8, 9, 10 and 11.

Using the data from this table and the equations above, the sensitivity equals 78% (128/164), the specificity equals 99% (465/470), the false-negative proportion equals 22% (36/164), and the false-positive proportion equals 1% (5/470). In addition, the PPV equals 96% (128/133) and the NPV equals 93% (465/501). The results here follow the intuition we would likely have about how accurately smoking during pregnancy might be captured on a birth certificate. As might be expected, true nonsmokers had a very high probability of reporting on the birth certificate that they were nonsmokers (nearly perfect specificity, very few false positives), while true smokers were less likely to report smoking (moderate sensitivity). We note further that two different misclassification mechanisms may be occurring here. One relates to social desirability, which would lead to understating or denying smoking when it occurred, reducing sensitivity, while others may simply have misheard or misunderstood the question (leading to very high but not perfect specificity).

As mentioned above, postive predictive value and negative predictive value depend on the prevalence of true exposure in the population. Equations 6.7 and 6.8 use Bayes theorem to show the relation of sensitivity, specificity, and prevalence of exposure with positive predictive value and negative predictive value, respectively. In these equations, $P(E_1)$ is the proportion of the population or subpopulation (e.g., cases) that is truly exposed to the dichotomous exposure E and $P(E_0)$ is the proportion of the population or subpopulation that is truly unexposed to E. For studies of disease or covariate misclassification, these equations can be adapted by using the prevalence of disease or the covariate instead of the prevalence of the exposure.

$$PPV = \frac{SE \cdot P(E_1)}{SE \cdot P(E_1) + (1 - SP)P(E_0)} \tag{6.7}$$

$$NPV = \frac{SP \cdot P(E_0)}{SP \cdot P(E_0) + (1 - SE)P(E_1)} \tag{6.8}$$

To demonstrate the relationship between predictive values and sensitivity and specificity, consider a hypothetical study that reports a survey of cigarette smoking in the United States as well as in Egypt. Assume that the two populations are administered identical questionnaires regarding self-reported smoking status (Table 6.3). The questionnaire has identical sensitivity (85%) and specificity (99%) of exposure classification in the two countries; however, Egypt has a far higher prevalence of true smokers than the US population (49.9% vs. 19.5%, respectively). Because Egypt has a higher prevalence of true smoking, the positive predictive value is higher in Egypt than the US (99% vs. 95%, respectively) while the negative predictive value is lower in Egypt than the US (87% vs. 96%, respectively).

Table 6.3 Estimation of positive and negative predictive values and sensitivity and specificity of smoking among hypothetical surveys conducted in Egypt and the United States.

	Egypt			United States	
	E* = 1	E* = 0		E* = 1	E* = 0
E = 1	424	75	E = 1	166	29
E = 0	5	496	E = 0	8	797
Pr(E) = (424 + 75)/1000 = 0.499			Pr(E) = (166 + 29)/1000 = 0.195		
Sensitivity = 424/(424 + 75) = 0.85			Sensitivity = 166/(166 + 29) = 0.85		
Specificity = 496/(496 + 5) = 0.99			Specificity = 797/(797 + 8) = 0.99		
PPV = 424/(424 + 5) = 0.99			PPV = 166/(166 + 8) = 0.95		
NPV = 496/(496 + 75) = 0.87			NPV = 797/(797 + 29) = 0.96		

Table 6.4 Options for design of an internal validation study.

	Participant selection strategy	Directly estimated bias parameters
1	Select individuals based on misclassified exposure measurement	PPV, NPV
2	Select individuals based on true exposure status	SE, SP
3	Select a random sample of individuals	SE, SP, PPV, NPV

Sources of Data

As described in Chapter 3, to conduct bias analyses, it is necessary to have or estimate bias parameters. For misclassification bias analysis, the classification method used in the study is compared to a more accurate method of classifying the exposure, disease, or covariate that is either too logistically difficult or too expensive to collect in the entire study population (as in Table 6.2, for example). The analyst has the option to conduct an internal validation study or, if that is not feasible, to use estimates calculated for a different study in a similar population. When an internal validation study is feasible, then the analyst has three general options for selecting individuals to be included in the validation study (see Chapter 3). This choice will affect the classification bias parameters that can be directly estimated, as discussed more completely in Chapter 3, and as summarized in Table 6.4.

To place these strategies in context, consider a study investigating maternal antibiotic use during pregnancy and the risk of orofacial clefts in the child. An analyst has enough funds to enroll 200 women into a validation study and needs to decide between the three strategies. For the first validation study strategy, the analyst could select 100 women who reported using antibiotics during pregnancy and 100 women who did not report using antibiotics during pregnancy. These self-reported (potentially misclassified) responses (E^*) could be compared to pharmacy records (the true exposure value or gold standard, E). With this strategy, the number of self-reported antibiotic users and nonusers (marginal row totals in Table 6.1) are fixed and therefore positive predictive value and negative predictive value can be calculated, but not SE and SP. To see why it is impossible to directly estimate SE and

SP using this study design, notice that the analyst could have artificially inflated the SE by enrolling 150 self-reported antibiotic users and only 50 non users. This would have no effect on the estimate of positive predictive value, in expectation, but would increase the number of people in the A cell in Table 6.1 and therefore increase the SE.

With the second strategy, the pharmacy records are searched and 100 individuals with antibiotic use and 100 individuals without antibiotic use are selected. In this method, the number of true antibiotic users and nonusers (the marginal column totals in Table 6.1) are fixed. Because we have sampled based on a subject's true exposure status, we can calculate SE and SP, but we cannot directly estimate positive predictive value or negative predictive value using this design. Of course, as noted in Chapter 3, this strategy is usually impossible since we would need the true exposure on everyone in the population, in which case we would have little need for the misclassified version. This option is only practical when a naturally occurring subset, such as one of a study's enrollment sites, has data on both the gold-standard and misclassified measurement.

The last strategy involves selecting 200 individuals from the study population at random without regard to their self-reported antibiotic use (E^*) or pharmacy records (E). Because this method does not fix any of the margin totals, all four of the classification parameters can be calculated. However, when using this third strategy, caution should be exercised if the exposure is rare, since it is possible to end up with no (or few) exposed individuals in the validation study, making parameter estimates imprecise or impossible to calculate (see Chapter 3). A modification to the third strategy involves using known sampling weights to select the 200 people to ensure adequate numbers of exposed and unexposed individuals in the validation study. The observed cell counts in the validation study table are multiplied by the inverse of the sampling weights to yield a validation study in which sensitivity, specificity, positive predictive value and negative predictive value can be estimated. Similar validation studies could be used for the outcome, comparing self-report with medical record data, or to assess the potential confounding effects by smoking comparing self-report with hair samples bioassayed for cotinine levels (a biomarker of tobacco smoke). More recently, adaptive designs have been proposed that evaluate validation data as they are collected, and stop the validation study when some goal has been met [9, 10].

When validation data are collected, careful consideration needs to be given to whether to collect sufficient data to be able to stratify the measures within levels of other variables for analysis. For example, in the case of exposure misclassification, care should be given as to whether the data will need to be stratified by the outcome to estimate the bias parameters within levels of the outcome. If a design is used that allows estimation of SE and SP and the outcome is ignored, the analysts are assuming that the misclassification is nondifferential with respect to the outcome and results of bias analyses using these parameters will reflect this. If estimation of SE and SP is done within levels of the outcome but sufficient sample size is not recruited to allow reasonably precise estimate of SE and SP within levels of the outcome, the estimates can differ within levels of the outcome due to sampling error that would make it appear the misclassification mechanism is differential by design when in fact it may be nondifferential (and vice versa).

When estimating positive predictive value and negative predictive value, additional care is needed. In this case, we must estimate the parameters within levels of the outcome. To see why, remember that positive predictive value and negative predictive value are a function of SE and SP, and also prevalence. Thus for any SE and SP, the values of positive predictive value and negative predictive value between two groups will differ if the prevalence differs between those groups. Prevalence of the exposure will differ within levels of the outcome, if the exposure affects outcome occurrence. This implies that if the classification of exposure is nondifferential with respect to the outcome, positive predictive value and negative predictive value will differ between those with and without the disease, if there is an effect of the exposure on the outcome. Estimating a common positive predictive value and negative predictive value pooled across levels of the outcome will likely lead to modelling a differential misclassification mechanism, even if the true underlying mechanism is nondifferential. Even if the true misclassification is differential, estimating a common positive predictive value and negative predictive value will typically lead an analyst to modelling the *wrong* differential misclassification mechanism.

Ultimately the decision around which design to use and what parameters to estimate will be a balance of getting valid and precise estimates of the bias parameters to be used and the costs to acquire the information in terms of time and money. Because analysts are often in a situation where only the misclassified data are currently available and new data are required on the gold standard to validate, and because the gold standard measure is often more expensive and time consuming to collect, often the first and third designs in Table 6.4 are the only feasible possibilities. As noted, if this decision is taken and negative predictive value and positive predictive value are to be used, care should be exercised in ensuring that negative predictive value and positive predictive value are estimated with reasonable precision and within levels of other key variables. However, when this approach is used, while estimating the bias parameter requires care, the calculations to bias-adjust the data using these parameters is very simple (see below). When sensitivity and specificity are estimated, the calculations are only slightly more complicated, but the estimates are often more likely to apply to other populations and thus can be of greater value to the wider research community.

If the resources to conduct an internal validation study are unavailable, then the literature should be searched for published validation studies that can be applied to the data. As discussed earlier, the previously published validation study should be conducted in a similar population to the current study to improve the external validity (transportability) of the bias parameter estimates. If multiple studies are found that inform bias parameters, a meta-analysis can be used to combine bias parameter estimates across the multiple validation studies or the results can be combined to form a distribution of values that can be used with a probabilistic bias analysis (Chapters 7, 8 and 9) or with a Bayesian bias analysis (Chapter 11). If no published data are available, then the analyst will need to estimate the classification parameters based on their experience (i.e., as an educated guess). This approach should only be taken as a last resort and can be seen as equivalent to assigning a prior distribution based on expert knowledge in a Bayesian inference framework. As such, relying on techniques for elicitation of expert opinion may be fruitful [11, 12]. However, in this case, multidimensional bias analysis (discussed later in this chapter), probabilistic bias analysis (Chapters 7, 8 and 9), or Bayesian bias analysis (Chapter 11) are better approaches because they allow the analyst to see the change in the estimate of association from using different values of the classification parameters. The analyst should also note that collecting validation data would be a productive item to add to the research agenda, because such data would potentially reduce the uncertainty about the estimate of association.

Regardless of the approach used, analysts should always be aware of the limitation of their own validation data. As described in Chapter 3, even internal validation data can suffer from systematic errors like selection bias (e.g., if the validation study required re-consenting participants and only a subset agreed, and the bias parameters estimated in the subset were not the same as would have been estimated in the full sample), misclassification (e.g., if the gold standard measure was not a true gold standard), and random error. It is important when presenting the results of the bias analysis to fully document and present the source for the bias parameters and to justify the values used if they come from external data or expert opinion. This documentation allows critical review of the values chosen and rigorous debate about their validity. Should others disagree with the choices the analyst has made, they should be able to recreate the bias analysis with their own parameters to reach their own conclusions, which is made easier when the bias analyst provides the computing code or tools used for the bias analysis [13].

Bias Analysis of Exposure Misclassification

The classification parameters obtained for the exposure classification scheme can now be used to conduct a simple bias analysis to bias-adjust for misclassification. As outlined above, the design of a validation study will determine which measures of classification can be calculated. First, we will outline the methods for misclassification bias-adjustments for situations in which the sensitivity and specificity are available and then methods where the predictive values are available. To develop the method we will use to bias-adjust for misclassification, it is useful to begin with the hypothetical situation in which we know the true classification status of study participants and would like to calculate the expected observed data using an imperfect classification scheme with known sensitivity and specificity. Considering this scenario helps to develop an understanding of the impact of misclassification and also explains the process for deriving the equations used to calculate the expected true estimate of effect given misclassified observed data (i.e., the situation ordinarily encountered).

In Table 6.5, the equations for calculating the cell frequency in the 2 × 2 table are presented. As throughout, we will use upper case letters to designate true

Table 6.5 Equations for calculating the expected observed data from the true data given sensitivity and specificity.

	Truth		Expected observed	
	E_1	E_0	E_1	E_0
D_1	A	B	$a = A(SE_{D1}) + B(1 - SP_{D1})$	$b = A(1 - SE_{D1}) + B(SP_{D1})$
D_0	C	D	$c = C(SE_{D0}) + D(1 - SP_{D0})$	$d = C(1 - SE_{D0}) + D(SP_{D0})$
Total	$A + C$	$B + D$	$a + c$	$b + d$

classification frequencies and lower case letters to designate expected observed or actual observed frequencies. Although there is overlap in the characters used, the notation is not the same as used for Equations 6.1 to 6.6 and the text and tables of that accompanying section. E_1 and D_1 represent the exposed and diseased populations, respectively, and E_0 and D_0 represent the unexposed and undiseased populations, respectively. In the first two columns, we present the true (not misclassified) data and in the last two columns we present the equations that relate the true data to the misclassified values we would expect to observe, given the bias parameters of sensitivity and specificity. For this example, we will assume that the mechanism of misclassification of exposure is nondifferential with respect to disease (i.e., $SE_{D1} = SE_{D0}$ and $SP_{D1} = SP_{D0}$) and that disease is correctly classified.

The equations for the expected cell count that would be observed divide the individuals in each cell into two groups: those who will remain in the cell (i.e., are correctly classified) and those who will be shifted to the adjacent cell in the row (i.e., are misclassified). For disease and covariate misclassification, the equations are essentially the same as presented above, with the difference being the direction in which misclassified subjects are shifted. For exposure misclassification, subjects always stay in the same row (because their disease is assumed to be correctly classified) but misclassified subjects are shifted from the exposed to unexposed column or vice versa. For disease misclassification, subjects always stay in the same column (because their exposure is assumed to be correctly classified), but misclassified subjects are shifted from the diseased to nondiseased row or vice versa. For covariate misclassification, if we can assume that the exposure and disease are classified correctly, then the misclassified data remain in the same exposure and disease cell but are shifted from one stratum of the confounder to another.

To calculate the "a" cell (i.e., the number of subjects we would expect to be classified as exposed and diseased given the bias parameters), we take the frequency of truly exposed individuals who were classified correctly ($A \cdot SE_{D1}$) and add the frequency of truly unexposed individuals who were incorrectly classified as exposed $[B \cdot (1 - SP_{D1})]$. The remaining cells in the table are calculated in a similar manner, using the equations in Table 6.5.

Table 6.6 presents hypothetical data to demonstrate how to calculate the expected observed cell frequencies. We will begin with an example of nondifferential exposure misclassification in which the classification scheme has a sensitivity of 85% and a specificity of 95%. The odds ratio (OR), risk ratios (RR), and risk difference (RD) are calculated for the true and expected observed data. In this example, the

Table 6.6 Hypothetical example for calculating the expected observed data given the true data assuming nondifferential misclassification of dichotomous exposure with sensitivity = 85% and specificity = 95%.

	Truth		Expected observed	
	E_1	E_0	E_1	E_0
D_1	200	100	200(0.85) + 100(0.05) = 175	200(0.15) + 100(0.95) = 125
D_0	800	900	800(0.85) + 900(0.05) = 725	800(0.15) + 900(0.95) = 975
Total	1000	1000	900	1100
OR	2.3		1.9	
RR	2.0		1.7	
RD	0.10		0.08	

Table 6.7 Hypothetical example for calculating the expected observed data given the true data assuming differential misclassification of dichotomous exposure with sensitivity = 100% for cases and 70% for non-cases and specificity = 90% for both cases and non-cases.

	Truth		Expected observed	
	E_1	E_0	E_1	E_0
D_1	200	100	200(1.0) + 100(0.1) = 210	200(0) + 100(0.9) = 90
D_0	800	900	800(0.7) + 900(0.1) = 650	800(0.3) + 900(0.9) = 1050
Total	1000	1000	860	1140
OR	2.3		3.8	
RR	2.0		3.1	
RD	0.10		0.17	

expected (misclassified) data are biased toward the null by the nondifferential exposure misclassification mechanism (in this case, a true risk ratio of 2.0 appears as 1.7 with misclassification). It is this calculation that is the basis for the (sometimes mistaken) generalization that nondifferential misclassification will bias associations toward the null, which we discuss in detail below.

Using the same data, we now assume a classification scheme with a specificity of 90% for both the diseased and nondiseased but with a sensitivity of exposure classification that is 100% for the diseased and 70% for the nondiseased. Table 6.7 presents the expected observed data assuming this differential misclassification mechanism. In this example, the differential misclassification resulted in measures of association that were larger than the true result for all three measures of association (i.e., bias away from the null). However, this increased measure of association is not the rule with differential misclassification; differential misclassification can bias either toward or away from the null. For any given problem, the expected direction of the bias is difficult to predict without quantitative bias analysis.

Bias-Adjusting for Exposure Misclassification Using Sensitivity and Specificity: Nondifferential and Independent Errors

The preceding examples show how misclassification can affect results. However, as epidemiologists, we are usually faced with a different problem. Our data classification scheme is imperfect, and we wish to estimate what data arrangement we would have observed had participants been correctly classified. To understand how to deal with this problem, we will algebraically rearrange the equations above to reflect what we would expect the true data to be, given the observed data and the bias parameters. We will demonstrate the approach using the A cell.

Equations 6.9 and 6.10, below, (from Table 6.5) show the relation of the true A and B cells and the bias parameters (SE an SP) to the observed "a" cell. $D_{1\ Total}$ is the total number of subjects with the disease $(a + b)$, and $D_{0\ Total}$ is the total number of subjects without the disease $(c + d)$.

$$a = A(SE_{D1}) + B(1 - SP_{D1}) \qquad (6.9)$$

$$B = D_{1Total} - A \qquad (6.10)$$

We can substitute Equation 6.10 into Equation 6.9 to have only one unknown (A) in the resulting equation:

$$a = A(SE_{D1}) + (D_{1Total} - A)(1 - SP_{D1})$$

We can then algebraically rearrange this equation to solve for A (the expected number of true exposed cases) as follows:

$$a = A(SE_{D1}) + D_{1Total} - A - D_{1Total}(SP_{D1}) + A(SP_{D1})$$

$$a - D_{1Total} + D_{1Total}(SP_{D1}) = A(SE_{D1}) - A + A(SP_{D1})$$

$$a - D_{1Total}(1 - SP_{D1}) = A(SE_{D1} - 1 + SP_{D1})$$

$$A = \frac{a - D_{1Total}(1 - SP_{D1})}{(SE_{D1} - 1 + SP_{D1})}$$

We can similarly rearrange the equations for C in Table 6.5 to calculate c. Table 6.8 shows all the equations for calculating the expected true table from observed data,

Table 6.8 Equations for calculating expected true data given the observed data with exposure misclassification.

	Observed		Misclassification adjusted data	
	E_1	E_0	E_1	E_0
D_1	a	b	$[a - D_{1\ Total}(1 - SP_{D1})]/(SE_{D1}-1 + SP_{D1})$	$D_{1\ Total} - A$
D_0	c	d	$[c - D_{0\ Total}(1 - SP_{D0})]/(SE_{D0}-1 + SP_{D0})$	$D_{0\ Total} - C$
Total	$a + c$	$b + d$	$A + C$	$B + D$

Table 6.9 Example of bias-adjusting for misclassification of smoking in a study of the effect of smoking during pregnancy on breast cancer risk assuming nondifferential misclassification of smoking information on the birth certificate.

SE = 78%, SP = 99%	Observed		Misclassification Adjusted data	
	Smokers	Nonsmokers	Smokers	Nonsmokers
Cases	215	1449	256.3	1407.7
Controls	668	4296	799.1	4164.9
OR	0.954		0.949	

Observed data from Fink and Lash, 2003 [14].

given estimates of the sensitivity and specificity. Once we have estimated the expected true table, given the sensitivity and specificity, we can use that table to estimate any measure of effect allowed by the initial study design. Note that using a sensitivity of 85%, specificity of 95% and assuming nondifferential misclassification of the exposure, we can use the equations in Table 6.8 to back calculate the true data in Table 6.6 from the misclassified observed data in Table 6.6.

We can now apply these equations to bias-adjust misclassified data using the validation data presented earlier. In a study of the effect of smoking during pregnancy on breast cancer risk, information on maternal smoking during pregnancy was ascertained from the mother's self-report on the birth certificate. There was concern that mothers may not accurately report their smoking history when completing a birth certificate, but since the pregnancy was before their breast cancer diagnosis, the misclassification mechanism was expected to be nondifferential. This study observed no association between smoking during pregnancy and risk of breast cancer (OR = 0.95 95%; CI = 0.81, 1.13) [14]. However, it was hypothesized that the lack of an observed association could have been caused by nondifferential misclassification of smoking. This is a reasonable critique of the study, since even without validation data we would expect that smoking during pregnancy as measured on a birth certificate might be poorly classified, and one might expect that had the issue not been addressed, that reviewers would raise the issue as part of the peer review process. Indeed, without quantification of the magnitude and direction of the bias, a reviewer might simply reject the analysis on the basis of concern about the misclassification. Quantitative bias analysis allows us to determine whether our intuition is correct, at least assuming a valid bias model.

In Table 6.2, we used data from a validation study to calculate a sensitivity of $128/164 = 78\%$ and a specificity of $465/470 = 99\%$ for the accuracy of smoking information on the birth certificate compared with medical record. We will use these values and the observed data from the study to estimate the association between smoking during pregnancy and breast cancer risk had smoking not been misclassified, conditional on the accuracy of the sensitivity and specificity.

Table 6.9 shows the observed and bias-adjusted data given the bias parameters. In keeping with good practice, we do not round any estimates until necessary. Sensitivities and specificities in this example are 128/164 and 465/470, respectively. In this example, the nondifferential misclassification had a minimal effect on the odds

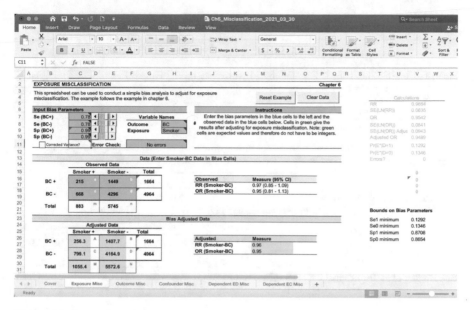

Figure 6.4 Screenshot of Excel spreadsheet to perform simple bias analysis for nondifferential misclassification of exposure.

ratio, suggesting it is unlikely that the observed null effect was the result of inaccurate reporting of smoking status on the birth certificate, if these estimates of the sensitivity and specificity were accurate.

Below is a screenshot from the Excel spreadsheet available on the text's website showing the same results (Figure 6.4). In the spreadsheet, it is necessary to input the sensitivity and specificity for the exposed and unexposed and then the a, b, c, and d cell frequencies from the observed 2×2 table. The spreadsheet will calculate the bias-adjusted cell frequencies and the bias-adjusted odds ratio and risk ratio.

A simple alternative formula is available when the OR is the desired measure of association [15, 16]. Let P_1 be the proportion who are observed to be exposed among the diseased and let P_0 be the proportion who are observed to be exposed among the nondiseased. These quantities can be estimated as $a/(a + b)$ and $c/(c + d)$ in Table 6.8. A simple formula for the OR bias-adjusted for exposure misclassification is:

$$OR = \frac{(P_1 + SP_1 - 1)(SE_0 - P_0)}{(P_0 + SP_0 - 1)(SE_1 - P_1)}$$

Following the previous example, we estimate $P_1 = 215/(215 + 1449) = 13\%$ and $P_0 = 668/(668 + 4296) = 13\%$. Inserting the appropriate sensitivity and specificity values, we compute the OR bias-adjusted for misclassification of 0.95, exactly as we did above (Table 6.9 and Figure 6.4).

Table 6.10 Equation for calculating expected true frequencies given the observed frequencies and classification predictive values.

	Observed		Misclassification adjusted data	
	E_1	E_0	E_1	E_0
D_1	a	b	$a(\text{PPV}_{D1}) + b(1 - \text{NPV}_{D1})$	$D_{1\ \text{Total}} - A$
D_0	c	d	$c(\text{PPV}_{D0}) + d(1 - \text{NPV}_{D0})$	$D_{0\ \text{Total}} - C$
Total	$a + c$	$b + d$	$A + C$	$B + D$

Table 6.11 Example of bias-adjustment for misclassification of BMI category using positive and negative predictive values in a study of the effect of BMI category on early preterm birth.

$\text{PPV}_{\text{D1}} = 65\%, \text{PPV}_{\text{D0}} = 74\%,$ $\text{NPV}_{\text{D1}} = 100\%, \text{NPV}_{\text{D0}} = 98\%$	Observed		Adjusted data	
	Underweight	Normal weight	Underweight	Normal weight
Preterm	599	4978	389.0	5188.0
Term	31,175	391,851	29,687.7	393,338.3
OR	1.5		1.0	

See Bodnar et al. for original data [18–20].

Bias-Adjusting for Exposure Misclassification Using Predictive Values

The methods presented to this point use sensitivity and specificity to complete the bias analysis. However, when positive predictive values and negative predictive values are available, a second method can be used for simple bias analysis to bias-adjust for misclassification. Table 6.10 displays the formulas for bias-adjustment of cell counts using positive predictive values and negative predictive values [17].

As an example of the approach, Bodnar et al. conducted a cohort study examining the association between maternal pre-pregnancy BMI and risk of early preterm (<34 weeks) delivery [18–20]. Prepregnancy BMI was estimated from self-reported pre-pregnancy height and weight from maternal birth certificates. Prepregnancy BMI was categorized as underweight, normal weight, overweight, and obese using World Health Organization guidelines. For the purposes of this example, we only examine the association between underweight (versus normal weight) and early preterm birth. The data from the study are shown in Table 6.11. Since BMI was collected from self-report data that were ascertained after women had a term or preterm birth, there was potential for differential misclassification. For instance, women who experienced a preterm birth may have been more likely to accurately report their BMI than women who had a term birth. Regardless of whether there had been differential misclassification, Bodnar et al. were interested in estimating positive predictive values and negative predictive values so needed to conduct the first type of valida-tion study shown in Table 6.4 by randomly sampling women conditional on their reported BMI category and term/preterm status [18]. For the women who were sampled, BMI was obtained from medical records from the woman's first prenatal

Table 6.12 Validation data comparing birth certificate (misclassified) BMI category to medical record review (gold standard) by term birth or early preterm birth [19].

	Term		Preterm	
	Medical Record			
Birth certificate	Underweight	Normal weight	Underweight	Normal weight
Underweight	132	47	50	27
Normal weight	2	115	0	120

visit (used as a gold standard for BMI category). A reduced version of the validation data are presented in Table 6.12. Because of the design strategy implemented by the authors, only positive predictive values and negative predictive values could be estimated. From the data in Table 6.12, we calculated a PPV_{D1} of 65%, a NPV_{D1} of 100%, a PPV_{D0} of 74%, and a NPV_{D0} of 98%.

Table 6.10 shows the equations to bias-adjust for misclassification using these bias parameters. Notice that the bias-adjustments are even simpler than when using SE and SP. As an example, the bias-adjusted A cell is just the positive predictive value among the cases times the observed "a" cell (those correctly classified among those classified as exposed) plus one minus the positive predictive value among the cases times the observed "b" cell (those incorrectly classified as unexposed). A similar logic can be applied to the three remaining cells.

Using the bias parameters from the Bodnar et al. example and applying the equations in Table 6.10 to the observed data, we estimate the data we would expect to have observed without misclassification in Table 6.11. Using these adjusted cell frequencies, we estimate that pre-pregnancy underweight women have approximately the same odds of early preterm delivery as pre-pregnancy normal weight women. The OR we observed after bias-adjusting for misclassification is quite attenuated from the crude OR of 1.5 in the misclassified data, suggesting the misclassification biased the estimates away from the null, conditional on the accuracy of the bias model.

As we noted above, using positive predictive values and negative predictive values to bias-adjust for misclassification can be perilous. Because positive predictive values and negative predictive values can change dramatically as a function of the prevalence of the true exposure status, the optimal situation is for an analyst to use positive predictive values and negative predictive values that are estimated from a representative sample of the study population, possibly conditional on sampling within categories of observed disease and exposure status (see above and Chapter 3). Alternatively, an analyst can choose a population that is as similar as possible to the source population for her study. However, this will require the assumption that the prevalence of the true exposure is the same in the two populations, including when stratified by disease status. This assumption will often be impossible to verify.

Bias-Adjustment for Nondifferential Outcome Misclassification Using Positive Predictive Values for the Risk Ratio Measure of Association

Brenner and Gefeller (1993) illustrate a special instance in which outcome misclassification can be adjusted using only positive predictive values with no knowledge of negative predictive values [21]. However, this method can only be implemented when there is nondifferential misclassification of the outcome with respect to the exposure and the risk ratio is the measure of association. To demonstrate their method, we assume that we want to estimate the risk ratio association without any disease misclassification:

$$RR = \frac{\Pr(D = 1|E = 1)}{\Pr(D = 1|E = 0)}$$

Because misclassification is nondifferential, we can multiply the right hand side of the equation by SE_1/SE_0 without altering the RR. We note that, technically, only the sensitivities need to be independent of exposure and specificities could differ across exposure groups with no impact on these results:

$$
\begin{aligned}
RR &= \frac{\Pr(D = 1|E = 1)}{\Pr(D = 1|E = 0)} \\
&= \frac{\Pr(D = 1|E = 1)}{\Pr(D = 1|E = 0)} \cdot \frac{SE_1}{SE_0} \\
&= \frac{\Pr(D = 1|E = 1)}{\Pr(D = 1|E = 0)} \cdot \frac{\Pr(D^* = 1|D = 1, E = 1)}{\Pr(D^* = 1|D = 1, E = 0)} \\
&= \frac{\Pr(D^* = 1, D = 1|E = 1)}{\Pr(D^* = 1, D = 1|E = 0)}
\end{aligned}
$$

This final term can also be derived using the misclassified RR (labeled RR_{obs}) and positive predictive values:

$$
\begin{aligned}
RR_{obs} \cdot \frac{PPV_1}{PPV_0} &= \frac{\Pr(D^* = 1|E = 1)}{\Pr(D^* = 1|E = 0)} \cdot \frac{\Pr(D = 1|D^* = 1, E = 1)}{\Pr(D = 1|D^* = 1, E = 0)} \\
&= \frac{\Pr(D^* = 1, D = 1|E = 1)}{\Pr(D^* = 1, D = 1|E = 0)}
\end{aligned}
$$

Combining these results, we arrive at Brenner and Gefeller's result that one can bias-adjust a misclassified RR using positive predictive values, under the assumption of nondifferential misclassification:

$$RR = \frac{\Pr(D^* = 1|E = 1)}{\Pr(D^* = 1|E = 0)} \cdot \frac{PPV_1}{PPV_0}$$

Table 6.13 The association between family history of hematopoietic cancer and risk of lymphoma before and after bias-adjustment for differential misclassification of family history Observed data from Chang et al., 2005 [22].

	Observed data		Misclassification adjusted data	
	Family history	No family history	Family history	No family history
Cases	57	1148	56.7	1148.3
Controls	26	1201	37.1	1189.9
OR	2.3		1.6	

This result is likely to be most useful in studies in which the analyst has reason to believe that only the disease is misclassified and that the misclassification is nondifferential. In that case, the analyst could design a validation study that only estimates positive predictive values, saving the cost of estimating the accuracy among those who test negative for the disease.

Bias-Adjustments Using Sensitivity and Specificity: Differential Independent Errors

The following section works through an example in which the scheme for classifying the exposure exhibits *differential* misclassification. This example uses data from a Swedish case-control study measuring the association between a family history of hematopoietic cancer on an individual's risk of lymphoma [22]. In this study, the analysts validated self-report of a family history by searching for family members' diagnosis in the Swedish Cancer Registry and examined whether the validity of self-report differed between cases and controls. Specificity was nearly perfect in both groups (98% and 99% among cases and controls, respectively), but sensitivity was low among cases (60%) and even lower among controls (38%), as might be expected by design. Table 6.13 presents the observed data from the study and the bias-adjusted data accounting for the sensitivity and specificity of report in the cases and controls.

In this example, differential misclassification of family history of hematopoietic disease had a substantial effect on the observed estimate of association (assuming a valid bias model), with a

$$\frac{(2.3 - 1) - (1.6 - 1)}{1.6 - 1} = 117\%$$

117% increase in the relative effect away from the null. We note here that the benefit of the bias analysis is not only in confirming the direction of the bias, but also the magnitude. Unlike in the example of misclassification of smoking during pregnancy, here the impact is substantial.

Table 6.14 The association between self-reported maternal residential proximity to agricultural fields and risk of neural tube defects in their offspring.

	Total population		Observed data among the validated subset		Observed data among the remainder	
	<0.25 miles	0.25+ miles	<0.25 miles	0.25+ miles	<0.25 miles	0.25+ miles
Cases	82	177	64	163	18	14
Controls	110	360	91	333	19	27
OR	1.5		1.4		1.8	

Adapted from Shaw et al., 1999 [23].

Table 6.15 The association between residential proximity to agricultural fields and risk of neural tube defects.

	Total		Data among the validated subset		Bias-adjusted data among the remainder	
	<0.25 miles	0.25+ miles	<0.25 miles	0.25+ miles	<0.25 miles	0.25+ miles
Cases	93.3	165.7	67	160	26.3	5.7
Controls	151.8	318.2	116	308	35.8	10.2
OR	1.2		1.1		1.3	

Adapted from Shaw et al., 1999 [23] and Rull et al., 2006 [24].

Bias-Adjustments Using Sensitivity and Specificity: Internal Validation Data

Finally, we present an example of bias-adjusting for exposure misclassification using internal validation data. Here we use data from a study of the relation between self-reported maternal residential proximity to agricultural fields during pregnancy (as a proxy for pesticide exposure) and risk of neural tube defects in their offspring [23]. For a large subset of the population, the analysts were able to geocode addresses and compare self-report to land-use maps. Table 6.14 below shows the self-reported data for the full sample, the subset that was validated, and the remainder of the population.

Table 6.15 displays the data for the validated subset and the bias-adjusted data for the remainder of the population. In the validation subset, the sensitivity of exposure classification was 65.7% among the cases and 50.0% among the controls, and specificity was 87.5% among the cases and 89.3% among the controls [24]. In this case, because some of the subjects in the study had their results validated, and therefore we already know their gold-standard exposure status, we do not need to apply any bias-adjustments to these subjects. Here the values of SE and SP were used to bias-adjust the data only for the subset of the population without validated data. For subjects in the validation study, we can simply use their validated exposure measure. We can then combine the bias-adjusted and validated data to obtain an overall bias-adjusted measure of association.

There are two options for calculating an overall odds ratio. First, we can add the cells in the validated and bias-adjusted groups to obtain total frequencies (the "Total" columns in Table 6.15). A second approach is to pool the ln of the odds ratios for the two strata of bias-adjusted and validated data, weighting by the inverse of their variances [25]:

$$\ln(OR_{total}) = \frac{\ln(OR_V)V_R + \ln(OR_R)V_V}{V_V + V_R}$$

where V_V is the variance of the odds ratio among the validated group, using the standard calculation of the sum of the inverse of each cell in the 2×2 table and V_R is the variance of the odds ratio in the remainder, bias-adjusted strata. The equation for V_R is postponed until Chapter 10, where we discuss direct estimation of variances for bias-adjusted data. Applying this equation to these data yields a pooled odds ratio of 1.1.

Note that using the validated measure of exposure decreased the estimate of effect from 1.4 to 1.1 in the validated subset and from 1.8 to 1.3 in the nonvalidated group. The higher estimate of effect in the nonvalidated group suggests that this group may not be representative of the total and that it is necessary to consider whether it is appropriate to apply the sensitivity and specificity from the validated sample to the nonvalidated sample. Alternatively, given the small sample size of the remainder sample, the discrepancy may arise in large part due to chance variation.

Overreliance on Nondifferential Misclassification Biasing Toward the Null

Epidemiologists have long relied on the mantra that "nondifferential misclassification biases observed results toward the null." By this, we mean that if the data collection methods preclude the expectation that the rate of errors in classifying one axis of an association (e.g., exposure status) may depend on the status of the second axis of an association (e.g., disease status), then the classification errors are expected to attenuate the strength of the association (i.e., bias toward the null). Reliance on this expectation has given epidemiologists a false sense of security regarding their results with typical discussion noting that even if the misclassification bias existed, the true association would be even greater than what was observed. The typical superficial approach to assessing misclassification bias prevents a thorough understanding of the extent to which the true association might differ from the observed association, and also sometimes results in a misinterpretation of the inferences that can be made from the bias toward the null [26]. It is not the job of an epidemiologist to simply determine if an exposure has an effect on an outcome as a binary (as is propagated by null hypothesis testing approaches), but rather it is to estimate the magnitude of the effect in a valid and precise manner. Thus, even if nondifferential misclassification did bias results towards the null, it matters how much because the decision made about the importance of the public

Table 6.16 Association between body mass index and the prevalence of diabetes using BRFSS data.

	<25	25–30	30+
Diabetics	3053	5127	6491
Nondiabetics	81,416	65,104	33,814
OR	Reference	2.1	5.1

Observed data from Mokdad et al., 2003 [32].

health consequences of intervening on the exposure depend on the magnitude of the effect (typically measured on the absolute scale). Simple bias analysis provides an accessible tool to easily quantify the expected impact of the misclassification, conditional on the accuracy of the bias model. Furthermore, as will be shown in this section, there are situations when the rule does not hold and this can easily be observed using the bias analysis techniques described in this chapter.

Even when the misclassification mechanism is truly nondifferential, there are a number of subtleties and exceptions to the rule that nondifferential misclassification leads to a bias toward the null. First, nondifferential misclassification guarantees that the bias will be toward the null only in *expectation*. If we assume that classification is a random process, any realization of that classification process, in a particular study, need not be biased toward the null even if the mechanism of misclassification was nondifferential. The bias in any particular study can be toward or away from the null, since the actual misclassification arising from the nondifferential mechanism can result in a data arrangement far from what is expected. Put another way, even if the mechanism of misclassification was nondifferential such that sensitivity among the exposed was expected to equal sensitivity among the unexposed, through random error, we could see deviation from this expectation creating differential misclassification in the realization of the study. This potential for departure from expectation has been illustrated multiple times [27–29] and quite extensively in a simulation study by Jurek et al. [28]. Their results showed that, if a misclassification mechanism is nondifferential, the average effect estimate across a large number of studies will be biased toward the null. For any individual study, however, the misclassified result could by chance be further from the null than the true result. In general, a relatively large proportion of misclassified results for individual studies were likely to be larger than the true results for weak associations (OR = 1.5), rare outcomes (1%), and unbalanced exposure prevalence (10%). While sample size was not varied in this study, one would expect that smaller studies would be more likely to yield results some distance from the expectation than larger.

Second, when the exposure variable has more than two exposure levels, in some cases, misclassification of exposure can bias the estimate of effect away from the null [30, 31]. Table 6.16 shows the association between body mass index and the prevalence of diabetes estimated using cross-sectional data from the Behavioral Risk Factor Surveillance System (BRFSS) [32]. The BRFSS surveys the US population using telephone interviews and collects height and weight information by self-report. Body mass index (BMI) was categorized into three groups— < 25, 25–30, and 30 + —representing healthy weight, overweight, and obese, respectively. A second study compared the national prevalence of body mass index in the BRFSS and the National Health and Nutrition Examination Survey (NHANES), which

Table 6.17 Bias-adjusted data for the association between body mass index and the prevalence of diabetes using BRFSS data.

	<25	25–30	30+
Diabetics	2,545.9	2607.5	9517.6
Nondiabetics	74,977.1	55,776.3	49,580.6
OR	Reference	1.4	5.7

Observed data from Mokdad et al., 2003 [32] and validation data from Yun et al., 2006 [33].

collects height and weight information during a physical examination by a trained health technician [33]. This study suggested that the prevalence of being overweight and obese were underestimated in the BRFSS study. For illustrative purposes, we will assume that there was no overreporting of body mass index and that misclassification of BMI was independent of diabetes status (i.e., independent and nondifferential misclassification). Extrapolating from the data provided in the man-uscript of the comparison study [33], we will assume that 9% of those overweight had been incorrectly classified as healthy and that 68.2% of those obese had been incorrectly classified as overweight. We assume that no one who was obese had been misclassified as having a healthy body mass index. Table 6.17 displays the recalculated table cells and odds ratios after applying the estimates of classification errors assuming that misclassified data only shifted one column. Note that the odds ratio for the obese group compared to the healthy group was biased toward the null, but the estimate of association for the overweight group compared to the healthy group was substantially biased *away* from the null, despite the fact that the misclassification was nondifferential. This pattern occurs because the group with the highest prevalence of diabetes (obese) mixed substantially with the group with the moderate prevalence of diabetes (overweight). When this group is removed from the moderate prevalence of diabetes group (overweight) and reallocated to the high prevalence of diabetes group (obese), the estimate of the strength of association comparing the overweight group to the healthy weight group decreased from 2.1 to 1.4.

Finally, although nondifferential misclassification of dichotomous categorical variables produces a bias toward the null, in the expectation, if the sum of SE and SP is below 1 the bias can continue past the null. For example, in Table 6.6, if the nondifferential misclassification process led to a SE = 15% and SP = 70%, the expected OR resulting from this misclassification would be 0.8. The true OR is 2.3, so the direction of this bias is, indeed, toward the null. However, an analyst, upon seeing an OR = 0.8 and suspecting nondifferential misclassification, might be tempted to posit the true OR is less than 0.8. In this case, that would be incorrect, since the combination of SE and SP being below 1 has produced a bias that reverses the direction of the effect. Such events are hopefully rare, though may occur in practice [34].

To summarize, the generalization that nondifferential misclassification biases results toward the null should be refined to say, "on average, nondifferential misclassification of a dichotomous independent variable will bias results toward the null." Further refinements to this statement will be made throughout this chapter.

Table 6.18 Equations for calculating the expected observed data from the true data when there is disease misclassification.

	Truth		Expected observed	
	E_1	E_0	E_1	E_0
D_1	A	B	$A(SE_{E1}) + C(1 - SP_{E1})$	$B(SE_{E0}) + D(1 - SP_{E0})$
D_0	C	D	$C(SP_{E1}) + A(1 - SE_{E1})$	$D(SP_{E0}) + B(1 - SE_{E0})$
Total	$A + C$	$B + D$	$a + c$	$b + d$

Table 6.19 Equations for calculating the expected observed data from the true data given sensitivity and specificity[a].

	Observed		Misclassification adjusted data	
	E_1	E_0	E_1	E_0
D_1	a	b	$[a - E_{1\ Total}(1 - SP_{E1})]/$ $[SE_{E1} - (1 - SP_{E1})]$	$[b - E_{0\ Total}(1 - SP_{E0})]/$ $[SE_{E0} - (1 - SP_{E0})]$
D_0	c	d	$E_{1\ Total} - A$	$E_{0\ Total} - B$
Total	$a + c$	$b + d$	$A + C$	$B + D$

[a]$E_{1\ Total}$ is the total number of exposed subjects $(a + c)$ and $E_{0\ Total}$ is the total number of unexposed subjects $(b + d)$.

Disease Misclassification

Bias-Adjustments with Sensitivity and Specificity: Nondifferential and Independent Errors

Returning to our descriptions of simple bias analysis to bias-adjust for misclassification, we will now follow the same process we used for exposure misclassification to bias-adjust for disease misclassification. In Table 6.18, we have modified the equations from Table 6.5 to be used to address disease misclassification; the major difference is that instead of misclassified data staying in the same row but moving columns, the data remain in the same column but shift rows. Following the same process we used with exposure misclassification, the equations in Table 6.18 are used to derive the equations for calculating expected truth cell frequencies given observed data. These equations are shown in Table 6.19.

To illustrate bias-adjustments for disease misclassification, we will apply these equations to data exploring the relation between sex and death from acute myocardial infarction (AMI) in Belgium in 1988. Note that this 1988 study made no distinction between natal sex and gender identity; we use "men" and "women" here as the terms used by the original authors, and acknowledge this shortcoming of the original study. Using mortality statistics provided by the vital statistics registry, the data indicate that, among Belgian citizens who died in 1988, men had 1.3 times the risk of dying from an AMI as women (Table 6.20). However, because the cause of death information came from death certificates, there is potential for

Table 6.20 Observed and bias-adjusted data for the association between sex and death from acute myocardial infarction (AMI).

	Observed data		Bias-Adjusted data	
	Men	Women	Men	Women
AMI deaths	4558	3428	7,363.9	5,210.3
Other deaths	46,305	46,085	43,499.1	44,302.7
Total	50,863	49,513	50,863	49,513
Risk (%)	9.0	6.9	14.5	10.5
Risk ratio	1.3	Reference	1.4	Reference

Table 6.21 Validation study of acute myocardial infarction recorded on the death certificate.

		MONICA registry		
		AMI+	AMI−	Total
Death certificate	AMI+	1750	184	1934
	AMI−	1550	12,075	13,625
	Total	3300	12,259	15,559

Validation data from De Henauw et al., 1998 [35].

outcome misclassification. A validation study estimated the extent of this misclassification by comparing death certificate data with the Multinational Monitoring of Trends and Determinants in Cardiovascular Diseases acute coronary event registry [35]. Table 6.21 shows the validation data, and from these data the sensitivity and specificity of AMI classification can be calculated. We can then bias-adjust for the outcome misclassification by applying the bias parameters (sensitivity and specificity) to the sex-specific misclassified vital statistics registry data using the equations in Table 6.19.

The validation data show that death certificates have a sensitivity of 53.0% (1750/3300) and a specificity of 98.5% (12,075/12,259) for AMI classification. These values for sensitivity and specificity will be applied to the data in Table 6.20, examining the relation between death from acute myocardial infarction and sex. We will assume that the disease misclassification was nondifferential. Assuming a valid bias-model, the observed result was biased toward the null (see Table 6.20 for bias-adjusted results). The bias of the relative effect was

$$\frac{(1.4 - 1) - (1.3 - 1)}{1.3 - 1} = 33\%$$

Figure 6.5 shows a screenshot of the Excel spreadsheet to conduct the same bias-adjustment for disease misclassification. Once the user inputs the sensitivity and specificity of disease misclassification (by exposure status if necessary), the program computes the bias-adjusted cell frequencies and measures of effect.

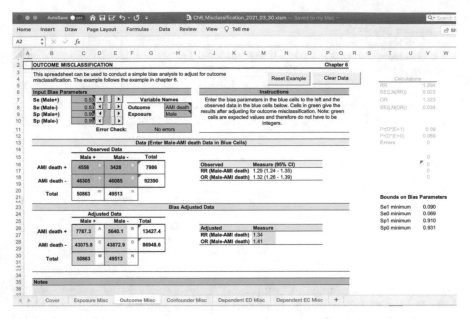

Figure 6.5 Screenshot of Excel spreadsheet for bias-adjusting estimates of effect for nondifferential disease misclassification. Note that the spreadsheets are designed to use inputs for SE and SP in increments of 0.01, so to get the exact values used in the spreadsheet above and in the example, one would need to overwrite the formulas in the cells for inputs and type in the values desired.

Disease Misclassification in Case-Control Studies

It would be a mistake to use the disease-misclassification adjustments in Table 6.19 for case-control data if (as is almost always true) SE and SP were determined from other than the study data themselves [36], because the use of different sampling probabilities for cases and controls alters the sensitivity and specificity within the study relative to the source population [37]. The study specificity could therefore be far from the population specificity. For example, if SE = SP = 0.90, all apparent cases are selected, and controls are 1% of the population at risk, the study specificity will be $0.01(0.90)/[0.1 + 0.01(0.90)] = 0.08$. Use of the population specificity (0.90) instead of the study specificity 0.08 in a bias analysis would produce extremely distorted results. Instead, following the notation in Table 6.19, one can obtain the bias-adjusted estimate of cases in an exposure group indexed by j as [37]:

$$A_j = \frac{a_j - \left(1 - SP_j\right)\left(a_j + \frac{E_{j\,total} - a_j}{f}\right)}{\left(Se + Sp - 1\right)}$$

and the bias-adjusted estimate of non-cases as:

$$B_j = \frac{\frac{E_{j\ total}-a_j}{f} - (1 - Se)\left(a_j + \frac{E_{j\ total}-a_j}{f}\right)}{(Se + Sp - 1)}$$

where f is the control sampling fraction and the bias-adjusted estimate of controls would be $f \cdot B_j$. In case-control studies with secondary-base design, f may be unknown, but may be possible to estimate given subject matter knowledge. A multidimensional analysis with a range of values assigned to f would be advisable.

Overreliance on Nondifferential Misclassification Biasing Toward the Null

When conducting case-control studies in which respondents are selected based on their disease status, disease misclassification can affect which participants are included in the study. This differs from exposure misclassification in case-control studies, in which respondents will still be included in the study, but may be classified into the incorrect exposure group. For many diseases, there are multiple clinical, laboratory, and pathological tests that can be used to diagnose disease. The analyst must define the disease for the research study, and this definition can and often should differ from a clinical definition. The research case definition must balance sensitivity and specificity. High specificity limits the false-positives but will often result in a lower sensitivity and may limit the number of available true cases and reduce the power of the study. On the contrary, high sensitivity allows harvesting all possible cases, but will often result in a lower specificity, which will also allow noncases to be included, and could bias the results and be cost-inefficient since controls need to be recruited for these false positive cases.

In the section on exposure misclassification, a number of subtleties and exceptions were made to the generalization of nondifferential misclassification biasing results toward the null. With disease misclassification, there are a number of conditions, which, if met, will allow unbiased estimates of effect even in the presence of disease misclassification. If the disease can be defined such that there are no false-positives (i.e., specificity is 100%) and the misclassification of true positives (i.e., sensitivity) is nondifferential with respect to exposure, then risk ratio measures of effect will not be biased, in the expectation [38]. Odds ratio and incidence rate ratio measures may be approximately unbiased in this setting if they are suitable approximations of the risk ratio (which usually requires risk less than about 10% in all combinations of independent variables used in the analysis). However, the risk ratio estimate with misclassification will generally be less precise than the risk ratio estimate without misclassification due to the loss of true cases by imperfect sensitivity.

To see why ratio measures of association are unbiased when specificity is perfect and the sensitivity is the same among the exposed and unexposed, consider the following equations from Table 6.18 that relate observed cell counts to true cell counts.

$$a = A(SE_{E1}) + C(1 - SP_{E1})$$
$$b = B(SE_{E0}) + D(1 - SP_{E0})$$

Because the specificity is 100%, the second half of each of the above equations $(1 - SP)$ is zero and can be eliminated. Therefore, the observed risk ratio would be

$$RR_{observed} = \frac{a/(a+c)}{b/(b+d)} = \frac{A \cdot SE_{E1}/(a+c)}{B \cdot SE_{E0}/(b+d)}$$

Because the sensitivity of disease classification is nondifferential, SE_{E1} will equal SE_{E0} and those terms will cancel in the equation. In addition, because misclassified subjects are not changing their exposure status, a + c will equal A + C. Therefore, the equation for the observed risk ratio simplifies to the equation for the true risk ratio:

$$RR_{observed} = \frac{a/(a+c)}{b/(b+d)} = \frac{A/(A+C)}{B/(B+D)} = RR_{true}$$

The confidence interval will, generally, be wider than had the classification been perfect, however, because of the reduction in the number of cases caused by the imperfect sensitivity (i.e., a reduced from A and b reduced from B by a factor SE_{E+}, such that $a/b = A/B$ but $a + b < A + B$).

A similar approach can be applied to examine the bias in the risk difference under this scenario. Here the risk difference will be biased towards the null by a factor of the sensitivity.

$$
\begin{aligned}
RD_{observed} &= \frac{a}{a+c} - \frac{b}{b+d} \\
&= \frac{A \cdot SE}{A+C} - \frac{B \cdot SE}{B+D} \\
&= SE\left(\frac{A}{A+C} - \frac{B}{B+D}\right) \\
&= SE \cdot RD_{true}
\end{aligned}
$$

Although the observed RD is biased in this case, if one had an estimate of the sensitivity, an estimate of RD_{TRUE} could be obtained as $\frac{RD_{observed}}{SE}$.

In Table 6.22, data are presented for a hypothetical study with 2000 participants, half with the exposure and the other half without. The true association between exposure and disease is a risk ratio of 2 and the true risk difference is 0.1. Using these data, Figure 6.6 shows the relation between sensitivity, specificity, and the expected observed risk ratio and risk difference. The black curves refer to the risk ratios. When specificity is held at 100% and sensitivity ranges from 50% to 100%, the analysis yields the unbiased risk ratio of 2 regardless of the sensitivity and the risk difference

Table 6.22 Hypothetical true association between exposure and disease.

	E_1	E_0
D_1	200	100
D_0	800	900
Total	1000	1000
Risk	0.2	0.1
Risk difference	0.1	
Risk ratio	2	

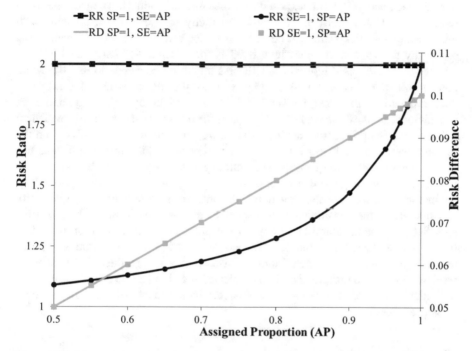

Figure 6.6 Effect of disease misclassification on expected estimates of the risk ratio and risk difference, with original data as shown in Table 6.22. The assigned proportion (AP) refers to the value assigned to the sensitivity or specificity not held constant, as shown in the figure legend. Note that the two lines overlap for the two risk difference scenarios, so one is shown with a marker and the second with a line.

is biased by a factor equal to the value of the nondifferential imperfect sensitivity. Conversely, when sensitivity is held at 100% and specificity ranges from 50% to 100%, there is an inverse relation between the bias and the specificity. Even a specificity of 99% will yield a risk ratio of 1.9, very close to the truth (2.0) though not completely unbiased. The gray curves refer to risk differences. When specificity is held at 100% and the sensitivity ranges from 50% to 100%, the risk difference approaches the truth linearly as the sensitivity approaches 100%.

In the example of death from myocardial infarction and sex (Tables 6.20 and 6.21), a risk ratio of 1.3 was observed and, after conducting a bias analysis for non-differential misclassification, a bias-adjusted risk ratio of 1.4 was obtained. The validation data indicated that the specificity of disease was close to perfect (98.5%), but, as we see in Figure 6.6, this near-perfect specificity is not always sufficient to produce an unbiased estimate. However, the high specificity does explain why the observed estimate was close to the bias-adjusted estimate despite the poor sensitivity (53%), in this example. There is, however, no general guarantee that having an approximately perfect specificity will result in an approximately unbiased risk ratio. For example, consider the data in Table 6.22. Suppose that, instead of 800 exposed non-diseased and 900 unexposed non-diseased individuals, we had 100,800 exposed non-diseased and 100,900 unexposed non-diseased individuals. The true risk ratio remains 2.0. If sensitivity is 53% and specificity is 100%, the estimated (with misclassification) risk ratio will be 2.0, in the expectation. However, even if the sensitivity is 53% and the specificity if 99%, the misclassification of such a large number of non-diseased individuals will bias the observed risk ratio to 1.05. More generally, if the disease is rare (or the exposure prevalence is low, for exposure misclassification problems), the strength of bias is driven by specificity, and even specificity near 100% can lead to strong bias if the outcome (or exposure prevalence) is quite rare. For this reason, analysts should be very cautious when relying on the assumption that a nearly perfect specificity will result in little bias and instead use bias analysis methods to quantify the actual expected magnitude of the bias.

Figure 6.6 does suggest that when conducting a study in which a risk ratio will be the measure of effect, it is useful to define the outcome with high (perfect) specificity and nondifferential sensitivity with respect to disease classification. Near-perfect specificity can be accomplished by requiring pathological confirmation of disease (rather than a clinical confirmation), for example, and serves to remind us that for research purposes, clinical definitions of disease may not produce the most valid research results. Nondifferential disease misclassification can be achieved by using individuals blinded to exposure status or independent sources for disease classification.

Covariate Misclassification

Bias-Adjustments with Sensitivity and Specificity: Nondifferential and Differential Misclassification with Independent Errors

Methods for conducting bias analysis for a misclassified covariate follow the same approach as for exposure misclassification, but the misclassified data stay in the same exposure and disease cells and move from one stratum of the covariate to another. The covariate could be a potential confounder, effect measure modifier, or both. Whereas nondifferential misclassification of exposure (outcome) implies that the sensitivity and specificity does not vary by disease (exposure) status, when it comes to nondifferential misclassification of other covariates, the term "nondifferential" implies that the sensitivity and specificity of covariate

Table 6.23 Equations for bias-adjusted observed data that are biased by nondifferential covariate misclassification.

			Observed data			
	Total		C_1		C_0	
	E_1	E_0	E_1	E_0	E_1	E_0
D_1	a	b	a_{C1}	b_{C1}	a_{C0}	b_{C0}
D_0	c	d	c_{C1}	d_{C1}	c_{C0}	d_{C0}
			Bias-adjusted data			
			C_1		C_0	
	E_1	E_0	E_1	E_0	E_1	E_0
D_1	A	B	$[a_{C1} - a(1 - \text{SP})]/$ $[(\text{SE} - (1 - \text{SP})]$	$[b_{C1} - b(1 - \text{SP})]/$ $[(\text{SE} - (1 - \text{SP})]$	$a - A_{C1}$	$b - B_{C1}$
D_0	C	D	$[c_{C1} - c(1 - \text{SP})]/$ $[(\text{SE} - (1 - \text{SP})]$	$[d_{C1} - d(1 - \text{SP})]/$ $[(\text{SE} - (1 - \text{SP})]$	$c - C_{C1}$	$d - D_{C1}$
	Total		$A_{C1} + C_{C1}$	$B_{C1} + D_{C1}$	$A_{C0} + C_{C0}$	$B_{C0} + D_{C0}$

classification do not vary as a function of either disease *or* exposure status. Differential misclassification of covariates occurs whenever either sensitivity or specificity (or both) of covariate classification differ by either exposure or disease status (or both).

We will first introduce bias-adjustments for nondifferential misclassification and then discuss the more general setting of differentially misclassified covariates. The equations for bias-adjusting nondifferentially misclassified covariate data (Table 6.23) are derived in the same manner as in the sections on exposure and disease misclassification.

To illustrate a bias-adjustment for covariate misclassification, we will use an example of a study that examined whether there was an association between prenatal folic acid vitamins and having twins. In unadjusted analyses, those who took vitamins were 2.4-fold more likely to have twins than those who did not. Use of *in vitro* fertilization (IVF) procedures increases the risk of twins and it was suspected that women undergoing IVF would be more likely to take folic acid supplements. Thus, IVF could be a strong confounder of the vitamins and twins relation. This study did not have data on use of IVF but used as a proxy "a period of involuntary childlessness." Adjustment for this variable yielded an adjusted risk ratio of 2.3, suggesting that IVF treatment was not a strong confounder of the relation between folic acid supplements and having twins. However, further data made available after the completion of this study indicated that involuntary childlessness was not a good proxy for IVF use. This proxy had a sensitivity of 60% and a specificity of 95% for IVF treatment [39], and these bias parameters were assumed to be nondifferential with respect to both vitamin use and having twins. The observed data are presented in Table 6.24 and the data bias-adjusted for the misclassification of IVF treatment by use of its proxy variable, compared with actual IVF treatment, are presented in Table 6.25. The observed date (Table 6.24) suggested an association between use of folic acid supplements and twin births in both the IVF+ stratum (RR = 4.0) and the IVF− stratum (RR = 1.7). In these strata, IVF+ was assigned by the proxy of a

Table 6.24 Observed data using a proxy for use of IVF treatment in a study estimating the effect of folic acid supplements on having twins.

| | | Total | | Proxy IVF+ | | Proxy IVF− | |
| | | Folic acid | | Folic acid | | Folic acid | |
		Yes	No	Yes	No	Yes	No
Twins	Yes	1319	5641	565	781	754	4860
	No	38,054	405,545	3583	21,958	34,471	383,588
	Total	39,373	411,186	4148	22,739	35,225	388,448
Risk ratio		2.4		4.0		1.7	
Standardized morbidity ratio				2.3			
Mantel-Haenszel risk ratio				2.2			

Data adapted from Berry et al., 2005 [39].

Table 6.25 Bias-adjusted data use of IVF treatment in a study estimating the effect of folic acid supplements on having twins.

| | | IVF+ | | IVF− | |
| | | Folic acid | | Folic acid | |
		Yes	No	Yes	No
Twins	Yes	907.4	907.2	411.6	4733.8
	No	3055.1	3055.8	34,998.9	402,490.2
	Total	3962.5	3963.0	35,410.5	407,224.0
Risk ratio		1.0		1.0	

Data adapted from Berry et al., 2005 [39].

period of involuntary childlessness and IVF− was assigned to those without such a period of involuntary childlessness. After bias-adjusting for the misclassification of IVF, the risk ratios in both strata of IVF were null. Assuming the values assigned to the bias parameters are accurate, misclassification of the IVF confounder created the appearance of an effect between folic acid and twins when there was truly no association. Figure 6.7 shows the screenshot from the Excel spreadsheet with this example.

Bias-adjustment for differential misclassification of covariates is conceptually the same as for nondifferential misclassification. However, in this case, sensitivity and specificity values must now be obtained for all possible combinations of exposure and disease. If exposure and disease are both dichotomous, there are up to eight bias parameters that must be specified—four sensitivities and four specificities:

$$SE_{Ex,Dy} = \Pr(C^* = 1 | C = 1, E = x, D = y)$$
$$SP_{Ex,Dy} = \Pr(C^* = 0 | C = 0, E = x, D = y)$$

where x and y can both be either 0 or 1 and C^* indicates the misclassified version of C. It is possible that the sensitivities and specificities depend on only the confounder and either exposure or disease, but not both in which case only four parameters would be needed. If sufficient data exist to specify values for these bias parameters, the equations in Table 6.23 can be employed to bias-adjust for misclassification of the covariate as in Table 6.26.

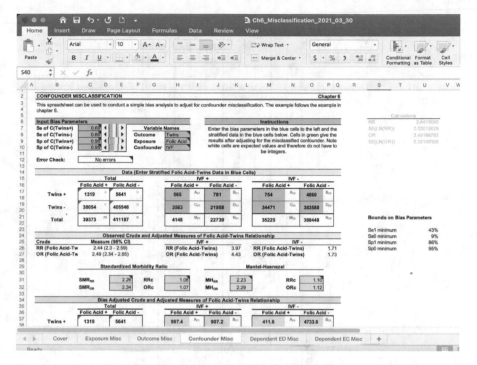

Figure 6.7 Screenshot of Excel spreadsheet for conducting simple bias analysis for misclassified covariates.

Overreliance on Nondifferential Misclassification of Covariates Biasing Toward the Null

As discussed in Chapter 5, the effect of a confounder on an association between an exposure and disease can be measured by the relative risk due to confounding (RR_{conf}). This bias parameter can be calculated by comparing the risk ratio with the risk ratio adjusted for the confounder by standardization (sRR) or pooling (such as with Mantel-Haenzel estimators, so long as these provide a reasonable estimate of the standardized result). Choice of the appropriate standardized estimate is important to compute a valid estimate of the strength of confounding (relative risk due to confounding, RR_{conf}, which was introduced in Chapter 5):

$$RR_{Conf} = \frac{RR_{crude}}{RR_{adj}}$$

In the example above (Table 6.24), the crude risk ratio relating folic acid intake to giving birth to twins was 2.4 and the RR adjusted for the proxy for IVF was 2.3. This result suggests that IVF was not an important confounder in the association between

Table 6.26 Equations for bias-adjusting observed data that are biased by differential covariate misclassification.

Observed data

	Total E_1	Total E_0	C_1 E_1	C_1 E_0	C_0 E_1	C_0 E_0
D_1	a	b	a_{C1}	b_{C1}	a_{C0}	b_{C0}
D_0	c	d	c_{C1}	d_{C1}	c_{C0}	d_{C0}

Adjusted data

	C_1 E_1	C_1 E_0	C_0 E_1	C_0 E_0
D_1	B $[a_{C1} - a(1 - SP_{E1D1})]/[(SE_{E1D1} - (1 - SP_{E1D1})]$	$[b_{C1} - b(1 - SP_{E0D1})]/[(SE_{E0D1} - (1 - SP_{E0D1})]$	$a - A_{C1}$	$b - B_{C1}$
D_0	$[c_{C1} - c(1 - SP_{E1D0})]/[(SE_{E1D0} - (1 - SP_{E1D0})]$	$[d_{C1} - d(1 - SP_{E0D0})]/[(SE_{E0D0} - (1 - SP_{E0D0})]$	$c - C_{C1}$	$d - D_{C1}$
Total	$A_{C1} + C_{C1}$	$B_{C1} + D_{C1}$	$A_{C0} + C_{C0}$	$B_{C0} + D_{C0}$

use of folic acid during pregnancy and having twins ($RR_{conf} = 1.08$). After bias-adjusting for the misclassification of IVF, the RR_{adj} equals 1.0, so $RR_{conf} = 2.4$. This change in the relative risk due to confounding with the bias-adjustment for misclassification indicates that the entire observed association between folic acid use and having twins may have been due to confounding by IVF use. In this situation, misclassification of the confounder prevented the adjustment from removing the confounding, so residual confounding bias remained. As a general rule, nondifferential misclassification of a confounder leaves residual confounding after adjusting for the misclassified confounder, and biases the relative risk due to confounding, not the adjusted estimate itself, toward the null [40]. Therefore, an estimate of association adjusted for a misclassified covariate will lie somewhere between the crude result (i.e., the estimate of association not adjusted for the confounder) and the true adjusted result (i.e., the estimate of association that one would obtain when adjusting for the correctly classified confounding). The estimate adjusted for the misclassified confounder is therefore biased away from the truth in the direction of the confounding. If the correctly classified confounder was available, adjusting for it would continue to move the estimate in the same direction as adjustment for the misclassified confounder; however, it is impossible to know how much further. In this example, the crude risk ratio was 2.4, the estimate adjusted using the bias-adjusted measure of IVF was a risk ratio of 1.0, and the estimate adjusted for the imperfect measure of IVF was a risk ratio of 2.3, which lies between the first two estimates. The relative risk due to confounding was biased toward the null from 2.4 to 1.08.

Unlike nondifferential misclassification of a covariate, differential misclassification leads to an unpredictable bias of the relative risk due to confounding, which can still be bias-adjusted using the methods described above. Quantitative bias analysis is a particularly useful tool in this setting, because of the unpredictable direction of the residual confounding.

To this point, we have considered the impact of a covariate that is acting as a confounder. If the covariate is an effect measure modifier, then nondifferential misclassification of either the exposure or the modifier in an analysis of interaction can create the appearance of interaction when none truly exists or can mask interaction when it truly exists [40]. In the example of folic acid and twins, if we hypothesized that IVF treatment could modify the effect of folic acid intake on risk of giving birth to twins, then relying on the data from the misclassified variable for IVF would have suggested that IVF was a modifier, while in truth the risk between folic acid and twins does not depend on IVF status.

Dependent Misclassification

As discussed earlier, dependent misclassification occurs when study participants misclassified with respect to one axis of the association or analysis (e.g., exposure) are more likely than those not misclassified with respect to that axis to also be

Table 6.27 Example of dependent misclassification.

D = 1, E = 1	D* = 1	D* = 0	D = 1, E = 0	D* = 1	D* = 0
E* = 1	0.95	0.02	E* = 1	0.02	0.01
E* = 0	0.02	0.01	E* = 0	0.95	0.02
D = 0, E = 1	D* = 1	D* = 0	D = 0, E = 0	D* = 1	D* = 0
E* = 1	0.02	0.95	E* = 1	0.01	0.02
E* = 0	0.01	0.02	E* = 0	0.02	0.95

The classification matrix indicates the probability of observations being classified (*) in different cells conditional on their true exposure (E) and disease (D) status.

misclassified with respect to a second axis of the association (e.g., disease). That is, those whose disease status is misclassified are more likely to have a misclassified exposure status than those who have a correctly classified disease status. Such misclassification can be either differential or nondifferential. The important distinction between differential versus nondifferential and dependent versus independent misclassification is that with the former we are focused on whether the probability of misclassification on one variable differs with the true *value* of a second. With the latter, we are concerned with whether the probability of misclassification of one variable depends on the *misclassified value* of the second variable. So misclassification of an exposure (outcome) can be nondifferential and dependent if overall SE and SP of exposure do not differ within levels of the true outcome (exposure) but the probability of being misclassified on both exposure and outcome is greater than we would expect based on the independent values of SE and SP of the exposures and outcomes if they had been independent.

Suppose for example, that the pattern of misclassification is as in Table 6.27, which shows how subjects are classified with respect to their true exposure-outcome combination. The probability of correct classification for each cell of the 2×2 table is $pr(E^* = x, D^* = y | E = x, D = y)$ where x and y can take on the value 0 or 1. In this case, we can calculate the expected sensitivity of exposure among those with the disease as $pr(E^* = 1, D^* = 1 | E = 1, D = 1) + pr(E^* = 1, D^* = 0 | E = 1, D = 1) = 0.95 + 0.02 = 0.97$ and the expected sensitivity of exposure among those without the disease is $pr(E^* = 1, D^* = 0 | E = 1, D = 0) + pr(E^* = 1, D^* = 1 | E = 1, D = 0) = 0.95 + 0.02 = 0.97$. The same approach can be used to calculate the specificity of exposure classification within levels of the outcome, and we again see that specificity is 0.97 among those with and without the outcome. Thus in this case, exposure is nondifferentially misclassified with respect to the outcome. Following the same logic, we can see that the outcome is also nondifferentially misclassified with respect to the exposure, with both sensitivity and specificity of outcome classification equal to 0.97. Given that we have misclassification of the exposure and the outcome, we can calculate the four probabilities of joint misclassification with respect to exposure and outcome, $pr(E^* = 1, D^* = 1 | E = 0, D = 0)$, $pr(E^* = 1, D^* = 0 | E = 0, D = 1)$, $pr(E^* = 0, D^* = 1 | E = 1, D = 0)$ and $pr(E^* = 0, D^* = 0 | E = 1, D = 1)$, that would be expected under independence. The expected values of these probabilities can be calculated by multiplying the probability of misclassification on one axis and the other. In all cases, the expected probability is 0.0009 (0.03 × 0.03). In reality we see that the probability of being doubly misclassified regardless of the true values of exposure and disease is 0.01. Thus, in this case we have dependent, but nondifferential misclassification, of exposure and disease.

Small amounts of dependent misclassification can have a large impact on estimates of effect [4], and the direction of the bias is not predictable without formal quantitative bias analysis. Unlike the misclassification errors previously discussed in this chapter, errors from dependent misclassification are more difficult to model in the analysis phase. However, since a major source of dependent errors arises when the same method is used to ascertain information on more than one variable, such as exposure and disease status, the best option is to design studies that prevent the risk of this error. For example, an interview used to collect data on exposure and disease would be susceptible to dependent errors, but if possible, the disease information could be collected by medical record review to remove the risk of dependent errors.

In the situations where there is no option of preventing this error in the design, Kristensen (1992) describes a set of equations to calculate bias-adjusted risk ratios [4]. One barrier to the utility of these equations is that they require estimates of the probability of misclassification in many different directions. For example, what proportions of respondents who are truly unexposed and undiseased are classified as (1) unexposed and diseased, $pr(E^* = 0, D^* = 1 | E = 0, D = 0)$, (2) exposed and undiseased, $pr(E^* = 1, D^* = 0 | E = 0, D = 0)$, and (3) exposed and diseased, pr $(E^* = 1, D^* = 1 | E - 0, D = 0)$? There are seldom validation data to provide estimates of these probabilities.

In the absence of estimates of the complete set of probabilities of misclassification, one option is to calculate the extent of misclassification that would be necessary to produce a specific odds ratio, for example, a null result. Then the analyst can consider whether it is plausible that the calculated misclassification occurred in the study. As an example, in a study of self-reported neighborhood problems and functional decline, individuals with one or more neighborhood problems were 2.2-fold more likely to report functional decline over a year of follow-up [41]. Both the neighborhood characteristics and functional decline were assessed by an interview with the participants, so it is possible that the personality characteristics that would lead an individual to overstate the problems in their neighborhood may also lead them to overstate their functional decline. We used the equations described by Kristensen (1992) [4] to find the minimum change in cell count necessary to remove the association between neighborhood characteristics and functional decline [42]. If there was truly no association between neighborhood problems and functional decline, it would only take 1.6% of the study population to have dependent errors—such that they were misclassified as having neighborhood problems and functional decline—to create the observed effect. Table 6.28 displays the association extrapolated from information in the original paper and the hypothetical 2 × 2 table if 1.6% of the population had dependent errors. While this approach can be compelling, it is complicated by the fact that many different combinations of movement between cells could lead to a null result when bias-adjusted and it is often difficult to determine which are plausible.

Table 6.28 Observed and hypothetical data from a study of neighborhood characteristics and functional decline.

	Observed data		Hypothetical data	
	Neighborhood problems		Neighborhood problems	
	Yes	No	Yes	No
FD	20	28	7.0	28.9
No FD	152	639	156.9	646.7
Total	172	667	163.9	675.6
Risk	0.12	0.042	0.043	0.043
Risk ratio	2.8	Reference	1.0	Reference

Original data from Balfour and Kaplan, 2002 [41]. Hypothetical data from Lash and Fink, 2003 [42]. FD = functional decline.

Identifying the structure of dependent error that would produce a null result is complex because the number of bias parameters increases to 12 (for each of the 4 cells in the 2 × 2 table, the 3 probabilities of being misclassified with respect to exposure and outcome). As a result, we recommend using iterative methods, such as those discussed in Chapters 7 and 8, in which small amounts of dependent errors are simulated to see if even these small errors can make a variable with no association with the outcome appear to be a strong risk factor. Figure 6.8 shows a screenshot from an Excel spreadsheet for the preceding example. The table in Figure 6.8 titled "Single Simulated Corrected Data" is the result of one such random simulation, so does not match the results in Table 6.28.

The example above examined the effect of dependent exposure–disease misclassification on observed estimates of association. However, dependent exposure–covariate misclassification can also create unpredictable and substantial bias to observed study results and may be more common than exposure-disease dependent misclassification in longitudinal studies. Often, both exposure and covariate information are taken from a single data collection instrument that is based on respondent self-report [43]. The risk of dependent misclassification should be considered for studies with known strong confounders. In addition, for variables with a strong association to the outcome where there is also a plausible mechanism for respondents to over-report both the exposure and covariate, there would now be the appearance of confounding and therefore the covariate would be included in the analysis, thereby biasing the observed results. When possible, different sources of information should be used to collect information on the exposure or confounder, or a validation study should be conducted on the two variables of concern to estimate the parameters needed to conduct sensitivity analysis to be able to bias-adjust the observed estimate of association. Given that this may not be possible in all cases, it is worth thinking through what the most influential confounders are likely to be for any analysis and at a minimum use different methods from the exposure classification scheme to capture information on the key confounders.

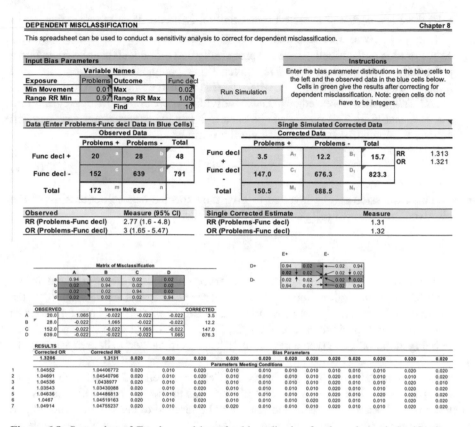

Figure 6.8 Screenshot of Excel spreadsheet for bias-adjusting for dependent misclassification. (Example from Table 6.28).

Matrix Method for Misclassification Adjustment

The methods presented above assume that the exposure, disease, and covariates were each dichotomous variables and the methods could only deal with one type of misclassification at a time. Here, we present a flexible method for dealing with polytomous variables, with misclassification patterns that can be differential and/or dependent and that allows simultaneous adjustment of misclassification of outcome, exposure and covariates [36, 44, 45]. The flexibility of this method is offset by the necessity of using a small amount of matrix algebra. However, the matrix method is more general than the approaches outlined above, while simplifying to be equivalent to those methods in simple cases.

To begin the matrix method for misclassification adjustment, we enumerate the cells of our observed data from 1 to K where order is not important but must be consistently maintained. For example, in a simple 2×2 table, the cells are: (1) exposed and diseased, (2) the unexposed and diseased, (3) the exposed and undiseased, and (4) the unexposed and undiseased, giving K = 4. For each of the K cells there is a vector of observed (potentially misclassified) cell frequencies, $F^* = [F_1^*, \ldots, F_K^*]'$, and a corresponding vector of true cell frequencies $F = [F_1, \ldots, F_K]'$. For each of the K cells, there is a probability that individuals are classified into a potentially misclassified cell m given that they truly belong in cell t, $p_{mt} = \Pr\,(Cell^* = m|\,Cell = t)$, where $Cell$ is the cell an individual truly belongs to and $Cell^*$ is their observed cell. These individual probabilities are aggregated into a $K \times K$ matrix, P. This matrix can be used to relate the expected misclassified cell counts, F^*, given true cell counts, F, via the formula:

$$F^* = PF \qquad\qquad (6.11)$$

This expression can be inverted to give the expected true cell counts given the observed misclassified cell counts, via the formula:

$$F = P^{-1}F^* \qquad\qquad (6.12)$$

For exposure misclassification, the observed cell counts are aggregated in the vector $F^* = [a, b, c, d]'$ and the true but unobserved cell counts are represented in the vector $F = [A, B, C, D]'$. The probabilities $\Pr(Cell^* = m|\,Cell = t)$ are functions of sensitivities and specificities. For example, $\Pr(Cell^* = 1|Cell = 1)$ is the probability of being in the "a" cell given a person truly belongs to the "A" cell, or sensitivity. Pr $(Cell^* = 1|Cell = 2)$ is the probability of being in the "a" cell given a person truly belongs to the "B" cell, or 1-specificity. $\Pr(Cell^* = 1|Cell = 3)$ is the probability of being in the "a" cell given a person truly belongs to the "C" cell, which is 0 since there is no disease misclassification and it is impossible to classify an exposed and undiseased person as exposed and diseased. The P matrix, in the example of exposure misclassification, can be filled out as:

$$P = \begin{bmatrix} SE_{D1} & 1 - SP_{D1} & 0 & 0 \\ 1 - SE_{D1} & SP_{D1} & 0 & 0 \\ 0 & 0 & SE_{D0} & 1 - SP_{D0} \\ 0 & 0 & 1 - SE_{D0} & SP_{D0} \end{bmatrix}$$

The matrix method seems complex; however, for simple settings, it delivers familiar results. The matrix method in Equation 6.11 relates true data to misclassified data and is equivalent to the equations in Table 6.5. To see this, we can vector multiply the first row of P by F and obtain $A\,SE_{D1} + B\,(1 - SP_{D1})$ which is the equation for the "a" cell in Table 6.5. The remaining cells in the matrix method of 6.11 match the remaining 3 cells in Table 6.5. Further, Equation 6.12 is simply the matrix equivalent of the equations in Table 6.8. If we invert the matrix P and vector-multiply the first row of the inverted matrix by F^* we obtain $A = \frac{a - D(1 - SP_{D1})}{SE_{D1} + SP_{D1} - 1}$, which is identical to the "A" cell in Table 6.8. To demonstrate the equivalence of the matrix method to the equations in Table 6.5, we use this method to adjust for the exposure misclassification in the data in Table 6.9. The four cells of the contingency table comprise our cell counts, with $K = 4$. The observed data is aggregated in the vector $F^* = (a = 215, b = 1449, c = 668, d = 4296)'$. Using the sensitivity=128/ 164 = 0.78 and specificity=465/470 = 0.99 for both cases and controls, we complete the P matrix as:

$$P = \begin{bmatrix} 0.78 & 0.01 & 0 & 0 \\ 0.22 & 0.99 & 0 & 0 \\ 0 & 0 & 0.78 & 0.01 \\ 0 & 0 & 0.22 & 0.99 \end{bmatrix}$$

Using Equation 6.12, we estimate the true frequency distribution of exposure and outcome as: $F = P^{-1}F^* = (256.3, 1407.7, 799.1, 4164.9)'$, which is identical to the results we obtained in Table 6.9. This calculation is quickly carried out in R (with the syntax: solve(P)%*%Fstar). As in that table, we did not round until the final step. We note that for all P matrix designed with this formatting, the columns must sum to 1, which is a helpful check on whether the matrix is constructed correctly.

The matrix method can be used to adjust for outcome misclassification with only minor changes. Instead of the upper-left and lower-right quadrants of the P matrix being filled in, the upper-right and lower-left quadrants are filled in to allow people to be transferred between disease and undiseased cells but not change their exposure status. The P matrix for the disease misclassification in Table 6.20, where sensitivity=1750/3300 = 53.0% and specificity=12,075/12,259 = 98.5% is:

$$P = \begin{bmatrix} 0.53 & 0 & 0.015 & 0 \\ 0 & 0.53 & 0 & 0.015 \\ 0.47 & 0 & 0.985 & 0 \\ 0 & 0.47 & 0 & 0.985 \end{bmatrix}$$

Letting $F^* = (a = 4558, b = 3428, c = 46,305, d = 46,085)'$, then $F = P^{-1}F^* = (7363.9, 5210.3, 43499.1, 44302.7)'$, which is identical to the bias-adjusted result in Table 6.20.

The power of the matrix method is that it can be extended, in a mathematically simple way, to accommodate categories with more than two levels and can simultaneously account for misclassification of exposure and disease. For example, with the same data as in Table 6.20, suppose exposure is misclassified as well as the outcome. Outcome misclassification remains nondifferential, but we need new notation do distinguish exposure and outcome misclassification so we say sensitivity and specificity of outcome classification are: $SE_1^D = SE_0^D = 1750/3300 = 0.53$ and $SP_1^D = SP_0^D = 12,075/12,259 = 0.985$. We also want to adjust for differential exposure misclassification of gender (male = 1, female = 0). Record abstraction may be more thorough among cases, resulting in $SE_1^e = 0.99$ and $SP_1^e = 0.97$, while among controls $SE_0^e = 0.97$ and $SP_0^e = 0.96$. Our P matrix is filled in by considering each of the 16 p_{mt}. For instance, $\Pr(Cell^* = 1 | Cell = 1)$ is the probability someone remains in the "a" cell if they are truly in the "A" cell. If we assume independent misclassification, this results in $SE_1^E \cdot SE_1^D = 0.99 \cdot 0.53 = 0.52$. Each of the cells are evaluated in the same way resulting in a P matrix of:

$$P = \begin{bmatrix} 0.99 \cdot 0.53 & 0.03 \cdot 0.53 & 0.97 \cdot 0.015 & 0.04 \cdot 0.015 \\ 0.01 \cdot 0.53 & 0.97 \cdot 0.53 & 0.03 \cdot 0.015 & 0.96 \cdot 0.015 \\ 0.99 \cdot 0.47 & 0.03 \cdot 0.47 & 0.97 \cdot 0.985 & 0.04 \cdot 0.985 \\ 0.01 \cdot 0.47 & 0.97 \cdot 0.47 & 0.03 \cdot 0.985 & 0.96 \cdot 0.985 \end{bmatrix}$$

Letting $F^* = (a = 4558, b = 3428, c = 46{,}305, d = 46{,}085)'$, as above, we have $F = P^{-1}F^* = (7277.8, 5296.4, 42996.8, 44805.0)'$ resulting in a RR $= 1.37$, somewhat attenuated from the RR attained when we only adjusted for exposure misclassification. As before, we did not round any results until the final step.

Finally, if an analyst has information on dependent misclassification, it can be seamlessly integrated into the P matrix. While the equations are relatively simple, the difficult issue is whether the analyst has sufficient knowledge to estimate the probability of misclassification for each cell of the table. The examples so far have implied that a study only had one source of bias from misclassification. For example, the study of smoking and the risk of breast cancer used the cancer registry to identify cases. The cancer registry requires histological confirmation of cancer diagnoses, so disease misclassification was thought to be minimal and there was no evidence of substantial confounder misclassification. Therefore, in this example, exposure misclassification was the source of bias that was most likely to have affected the results.

When an analysis is susceptible to more than one set of classification errors, the corrections above can be performed sequentially, so long as the classification errors are independent, or performed at the same time using the matrix method above. However, if the errors are dependent, then these methods will not be valid when applied sequentially and the method described by Equation 6.12 must be implemented.

Multidimensional Bias Analysis for Misclassification

As discussed in previous chapters, when implementing quantitative bias analysis, there will be uncertainty in what value should be assigned to the bias parameters. Earlier, we introduced a case-control study of maternal smoking and breast cancer with an observed conventional OR $= 0.95$ (Table 6.9). Concern was raised about the potential for misclassification of self-reported maternal smoking. A quantitative bias analysis was conducted using a sensitivity of 0.78 and specificity of 0.99, and we found a bias-adjusted OR $= 0.95$. However, there was still concern regarding the uncertainty in the values assigned to the sensitivity and specificity parameters of the

Table 6.29 Multiple bias analysis with various values of sensitivity and specificity of classification of maternal smoking status using the results of a study of the association between maternal smoking during pregnancy and breast cancer risk.

Sensitivity	Specificity	Conventional estimate	Bias adjusted estimate
0.68	0.99	0.95	0.95
0.78	0.99	0.95	0.95
0.88	0.99	0.95	0.95
0.68	0.95	0.95	0.93
0.78	0.95	0.95	0.93
0.88	0.95	0.95	0.93
0.68	0.90	0.95	0.84
0.78	0.90	0.95	0.84
0.88	0.90	0.95	0.84

Original data from Fink and Lash, 2003 [14].

bias model. In Chapters 7, 8, 9, 10 and 11, we will illustrate how to incorporate that uncertainty in a comprehensive fashion. A simpler approach, however, is to conduct a multidimensional bias analysis, which repeats the previous bias analysis at various sensitivity and specificity values. To illustrate the process, we choose sensitivities of 0.68, 0.78, and 0.88, as well as specificities of 0.99, 0.95 and 0.90, and compute bias-adjusted ORs for the 9 combinations of these parameters, using the formulae in Table 6.8. Results are shown in Table 6.29. The result of this simple multidimensional bias analysis demonstrate the importance of the specificity parameter in this example. When the specificity is 0.99, the bias-adjusted OR is 0.95 regardless of which sensitivity was chosen. When the specificity was 0.95, the bias-adjusted OR = 0.93 regardless of which sensitivity was chosen and when the specificity is 0.90, the bias-adjusted OR = 0.84 regardless of sensitivity. Because it is unlikely that women who truly do not smoke report having smoked, OR = 0.95 is more likely and the bias-adjusted result varies only a little over the values of sensitivity chosen. This type of multidimensional bias analysis can help address the sensitivity of the bias analysis to the values assigned to the bias parameters. To show this approach graphically, we bias-adjusted the observed odds ratio for a range of sensitivity values of 0.5 to 1.0, while holding the specificity constant at 1, 0.99, 0.95 and 0.90. Figure 6.9 displays the results of this multiple bias analysis.

Figure 6.10 displays a screenshot of the table generated by this same multiple bias analysis result in an Excel spreadsheet available at the text's website. The Figure expands the values assigned to the bias parameters by allowing differential misclassification. This aspect shows that if differential misclassification was likely (contrary to the study's design) and the sensitivity of smoking classification was higher for the cases than the controls (one possible manifestation of recall bias), then it is possible that there was a weak positive association between maternal smoking and breast cancer risk that was masked by the smoking misclassification. Given the study's design protection against differential misclassification, however, this scenario seems unlikely.

Figure 6.9 Multiple bias analysis resulting from assigning values ranging from 0.5 to 1 to the sensitivity of classification of maternal smoking status, holding specificity equal to 1 (black line), 0.99 (red line), 0.95 (green line), 0.9 (blue line), using the results of a study of the association between maternal smoking during pregnancy and breast cancer risk. (Original data from Fink and Lash, 2003 [42]).

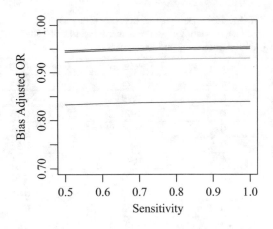

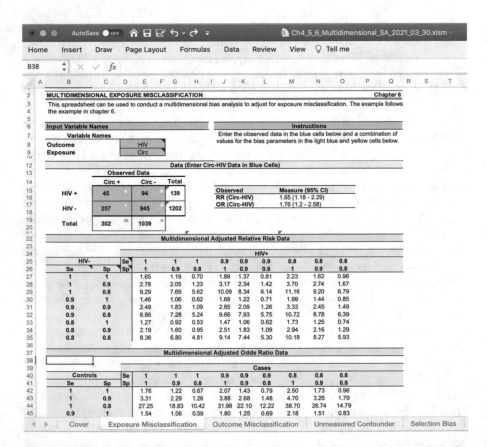

Figure 6.10 Screenshot of Excel spreadsheet to perform multidimensional bias analysis for misclassification of exposure. Original data from Fink and Lash, 2003 [42]. Similar rationale and methodologies could be applied to research studies with potential misclassification of outcome or a confounder.

Limitations

Negative Expected Cell Frequencies

One possible outcome of applying the equations that use sensitivity and specificity for any of the bias adjustments above is that they can result in negative expected cell counts. There are combinations of sensitivity and specificity for any observed table that will produce negative expected cell counts. We refer to sensitivity and specificity values that produce negative expected cell counts as inadmissible and those that produce non-negative expected cell counts as admissible. It is possible to derive bounds around admissible sensitivity and specificity values [15, 46]. For exposure misclassification, let $Pr(E^* = 1|D = d)$ be the observed exposure probability within level d of the disease status. Two sets of bounds can be derived for admissible values of sensitivity and specificity among cases and controls:

$$\Pr(E^* = 1|D = d) \le SE_d \le 1 \ \ and \ \ 1 - \Pr(E^* = 1|D = d) \le SP_d \le 1$$

or

$$0 \le SE_d \le \Pr(E^* = 1|D = d) \ \ and \ \ 0 \le SP_d \le 1 - \Pr(E^* = 1|D = d)$$

The above bounds are obtained under differential misclassification and there are separate bounds for cases and controls. However, if there is nondifferential misclassification, then we have one overall set of bounds that is the intersection of the two bounding sets above:

$$\max \left[\Pr(E^* = 1|D = 1), Pr(E^* = 1|D = 0) \right] \le Se \le 1$$
$$1 - \min \left[\Pr(E^* = 1|D = 1), Pr(E^* = 1|D = 0) \right] \le Sp \le 1$$

or

$$0 \le Se \le \min \left[\Pr(E^* = 1|D = 1), Pr(E^* = 1|D = 0) \right]$$
$$0 \le Sp \le 1 - \max \left[\Pr(E^* = 1|D = 1), Pr(E^* = 1|D = 0) \right]$$

In both cases, the second expression involves low sensitivity and specificity estimates and is less likely to pertain to applied research. Typically, values of sensitivity and specificity that fall within the first set of bounds will be most applicable. Similar bounds can be derived for disease misclassification. The important aspect of these bounds is that at the limit (say, when $SP_d = 1 - Pr(E^*|D = 1)$), one of the cells in the adjusted 2×2 table will equal 0. For example, in Table 6.9, the probability of smoking among cases, $Pr(E^* = 1|D = 1)$, is 215/1664 or approximately 12.9%. If the specificity among cases was $100 - 12.9 = 87.1\%$, there would be 0 exposed cases in the bias-adjusted table. This result corresponds to the left hand bound on the specificity in the first set of expressions above. If the specificity among cases was less than 87.1%, the number of exposed cases in the bias-adjusted table would be less than 0. Earlier we noted that specificity could have a substantial impact on model results when exposure prevalence is low. Notice in the equations above that the lower bound on the specificity is typically 1 minus the exposure prevalence. A low exposure prevalence equates to a high bound and, therefore, a small change in the specificity could result in a bias-adjusted cell count that approaches zero and a large change in the effect estimate.

The occurrence of negative cell counts signals a fundamental incompatibility between the observed data and the specified bias parameters. An analyst who obtains negative bias-adjusted cell counts from a bias model should carefully consider two possible explanations. First, it is possible that the chosen values of sensitivity and specificity were inappropriate for the target population of interest. Consider a hypothetical study of the effect of self-reported maternal smoking on birth defects in the United States in 2017. If estimates of sensitivity and specificity were derived from the published literature on reliability of self-reported smoking versus a gold standard in Europe in 1987, this transport of validation results could hinder the external validity of the estimates, since attitudes regarding smoking were so different in the two populations. In this instance, analysts should attempt to find more valid estimates of sensitivity and specificity for their population. Second, it is possible that the data the analyst has collected are invalid in some important way. For example, there could be confounding or selection bias that needs to be taken into account to apply the estimates of sensitivity and specificity. The data may also have deviated substantially from expectation by chance, as described above. This concern would apply most often to relatively small studies or studies with sparse data in some combination of exposure and disease. In this circumstance, one might have valid estimates of sensitivity and specificity, but their application to classifications in the single study sample deviated from expectation, so when the valid estimates are applied against the observed data, negative cell frequencies arise. Unfortunately, there is no diagnostic signal for which of the two possibilities best explains the bias-adjusted negative cell frequency, and the two can work in conjunction with one another. Concerns about transportability of the validation data favor the first explanation and small sample sizes or threats to validity favor the second explanation, but one can never be sure of the explanation in practice. We address these topics in greater detail in Chapters 7, 8 and 11.

Other Considerations

The principal limitation surrounding the methods described in this chapter to bias-adjust for exposure misclassification is that they rely upon the analyst to correctly specify the values assigned to the bias parameters. Probabilistic bias analysis, which will be presented Chapters 7, 8 and 9, will relax this assumption. While it may seem reasonable to assume that a small deviation between the specified bias parameter and the true bias parameter would have little impact on the bias-adjusted effect estimate, this is not always the case. We earlier showed how values assigned to specificity dramatically affect the strength of the modeled bias when the exposure prevalence (or disease occurrence, for disease misclassification problems) is low. In that circumstance, the value assigned to the sensitivity may have little effect on the strength of the modeled bias. As a second example, when the sensitivity or specificity values are close to the bounds given above, a small change in the bias parameter could result in a very large change in the bias-adjusted effect estimate. The reason for this is that while a small change in the bias parameter may increase (or decrease) the number of people in a cell in the expected true table by only a small amount, if the number of people in a cell is already small, adding (or subtracting) a small number could have a substantial impact. For this reason, analysts should be very careful about specifying values assigned to bias parameters, particularly if they are close to one of the bounds given above (or, alternatively, if their bias-adjusted contingency table has a small number in a cell). Good practice for all exposure misclassification analyses will also vary the values assigned to the bias parameters in a sensitivity analysis to gauge what impact it has on the resulting adjusted estimate [13], as we discuss throughout this text.

References

1. Willett WC, Sampson L, Stampfer MJ, Rosner B, Bain C, Witschi J, et al. Reproducibility and validity of a semiquantitative food frequency questionnaire. Am J Epidemiol. 1985;122:51–65.
2. Bodnar LM, Siega-Riz AM, Simhan HN, Diesel JC, Abrams B. The impact of exposure misclassification on associations between prepregnancy body mass index and adverse pregnancy outcomes. Obes. 2010;18:2184–90.
3. MacLehose RF, Olshan AF, Herring AH, Honein MA, Shaw GM, Romitti PA. Bayesian methods for correcting misclassification an example from birth defects epidemiology. Epidemiol. 2009;20:27–35.
4. Kristensen P. Bias from nondifferential but dependent misclassification of exposure and outcome. Epidemiology. 1992;3:210–5.
5. Hernan MA, Cole SR. Causal diagrams and measurement bias. Am J Epidemiol. 2009;170:959–62.
6. Agresti A, Caffo B. Simple and effective confidence intervals for proportions and differences of proportions result from adding two successes and two failures. Am Stat. 2000;54:280–8.
7. Agresti A, Coull BA. Approximate is better than "exact" for interval estimation of binomial proportions. Am Stat. 1998;52:119–26.
8. Piper JM, Mitchel EF Jr, Snowden M, Hall C, Adams M, Taylor P. Validation of 1989 Tennessee birth certificates using maternal and newborn hospital records. Am J Epidemiol. 1993;137:758–68.
9. Collin LJ, MacLehose RF, Ahern TP, Nash R, Getahun D, Roblin D, et al. Adaptive Validation Design. Epidemiol. 2020;31:509–16.
10. Collin LJ, Riis AH, MacLehose RF, Ahern TP, Erichsen R, Thorlacius-Ussing O, et al. Application of the adaptive validation substudy design to colorectal cancer recurrence. Clin Epidemiol. 2020;12:113–21.
11. Johnson SR, Tomlinson GA, Hawker GA, Granton JT, Feldman BM. Methods to elicit beliefs for Bayesian priors: a systematic review. J Clin Epidemiol. 2010;63:355–69.

12. Kadane J, Wolfson LJ. Experiences in elicitation. J R Stat Soc Ser Stat. 1998;47:3–19.
13. Lash TL, Fox MP, MacLehose RF, Maldonado G, McCandless LC, Greenland S. Good practices for quantitative bias analysis. Int J Epidemiol. 2014;43:1969–85.
14. Fink AK, Lash TL. A null association between smoking during pregnancy and breast cancer using Massachusetts registry data (United States). Cancer Causes Control. 2003;14:497–503.
15. Gustafson P, Le ND, Saskin R. Case–control analysis with partial knowledge of exposure misclassification probabilities. Biometrics. 2001;57:598–609.
16. Chu H, Wang Z, Cole SR, Greenland S. Sensitivity analysis of misclassification: a graphical and a Bayesian approach. Ann Epidemiol. 2006;16:834–41.
17. Marshall RJ. Validation study methods for estimating exposure proportions and odds ratios with misclassified data. J Clin Epidemiol. 1990;43:941–7.
18. Bodnar LM, Abrams B, Bertolet M, Gernand AD, Parisi SM, Himes KP, et al. Validity of birth certificate-derived maternal weight data. Paediatr Perinat Epidemiol. 2014;28:203–12.
19. Lash TL, Abrams B, Bodnar LM. Comparison of bias analysis strategies applied to a large data set. Epidemiol. 2014;25:576–82.
20. MacLehose RF, Bodnar LM, Meyer CS, Chu H, Lash TL. Hierarchical semi-Bayes methods for misclassification in perinatal epidemiology. Epidemiology. 2018;29:183–90.
21. Brenner H, Gefeller O. Use of the positive predictive value to correct for disease misclassification in epidemiologic studies. Am J Epidemiol. 1993;138:1007–15.
22. Chang ET, Smedby KE, Hjalgrim H, Porwit-MacDonald A, Roos G, Glimelius B, et al. Family history of hematopoietic malignancy and risk of lymphoma. J Natl Cancer Inst. 2005;97:1466–74.
23. Shaw GM, Wasserman CR, O'Malley CD, Nelson V, Jackson RJ. Maternal pesticide exposure from multiple sources and selected congenital anomalies. Epidemiology. 1999;10:60–6.
24. Rull RP, Ritz B, Shaw GM. Validation of self-reported proximity to agricultural crops in a case-control study of neural tube defects. J Expo Sci Env Epidemiol. 2006;16:147–55.
25. Greenland S. Variance estimation for epidemiologic effect estimates under misclassification. Stat Med. 1988;7:745–57.
26. Greenland S, Gustafson P. Accounting for independent nondifferential misclassification does not increase certainty that an observed association is in the correct direction. Am J Epidemiol. 2006;164:63–8.
27. Wacholder S, Hartge P, Lubin JH, Dosemeci M. Non-differential misclassification and bias towards the null: a clarification. Occup Env Med. 1995;52:557–8.
28. Jurek AM, Greenland S, Maldonado G, Church TR. Proper interpretation of non-differential misclassification effects: expectations vs observations. Int J Epidemiol. 2005;34:680–7.
29. Loken E, Gelman A. Measurement error and the replication crisis. Science. 2017;355:584–5.
30. Dosemeci M, Wacholder S, Lubin JH. Does nondifferential misclassification of exposure always bias a true effect toward the null value? Am J Epidemiol. 1990;132:746–8.
31. Weinberg CR, Umbach DM, Greenland S. When will nondifferential misclassification of an exposure preserve the direction of a trend? Am J Epidemiol. 1994;140:565–71.
32. Mokdad AH, Ford ES, Bowman BA, Dietz WH, Vinicor F, Bales VS, et al. Prevalence of obesity, diabetes, and obesity-related health risk factors, 2001. JAMA. 2003;289:76–9.
33. Yun S, Zhu BP, Black W, Brownson RC. A comparison of national estimates of obesity prevalence from the behavioral risk factor surveillance system and the National Health and Nutrition Examination Survey. Int J Obes. 2006;30:164–70.
34. Ogilvie RP, MacLehose RF, Alonso A, Norby FL, Lakshminarayan K, Iber C, et al. Diagnosed Sleep Apnea and Cardiovascular Disease in Atrial Fibrillation Patients: The Role of Measurement Error from Administrative Data. Epidemiology. 2019;30:885–92.
35. De Henauw S, de Smet P, Aelvoet W, Kornitzer M, De Backer G. Misclassification of coronary heart disease in mortality statistics. Evidence from the WHO-MONICA Ghent-Charleroi Study in Belgium. J Epidemiol Community Health. 1998 Aug;52(8):513–9.
36. Greenland S, Kleinbaum DG. Correcting for misclassification in two-way tables and matched-pair studies. Int J Epidemiol. 1983;12:93–7.

37. Jurek AM, Maldonado G, Greenland S. Adjusting for outcome misclassification: the importance of accounting for case-control sampling and other forms of outcome-related selection. Ann Epidemiol. 2013;23:129–35.
38. Brenner H, Savitz DA. The effects of sensitivity and specificity of case selection on validity, sample size, precision, and power in hospital-based case-control studies. Am J Epidemiol. 1990;132:181–92.
39. Berry RJ, Kihlberg R, Devine O. Impact of misclassification of in vitro fertilisation in studies of folic acid and twinning: modelling using population based Swedish vital records. BMJ. 2005;330):815.
40. Greenland S. The effect of misclassification in the presence of covariates. Am J Epidemiol. 1980;112:564–9.
41. Balfour JL, Kaplan GA. Neighborhood environment and loss of physical function in older adults: evidence from the Alameda County Study. Am J Epidemiol. 2002;155:507–15.
42. Lash TL, Fink AK. Re: "Neighborhood environment and loss of physical function in older adults: evidence from the Alameda County Study". Am J Epidemiol. 2003;157:472–3.
43. Brennan AT, Getz KD, Brooks DR, Fox MP. An underappreciated misclassification mechanism: implications of nondifferential dependent misclassification of covariate and exposure. Ann Epidemiol. 2021;58:104–23.
44. Barron B. The effects of misclassification on the estimation of relative risks. Biometrics. 1977;33:414–8.
45. Greenland S, Lash TL. Bias Analysis. In: Rothman KJ, Greenland S, Lash TL, editors. Modern Epidemiology. 3rd ed. Philadelphia: Lippincott Williams & Wilkins; 2008. p. 345–80.
46. MacLehose RF, Gustafson P. Is probabilistic bias analysis approximately Bayesian? Epidemiology. 2012;23:151–8.

Chapter 7
Preparing for Probabilistic Bias Analysis

Introduction

To this point, we have assigned only a single set of fixed values to the bias parameters of bias models (*i.e.,* simple bias analysis, see Chapters 4, 5, and 6) or combinations of fixed values (*i.e.,* multidimensional bias analysis, also in Chapters 4, 5, and 6). Simple bias analysis is an improvement over conventional analyses, which implicitly assume that all the bias parameters are fixed at values that imply that there is no bias. However, simple bias analysis is limited by its assumption that the values of the bias parameters are known without error, which is never a valid assumption (1). Even when internal validation studies are conducted, the resulting bias parameters are measured with random error in a subsample, and often have their own sources of systematic error (see Chapter 3). Multidimensional bias analysis improves on simple bias analysis by examining the impact of more than one set of values for the bias parameters, but even this approach only examines the bias conferred by a limited set of values for the bias parameters. For any analysis, many other possible combinations of values are plausible, and a multidimensional analysis will not describe the impact of these possibilities. More important, multidimensional analysis gives no sense of which bias-adjusted estimate of effect is the most likely under the assumed bias model, which can make interpretation of the results challenging.

One strategy to address the limitations of simple and multidimensional bias analysis is to use probabilistic bias analysis (2–7). With probabilistic bias analysis, rather than specifying one set of values for the bias parameters, or a limited number of sets, the analyst assigns probability distributions for each of the bias parameters and then uses Monte Carlo methods to generate a distribution of bias-adjusted estimates of effect. By "Monte Carlo methods" we mean that we randomly sample values for the bias parameters from a probability distribution specified by the analyst (8), and then use these chosen values to conduct a single simple bias analysis. We then repeat this process many times, each time randomly sampling a set of values for

© Springer Nature Switzerland AG 2021
M. P. Fox et al., *Applying Quantitative Bias Analysis to Epidemiologic Data*,
Statistics for Biology and Health, https://doi.org/10.1007/978-3-030-82673-4_7

the bias parameters and bias-adjusting the estimate of association. The validity of the analysis therefore relies substantially on the validity of the probability distributions assigned to the bias parameters. After we have computed a large number of bias-adjusted estimates, inference can proceed by examining the distribution of these estimates. For instance, it is common to report the median of this distribution as the point estimate of effect and the 2.5^{th} and 97.5^{th} percentile as limits of the "simulation intervals." This process will be described in Chapters 8 and 9.

Probabilistic bias analysis therefore extends the methods explained in the earlier chapters on simple and multidimensional bias analysis by repeatedly sampling from distributions of values assigned to bias parameters. Probabilistic bias analysis can be used for any of the simple bias analysis methods discussed in Chapters 4, 5, and 6. Misclassification, uncontrolled confounding, and selection bias problems can all be approached with a straightforward probabilistic bias analysis extension to the simple bias analysis methods that have already been discussed. In this chapter, we focus on the steps needed to prepare for a probabilistic bias analysis and the decisions that need to be made to complete one. The next two chapters focus on the steps needed to implement probabilistic bias analysis for either summary level data (Chapter 8) or record level data (Chapter 9).

Preparing for Probabilistic Bias Analysis

The most important first step is the same for all data analysis projects. Before modeling the data using probabilistic bias analysis, one should have a thorough understanding of the data including where missing data exist, where exposure and outcome data become sparse when stratified by other variables, and what measured variables are essential to adjust for in the analysis. This step, and especially the selection of variables for control (9,10), can be aided by use of directed acyclic graphs (see Chapter 2), but there is no replacement for the due diligence of examining simple cross tabulations of relevant variables.

In conventional analyses, analysts compute crude effect estimates before using regression or other methods to control for what they posit as a sufficient set of confounders. Analysts should be able to explain differences between crude and adjusted estimates of effect, presumably due to the analytic control for confounding. This advice extends beyond just the conventional methods for confounder control. Before engaging in probabilistic bias analysis, the analyst should first conduct a series of simple bias analyses (11). By conducting simple bias analyses, the analyst can anticipate what to expect for a median value from a probabilistic bias analysis. Deviations from this expectation may be valid but should at least trigger an evaluation to be sure there is a sound explanation. A preliminary simple bias analyses can be implemented by first assigning the median values of the bias distributions (as described below) to the parameters of the bias model and then applying the methods for simple bias-adjustment described in earlier chapters. If the analyst then implements a probabilistic bias analysis and finds that the median bias-adjusted value differs substantially from the result of the preliminary simple bias analysis, this difference could indicate that there is an error in coding, that the input distributions are producing illogical values (see the discussion of negative cell frequencies in Chapter 6), or that the distribution of the data is leading to realizations different from expectation. Whatever the explanation, the analyst will know to seek an explanation if she first has an expectation for the bias analysis from a series of simple bias analyses. Completing a series of simple bias analyses (multidimensional bias analysis) may also reveal what bias parameters or values assigned to bias parameters drive the bias analysis. For example, in a dataset with a rare outcome, small deviations in specificity of outcome classification can lead to dramatically different bias-adjusted results (see Chapter 6), whereas large changes in sensitivity might have little effect (12). Knowing these dependencies can help analysts explain results seen in the subsequent probabilistic bias analyses.

Probabilistic bias analysis requires that the analyst assign a distribution for the bias parameters. We discuss the technical details of specifying a bias parameter distribution below, and several important considerations need to be kept in mind in the planning phase. The final quality of a probabilistic bias analysis will be determined, to a large extent, by the plausibility of the bias distribution. Substantial thought should go into informing these distributions and it should be guided by a thorough literature review and expert knowledge, whenever possible. Further, the bias parameter generally should not depend on the observed data in the study under consideration, except for internal validation studies. This guideline implies that an excellent time to specify a bias distribution is prior to observing the results of a study. Unfortunately, this practice will seldom be possible as an analyst may not suspect the importance of a bias until seeing an unexpected conventional result. However, it is paramount that analysts not allow the desire to observe a particular result guide the specification of the bias distribution (13). The cognitive biases outlined in Chapter 1 make it all too easy to fall into this trap and awareness of these biases may be the best remedy to the problem.

Statistical Software for Probabilistic Bias Analysis

Probabilistic bias analysis can be implemented using freely available software in some cases, while in others, the analyst will want to adapt code we or others provide or write their own to create a bias analysis that best suits the analyst's needs. In some cases, because the approach requires running many (as many as 1 million) simulations, such analyses can require substantial computing time (14). When access to a computing cluster is available, this access can vastly reduce the amount of computing time. However, as computing speed increases, this has become less necessary. Many probabilistic bias analysis computing tools, particularly for summary level data, are available in common statistical packages. In Stata, the package *episens* will implement many of the analyses we outline here, as will the package *episensr* in R. For summary level bias analyses, we also provide Excel spreadsheets and R code.

When implementing probabilistic bias analysis for multiple sources of bias, there will rarely be existing software and the analyst will be required to write their own program. Often this can be accomplished with relatively little difficulty by "nesting" programs to implement bias-adjustment within one another. Similarly, bias analyses at the record level will generally require analysts to write their own programs. At the website for this book, we include many programs to implement bias analyses that were written in R, a programming language that is amenable to adapting code and running large simulations.

Summary Level Versus Record Level Probabilistic Bias Analysis

An important decision the analyst must make is deciding what type of data they will work with, summary level data (*e.g.,* a collapsed 2×2 table, or similar) or a dataset containing records of individuals, with information on study participant exposure and outcome status and often additional information on confounders, potential effect measure modifiers, and other variables. When analysts have access to primary data, both approaches are often options. Bias analysis applied to published papers, for which the bias analyst seldom has access to primary data, are often constrained to the summary level approach.

When both approaches are feasible, there are several considerations to make when choosing between them. Summary level probabilistic bias analysis has many advantages that make it appealing. Probabilistic bias analysis using summary level data is often easier to implement than probabilistic bias analysis using record level data. One reason is that it can often be conducted using pre-written code or software packages. At this book's website, we provide Excel spreadsheets in which summary level probabilistic bias analysis can be conducted and point the reader to Stata, R and shiny apps in which summary level probabilistic bias analysis can be implemented. Because only summary level data are required, it can also be implemented when exploring the impact of bias on the results of a study identified in the literature (15) or when conducting bias analysis on data for a meta-analysis (16).

Summary level probabilistic bias analysis will often require less computing time and computing power than record level probabilistic bias analysis, although marked differences in duration will really only become apparent for large datasets analyzed with machines that have relatively modest computing power given current standards. Simulations typically run in seconds for summary level data and could take several minutes, or even hours, for record level data, depending on the complexity of the data and of the bias model and on the computational capacity of the machine and the size of the original dataset. The computational efficiency of summary level bias analysis accrues because adjustments to biases occur at the aggregate level (something we discuss in greater detail in Chapter 8) rather than at each individual record (Chapter 9) (14). When implementing a bias-adjustment at the record level in a dataset of 100,000 individuals, the adjustment needs to proceed separately for each of the 100,000 observations. On the other hand, at the most basic summary level for an association, the 100,000 observations can be aggregated into four cell counts and the adjustment only need proceed through those four cells. In addition, simple bias-adjustments like these can often be summarized over strata using an adjusted measure like a standardized morbidity ratio or inverse-variance weighted summary estimate. In this case, bias-adjustments are made within the G strata over which the data will be summarized, so there are Gx2x2 cells into which the data are collapsed. No regression modelling is needed and as such, this summary approach may make the most sense when the analyst does not need to adjust for many measured confounders. Alternatively, the overall strength of confounding can be estimated using regression modeling by comparing the crude estimate with the estimate adjusted for a sufficient set of variables. As a reasonable approximation, bias-adjustments for misclassification or selection bias can then be applied to the crude 2×2 table, and the resulting bias-adjusted estimate can be further adjusted by applying the estimate of the strength of confounding to the bias-adjusted crude estimate.

Given these advantages, it may seem that summary level probabilistic bias analysis is the obvious approach to implement. Indeed, it may be the only choice if record level data are inaccessible. When the record level data are accessible, record level probabilistic bias analysis may be preferred given the modeling limitations of summary level probabilistic bias analysis. With summary level probabilistic bias analysis, it is difficult to take full advantage of regression modeling techniques, such as time-varying analyses, censoring, competing risk adjustment, and so on. Record level probabilistic bias analysis can accommodate the full range of regression modeling advantages by simulating the bias-adjustments for each record in the dataset. Adjusted datasets can then be used in regression models in the same way as the original data.

This advantage does come at the price of greater complexity. When using record level probabilistic bias analysis, one must apply bias-adjustments to each record within the dataset. If we wish to obtain 100,000 bias-adjusted estimates from a dataset of 1,000 individuals, the total dataset size becomes 100,000,000 observations. The computing time can become quite long and computers with more computing power may be necessary. One might find, though, that some of the 1000 records are identical (except for a personal identifier). For example, there may be many people in the dataset with the same pattern of covariate information (*e.g.*, gender, age category, unexposed, and free of the outcome). In this case, these records can be collapsed and assigned a weight equal to their frequency of appearance. Unique records receive a weight of one. This simplification can dramatically reduce the computational burden without sacrificing the advantages of record level bias analysis (14) (see Chapter 9).

A second consideration when choosing between summary level and record level approaches is that the bias models required to conduct record level probabilistic bias analysis to address misclassification bias may be different from a bias model for summary level data. For example, when implementing a bias-adjustment for misclassification, the simplest summary level adjustments (which are common in practice, but not what we introduce in Chapter 8) rely only on estimates of the sensitivity and specificity of classification. For more advanced summary level bias analyses, and record level bias analysis, the bias model must use the probability that the record contains a classification error (*i.e.*, the positive and negative predictive values). When only estimates of sensitivity and specificity of classification are available to inform the assignment of probability distributions, the predictive values must be computed by combining the values for sensitivity and specificity drawn from their respective distributions, with an estimate of the prevalence of the classified variable, as described in Chapters 8 and 9.

A third consideration is that applying the simplest bias model to summary level data may understate the total uncertainty (see further discussion in Chapter 8 of a summary level model that does not understate the total uncertainty). For example, consider a disease that is imperfectly classified, with a positive predictive value of 95%, estimated in a validation study in which 95 of 100 observed cases were confirmed by medical record review. In 1000 observed cases, we would expect 950 to be true cases, but there is uncertainty in this estimate. The first source of uncertainty derives from the validation data, which were obtained with (at least) random error. We can model random error in the bias parameter by assigning a probability distribution to the estimate obtained from the validation data. Rather than setting the positive predictive value to exactly 95%, we can assign a beta distribution (described below) with alpha=95 and beta=5, following directly from the validation study. This distribution has mean of 95% and 95% confidence interval limits of 90.0% and 98.3%. In Excel, the formula "=BETA.INV(RAND(),95,5)" returns a draw from this beta distribution. In 1000 simulations, we obtained a median positive predictive value of 95.3% and a 95% simulation interval from 90.2% to 98.4%. As expected in a sample of 1000 simulations, the distribution of results corresponds well to the assigned density. If we simply multiply the positive predictive value for each simulation by 1000 observed cases to obtain the number of simulated true cases, we obtain the corresponding values of median 953 simulated true cases, with 95% simulation interval from 902 to 984 simulated true cases.

Note, however, that if we had a positive predictive value of exactly 95%, we would not expect that exactly 950 out of 1000 individuals would be cases. Rather, 950 is the number we would expect on average. Random variability would cause any single study to have slightly more or less than 950 true cases. This is a second source of uncertainty. The positive predictive value would act on each of the 1000 individual cases, each with a probability of being a true case equal to the PPV. We can conceptualize that the actual data were collected with a coin flip for each record, only instead of a 50:50 coin, it was a PPV:(1−PPV) coin. There is uncertainty in the positive predictive value because it was measured in only 100 people, and there is uncertainty in the realization of the data (as a function of positive predictive values).

We can also simulate this second source of uncertainty. Once we have a value for the PPV, we can sample the misclassification-adjusted disease status from a binomial distribution. We note that this step is simply a type of imputation technique, which we discuss in detail in Chapter 10. We draw the positive predictive value from the beta distribution to model the first source of uncertainty, then we use that positive predictive value on all 1000 records to simulate the second source of uncertainty. We do not have to have access to the individual records to simulate the second source of uncertainty. For example, in Excel, the formula "=BINOM.INV(1000,0.95,RAND())" returns the number of simulated true positive cases among 1000 observed cases with positive predictive value fixed at 0.95, thus modeling only this second source of uncertainty. It is possible to simultaneously model both sources of uncertainty. In Excel, the formula "=BINOM.INV(1000,BETA.INV(RAND(),95,5),RAND())" models the first and second sources of uncertainty simultaneously by replacing 0.95 in the preceding formula with the formula for drawing from a beta distribution to assign the positive predictive value. Simulating both sources of uncertainty, we obtained a median 953 simulated true cases, with 95% simulation interval from 901 to 985 simulated true cases, which maps well to the expectation, given the distributions and model (mean 950, 95% CI 897 to 985). The slightly wider simulation interval than above (902 to 984 simulated true cases, 900 to 983 expected true cases given only the uncertainty in the positive predictive value) arises because of the inclusion of the second source of uncertainty, which is only observed when the bias parameters are applied at the record level instead of at the summary level. Although the difference in the simulation interval limits in this example is small, the interval around the difference in the number of simulated true positive cases from each iteration when only the first source of uncertainty was included, versus when both sources of uncertainty were included, was much larger: median difference=0, 95% simulation interval -14 to 14 simulated true positive cases. More generally, bias models for a summary level bias analysis have often omitted the uncertainty arising from the fact that biases act on individual records, and thereby slightly understate the total uncertainty. We encourage the analyst to be careful to write a bias model that includes the total uncertainty, as explained here and in Chapters 8, 9, 10, and 11. The tools on this text's website incorporate both sources of uncertainty. When using tools provided by others, the analyst should evaluate whether both sources of uncertainty have been properly modeled.

   A final consideration when choosing between summary level and record level approaches is that computing code must collate and summarize the accumulated bias-adjusted estimates generated from the regression model (as will be described in detail in Chapter 8). For some software packages, this can be done quite simply and easily while in others it may involve a more sophisticated ability to code, including use of matrix methods. No software currently exists that allows implementation of record level probabilistic bias analysis using a pre-coded routine. Although the code necessary to conduct record level bias analysis is not overly complex, it often needs to be written for, or adapted to, the specific dataset and bias problem, requiring customized coding. Details of summary level and record level probabilistic bias analysis are provided in Chapters 8 and 9, respectively, and example and adaptable code is available at the website that accompanies this text.

Describing Uncertainty in the Bias Parameters

A key advantage of probabilistic bias analysis is that it allows an accounting for uncertainty in the values assigned to the bias parameters. The bias parameter values cannot be known for certain; if they were known for certain, there would often be no bias. For example, to know the sensitivity and specificity of exposure classification with certainty, one would have to measure the gold standard in every member of the study population, in which case there would be no information bias because the gold standard measure of exposure and not the misclassified measure would be used in the analysis. To know the strength of association of an unmeasured confounder with the exposure and the outcome with certainty, it would have to be measured (without error) in the study population, and it would no longer be an uncontrolled confounder. To know the participation proportions within exposure categories of a case-control study, exposure and disease status of every member of the source population would have to be known, which can only be realized when everyone participates in the study population, hence there could be no selection bias from this mechanism. In short, the very nature of bias requires that the values of the bias parameters are uncertain.

That said, there is often very sound information available to inform a reasonable range of values to assign to the bias parameters. For example, we may estimate the positive predictive value of outcome classification with a validation study conducted within a subset of the study population. Assuming the validation study was well conducted, we can be reasonably sure that the estimates apply to the remainder of the study population. However, even these will be measured with random error, such that we cannot be sure that the true estimate equals the point estimate from the validation substudy. If we estimated that the positive predictive value was 95.0%, but a 95% confidence interval calculated for the positive predictive value ranged from 90.0% to 98.3%, it would be unwise to assume with 100% certainty that the true positive predictive value is 95%. Different values of positive predictive value, all compatible with the validation data, may produce substantially different bias-adjusted results. Probabilistic bias analysis accounts for the uncertainty by assigning a probability distribution to the bias parameter to express our uncertainty, mathematically. Typically, the bias parameter distribution will be formally conceived of in a subjective fashion, as a statement about an analyst's uncertainty about the value of the bias parameter. As we point out in Chapter 11, this bias parameter specification has a distinctly Bayesian flavor. For instance, an analyst may think the best guess for positive predictive value is 95% and they are 95% certain that the positive predictive value lies between 90.0% and 98.3%. Below we introduce probability distributions that will allow the analyst to account for this uncertainty by creating and sampling from mathematical specifications of these subjective statements.

Probability Distributions

To conduct a probabilistic bias analysis, one needs to assign a probability distribution to all or some of the bias parameters. Most of the bias parameters we are concerned with are continuous (even though they may be bounded within some range) and, as such, we consider common density distributions for continuous variables. A *probability density function*, is a mathematical function that is defined over an infinite number of points, potentially constrained to an interval. The probability at any single point equals zero, so probabilities are measured over intervals. The integral of the probability density function between the two points of the interval equals the probability for that interval, and the integral over the range of all possible points equals one.

For any analysis, many choices for distributions are available, so the analyst will need to define a probability distribution that represents the best understanding of the possible values of the bias parameter and the likelihood that each value in the defined range equals the true value. Below we discuss several useful distributions and how to create them using standard statistical analysis software. Further information on distributions can be found in Vose (17). For all distributions, we assume the bias parameter is a continuous random variable, denoted by x.

Uniform Distribution

To introduce the concept of probability distributions we will use a very simple distribution, the uniform distribution. A random variable that follows a uniform distribution has a minimum value and a maximum value and every value of the random variable in the range has an equal probability density. This uniform probability density equals:

$$f(x) = \frac{1}{maximum - minimum}$$

Uniform distributions are easy to generate in most statistical analysis software such as SAS, R and Stata and can also be generated using Microsoft Excel or any software package that can generate a random number from a standard uniform distribution (*i.e.*, a uniform distribution with a minimum of 0 and a maximum of 1). A random draw from a distribution with a specified minimum and maximum value can be created from a random draw from a standard uniform distribution using the following formula:

$$random_{min,max} = min + u \cdot (max - min)$$

where $random_{min,max}$ is a random draw from a uniform distribution between some specified minimum (min) and some specified maximum (max), and u is a random draw from a standard uniform distribution. Microsoft Excel's "rand()" function generates a uniform random draw in the range 0–1 and its "randbetween(bottom, top)" function generates a uniform random draw between the values assigned to "bottom" and "top." In SAS, the "rand('UNIFORM')" function generates a uniform random draw, u, in the range 0–1 and the more general "quantile('UNIFORM',u,l, r)" function can also be used to draw from a uniform distribution with minimum of l and maximum of r. The default values for the location parameters l and r equal 0 and 1, respectively, so can be omitted to sample from the standard uniform. In Stata, the function "runiform(*min, max*)" draws a random variable from a uniform distribution with a minimum of *min* and maximum of *max*. In R, the function "runif (*n, min, max*)," where n is the number of random values to generate and *min* and *max* are the range of numbers to sample from, will also allow sampling from a uniform distribution.

To illustrate, if we sampled 50,000 estimates of sensitivity from a uniform distribution (min = 70%, max = 95%, so the uniform probability density f(x)=1/ (0.95 − 0.70) = 4), the resulting histogram of the chosen values would be as displayed in Figure 7.1. Note that each value within the minimum and maximum values is equally likely to be chosen, while values below and above the minimum and maximum values cannot be chosen. The uniform distribution is conceptually easy to understand, and the range of allowed values might seem easy to specify. This ease of understanding and specification, however, come at an important expense. One would seldom have sufficient information to completely disallow values just above or just below the allowed range, and one would often have some information to support values within the range as more likely than others. These circumstances

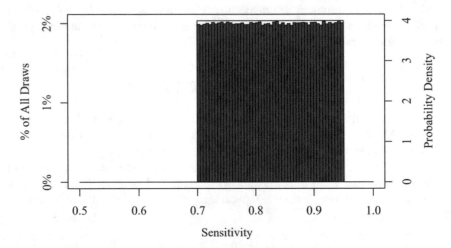

Figure 7.1 Example of a uniform probability density distribution for sensitivity (black) and the results of drawing randomly from that distribution (red) with a minimum value of 70% and a maximum value of 95%. While it is expected that each value in the allowed range will be selected equally, the observed data do not match our expectations exactly since this process was not repeated infinitely.

are not reflected in the probability density assigned by a uniform distribution. There are some bounds that make good sense, such as an upper limit on a proportion (*e.g.,* a classification parameter) equal to 1 or a lower limit on a proportion equal to 0 or equal to a value below which negative cell frequencies are returned by the bias model. These upper and lower bounds can also be incorporated into other probability density distributions, as we will discuss further below.

Generalized Method for Sampling from Distributions

Not all probability distributions can be randomly sampled in all statistical computing packages. However, one can readily sample from any continuous probability distribution, f(x). First, we need to define the cumulative distribution function of f(x) as the integral of that function up to some point x:

$$F(x) = \int_{-\infty}^{x} f(t)dt \qquad (7.1)$$

and note that $F(\infty) = 1$ for all distributions. By recognizing that the distribution of a random variable $U-F(x)$ is uniformly distributed over the interval $[0,1]$, one can choose a random number u from a standard uniform distribution and interpret it as a random value selected from $F(x)$. To obtain x [a random value selected from $f(x)$], one solves:

$$F(x) = \int_{-\infty}^{x} f(x)dx = u \qquad (7.2)$$

for x by making use of the inverse cumulative distribution function (programmed in most software packages for common distributions). Numeric integration and interpolation will ordinarily suffice if $f(x)$ is not integrable. This procedure is implemented in SAS by calling "quantile('function',u,parm-1,....,parm-k)" where 'function' is the density function and 'parm-1,....,parm-k' are optional shape and location parameters. In Stata and R, there are separate quantile functions (or inverse cumulative distribution functions) for each distribution. In Excel, only the uniform has a built-in density sampling function, but any of the density distributions with a built-in "INV" function can be sampled using this approach.

Trapezoidal Distribution

As noted above, the uniform distribution is appealing because it is easy to understand, but it is rarely the case that all values in the plausible range of a bias parameter are equally likely to be true, or that every value outside that range is impossible. As such, we do not recommend assigning a uniform distribution to bias parameters except when no other distribution can be credibly parameterized. To address the shortcomings of the uniform distribution, a trapezoidal distribution provides a more appropriate choice. The trapezoidal distribution has two modes, an upper and a lower mode. Each value between the modes has an equal probability density. A trapezoidal distribution also has a minimum and a maximum value, and the probability density decreases linearly to zero as one moves away from the nearest mode and towards the minimum or maximum values.

The trapezoidal distribution may often provide a more realistic distribution than the uniform distribution for values assigned to bias parameters. This is because one might be confident that the true value lies within a certain range (between the modes) but might not want to exclude the possibility that the value of the parameter lies outside that range. In such cases, a trapezoidal distribution gives an option in which the values between the modes have greater density than values outside the modes. Values outside the modes but within the specified range are, however, allowed. There is no large change in density at either the modes or the bounds. The trapezoidal density therefore allows for some values within a range to be characterized as more likely than others. In addition, the minimum and maximum can be specified outside the range one might be inclined to specify with a uniform distribution, which assigns some probability to values that would be disallowed by the uniform. The trapezoidal is also an easy distribution to comprehend, and often an easy distribution to parameterize, because it requires the analyst to specify only the minimum, maximum, and two modes. Further, because it has minimum and maximum values, it is easy to bound the trapezoidal distribution at the limits of logical values (*e.g.,* between 0 and 1 for proportions).

Despite these advantages, the trapezoidal retains some of the disadvantages of the uniform distribution. The minimum and maximum values establish sharp boundaries outside of which values are completely disallowed. While such boundaries may sometimes be justified (*e.g.,* at 0 and 1), in some cases these boundaries will not have a sound basis. Furthermore, all values within the range established by the two modes are given equal probability density, whereas there may be reason to support some values within that range more strongly than others.

An example of 50,000 values drawn from a trapezoidal distribution with a minimum value of 70%, a lower mode of 75%, an upper mode of 85%, and a maximum value of 95% is given in Figure 7.2.

Most statistical analysis software packages do not provide the ability to sample randomly from a trapezoidal distribution. The main exception to this is R, which has a trapezoid package that can be installed and has a function "rtrapezoid" that can be used to sample randomly from a trapezoidal distribution. However, for any statistical software package, sampling from a trapezoidal can be implemented using a standard uniform distribution random number generator and the inverse cumulative distribution function approach described in Equations 7.1 and 7.2. To generate a random draw from a trapezoidal distribution, first choose a random number u from a uniform (0,1) distribution. Then calculate:

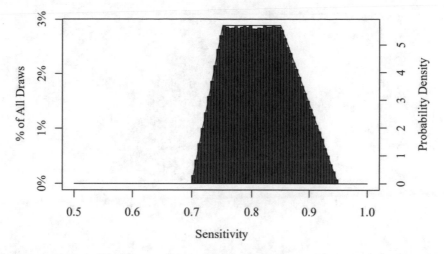

Figure 7.2 Example of a trapezoidal probability distribution for sensitivity (black) and the results of drawing randomly from that distribution (red) with a minimum value of 70%, a lower mode of 75%, an upper mode of 85%, and a maximum value of 95%.

$$s = \frac{[\min + \text{mod}_{low} + u \cdot (\max + \text{mod}_{up} - \min - \text{mod}_{low})]}{2}$$

where min, mod_{up}, mod_{low}, and max are the minimum, lower mode, upper mode and maximum value of the specified trapezoidal distribution, and u is the random draw from a standard (0,1) uniform distribution. Next, calculate the random value *trap* from the trapezoidal distribution as follows:

If the value of s is greater than or equal to the lower mode and less than the upper mode:

$$trap = s$$

If s is greater than or equal to the upper mode (and less than or equal to the upper limit) then:

$$trap = \max - \sqrt{(\max - \text{mod}_{up}) \cdot (-2 \cdot s + \max + \text{mod}_{up})}$$

If s is greater than or equal to the lower limit and less than the lower mode then:

$$trap = \min + \sqrt{(\text{mod}_{low} - \min) \cdot (2 \cdot s - \min - \text{mod}_{low})}$$

The trapezoidal distribution is also useful because it is a flexible distribution. One could create a uniform distribution from the trapezoidal distribution by setting the minimum equal to the lower mode and the maximum equal to the upper mode.

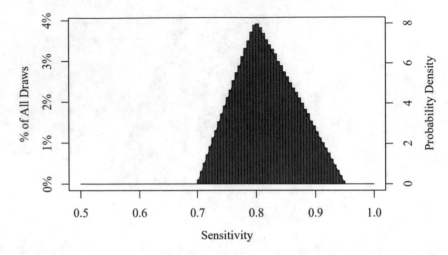

Figure 7.3 Example of a triangular probability distribution for sensitivity (black) and the results of drawing randomly from that distribution (red) with a minimum value of 70%, a mode of 80% and a maximum value of 95%.

Triangular Distribution

Another choice for a probability distribution is the triangular distribution; 50,000 samples drawn from a triangular distribution are shown below in Figure 7.3. The triangular is similar to the trapezoidal distribution in that the probability increases linearly from the minimum up to its modal value and then decreases linearly to the maximum, but differs in that it has only one mode. In many bias analyses, it would be difficult to support a single value as most likely, which is a disadvantage of the triangular distribution compared with the trapezoidal distribution. Other advantages and disadvantages discussed with regard to the trapezoidal distribution therefore apply in equal measure to the triangular distribution. The mode represents the most likely value with the probability trailing off to zero as one moves away from the mode and approaches the minimum or the maximum values. A triangular distribution can be sampled in SAS using the "rand('TRIANGLE', h)" function, which samples from a distribution from 0–1 with a mode at h. In Stata, there is no built-in function to generate a random variable from a triangular distribution but we include one (rtri.ado) posted to the website that accompanies this book. There is also no built-in function for sampling a triangular distribution in R and we also provide that in the online appendix (rtriangular.r). Note also that one can sample from a triangular distribution in R using the rtrapezoid package with the lower and upper modes set to the same value (which is how we created Figure 7.3).

Normal Distribution

For some analyses, a normal distribution might be a useful and familiar choice for assignment to a bias parameter, particularly because many statistical packages allow random sampling of a standard normal deviate and because it is a commonly known distribution. A normal distribution requires the analyst to specify a mean and a standard deviation. In some cases in this book, a specific type of standard deviation, the standard error will be used in conjunction with the normal distribution. In particular, in Chapter 8 we will sample from a normal distribution centered on a parameter value with an estimated standard error. Because the standard error is one type of standard deviation (the standard deviation of the sampling distribution), we remain general and refer to standard deviations in this chapter. Software packages that allow sampling of a random standard normal deviate (that is, a draw from a normal distribution with a mean of 0 and standard deviation of 1) can be used to create a draw from a normal distribution as:

$$random_{mean,sd} = mean - random_{0,1} \cdot sd \qquad (7.3)$$

where $random_{mean,sd}$ is a random draw from a normal distribution with a mean of *mean* and a standard deviation of *sd* and $random_{0,1}$ is a random draw from a standard normal distribution. In Microsoft Excel this draw can be done by setting "$random_{0,1}$ = NORMINV(RAND(),0,1)," which follows the general procedure outlined in Equations 7.1 and 7.2. Alternatively, one could substitute *mean* for 0 and *sd* for 1 in the Excel call to the norminv function. In SAS, the function "rannor(seed)" returns a random draw from the standard normal distribution as does "rand('normal')". Alternatively, the SAS function "quantile('normal',u,*mean*,*sd*)" returns a random draw from a normal distribution with mean equal to *mean* and standard deviation equal to *sd*, which is the equivalent of $random_{mean,sd}$ in Equation (7.3). The default values for these optional location and shape parameters are 0 and 1, respectively, and may be omitted if these values are desired, in which case the quantile function returns the equivalent of $random_{0,1}$ in Equation 7.3. In Stata the "rnormal (*mean*, *sd*)" function can be used while in R the "rnorm(*n*, *mean*, *sd*)" function can be used, where *n* is the number of draws. Figure 7.4 depicts 50,000 draws from a normal distribution for sensitivity with a mean of 80% and a standard deviation of 10%. This highlights a potential drawback to using normal distributions for bias parameters that are percentages; namely, that the sampled values can be larger than 100% or less than 0%. We discuss this in more detail, and offer solutions, further below.

Specifying the mean and standard deviation of the normal distribution is generally straightforward. The mean of the normal distribution is chosen as the value of the bias parameter that the analyst thinks is most likely. This selection can be informed by the prior literature or expert opinion (see Chapter 3). The standard deviation could be specified directly from a previous validation study, if the study gives the standard deviation. Alternatively, the standard deviation could be back

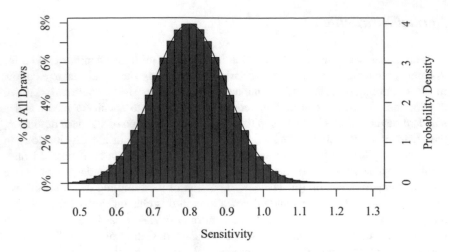

Figure 7.4 Example of a normal probability distribution for sensitivity (black) and the results of drawing randomly from that distribution (red) with a mean value of 80% and a standard deviation of 10%.

calculated from the 95% confidence interval limits if those are given in a validation or external study, using the formula:

$$sd = \frac{U - L}{2 \cdot 1.96} \qquad (7.4)$$

Where U is the upper 95% confidence limit and L is the lower limit. If intervals are not available for use, it may be acceptable to specify the extreme limits of the bias parameter that the analyst thinks is likely. The upper and lower limit of this range can be substituted for U and L in the formula above, respectively. If these are truly extreme values, a 99% interval might be more suitable than a 95% interval, so one would replace 1.96 with 2.58 in Equation 7.4. A transformation of the bias parameter may be necessary when using the normal distribution. For example, if the bias parameter is an odds ratio (OR) or risk ratio (RR) (e.g., if performing a bias-adjustment for uncontrolled confounding), a normal distribution would typically be placed on the ln(OR) or ln(RR). In this case U and L in Equation 7.4 would be the natural log of the upper and lower 95% confidence interval, respectively. Drawing a large number of random samples from the specified distribution, graphing it and examining the 2.5th, 50th and 97.5th percentiles of the parameter of interest on the scientifically meaningful scale is an excellent way to double check accuracy of computation and to ensure that the resulting distribution reflects prior belief. For instance, suppose an analyst was specifying a bias distribution for an odds ratio that was estimated external literature as 0.61 with 95% CI: 0.40, 0.92. Equation 7.4 can be used to estimate the standard deviation (also a standard error): $sd = \frac{ln(0.92) - ln(0.40)}{2 \cdot 1.96} = 0.21$. To sample from the bias distribution, we draw samples from a normal distribution with mean of ln(0.61) and standard deviation of 0.21: N-$(mean = \ln(0.61), sd = 0.21)$. We draw 10^6 samples from this distribution, exponentiate them, and estimate the 2.5th, 50th and 97.5th percentile which gives values of 0.404, 0.610, and 0.920 which match the values used to calculate the parameters of the normal distribution.

Unlike the trapezoidal distribution, the normal distribution does not have a range of values that are considered equally likely, and in some cases the analyst may feel this model is a better fit to their prior assumptions about the bias parameter. This might be the case when the analyst has data from a validation study that suggests a most likely value, but then wishes to depict a decreasing probability of values away from the modal value. However, Figure 7.4 demonstrates one important problem with using a normal distribution. Because the tails of the normal distribution extend past values that might be logically possible (*i.e.,* values outside the range of 0–1 required for sensitivity and specificity, which are proportions), this distribution might not be as useful as it first appears.

For parameters such as sensitivity, specificity, positive predictive value or negative predictive value (or other proportions), which are constrained to lie between 0 and 1, a simple logit transformation will suffice. If p is the constrained parameter value, $logit(p) = \ln\left(\frac{p}{1-p}\right)$ can take values from negative infinity to infinity. A normal distribution can be used as distribution for logit(p) rather than (p). After we sample logit(p), we can use the $expit(x) = \frac{\exp(x)}{1+\exp(x)}$ function to convert from the logit-probability back to the probability scale. One potential problem with this approach is that it can be slightly more complicated to specify a distribution for logit(p) than for p. We illustrate how this can be accomplished using a simple example. Suppose we wish to specify a bias distribution for the positive predictive value of outcome classification in a hypothetical validation study to estimate the positive predictive value = 95.0% (95%CI: 88.0%, 98.0%). Specifying the parameters of the normal distribution for the logit of the positive predictive value requires us to specify a mean and standard deviation. The mean can be specified easily as logit(0.95)=2.94. The standard deviation can be computed using the formula above with logit(L) and logit(U) substituted for L and U, respectively which, in this example gives an answer of sd=0.48 and a final specification of: $logit(PPV) \sim N$-$(mean = 2.94, sd = 0.48)$. We again emphasize that it is important to check the model specification by drawing random samples from the distribution. In this case, we draw 10^6 PPV samples and examine the 2.5th, 50th, and 97.5th quantiles of the expit of these samples, which are 0.88, 0.95, and 0.98, respectively, as expected. We also plot a histogram of the randomly sampled logit(PPV) as well as the $expit(logit(-PPV)) = PPV$. As shown in Figure 7.5, the distribution of the logit(PPV) is normal and centered at 2.94, as expected. The distribution of the positive predictive value, however, is not normal and has a longer left tail. The analyst should be careful to examine the distribution of the bias parameter's on the bias parameters natural scale (such as a percentage) to ensure it is in keeping with prior knowledge.

One could consider a different approach, in which a normal distribution is assigned to the bias parameter (rather than the transformed bias parameter), and

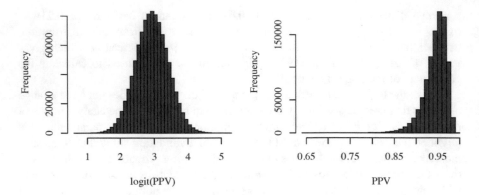

Figure 7.5 Histograms of logit(PPV) and PPV on the left and right panels, respectively. Samples were drawn from a Normal(($\mu = 2.94, \sigma = 0.48$) distribution on the logit scale and then transformed using the expit function.

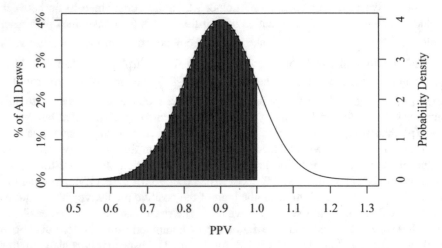

Figure 7.6 Example of a normal probability distribution for sensitivity (black) and the results of drawing randomly from that distribution (red) with a mean value of 90% and a standard deviation of 10% truncated at 0% and 100% Truncation results in a realized mean of the samples in red of 87% with standard deviation of 8%.

this normal distribution is truncated to the range of possible parameter values as in Figure 7.6. Sampling from truncated distributions is somewhat more difficult, however, and it changes the bias parameter distribution in ways that are difficult to intuit. For example, if we had prior reason to specify *PPV~N(mean = 0.9, sd = 0.1)*, we have made the decision that we think 90% is the most likely value for the positive predictive value. However, if we sample positive predictive value estimates from this distribution and discard all values greater than 1, we end up with a distribution that may not conform to our prior belief about the positive predictive; in this case, a mean PPV=0.87 and sd=0.08, rather than the values we intended to specify. For this reason, we suggest avoiding truncated distributions or being extremely careful and sampling from the distribution to make sure the specified distribution corresponds to prior knowledge.

Beta Distribution

The beta distribution provides a probability density function that is well-suited for assignment to proportions – such as sensitivity, specificity, or predictive values – because it is constrained to the interval [0,1]. It can be scaled and located elsewhere, though, so its flexible shapes can also be valuable for application to bias parameters that are not constrained to the interval [0,1]. The beta distribution is parameterized by two positive shape parameters, often denoted as α and β. The expected value of a random variable X drawn from a beta distribution and its variance are as follows.

$$E(X) = \frac{\alpha}{\alpha + \beta} \tag{7.5}$$

$$Var(X) = \frac{\alpha\beta}{(\alpha + \beta)^2(\alpha + \beta + 1)} \tag{7.6}$$

A second advantage of the beta distribution is its flexibility to model a wide range of probability density shapes. Figure 7.7 illustrates this point. With α and β both set equal to 1.1, the beta distribution yields a near uniform density, but with probability

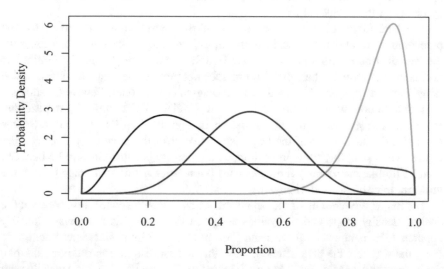

Figure 7.7 Examples of probability density functions generated by the beta distribution by setting different combinations of values for α and β. Blue density: α=1.1, β=1.1. Red density: α=7, β=7. Black density: α=3, β=7. Green density: α=15, β=2.

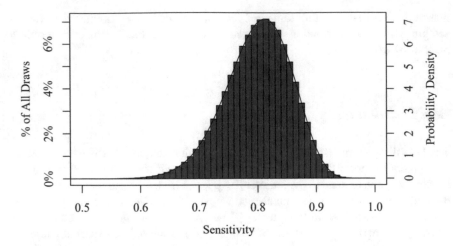

Figure 7.8 Example of a beta probability distribution for sensitivity (black) and the results of drawing randomly from that distribution (red) with α=40 and β=10, so mean at sensitivity=80%.

density that declines smoothly to 0 at the boundaries of 0 and 1. In fact, with $\alpha = 1$ and $\beta = 1$,then the beta distribution is exactly a uniform(0,1) distribution. With α and β both set equal to 7, the beta distribution yields a symmetric density similar to the normal distribution. With α and β set to different values, the beta distribution yields asymmetric densities centered on different means. In general, if α and β are both less than 1, then the density distribution will be u-shaped, so bimodal, giving more weight to values at the tails than values at the center. If α and β are both greater than 1, then the density distribution will be unimodal, giving more weight to values at the center than values at the tails.

The beta distribution addresses many of the shortcomings described for the preceding distributions. Unlike the uniform, trapezoidal, and triangular distributions, the beta distribution does not have sharp boundaries (except at 0 and 1). Unlike the normal distribution, the beta distribution does not yield values outside of an allowed range (such as proportions less than zero or greater than one). Figure 7.8 shows a beta distribution with location and spread similar to the uniform, trapezoidal, triangular and normal distributions shown in Figures 7.1, 7.2, 7.3 and 7.4. This illustrates the advantages of the beta distribution over those shown in these earlier figures when sampling for a proportion, as the density is constrained to lie between 0 and 1 (unlike the normal) and lacks hard boundaries within that range (unlike the uniform, trapezoidal, and triangular).

The major shortcoming of the beta distribution is that the connection between the desired density shape and the values assigned to α and β is not always directly apparent. The parameter values α and β do not have an immediate interpretation in the same way that the parameters of, say, the normal distribution do. For example, the beta density distribution in Figure 7.7 with $\alpha = 15$ and $\beta = 2$ yields a density with a mean of 0.88, a median of 0.90 and 2.5[th] and 97.5[th] percentiles of 0.70 and 0.98, respectively. It is not immediately apparent when specifying the values of $\alpha = 15$ and $\beta = 2$ that these numbers will be returned as indicators of the location and spread; the analyst will need to rely on a computer to generate them.

For a classification parameter (sensitivity, specificity, positive predictive value, or negative predictive value) one can use the results from a validation study to directly assign values to the beta distribution parameters. For example, to parameterize the sensitivity of classification, one can set α equal to the number of people who test positive and are truly positive, and set β equal to the number of people who test negative and are truly positive. For example, if a validation study of smoking found that among 17 women who truly smoke, 15 self-reported smoking and 2 self-reported not smoking, the sensitivity would be 88% with a Wald 95% confidence interval: 73%, 100%. If we used these data to directly inform a bias distribution it would be of the form:

$$sens \sim beta(\alpha = 15, \beta = 2)$$

Using Equations 7.5 and 7.6 for mean and variance above, we find this distribution has a mean of 15/(15+2)=88% and standard deviation of 0.08 resulting in a Wald 95% confidence interval of: 73%, 100%. However, a Wald interval may be a suboptimal choice since the distribution is skewed. The 2.5^{th} and 97.5^{th} percentiles of the distribution can be used instead: 70%, 98%. The beta specification matches the 95% confidence interval results very well. A similar approach is used to specify the distribution for specificity ($\alpha =$ number of test negative and truly negative, β=number of test positive and truly negative) for positive predictive value ($\alpha =$ number of truly positive and test positive, β=number of truly negative and test positive) and negative predictive value ($\alpha =$ number truly negative and test negative, β=number of truly positive and test negative). We can express this more generally as:

$$\theta \sim beta(\alpha = A, \beta = B)$$

where θ is the bias parameter of interest, A and B are the number of people meeting the conditions described above (*e.g.*, if θ is sensitivity, A is the number who are true positives and test positive while B is the number who are true positive and test negative). This method uses the results of the validation study to directly parameterize the beta distribution, so should be implemented only after consideration of the direct applicability of the validation study results to the classification problem at hand. This topic was discussed in detail in Chapter 3.

Alternative prior specifications for beta distributions have been suggested. In Chapter 11, we introduce the fundamentals of Bayesian inference for quantitative bias analysis. Some analysts (17) have advocated taking a Bayesian approach to estimating bias parameters in bias analyses (we discuss this approach further in Chapter 11). This approach requires placing a prior distribution on the bias parameter and a uniform distribution is often chosen if there is no relevant prior information on the parameter of interest. A more informative prior distribution may be justifiable if there are previous studies estimating the parameter. We can express a uniform prior using a beta distribution, as mentioned above:

$$\theta \sim beta(\alpha = 1, \beta = 1)$$

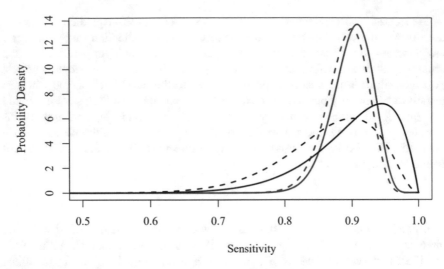

Figure 7.9 Shapes of beta distributions when specifying the parameters by adding 1 or not: dashed lines add 1 while solid lines do not. Red lines are for data with 90 true positives and 10 false negatives. Black lines are for data with 18 true positives and 2 false negatives: red solid line is beta ($\alpha=90$, $\beta=10$); red dashed line is beta($\alpha=91$, $\beta=11$); black solid line is beta($\alpha=18$, $\beta=2$); black dashed line is beta($\alpha=19$, $\beta=3$).

Adding the validation data, the distribution of the bias parameter of interest becomes:

$$\theta \sim beta(\alpha = A + 1, \beta = B + 1)$$

This distribution is very similar to the previous specification we had for this problem, only with 1 added to each parameter. For instance, the distribution assigned to sensitivity in the example above would be:

$$sens \sim beta(\alpha = 16, \beta = 3)$$

which would result in a mean of $16/(16+3)=84\%$ and a standard deviation of 0.08 resulting in a Wald 95% confidence interval of: 68%, 100% and 2.5th and 97.5th percentiles: 65%, 96%. Notice that this approach gives a result similar to above, though not identical; the mean is slightly lower using this alternative approach.

To further illustrate the differences in the two approaches (adding 1 to alpha and beta versus not), we consider specifying a beta distribution for the sensitivity parameter if a validation study found 90 people who tested positive and truly had the disease and 10 people who tested negative and truly had the disease. The solid line in Figure 7.9 shows the beta($\alpha=90$, $\beta=10$) distribution that would be specified by directly using this validation study while the dashed line shows the alternative beta($\alpha=91$, $\beta=11$) distribution. Because the sample size is reasonably large (n=100) there is relatively little difference between the two. Alternatively, when the sample is small the difference between the two approaches could differ more dramatically. If the validation study had been smaller and 18 of 20 people who truly had the disease tested positive, a beta(α=18, β=2) and a beta(α=19, β=3) distribution would be specified using the two approaches. A more dramatic difference is noticeable in these distributions. While the beta(α=90, β=10) and beta(α=18, β=2) distribution both specify means of 90%, their modal values are greater than 90% because of the skewness of the distribution. Alternatively, the +1 approach centers the mode at 90%. Centering the mode at 90% alters the variance of the beta distribution. We strongly encourage analysts to plot the proposed density distribution to ensure that they accurately represent the prior knowledge. In skewed distributions, in particular, analysts may be more interested in specifying a modal value for the bias parameter than the mean.

Deciding between which of the two approaches to adopt requires considerations of various factors. These considerations largely mirror factors that must be taken into account when deciding between frequentist and Bayesian approaches. Indeed, this topic pertains to all distributions, not only the beta distribution, but the beta distribution is one of the distributions for which it is particularly easy to incorporate prior information. The first approach above, that uses only the data in the validation study, is an unbiased approach that will produce a mean point estimate for the bias parameter that is equal to the truth on average, assuming a valid bias model. The second approach (adding 1) will produce a slightly biased estimate of the bias parameter, in general. The benefit of the second approach is that it can have a smaller variance than the first approach and, further, specifying the mode may be more intuitive. If we were only interested in estimating the bias parameter, the second approach could produce smaller mean squared error and might be preferred. However, the bias parameter is of secondary concern in this case because our goal is to estimate the bias-adjusted measure of association. We are aware of no literature suggesting that one approach is preferable to the other and, since small amounts of bias can have large effects, as we have seen in the specificity example (see Chapter 6), we generally opt for the first approach. There may, however, be a practical consideration that requires the second approach. If validation data measure perfect discordance or perfect concordance, then either alpha or beta will be zero. For example, if validation data find PPV $= 12/12 = 100\%$ (perfect concordance), then the corresponding beta distribution would have alpha $= 12$ and beta $= 0$ under the first approach. This is an improper beta distribution, so we advise adding 1 to both alpha and beta, in accordance with the second approach. In the example, the beta distribution for PPV would have alpha $= 13$ and beta $= 1$. For consistency, it would be advisable to add 1 to all validation data frequencies used in this setting, even if other validation data have no zero-frequency problem.

When one wishes to less directly parameterize the beta distribution, an approximate guide for choosing values for α and β is to specify a range of likely values. This range, call it $[a,b]$, should contain 95% of the parameter values that are deemed likely with only 2.5% probability of seeing a value lower than the low end and a 2.5% probability of seeing a value higher than the high end of the range. The ends of the range, a and b, can be substituted for the confidence limits in Equation 7.4 to solve for the standard deviation. Continuing with the example above, we could specify a=0.70 and b=0.98 along with x=0.88. Using Equation 7.4, we have $std = \frac{0.98-0.70}{2*1.96} = 0.07$. Given the estimated mean (x) and standard deviation (sd), we can solve for α and β with the following equations.

$$\alpha = x\left(\frac{x(1-x)}{sd^2} - 1\right)$$

$$\beta = (1-x)\left(\frac{x(1-x)}{sd^2} - 1\right)$$

Using these equations with this example yields α=17.3 and β=2.4, which are close to the actual shape parameters ($\alpha = 15$, $\beta = 2$).

Another shortcoming of the beta distribution is that some statistical computing software programs have no built-in function that allows one to sample from it. For example, Microsoft Excel has no such function, so one would have to adapt the general sampling strategy in Equations 7.1 and 7.2. This adaptation calls "BETA. INV(RAND(),a,b)," as we have seen above. In SAS, "rand("BETA", a, b)," in Stata, "rbeta(a, b)" and in R, "rbeta(n, a, b)" allow drawing from a beta distribution with shape parameters α=a and β=b. In SAS the function "quantile('beta',u,a,b,l,r)" returns a random draw from a beta distribution with shape parameters α=a and β=b, and with optional lower and upper bounds l and r, respectively. These last two are location parameters with default values of 0 and 1, respectively, and may be omitted if these values are desired.

As we emphasized above, it is good practice to draw random samples from the bias distribution, plot the distribution, and look at key percentiles to ensure that the specified distribution matches prior knowledge. Because the beta distribution is so flexible and can be highly skewed or even bimodal, it is particularly important to employ this strategy. Beta distributions can sometimes be difficult to match to prior knowledge if the best guess (mean or mode) of the parameter is near 0 or 1. For example, suppose a validation study is conducted in 91 truly negative people, 90 of whom test negative. In Figure 7.10, we plot two possible beta distributions, based on

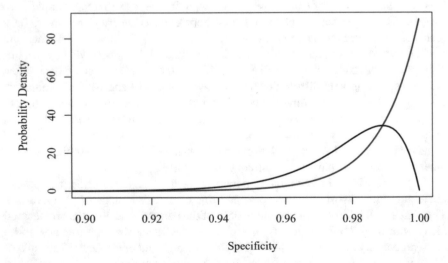

Figure 7.10 Probability density for beta distributions for a specificity parameter. Red density is for beta(α=90, β=1) and black is for beta(α=91, β=2).

the discussion above (either using the crude data or adding 1): beta($\alpha=90$, $\beta=1$) and beta($\alpha=91$, $\beta=2$). The 2.5th, 50th, and 97.5th quantiles from the two distributions are extremely similar with values of (96%, 99%, 100%) and (94%, 98%, 100%), respectively. However, examination of Figure 7.10 reveals a crucial distinction. In the beta($\alpha=90$, $\beta=1$) specification, the density uniformly increases as the specificity increases. In the beta($\alpha=91$, $\beta=2$) specification, the probability density increases up to a modal value of approximately 99% and then decreases rapidly from 99% to 100%. As we discussed in Chapter 6 small changes in specificity can have profound impacts on bias-adjusted estimates and the seemingly small difference between beta ($\alpha=90$, $\beta=1$) and beta($\alpha=91$, $\beta=2$) could have a meaningful impact since the former puts far greater weight on specificities near 100% than the latter. Prior knowledge will be crucial in deciding which of these distributions is more plausible. If one is struggling to identify the parameters for a beta distribution, an online tool may be of value (https://jhubiostatistics.shinyapps.io/drawyourprior/). This site allows the analyst to draw the distribution with the analyst's mouse and the shiny app will generate parameters for the closest beta distribution it can identify.

Bernoulli and Binomial Distributions

Bernoulli distributions are, in some ways, the simplest to understand. Unlike the continuous outcomes above, Bernoulli random variables are discrete and can attain values of 0 or 1. Bernoulli random variables are common in epidemiology. Outcomes (diseased=1, undiseased=0) are often modeled as Bernoulli random variables. In subsequent chapters, when introducing methods for exposure misclassification, we will often model these data as Bernoulli. For example, among those who are truly exposed a person might report being exposed (1) or report being unexposed (0). Bernoulli random variables can be conceived as the outcome of a potentially biased coin flip. If the biased coin attains a head, a truly exposed person reports being exposed. If the biased coin attains a tail, a truly exposed person reports being unexposed. The probability that the biased coin comes up heads in this example is simply the sensitivity.

We can define a Bernoulli distribution more formally by saying if the probability of seeing an event is p then a random variable X has a Bernoulli distribution if:

$$P(X = x) = p^x(1 - p)^{1-x}$$

That is, $P(X = 1) = p$ and $P(X = 0) = 1 - p$. The expectation of a Bernoulli random variable (the proportion of times we would expect to see an event) is $E(X) = p$ and the variance is $V(X) = p(1 - p)$.

We often see more than a single Bernoulli random variable at once. If N people experience the outcome in our study, we have N Bernoulli random variables. Rather than conceive of this as N individual Bernoulli trials, we can, equivalently, think of this as one binomial trial of N people. Now our random variable X is the number of people who experience the event of interest (*e.g.*, the number of people who report being exposed if they truly are exposed). This binomial variable X has a probability mass function

$$P(X = x) = \binom{N}{x} p^x (1 - p)^{N-x}$$

where $\binom{N}{x}$ is the binomial coefficient $\binom{N}{x} = \frac{N!}{(N-x)!x!}$. This probability function gives the probability of seeing any number of events from 0 to N. The mean of a binomial random variable is $E(X) = Np$ and its variance is $V(X) = Np(1 - p)$. We can see that if N=1, the binomial distribution reduces to a Bernoulli distribution. There is no mathematical difference between thinking of a study as containing N Bernoulli variables or 1 binomial variable of size N. In the simulations we present in Chapters 8 and 9, sampling from a single binomial distribution may have moderately faster computation time than sampling multiple times from a Bernoulli as it avoids looping through each observation in the dataset, for some software programs.

In much of epidemiologic data analysis, we observe an instance of $X = x$ and want to estimate a value of p, call it $\widehat{p}$, that is a function of the data. For example, if X_i is a Bernoulli outcome for the ith person in the study, we might want to estimate the probability that person has the outcome. Often that probability will depend on covariates in the dataset and we might model the probability as a function of these covariates as in a logistic regression model:

$$\text{logit}(p_i) = \beta_0 + \beta_1 Z_i$$

Here, our goal is to derive an estimate of $\widehat{\beta}_1$ given the observed data and standard software can easily fit this model.

In bias analysis we will sometimes have a different use of the Bernoulli and binomial distributions. We may have a study in which exposure is misclassified and have estimates of positive and negative predictive values. We will use a Bernoulli distribution in parts of Chapters 8, 9, 11, and 12 to sample (impute) the unobserved true exposure status conditional on their misclassified exposure value and the positive or negative predictive value. For a person who reports not being exposed, we can sample their true exposure status from $E_{0,i} \sim Bernoulli(1\text{-}npv)$. We note that 1 minus the negative predictive value is the probability of a person truly being exposed if they report not being exposed. For a person who reports being exposed, we can sample their true exposure status from $E_{1,i} \sim Bernoulli(ppv)$. We could repeat this for the total number of people who report not being exposed (M_0) and for the total number of people who report being exposed (M_1) in the study. Alternatively, we could resample the exposure status for all people who report not being unexposed as $E_0 \sim binomial(M_0, 1\text{-}npv)$ and all people who report being exposed $E_1 \sim binomial(M_1, ppv)$. The two approaches are equivalent, and we will use both in future chapters.

Other Probability Distributions

There are a number of probability distributions that could be assigned to bias parameters in a bias analysis. The ones we have presented above are flexible, familiar, and simple. Others that meet these criteria may also be implemented (*e.g.,* the Poisson distribution and the gamma distribution). Rather than catalog all the possible choices, we refer the reader to Vose (17). Readers can also use *The Engineering Statistics Handbook*, which is freely available on the US National Institute of Standards and Technology Web page (http://www.itl.nist.gov/div898/handbook/index.htm) and in particular the handbook's gallery of distributions available at http://www.itl.nist.gov/div898/handbook/eda/section3/eda366.htm).

Sensitivity to Chosen Distributions

It is important to choose probability distributions with sound underlying theoretical properties and to parameterize them so that they reflect the analyst's best understanding of the density supported by her review of the relevant evidence and her expert judgment. It is also important to realize that violations of the assumptions required to use certain distributions, or theoretical shortcomings of particular distributions, or small perturbations in the correspondence between the distribution's density and the analyst's optimally supported density, may have little consequence for the results of the bias analysis. Figure 7.11 shows five density distributions

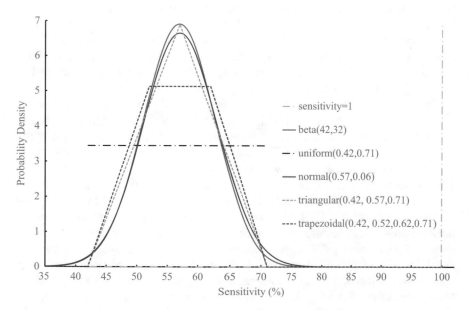

Figure 7.11 Five probability density functions centered on a proportion of 0.57 (location) and with the vast majority of the density lying between 0.42 and 0.71 (spread).

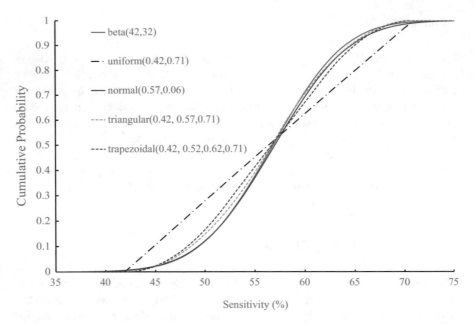

Figure 7.12 Five cumulative probability functions centered on a proportion of 0.57 and with the vast majority of the density lying between 0.42 and 0.71, corresponding to the five probability density functions depicted in Figure 7.11.

centered on a proportion of 0.57 and with the vast majority of the density lying between 0.42 and 0.71 [uniform(0.42, 0.71), trapezoidal(0.42,0.52,0.62, 0.71), triangular(0.42, 0.57, 0.71), normal($\mu = 0.57, \sigma = 0.06$), and beta($\alpha = 42, \beta = 32$)]. Notice that there is substantial similarity between these distributions, a fact that is more easily visualized by examining the corresponding cumulative probability functions in Figure 7.12. As a consequence, choosing between these five distributions might have little effect on the results of a probabilistic bias analysis.

This distribution for sensitivity derives from a validation study of self-reported history of taking antidepressants (18), the results of which were applied to a simple quantitative bias analysis of the association between self-reported use of antidepressants and breast cancer risk (19). The odds ratio associating ever-use of antidepressants, compared with never-use, with incident breast cancer was 1.2 (95% CI 0.9, 1.6). Boudreau et al. (2004) had conducted an internal validation study in a subset of participants, in which self-reported medication history was compared with electronic pharmacy records (18). 42 of 74 women (57%) with a pharmacy record of antidepressant prescription self-reported antidepressant use, which informed our assignment of the five density distributions described above and displayed in Figures 7.11 and 7.12. 274 of 280 women (98%) with no pharmacy record of antidepressant prescription self-reported no antidepressant use, which we used to inform the assignment of the following density distributions to the specificity of self-reported antidepressant use: uniform(0.95,1), trapezoidal(0.95,0.97,0.99,1), triangular

(0.95,0.98,1), normal(μ =0.98, ,σ =0.01), and beta(α =274, β =6). We used probabilistic bias analysis to bias-adjust using the simple misclassification bias models described in Chapter 6, with each density distribution for sensitivity assigned (specificity fixed equal to 1) and with each density distribution for specificity assigned (sensitivity fixed equal to 1). These density distributions adequately represent the evidence from the validation study regarding the sensitivity and specificity of classification of antidepressant use (18), particularly given that the emphasis is to illustrate differences between choice of density functions and not to conduct the optimal bias analysis for this study's results. We will, in fact, use this paper as a detailed example of multiple bias modeling in Chapter 12. The methods used to perform this probabilistic bias analysis will be explained in Chapter 8. The purpose here is only to explore the impact of the selected density distribution.

Figure 7.13 compares the bias analysis results obtained after 1000 iterations of the bias-adjustment equations, each with the sensitivity or specificity drawn from the respective density functions. The center of each line is the median and the error bar

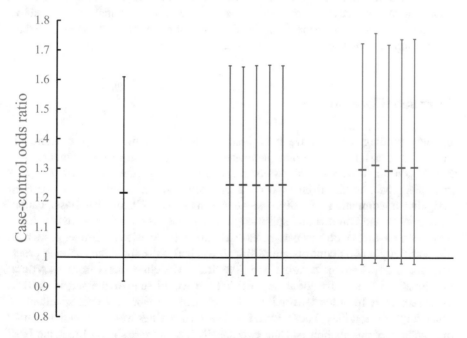

Figure 7.13 Probabilistic bias analysis results as a function of choice of density distribution applied to the sensitivity of exposure classification. For each density distribution, the midpoint depicts the median result after drawing 1000 values from the corresponding density distribution, and the error bars depict the 2.5 and 97.5 percentiles of the ranked bias-adjusted odds ratios. Black line: random error only. Red group: all have specificity=1; with left to right sensitivity~beta (α=42, β=32), ~uniform(0.42,0.71), ~normal((μ = 0.57,σ = 0.06), ~triangular(0.42, 0.57,0.71), ~trapezoidal(0.42,0.52,0.62,0.71). Blue group: all have sensitivity=0.57; with left to right specificity~beta(α=274, β=6), ~uniform(0.95,1), ~normal((μ = 0.98,σ = 0.01), ~triangular (0.95,0.98,1), ~trapezoidal(0.95,0.97,0.99,1) Original study data from Chien et al., 2006 (19) and validation data to inform the distributions are from Boudreau et al., 2004 (18).

limits are the 2.5 and 97.5 percentiles of the ranked bias-adjusted results. The figure illustrates that the choice of density distribution had very little impact on the median bias-adjusted results, which is as expected given that all of the distributions are centered on the same sensitivity (57%) or specificity (100% or 98%). The most notable, albeit small, difference is between bias-adjustment with sensitivity allowed to vary and specificity fixed at 100%, and the bias-adjustment with sensitivity fixed at 57% and specificity allowed to vary. The slightly stronger bias-adjustment when specificity is centered on 98% versus 100% is consistent with the expectation that strength of bias from exposure misclassification is driven by imperfect specificity when the exposure prevalence is low (see Chapter 6), as it was in this study (19).

Perhaps more surprisingly, there is very little difference in the width of interval between the 2.5% and the 97.5% regardless of the bias model. This observation derives from the fact that there is little difference in the shape or location of the probability distributions (Figures 7.11 and 7.12). The theoretical underpinnings of the choice of density distribution may, therefore, distinguish them more from one another than the difference in results yielded by the bias analysis. Nonetheless, it is good practice for analysts to evaluate the sensitivity of the bias analysis to assumptions about the assigned shape of the density distribution, particularly when different distributions are viewed as equally credible.

Correlated Distributions

In the preceding sections, we have assumed that the values sampled from the distributions assigned to the bias parameters are independent of one another. This assumption might be realistic in some cases, but in many other cases it is not. For example, modeling bias from exposure misclassification as non-differential implies assigning a correlation of 100% between sensitivity of exposure classification among cases and controls and specificity among cases and controls. That is, for a particular draw from the bias parameter distribution, the sensitivity among cases and the sensitivity among controls are identical. Similarly, the specificity among cases and specificity among controls are identical. Modeling bias from exposure misclassification as differential implies that we would sample the sensitivity for cases separately from the sensitivity among controls, or would sample specificities separately, or (most likely) both. Even if the two sensitivities were assigned the same bias parameter distribution and the two specificities were assigned the same bias parameter distribution, the misclassification would be differential since the two sensitivities and two specificities would not be equal for any draw from the distributions.

A simple method to specify these bias parameter distributions is to assume there is no correlation between the sensitivities (among cases and controls) or among specificities (among cases and controls). This assumption, though simple, will generally have little rationale support. We would want to avoid modeling the two distributions as independent (*i.e.,* a correlation of 0) (7). Even when the misclassification is differential, it is unrealistic to assume that there is no correlation between the sensitivity in cases and the sensitivity in controls and between the specificity in cases and the specificity in controls. Higher sensitivities among cases are likely associated with higher sensitivities among controls, even if they are not the same, presuming that the same classification method was used for both cases and controls (*e.g.*, a telephone interview) and that cases and controls were both selected from the same source population.

One way to think about the logic for including a correlation in distributions is to think about differential exposure misclassification. Suppose that an analyst expects a difference in the sensitivity of exposure misclassification in those with and without the outcome such that on average, those with the outcome have a sensitivity 30% higher than those without. Further suppose the analyst has a great deal of uncertainty in what the values of the sensitivity are such that they parameterize them with (unrealistic) uniform distributions ranging from 70% to 100% among those with the outcome and 40% to 70% among those without the outcome. If there were no correlation between the distributions, the analyst could draw a value of 100% for those with the outcome and 40% for those without (very large difference of 60%, much larger than the target difference of 30%) and this draw would be just as likely as drawing a value of 70% for those with the outcome and 70% for those without (no difference, much smaller than the target difference of 30%). By specifying a correlation between the two distributions, the analyst could specify that the target difference of 30% between the two values is the most likely difference but without a perfect correlation, so allow for variation in that difference. Still, extreme differences are far less likely than ones nearer to the target 30% difference. Note that this example makes it easy to see the point and works because the two distributions are uniform and have the same width for their ranges. Other distributions may not mimic this exact pattern of keeping a standard difference between chosen values but will follow the same principles.

A correlation can be included in a bias analysis by inducing a correlation between the random values selected from the standard uniform distribution (u in Equation 7.2). The following approach uses the inverse cumulative distribution function method of Equations 7.1 and 7.2 and is often referred to as a NORmal To Anything (NORTA) transformation (20,21). The NORTA transformation has three steps.

First one samples a pair of variables, Z_1 and Z_2, from a standard multivariate normal distribution with correlation ρ:

$$\begin{bmatrix} Z_1 \\ Z_2 \end{bmatrix} \sim MVN\left(\begin{bmatrix} 0 \\ 0 \end{bmatrix}, \begin{bmatrix} 1 & \rho \\ \rho & 1 \end{bmatrix} \right)$$ (7.7)

We would draw N such pairs, which would result in correlation of strength ρ between the sampled Z_1 and Z_2. The sampling in this step is easily accomplished using the function "mvrnorm" (in the MASS package of R), "drawnorm" in Stata, or the "randnormal(N, mean, cov)" function in SAS. In Excel, first draw "Z_1=NORMINV(RAND(),0,1)" and then "Z_2 =NORMINV(RAND(),ρZ_1,SQRT$(1-\rho^2)$)."

Second, each of the normal random variates, Z_1 and Z_2, is transformed using the *univariate* normal cumulative distribution function, ϕ:

$$U_1 = \phi(Z_1)$$

$$U_2 = \phi(Z_2)$$

After this second step, we have generated 2 vectors of length N, U_1 and U_2, that have a correlation of *approximately* ρ, both of which have uniform marginal distributions.

Third, because U_1 and U_2 are uniformly distributed, we can now use the inverse cumulative distribution function method described earlier (Equation 7.2) to generate final random variables:

$$X_1 = F_1^{-1}(U_1)$$

$$X_2 = F_2^{-1}(U_2)$$

where F_1^{-1} and F_2^{-1} are the inverse cumulative distribution functions for the first and second distributions of interest (such as the distribution of sensitivities among cases and controls, respectively). These variables, X_1 and X_2, are now random samples from the distribution of interest and will have a correlation of approximately ρ.

For instance, suppose an analyst wanted to generate 10^5 correlated sensitivities where the sensitivity among the controls, SE_0, and sensitivity among the cases, SE_1, had the following marginal distributions and had a correlation of $\rho = 0.8$:

$$SE_0 \sim \text{beta}(\alpha = 90, \beta = 10)$$

$$SE_1 \sim \text{beta}(\alpha = 80, \beta = 20)$$

First, the analyst would sample 10^5 pairs of random variables from a multivariate normal (Equation 7.7) with a correlation of $\rho = 0.8$. Second, the standard normal cumulative distribution function would be applied to each of the 2 vectors of 10^5 random variables to produce 2 vectors of uniform random variables. Third, and finally, the inverse beta cumulative distribution function with parameters $\alpha = 90$, $\beta = 10$ would be applied to the first vector of random variables and a cumulative inverse beta distribution function with parameters $\alpha = 80$, $\beta = 20$ would be applied to the second vector of random variables. This would result in samples of SE_0 that would have a marginal beta(α=90, β=10) distribution with mean of approximately 0.9 and samples of SE_1 that have a marginal beta(α=80, β=20) distribution with a mean of approximately 0.8. The two sensitivities will have a correlation of approximately $\rho = 0.8$. It is important to recognize that the final correlation will not be exactly what we specified in the beginning but, rather, slightly smaller. Corrections exist to account for this, but trial and error approaches of increasing the original correlation among Z may be quicker. In this example, specifying $\rho = 0.8$ for Z results in a correlation of 0.797 between the sensitivities whereas specifying

$\rho = 0.803$ for the correlation between Z would result in a final correlation of 0.8 between sensitivities. Since the strength of correlation (0.8) is a matter of judgment, small deviations such as these should have little consequence.

An alternative approach, which results in a slightly narrower distribution, was described in the first edition of this text and elsewhere (22). First, select three random variables, u_1, u_2, and u_3, from a standard uniform distribution and calculate the logit of each as:

$$g_i = \text{logit}(u_i) = \ln\left(\frac{u_i}{1 - u_i}\right)$$

To create two correlated random variables, calculate two variables, c_1 and c_2, as:

$$c_1 = \frac{\exp(\sqrt{\rho} \cdot g_0 + g_1 \cdot \sqrt{1 - \rho})}{1 + \exp(\sqrt{\rho} \cdot g_0 + g_1 \cdot \sqrt{1 - \rho})}$$

$$c_2 = \frac{\exp(\sqrt{\rho} \cdot g_0 + g_2 \cdot \sqrt{1 - \rho})}{1 + \exp(\sqrt{\rho} \cdot g_0 + g_2 \cdot \sqrt{1 - \rho})}$$

where ρ is a number between 0 and 1 that denotes the desired correlation between the two variables (e.g., $\rho=0.8$). These two random variables (c_1 and c_2) will be correlated with strength ρ and can be used as the random uniform variables in the general inverse cumulative distribution function approach (Equations 7.1 and 7.2) or specific approaches to sampling from probability distributions that were described above. We used this approach in a SAS macro to adjust for differential misclassification of a variable while retaining some correlation between the sensitivity and/or specificity of classification within levels of a second variable (7).

We have consistently found, however, that the realized distribution yields a realized correlation somewhat less than the specified correlation. For example, if one specifies an initial correlation of 0.80 and uses the first method, they will observe a correlation in the simulation of 0.797 but with the second method will observe a correlation of 0.79. The first method (starting with Equation 7.7) consistently generates an observed correlation closer to the specified correlation, so we now prefer that method. Either method can be fine-tuned by modifying the initial correlation to result in the desired final correlation. The difference is small, so choice of method is unlikely to substantially influence the results of a quantitative bias analysis, but it is good practice to use the method that generates results closer to what has been specified. A larger concern is that we know of no evidence base to inform the strength of correlation between distributions. We originally used 0.8 (7), and that value has consistently been assigned when these methods have been used, but there is only expert opinion and this historical practice to support it. It would be a good idea for analysts using these methods to evaluate the sensitivity of a bias analysis to different assigned values.

Conclusions

Probabilistic bias analysis can provide a substantial improvement over simple bias analysis approaches (or even a series of simple bias analyses used for multidimensional bias analysis) by assigning distributions to the bias parameters. However, designing a valid probabilistic bias analysis requires careful planning just as with any data analysis. Planning realistic distributions to summarize the analyst's uncertainty in the bias parameters takes careful thought and planning and ideally summarizes the totality of the available validation data and other external information. Because the values or distributions assigned to bias models can never be known with certainty, it is also good practice to plan a sensitivity analysis of the bias analysis that will evaluate the sensitivity of results and inferences to different assumption sets (11). Carefully weighing these decisions should improve the reliability of the results of the bias analysis. In the following two chapters, we will discuss how we can apply these decisions to the data to model the impact of bias on study results.

References

1. Greenland S. Multiple-bias modeling for analysis of observational data. J R Stat Soc Ser A. 2005;168:267–308.
2. Shlyakhter A, Mirny L, Vlasov A, Wilson R. Monte Carlo modeling of epidemiological studies. Hum Ecol Risk Assess Int J. 1996;2:920–38.
3. Lash TL, Silliman RA. A sensitivity analysis to separate bias due to confounding from bias due to predicting misclassification by a variable that does both. Epidemiology. 2000;11:544–9.
4. Lash T, Fink AK. Semi-automated sensitivity analysis to assess systematic errors in observational data. Epidemiology. 2003;14:451–8.
5. Phillips CV. Quantifying and reporting uncertainty from systematic errors. Epidemiology. 2003;14:459–66.
6. Greenland S. Interval estimation by simulation as an alternative to and extension of confidence intervals. Int J Epidemiol. 2004;33:1389–97.
7. Fox MP, Lash TL, Greenland S. A method to automate probabilistic sensitivity analyses of misclassified binary variables. Int J Epidemiol. 2005;34:1370–6.
8. Harrison RL. Introduction to Monte Carlo simulation. AIP Conf Proc. 2010;1204:17–21.
9. Greenland S, Pearl J, Robins JM. Causal diagrams for epidemiologic research. Epidemiology. 1999;10:37–48.
10. Robins JM. Data, design, and background knowledge in etiologic inference. Epidemiology. 2001;12:313–20.
11. Lash TL, Fox MP, MacLehose RF, Maldonado G, McCandless LC, Greenland S. Good practices for quantitative bias analysis. Int J Epidemiol. 2014;43:1969–85.
12. Brenner H, Savitz DA. The effects of sensitivity and specificity of case selection on validity, sample size, precision, and power in hospital-based case-control studies. Am J Epidemiol. 1990;132:181–92.
13. Lash TL, Ahern TP, Collin LJ, Fox MP, MacLehose RF. Bias analysis gone bad. Am J Epidemiol. 2021;190:1604–1612
14. Lash TL, Abrams B, Bodnar LM. Comparison of bias analysis strategies applied to a large data set. Epidemiology. 2014;25:576-82.

15. Lash TL, Schmidt M, Jensen AØ, Engebjerg MC. Methods to apply probabilistic bias analysis to summary estimates of association. Pharmacoepidemiol Drug Saf. 2010;19:638–44

16. Ahern TP, Hertz DL, Damkier P, Ejlertsen B, Hamilton-Dutoit SJ, Rae JM, et al. Cytochrome P-450 2D6 (CYP2D6) Genotype and Breast Cancer Recurrence in Tamoxifen-Treated Patients: Evaluating the Importance of Loss of Heterozygosity. Am J Epidemiol. 2017;185:75–85.

17. Vose D. Risk Analysis: A quantitative guide. 3 edition. Chichester, England; Hoboken, NJ: Wiley; 2008. 752.

18. Boudreau DM, Daling JR, Malone KE, Gardner JS, Blough DK, Heckbert SR. A validation study of patient interview data and pharmacy records for antihypertensive, statin, and antidepressant medication use among older women. Am J Epidemiol. 2004;159:308–17.

19. Chien C, Li CI, Heckbert SR, Malone KE, Boudreau DM, Daling JR. Antidepressant use and breast cancer risk. Breast Cancer Res Treat. 2006;95:131–40.

20. Cairo MC, Nelson BL. Modeling and generating random vectors with arbitrary marginal distributions and correlation matrix. Department of Industrial Engineering and Management Sciences: Northwestern University, Evanston, IL; 1997.

21. Li ST, Hammond JL. Generation of pseudorandom numbers with specified univariate distributions and correlation coefficients. IEEE Trans Syst Man Cybern. 1975;5:557–61.

22. Lash TL. Bias Analysis. In: Lash TL, VanderWeele TJ, Haneuase S, Rothman KJ, editors. Modern Epidemiology. 4rd ed. Philadelphia: Wolters Kluwer Health; 2020. 711–54.

Chapter 8
Probabilistic Bias Analysis for Simulation of Summary Level Data

Introduction

Summary level probabilistic bias analysis uses aggregate data, such as a 2 × 2 table describing the cell counts and cross-tabulating an exposure and an outcome, or the number of observed cases divided by a census estimate of the number at risk describing disease surveillance. Summary level probabilistic bias analysis applies many of the same equations we introduced in Chapters 4, 5, and 6. The main difference between simple and probabilistic bias analysis is that for probabilistic bias analysis, instead of assuming a fixed value (or a set of fixed values) to assign to the bias parameters, we specify probability distributions for the bias parameters. The distributions are meant to represent our uncertainty about the values assigned to the bias parameters. We then use Monte Carlo simulation methods to estimate the bias-adjusted estimates and simulation intervals that represent the uncertainty about the estimate deriving from the uncertainty about the values to assign to the bias parameters.

When adjusting for misclassification in Chapter 6, we analyzed data from a case-control study of smoking and breast cancer in which we were concerned that smoking during pregnancy, ascertained by self-report on a birth certificate, was measured with error. Based on an external validation study that used data from medical records as the gold standard, measuring smoking by self-report on a birth certificate had a sensitivity of 78%. However, if we use only this estimate of sensitivity to bias-adjust for the misclassification using a simple bias analysis, we would be expressing strong confidence in our knowledge that the actual sensitivity is exactly 78%. It is the same degree of confidence, in fact, that is inherently implied when the sensitivity and specificity are implicitly set to 100% in a conventional analysis. In reality, we have some uncertainty about the sensitivity of 78%, and this additional uncertainty can be incorporated into the bias analysis. Our uncertainty could come from various sources. These include: (a) the sampling error in the data reported in the validation study, (b) heterogeneity in the estimates of sensitivity,

© Springer Nature Switzerland AG 2021
M. P. Fox et al., *Applying Quantitative Bias Analysis to Epidemiologic Data*,
Statistics for Biology and Health, https://doi.org/10.1007/978-3-030-82673-4_8

especially, among all studies validating maternal self-report of smoking on birth certificates, and (c) the applicability of validation data obtained from one study population when transported to a second study population. Indeed, (b) is particularly important in this example because two similar validation studies reported estimates of sensitivity ranging from 68 to 88% [1]. There is no reason to believe that any one of the point estimates from these studies is more applicable to our study population than the others. A more complete approach to describing our belief about the sensitivity of smoking classification would be to aggregate these results and use them to assign a probability distribution to the sensitivity bias parameter (see discussion below for the rationale and literature supporting this range). These sources of uncertainty about the values to assign to the bias parameters, and possibly other sources of uncertainty, should be propagated in the final 95% interval estimate around the bias-adjusted estimate. These bias-adjusted estimates will typically have wider intervals than intervals that are not bias-adjusted, reflecting the additional uncertainty in the bias parameters.

Assigning a probability distribution to the sensitivity of smoking classification allows a more complete depiction of our belief in the potential impact of information bias in the study. If we consider the average estimate of effect, probabilistic bias analysis is a direct extension of the multidimensional approach, which produces bias-adjusted estimates for a range of fixed bias parameter values. However, the multidimensional approach does not quantitatively assess which of the bias parameter values is most likely. Heuristically, in probabilistic bias analysis, the distribution assigned to the bias parameter assigns a relative weight to each possible value of the bias parameter. The final estimate of effect produced by a probabilistic bias analysis can be thought of as an average of the multidimensional results using the weights assigned by the bias parameter distribution. Adjusted estimates that correspond to bias parameter values that are deemed to be more likely, based on prior evidence and topic matter context, are given higher weight, while estimates corresponding to bias parameters that are deemed relatively unlikely are given lower weight.

Unlike multidimensional bias analysis, probabilistic bias analysis provides both a summary estimate of effect and its distribution under the bias model, incorporating both random error and the uncertainty in the bias parameters. This distribution allows estimation of the average or median bias-adjusted effect and, given the distributions of values assigned to the bias parameters and the bias model, the limits of an interval encompassing some proportion of the bias-adjusted estimates (*e.g.,* the central 95% of the distribution). Knowing this information allows the analyst to summarize the results of a bias analysis incorporating many possibilities for the bias parameters and summarize the variability of possible bias-adjusted estimates, given the assigned probability distributions for the bias parameters and the bias model. Assuming a valid bias model, this summary gives a more complete picture of the impact of the bias on the direction, magnitude and uncertainty in the estimate, allows stronger statements about the impact that the biases might have had, and allows these statements to be made more compactly than with the multidimensional approach.

Analytic Approach for Summary Level Probabilistic Bias Analysis

To paraphrase Tolstoy, all standard epidemiologic analyses resemble one another, but every probabilistic bias analysis is unique in its own way. Each of the three biases we address arises from a different mechanism and as such, the way we probabilistically adjust the estimates will naturally be different for each bias. Selection bias results from differential selection, participation, or completion of follow up of study participants. Misclassification bias results from imperfect measurement of a true covariate value. Uncontrolled confounding bias results from failure to measure (or control through design) a confounding factor or from imperfect measurement of those confounding factors that are controlled analytically. The probabilistic bias analysis methods we present for each of these three biases differ from one another because the structures of the biases are different. Further, the bias mechanism is context specific and careful consideration of the analyst's particular study design and type of bias is necessary. Even within the seemingly narrow context of misclassification, as we show below, the algorithm we use to bias-adjust for misclassification if we have information on sensitivity and specificity is different than the algorithm we use if internal validation estimates of positive and negative predictive values are available. A conceptual model or, better, a formal statistical model of how the biases arise is essential for guiding the methods and analysts need to consider the ways in which multiple sources of uncertainty arise in the context of a specific study before proceeding with the analysis.

Regardless of the type of bias that is being assessed in the probabilistic bias analysis, there is a general framework that can be applied, and all analyses will need to incorporate common elements, such as conventional random error and uncertainty resulting from assuming the bias parameter follows a distribution. We focus here on summary level probabilistic bias analysis and will then discuss individual level or record level probabilistic bias analysis in Chapter 9. The eight steps we will take are summarized in Table 8.1.

The remainder of Chapter 8 details algorithms to implement probabilistic bias analyses for each of the three main sources of bias. The specific implementation and order of steps 4a–4c will depend on what type of bias is being addressed. When presenting these algorithms, we carefully describe the type of data to which we are applying them so readers may follow the examples. The algorithms we have presented were developed based on theory surrounding how biases operate in these applied examples. Many of the algorithms we present have a strong theoretical justification as approximately or, in one special case of misclassification with predictive values, exactly Bayesian modeling solutions. We defer this discussion until Chapter 11, however, where we describe Bayesian analysis and inference in detail. The performance of other algorithms that have a less obvious theoretical justification have been studied in detail [2].

In the following section on implementation, we first discuss each of the steps in Table 8.1 in detail using the example of misclassification of smoking status as

Table 8.1 General steps for a probabilistic bias analysis.

Step	Explanation
1	Identify the likely sources of important bias in the data
2	Identify the bias parameters needed to adjust for the bias
3	Assign probability distributions to each bias parameter
4	Use simple bias analysis methods to incorporate uncertainty in the bias parameters and random error
4a	Incorporate bias parameter uncertainty by randomly sampling from each bias parameter distribution
4b	Generate bias-adjusted data using simple bias analysis methods and the sampled bias parameters
4c	Incorporate conventional random error by sampling summary statistics
5	Save the bias-adjusted estimate and repeat steps 4a-c
6	Summarize the bias-adjusted estimates with a frequency distribution that yields a central tendency and simulation interval

measured by a birth certificate in a study of the relation between smoking during pregnancy and risk of breast cancer. Subsequent sections detail methods for bias analysis applied to uncontrolled confounding and selection bias problems. The algorithm to implement the specific bias analysis will differ in important ways for each of these biases, but we will apply the general outline in Table 8.1 for these problems as well.

Exposure Misclassification Implementation

Step 1: Identify the Source of Bias

As with any bias analysis, the first step is to identify the likely sources of important bias affecting the study results [3]. For our example, we have previously identified that the exposure—smoking status during pregnancy—was assessed by self-report of a mother on the child's birth certificate. One would expect that exposure was not likely to be perfectly classified, and several external validation studies support that expectation. The probabilistic bias analysis will explore the impact of this exposure misclassification on the odds ratio associating smoking during pregnancy with breast cancer occurrence. Note that the example will use the odds ratio as the parameter of interest; this approach is easily extended for the risk difference or risk ratio.

Step 2: Select the Bias Parameters

To adjust for the bias, the analyst must identify the requisite bias parameters. This identification would be accomplished just as in a simple bias analysis, following the approach in earlier chapters. When bias-adjusting for exposure misclassification, the requisite bias parameters are either sensitivity and specificity or positive and negative predictive values. This choice will typically be guided by availability of validation studies. Predictive values are more likely to be available in the case of internal validation studies (see Chapter 3). In this analysis we will use estimates of sensitivity and specificity of exposure classification to inform the probabilistic bias analysis, as they are available from external sources in the existing literature.

Step 3: Assign Probability Distributions to Each Bias Parameter

In the simple bias analysis (Chapter 6), we specified self-report of smoking as reported on a birth certificate to be classified with a sensitivity (*SE*) of 78% and a specificity (*SP*) of 99% and for the misclassification to be nondifferential with respect to the outcome (*i.e.*, independent of true breast cancer status). This simple bias analysis implicitly assumed that we knew the true sensitivity and specificity with perfect certainty and ignored results from other validation studies, which suggested a broader range of possible values (presented in Table 8.2). Given the results of available validation studies conducted at about the same time as the original case control study and in the United States, but in different source populations, we consider various distributions for assignment to the bias parameters. One option is a uniform distribution, which represents an assumption that every sensitivity value between 68 and 88% is equally likely to be the true sensitivity. However, because only one study finds sensitivities as low as 68%, we instead opt for a distribution that places less probability mass on the tail ends of this range. Using the approach outlined in Chapter 7, we specify 0.68 and 0.88 as the 2.5th and 97.5th percentiles of the bias distribution. We use these values to estimate a standard deviation, $sd = \frac{0.88-0.68}{2\cdot 1.96} = 0.05$. Using 78% as the prior mean, we use equations for specifying the parameters of a beta distribution that were presented in Chapter 7:

$$\alpha = x\left(\frac{x(1-x)}{sd^2} - 1\right) = 0.78\left(\frac{0.78(1-0.78)}{0.05^2} - 1\right) = 50.6$$

$$\beta = (1-x)\left(\frac{x(1-x)}{sd^2} - 1\right) = (1-0.78)\left(\frac{0.78(1-0.78)}{0.05^2} - 1\right) = 14.3$$

Table 8.2 Summary of studies comparing birth certificate smoking information with smoking information from other sources (standard).

References	Standard	Sensitivity	Specificity
Buescher et al. [4]	Medical record	0.87	0.99
Piper et al. [5]	Medical record	0.78	0.99
Dietz et al. [1]	Questionnaire	0.88	0.99
Dietz et al. [1]	Capture–recapture	0.68	1

where, as before, x is the mean of the bias distribution and sd is the standard deviation of the bias distribution. This procedure translates our prior knowledge about the bias distribution into a beta($\alpha = 50.6$, $\beta = 14.3$) distribution. We can check that this distribution closely matches our prior knowledge by plotting the distribution, as well as examining percentiles of the distribution. The 2.5th, 50th and 97.5th percentiles of the beta($\alpha = 50.6$, $\beta = 14.3$) distribution occur at 67%, 78% and 87%, which matches the values used to inform the beta parameters quite closely. This choice is not the only defensible one for a distribution of values to assign to the sensitivity of exposure classification. How bias-adjusted estimates change when different distributional assumptions are made should be tested, perhaps by choosing other distributions for assignment to the bias parameter. The same studies in Table 8.2 suggest it is unlikely that non-smoking women report that they do smoke on the birth certificates of their children. We believe it is likely that the true specificity is higher than 95%, and most likely near 99%, so we choose a beta ($\alpha = 70$, $\beta = 1$) distribution which has 2.5th, 50th and 97.5th percentiles of the distribution at 95%, 99% and 100%.

In specifying these bias distributions to be the same in cases and controls, with 100% correlations (*i.e.*, sensitivity and specificity drawn from their respective distributions are used in both cases and controls), we are assuming that the misclassification is non-differential with respect to the outcome. This assumption is defensible in this example because accuracy of self-reporting of smoking at the time of birth certificate completion is unlikely to depend on subsequent breast cancer status. However, if one believes that differential misclassification was possible, the approach discussed here can be easily extended to incorporate four bias distributions: the sensitivity among cases, the sensitivity among controls, the specificity among cases, and the specificity among controls. If the misclassification is differential, draws from the distributions of sensitivity for cases should be correlated with draws from the distributions of sensitivity for controls and draws from the distributions of specificity for cases should be correlated with draws from the distributions of specificity for controls. As described in Chapter 7, the NORTA algorithm can be used to induce a specified level of correlation between bias parameters.

Step 4: Use Simple Bias Analysis Methods to Incorporate Uncertainty in the Bias Parameters and Random Error

As described above, probabilistic bias analyses depend on the context of the problem. For exposure misclassification in case-control studies, there are multiple equivalent ways to conceive of a statistical model for the study (see, for example, Gustafson [6]). We describe one such model for the way data arise in a study. We have found, through simulation studies, that the probabilistic bias analysis algorithm based on this model performs well in terms of bias-adjustment and interval coverage.

We conceptualize this model as follows. We first collect a sample of cases and controls who, with some probability, have some true exposure status (call this random error RE_1). This source of random error is the same random error that we would describe in a conventional analysis using the frequentist 95% confidence interval if there were no misclassification. In a case-control study with no exposure misclassification, RE_1 is the only source of random error.

If there is exposure misclassification, a second source of random error (RE_2) is the misclassification process. Namely, for people who are truly exposed, we assume they report exposure with probability equal to the sensitivity of the classification scheme and for people who are truly unexposed we assume they report being unexposed with probability equal to the specificity of the classification scheme. These misclassification parameters apply to individuals in a stochastic or probabilistic manner. The classification scheme is applied to each individual in the study and has a probability of correct classification equal to the sensitivity or specificity. In the expectation, the proportion of truly exposed subjects classified as exposed will be equal to the sensitivity. In reality, this is a stochastic process that can lead to deviations from the expectation and RE_2 captures this random application of the classification scheme.

These two sources of random error, RE_1 and RE_2, are the total randomness that is inherent in the data generation process. In addition, we have uncertainty in the bias parameters (the sensitivity and specificity) which must also be incorporated. This is the uncertainty that we describe by assigning distributions, rather than fixed values, to the bias parameters (Step 3 in Table 8.1). We describe this model in more statistical terms and discuss its equivalence to a more commonly used model for exposure misclassification in the Appendix at the end of this chapter. We note here, however, that it is this source of uncertainty, the uncertainty in the values for the bias parameters themselves, that leads to the systematic error intervals that we describe below, because error sources RE_1 and RE_2 both approach 0 as the sample size approaches infinity, whereas the uncertainty in the bias parameters does not decrease as we increase the sample size.

To conduct a probabilistic bias analysis, the algorithm must incorporate the uncertainty in the bias parameter and both sources of random error (RE_1 and RE_2). This bias model, which adjusts for misclassification error while also incorporating uncertainty and random error, is used to estimate a distribution of the bias-adjusted estimate using Monte Carlo techniques in which we draw random samples from distributions to propagate the uncertainty and random error into the final bias-adjusted effect estimate. The first random sample we draw (step 4a in Table 8.1) propagates the uncertainty in the bias parameters by sampling from the bias parameter distributions. The second random sample (relating to step 4b in Table 8.1) imputes the unknown true exposure status and propagates the randomness of the misclassification process (RE_2), conditional on the sensitivity and specificity parameters. The third random sample (relating to step 4c in Table 8.1) propagates random error associated with sampling truly exposed individuals from the population of cases and controls (RE_1).

The algorithm we present implements a probabilistic bias analysis that is approximately Bayesian. For that reason, we wait until Chapter 11 on Bayesian bias analysis to offer a formal justification for the algorithm we outline here.

Step 4a: Sample from the Bias Parameter Distributions

In the simple bias analysis (Chapter 6), we specified a single value for each of the bias parameters. In this probabilistic bias analysis, we have now assigned probability distributions to both the sensitivity and the specificity of smoking classification. To begin making bias-adjustments to the data set, we take an initial draw from both of the probability distributions. We then use these values in subsequent steps of the algorithm.

Drawing samples from standard probability distributions is straightforward with modern statistical software such as R, Stata or SAS (see Chapter 7). We first sampled a sensitivity value from a beta($\alpha = 50.6$, $\beta = 14.3$) distribution and obtained a value of 71.22%. Sampling a specificity from a beta($\alpha = 70$, $\beta = 1$) distribution yielded a value of 95.78%. We note, again, that because we assumed non-differential misclassification, we use the same sensitivity values for cases and controls and the same specificity values for cases and controls rather than sampling them separately. These sampled values are treated as fixed quantities in the next two steps of the algorithm.

Step 4b: Generate Bias-Adjusted Data Using Simple Bias Analysis Methods and the Sampled Bias Parameters

We now have an estimated value of sensitivity and specificity with which to work. In this step we will use those bias parameters to bias-adjust the data and account for the random error associated with RE_2; namely, randomness associated with the misclassification process. Our goal is to use the sampled sensitivity and specificity to impute data on the true exposure status given these parameters (using variants of the simple bias analysis methods we used in Chapter 6). We can impute the non-misclassified data very simply, while also accounting for random error (in RE_2) if we know the positive and negative predictive values (Table 8.3). To do this, we need to change the sensitivity and specificity parameters from step 4a into positive and negative predictive values. However, as we have shown in Chapter 6, this manipulation requires not only the sensitivity and specificity, but also the true exposure prevalence among cases and controls.

The first step in this process, therefore, is to generate an estimate of the true prevalence of the exposure while also propagating the conventional random error— namely, that we have assumed that truly exposed individuals are randomly sampled from the population of cases and controls (which we called RE_1 above). We need to sample the proportion of those who are exposed among cases and controls, $P(E_{D1})$ and $P(E_{D0})$, respectively, to allow us to estimate positive and negative predictive

Table 8.3 Equation for calculating expected true frequencies given the observed frequencies and classification predictive values.

	Observed		Misclassification adjusted data	
	E_1	E_0	E_1	E_0
D_1	a	b	$a(\text{PPV}_{D1}) + b(1 - \text{NPV}_{D1})$	$D_{1\ \text{Total}} - A$
D_0	c	d	$c(\text{PPV}_{D0}) + d(1 - \text{NPV}_{D0})$	$D_{0\ \text{Total}} - C$
Total	$a + c$	$b + d$	$A + C$	$B + D$

Table 8.4 Equations for calculating expected true data given the observed data with exposure misclassification.

	Observed		Misclassification adjusted data	
	E_1	E_0	E_1	E_0
D_1	a	b	$[a - D_{1\ \text{Total}}(1 - \text{SP}_{D1})] / [\text{SE}_{D1}{-}1 + \text{SP}_{D1}]$	$D_{1\ \text{Total}} - A$
D_0	c	d	$[c - D_{0\ \text{Total}}(1 - \text{SP}_{D0})] / [\text{SE}_{D0}{-}1 + \text{SP}_{D0}]$	$D_{0\ \text{Total}} - C$
Total	$a + c$	$b + d$	$A + C$	$B + D$

values. Sampling values for these two parameters is made difficult by the fact that we do not directly observe true exposure status. Instead, we can use the sampled sensitivity ($SE_{D1} = 71.22\%$) and specificity ($SP_{D1} = 95.78\%$) to estimate these counts using the same simple bias analysis methods described in Chapter 6. We use these sampled values and the equations in Chapter 6 for estimating the number of exposed cases, and the data arranged as in Table 8.4 to obtain an initial estimate of exposed cases, A, which we denote A^0:

$$A^0 = \frac{a - D_{1total}(1 - SP_{D1})}{SE_{D1} - 1 + SP_{D1}}$$
$$B^0 = D_{1total} - A_0$$

Using $SE_{D1} = 71.22\%$, $SP_{D1} = 95.78\%$, $a = 215$, and $D_{1total} = 1664$, we obtain an initial estimate of $A^0 = 216.09$. If there were 216.09 misclassification-adjusted exposed cases, there would have to be $B^0 = 1664{-}216.09 = 1447.91$ misclassification-adjusted unexposed cases. The distribution for $P(E_{D1})$ can be expressed as a beta distribution:

$$P(E_{D1}) \sim \text{beta}(\alpha = A^0, \beta = B^0)$$

For this example, $P(E_{D1})\sim$ beta($\alpha = 216.09$, $\beta = 1447.91$) which has median of approximately 12.97%. We can draw a single value from this distribution using R, Stata, SAS or Excel as shown in Table 8.5. Drawing a single value for $P(E_{D1})$ from this distribution yielded $P(E_{D1}) = 13.28\%$.

The simulation must then also be applied to the controls. We should not expect $P(E_{D0}) = P(E_{D1})$; in fact, this equality corresponds to the null hypothesis that there is

Program	Syntax
R	*rbeta(1,216.09,1447.91)*
Stata	*g x = rbeta(216.09,1447.91)*
SAS	*x = rand('beta', 216.09,1447.91)*
Excel	*=BETA.INV(RAND(),216.09, 1447.91)*

Table 8.5 Examples of how to randomly sample the exposure prevalence from a beta distribution (beta(α = 216.09, β = 1447.91)).

no association between the exposure and outcome, which is the very association that we aim to estimate. We repeat for controls, the same process we used for cases. The draws for sensitivity and specificity are held constant for the application to controls in the first iteration ($SE_{D0} = SE_{D1} = 71.22\%$ and $SP_{D0} = SP_{D1} = 95.78\%$) because the bias model assumes non-differential exposure misclassification. Were the bias model to allow for differential exposure misclassification, the sensitivity and specificity for controls would be drawn from distributions assigned to them and may be correlated with the sensitivity and specificity for cases, as described in Chapter 7.

Repeating the process we used for cases, we first estimate the number of misclassification-adjusted exposed controls using the sensitivity and specificity to obtain an initial estimate of C^0:

$$C^0 = \frac{c - D_{0total}(1 - SP_{D0})}{SE_{D0} - 1 + SP_{D0}}$$
$$D^0 = D_{0total} - C_0$$

Using $SE_{D0} = 71.22\%$, $SP_{D0} = 95.78\%$, $c = 668$, and $D_{0total} = 4964$, we obtain an initial estimate of $C^0 = 684.36$, which implies $D^0 = 4964–684.36 = 4279.64$. Using this result to specify a beta distribution, as above, we have $P(E_{D0})$ ~ beta ($\alpha = 684.36$, $\beta = 4279.64$) that has a median of approximately 13.78%. Drawing a single value for $P(E_{D0})$ from this distribution yielded $P(E_{D0}) = 13.99\%$.

As we noted previously, our goal is to simulate the misclassification-adjusted cell counts, but to do so we need positive and negative predictive values. Recall the equation for estimating the bias-adjusted number of exposed cases using predictive values (see Table 8.3):

$$A = a \cdot PPV_{D1} + b(1 - NPV_{D1}) \tag{8.1}$$

To estimate the bias-adjusted number of exposed cases (A), we must compute PPV_{D1} and NPV_{D1} using equations presented in Chapter 6:

$$PPV_{D1} = \frac{SE_{D1}P(E_{D1})}{SE_{D1}P(E_{D1}) + (1 - SP_{D1})(1 - P(E_{D1}))} \tag{8.2}$$

$$NPV_{D1} = \frac{SP_{D1}(1 - P(E_{D1}))}{(1 - SE_{D1})P(E_{D1}) + SP_{D1}(1 - P(E_{D1}))} \tag{8.3}$$

We cannot know $P(E_{D1})$ with certainty—it is a component of the target estimand. Instead, we substitute the estimate of $P(E_{D1})$ that was sampled in step 4b along with the sampled sensitivities and specificities into Equations 8.2 and 8.3. Using our sampled values, we find $PPV_{D1} = 72.10\%$ and $NPV_{D1} = 95.60\%$.

To complete this step and impute the misclassification-adjusted number of exposed, we simulate a binomial trial for the number of truly exposed cases among those cases who report being exposed using $PPV_{D1} = 72.10\%$. Conceptually, this simulation is like flipping a coin for each observed exposed case to decide whether to retain the exposure status as exposed or to reclassify the exposure status as unexposed. This simulation parallels the approach we will use for record level bias-adjustments in Chapter 9. The flipped coin does not have a 50% probability of coming up heads though, rather it has probability of retaining the exposure status as exposed of 72.10% (the positive predictive value in cases) and probability of reclassifying the exposure status as unexposed of 1–72.10% or 27.90%. The expectation of this simulation, summed over all observed exposed cases, is $a \cdot PPV_{D1}$, which is the first term on the right side of Equation 8.1. We then also simulate a binomial trial for the observed count of misclassification-adjusted exposed cases among those cases who reported being unexposed with probability $1 - NPV_{D1} = 1$–95.60% = 4.40%. Conceptually, this simulation is like flipping a coin for each observed unexposed case to decide whether to reclassify the exposure status as exposed or to retain the exposure status as unexposed. Again, the flipped coin does not have a 50% probability of coming up heads, it has a probability of reclassifying the exposure status as exposed of 4.40% (1 minus the negative predictive value in cases) and probability of retaining the exposure status as unexposed of 95.60% (the negative predictive value). The expectation of this simulation is $b \bullet (1 - NPV_{D1})$, which is the second term on the right side of Equation 8.1. The new estimate of A is:

$$W_1 \sim \text{binomial}(a, ppv_1)$$
$$Y_1 \sim \text{binomial}(b, (1 - npv_1))$$
$$A = W_1 + Y_1$$
$$B = D_{1total} - A$$

These binomial trials can be simulated in R, Stata, SAS or Excel using the data in the current example with the commands in Table 8.6.

Sampling in this way yields a value of $W_1 = 153$ and $Y_1 = 77$ to give a value of $A = 153 + 77 = 230$, whereas in expectation $A = 218.8$. The difference between 230 and 218.8 results from modeling the second source of random error, RE_2. We compute B (the number of unexposed cases) by difference ($B = D_{1total} - A$), so $B = 1664$–$230 = 1434$ in the first iteration of the simulation while in expectation, $B = 1664$–$218.8 = 1445.2$. Note that it is critical that we sample the frequency of exposed and then set the number of unexposed to be the total minus the simulated frequency of exposed (i.e., $A = W_1 + Y_1$ and $B = D_{1Total} - A$) because the number of cases is fixed. Simulating both the A and B cell, instead of computing B by

Table 8.6 Examples of how to randomly assign the misclassification-adjusted frequency of exposed from the binomial distributions: *binomial(n = 215, p = 0.721)* and *binomial(n = 1449, p = 0.044)*.

Program	Syntax
R	*rbinom(1,215,0.721)+ rbinom(1,1449,0.044)*
Stata	*g x = rbinomial(215,0.721) + rbinomial(1449,0.044)*
SAS	X = rand('binomial',215,0.721) + rand('binomial',1449,0.044)
Excel	=BINOM.INV(215,0.721,RAND()) + BINOM.INV(1449,0.044,RAND())

difference, would usually lead to a different total number of cases then was actually observed.

The same procedure is used to impute the true exposure status among controls. The equation for estimating the bias-adjusted number of exposed controls from Table 8.3 is:

$$C = c \cdot PPV_{D0} + d(1 - NPV_{D0})$$

To impute the number of misclassification-adjusted exposed controls (C), we must compute PPV_{D0} and NPV_{D0}, which, as we have noted, are a function of sensitivity, specificity, and the prevalence of the exposure in controls, $P(E_{D0})$. Even under a non-differential misclassification model that holds sensitivity and specificity constant for cases and controls, we should not expect PPV_{D0} and PPV_{D1} to be the same, nor should we expect NPV_{D0} and NPV_{D1} to be the same. The equations to estimate PPV_{D0} and NPV_{D0} are:

$$PPV_{D0} = \frac{SE_{D0}P(E_{D0})}{SE_{D0}P(E_{D0}) + (1 - SP_{D0})(1 - P(E_{D0}))}$$

$$NPV_{D0} = \frac{SP_{D0}(1 - P(E_{D0}))}{(1 - SE_{D0})P(E_{D0}) + SP_{D0}(1 - P(E_{D0}))}$$

where $P(E_{D0})$ is the proportion of the controls that is truly exposed to the dichotomous exposure. Using estimates of sensitivity and specificity from step 4a and estimates of $P(E_{D0})$ from step 4b, yields estimates of $PPV_{D0} = 73.30\%$ and $NPV_{D0} = 95.34\%$.

To complete the model for the second source of uncertainty among controls, we simulate a binomial trial for the observed exposed controls with $PPV_{D0} = 73.30\%$ and for the observed unexposed controls with $1 - NPV_{D0} = 1-95.34\% = 4.66\%$. The number of misclassification-adjusted exposed controls, C, can be sampled from the following distributions:

$$W_0 \sim \text{binomial}(c, ppv_0)$$
$$Y_0 \sim \text{binomial}(d, (1 - npv_0))$$
$$C = W_0 + Y_0$$
$$D = D_{0\,total} - C$$

In this example, we sample from $W_0 \sim \text{binomial}(668, 0.733)$ and $Y_0 \sim \text{binomial}$ (4296, 0.0466). Doing so resulted in $W_0 = 482$ and $Y_0 = 189$ and, therefore, $C = 482 + 189 = 671$, whereas in expectation $C = 689.9$. The difference between 671 and 689.9 results from modeling the second source of uncertainty. We compute D (the number of unexposed controls) by difference ($D = D_{0total} - C$), so $D = 4964$–$671 = 4293$ in the first iteration as opposed to the expected value of $D = 4964$–$689.9 = 4274.1$.

Step 4c: Incorporate Conventional Random Error by Sampling Summary Statistics

Our final task is to propagate the random error due to assuming that truly exposed individuals are randomly sampled from the population of cases and controls (RE_1, above). An astute reader may comment that we are double counting RE_1 since we have already sampled $P(E_{D1})$ and $P(E_{D0})$ in step 4b. We leave a formal justification for why this is acceptable until Chapter 11. For now, it suffices to think of the sampling $P(E_{D1})$ and $P(E_{D0})$ in step 4b as a starting point to generate the necessary PPV and NPV.

To simulate the final source of random error (RE_1), we choose a standard normal deviate (z_i) for each iteration (i) and multiply it by the standard error calculated from the imputed data for the misclassification-adjusted exposure (SE_i^{adj}) to simulate the conventional random error in the study. Then, for each simulation, we combine the systematic error and random error as:

$$estimate_i^{total} = estimate_i^{adj} - z_i \cdot SE_i^{adj} \tag{8.4}$$

where $estimate_i^{total}$ is a single simulated estimate of association that incorporates both sources of random error as well as the uncertainty in the bias parameters, $estimate_i^{adj}$ is a single simulated estimate of association generated using the misclassification-adjusted data from step 4b, and SE_i^{adj} is the standard error based on the bias-adjusted exposure status from step 4b. If the estimate is generally more normally distributed on the ln scale, such as for relative measures like the odds ratio or risk ratio, then Equation 8.4 would be modified as:

$$OR_i^{total} = e^{\ln(OR_i^{adj}) - z_i \cdot SE_i^{adj}} \tag{8.5}$$

Table 8.7 Examples of how to randomly sample from a standard normal distribution.

Program	Syntax
R	*rnorm(1,0,1)*
Stata	*g z = rnormal(0,1)*
SAS	*z = RAND('NORMAL')*
Excel	*NORM.S.INV(RAND())*

Using this approach, it is easy to augment statistical computing code created for probabilistic bias analysis to generate a total error interval by sampling z_i from a standard normal distribution using functions built-in to most statistical packages, as shown in Table 8.7.

In the example above, we sample z_1 from a normal($\mu = 0$, $\sigma = 1$) distribution and obtained $z_1 = 0.231$. From step 4b, we have A $= 230$, B $= 1434$, C $= 671$ and D $= 4293$, which gives a bias-adjusted $OR_1^{adj} = 1.03$ and $SE_1^{adj} = \sqrt{\left(\frac{1}{230} + \frac{1}{1434} + \frac{1}{671} + \frac{1}{4293}\right)} = 0.082$. Using this standard normal deviate in Equation 8.5 with the bias-adjusted estimate we obtain:

$$OR_1^{total} = e^{\ln(1.03) - 0.231 \cdot 0.082} = 1.01$$

Step 4c (Alternate): Resample the Prevalence of Misclassification Adjusted Exposure

We offer an alternative way to incorporate the conventional random error (RE_1). Results obtained using step 4c or step 4c (alternate) will typically be similar; however, we have found through simulation that step 4c (alternate) may have slightly higher bias. Up to this step in the algorithm, we have sampled (a) the bias parameters, sensitivity and specificity and (b) the prevalence of misclassification-adjusted exposure among cases and controls and a bias-adjusted estimate of the number of truly exposed cases and controls. We could use the bias-adjusted estimate of the count of truly exposed cases and controls to compute a bias-adjusted odds ratio; however, the performance of the algorithm is improved (in terms of coverage probability in simulation studies) if we treat $P(E_{D1})$ and $P(E_{D0})$ from step 4b as initial guesses that were necessary to obtain estimates of A, B, C and D. Now that we have estimates of those counts, we can obtain a final estimate of $P(E_{D1})$ and $P(E_{D0})$ by repeating the first part of step 4b, above. Namely, given the estimate of the misclassification-adjusted case counts from step 4c, we resample the proportion of misclassification adjusted exposure among cases and controls:

$$P(E_{D1}) \sim \text{beta}(\alpha = A, \beta = B)$$
$$P(E_{D0}) \sim \text{beta}(\alpha = C, \beta = D)$$

This sampling step gives a final estimate of the prevalence of misclassification-adjusted exposure among the cases and controls in this iteration of the simulation, and allows computation of the bias-adjusted odds ratio as:

$$OR_1^{total} = \frac{P(E_{D1})}{1 - P(E_{D1})} \Big/ \frac{P(E_{D0})}{1 - P(E_{D0})}$$

where the superscript "total" refers to the fact that this estimate incorporates the uncertainty in the bias parameters as well as the two sources of random error and the subscript 1 indexes the iteration number. From step 4c, we have $A = 230$, $B = 1434$, $C = 671$ and $D = 4293$, so we sample from $P(E_{D1}) \sim \text{beta}(\alpha = 230, \beta = 1434)$ and $P(E_{D0}) \sim \text{beta}(\alpha = 671, \beta = 4293)$. The first iteration yielded values of $P(E_{D1}) = 14.16\%$ and $P(E_{D0}) = 13.84\%$, and therefore $OR_1^{total} = 1.03$.

We note that, if this were not data from a case-control study and a different summary measure was desired, we could calculate the risk ratio as $RR_1^{total} = P(E_{D1})/P(E_{01})$ or the risk difference as $RD_1^{total} = P(E_{D1}) - P(E_{01})$.

Step 5: Save the Bias-Adjusted Estimate and Repeat Steps 4a-c

The bias-adjusted odds ratio of 1.01 (from Step 4c) is the result of a single simulation to bias-adjust for smoking misclassification, taking account of uncertainty about the values to assign to the sensitivity and specificity, the uncertainty from applying the values to the dataset, and the conventional random error. It represents only one possible estimate bias-adjusted for misclassification. Making inference based on this first adjustment alone would inadequately incorporate our uncertainty about the sensitivity and specificity values and the other sources of uncertainty. Repeating steps 4a–4b would yield different estimates of sensitivity and specificity of exposure classification and would therefore give a different bias-adjusted estimate of association. Repeating step 4c would yield a different influence of conventional random error, so would also change the bias-adjusted estimate of association. Ideally, we would like to know the distribution for the bias-adjusted estimate of association that results from combining the probability distributions assigned to the bias parameters and the distributions assigned to the random errors associated with the estimate. This distribution can be approximated by repeating steps 4a-4c many times, saving each bias-adjusted estimate of association obtained from each iteration. This series of bias-adjusted estimates can then be used to create a histogram, which approximates the desired distribution. The more iterations that are conducted, the better the approximation. Comparison of the distribution with the conventional distribution indicates the direction, magnitude, and uncertainty of the bias arising from exposure misclassification, assuming a valid bias model.

It is difficult to implement probabilistic bias analysis without using statistical software that has functions to generate random numbers and the ability to bias-adjust the data many times while each time saving the bias-adjusted estimates. This can be accomplished without much difficulty in statistical analysis programs like SAS, Stata and R, and is slightly more complicated in programs such as Microsoft Excel. For any probabilistic bias analysis, the user must decide how many simulations to run, such that the resulting frequency distribution is sufficiently precise to satisfy the inferential objective. This number can be tens or hundreds of thousands of simulations if there are many bias parameters and the probability distributions assigned to those parameters are wide. With modern computing power a million iterations can be run in a matter of seconds in many statistical packages.

Step 6: Summarize the Bias-Adjusted Estimates with a Frequency Distribution that Yields a Central Tendency and Simulation Interval

We repeated the simulation described above 10,000 times using Excel and 100,000 times using R, each time choosing new estimates of sensitivity and specificity, adjusting for the bias, reintroducing conventional random error using SE_i in Equation 8.5, and then saving the bias-adjusted odds ratios. The distribution of OR_i^{total} yielded by these simulations is shown in Figure 8.1. The histogram shape is similar for both sets of simulations. However, the distribution of OR_i^{total} generated by R extends above and below the distribution generated by Excel, presumably because of

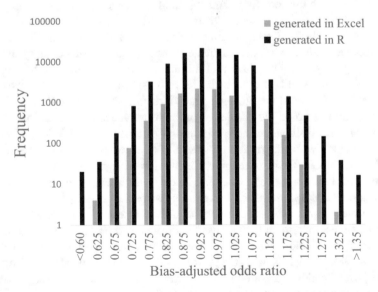

Figure 8.1 Histogram of bias-adjusted odds ratios $\left(OR_i^{total}\right)$, adjusting for misclassification and incorporating random error using SE_i^{adj} in Equation 8.5. 10,000 iterations generated using Excel and 100,000 iterations generated using R. Note the frequency is on a logarithmic scale.

Table 8.8 Results of a probabilistic bias analysis of the relation between smoking and breast cancer bias-adjusting for nondifferential misclassification of smoking using 100,000 iterations in R.

Analysis	Median OR	95% Interval
Conventional result, assuming no bias	0.95	(0.81, 1.13)[a]
Systematic error only	0.95[b]	(0.93, 0.95)[b]
Systematic and random error	0.95[b]	(0.78, 1.14)[b]

Original data from Fink and Lash, (2003) [7].
[a]Computed using exact confidence intervals.
[b]Computed using simulation.

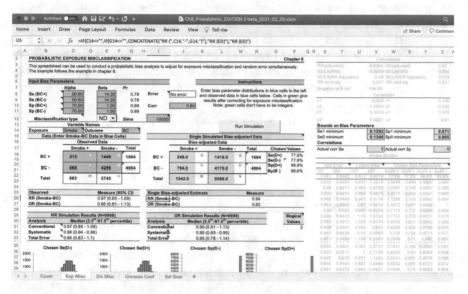

Figure 8.2 Screenshot of Excel spreadsheet to conduct probabilistic bias analysis for exposure misclassification. Original data from Fink and Lash, 2003 [7].

the tenfold greater number of iterations, which allows greater chance for rare combinations of parameters to be observed. Negative cell frequencies were generated by one iteration out of 10,000 in Excel and five iterations out of 100,000 in R; these iterations were discarded.

These distributions can be summarized by creating a 95% simulation interval. Conceptually, we begin by ranking the bias-adjusted estimates from lowest to highest. Functionally, this ranking is not necessary to implement because the software will rank the results as part of the computation. To summarize the results, we use the median value of the bias-adjusted estimates as our measure of central tendency. We then take the 2.5th percentile and 97.5th percentile of the distributions as the limits between which 95% of the bias-adjusted odds ratios lie in this set of simulations. We can create intervals like this for the series of estimates bias-adjusted for each source of uncertainty. The results are presented in Table 8.8 for the 100,000 iterations generated in R. Figure 8.2 shows a Screenshot of the Excel spreadsheet available on the text's website to conduct probabilistic bias analysis for exposure misclassification.

Without accounting for the classification errors, the result showed that women who smoked during pregnancy had 0.954 times the odds of breast cancer as those who did not with a conventional frequentist 95% confidence interval from 0.806 to 1.129. We refer to this as the conventional result because it only includes an assessment of random error and assumes no misclassification.

To summarize the extent of the systematic error, we first present results for a misclassification-adjusted OR that accounts for uncertainty in the values of sensitivity and specificity but does not include any random error. We accomplish this by sampling sensitivities and specificities as normal in step 4a. We also proceed through step 4b and 4c as normal but without any random sampling (as these steps propagate the random error in the estimate). Instead of randomly sampling in steps 4b or 4c, we set the parameter to the mean of the distribution we would have sampled from. For example, instead of sampling the number of truly exposed cases in step 4b using:

$$W_1 \sim binomial(a, ppv_1)$$
$$Y_1 \sim binomial(b, (1 - npv_1))$$
$$A = W_1 + Y_1$$

we instead set $A = a\, ppv_1 + b(1 - npv_1)$. In step 4c, rather than sampling $z \sim N(0, 1)$, we set z = 0. This procedure results in a misclassification-adjusted OR that allows for uncertainty in sensitivity and specificity but no random error. This yielded a median odds ratio of 0.949 with a 95% simulation interval ranging from 0.928 to 0.953. Even though there was evidence of substantial misclassification due to low sensitivity, the median of the bias-adjusted estimates was nearly identical to the conventional odds ratio. In addition, 95% of the distribution of this OR lies between approximately 0.93 and 0.95, suggesting negligible impact of uncertainty in the extent of misclassification. The extent of bias due to systematic errors can be difficult to predict in practice and conducting a formal bias analysis like this will often be the only way to determine whether systematic biases are important.

The last row of Table 8.8 shows the point estimate and 95% simulation interval bias-adjusting for misclassification and accounting for random error. The median bias-adjusted estimate was 0.947, extremely close to the median from the systematic bias alone OR. The simulation interval is wider still, with limits extending from 0.780 to 1.140. We can evaluate the importance of each source of uncertainty by computing and comparing the widths of their respective intervals. Using the results in Table 8.8, the width of the interval corresponding to systematic bias and uncertainty in the bias parameters alone is 0.953/0.928 = 1.027. The width of the interval corresponding to all sources of uncertainty and random error is 1.140/0.780 = 1.462. The uncertainty added by the random error, beyond the uncertainty in the sensitivity and specificity was 1.462/1.027 = 1.424. The largest component of the total uncertainty, therefore, arises from the random error, not from the uncertainty in bias parameters. That has implications for future studies. More precise information on sensitivities and specificities, collected through future validation studies, is unlikely to reduce the width of the uncertainty intervals. Instead, future researchers would be wise to consider larger case-control studies, presuming that the interval from the present study is too wide to meet the inferential goal. As always, these results and inferences assume that the specifications of the bias distributions are valid and there are no other sources of error.

When interpreting a simulation interval, we recommend focusing on three things: magnitude, direction and uncertainty. By magnitude we mean how far from the bias-adjusted estimate was the conventional point estimate, an indicator of how much bias the conventional estimate suffered given the assumed bias model. By direction we mean noting where the median bias-adjusted estimate lies compared with the conventional point estimate. Was the bias, on average, towards or away from the null, or even past the null? And by uncertainty, we mean how wide was the bias-adjusted interval (or the total error interval) compared with the conventional 95% confidence interval. This comparison gives a sense for whether there was more uncertainty from the systematic error or the random error. These three allow a full description of the impact of the bias.

This application of bias-adjustment for misclassification dealt with exposure misclassification in a case-control study. The same algorithm can be used for cohort studies, in which case other measures of effect might be preferred and could be sampled in step 4c. Disease misclassification in cohort studies can also be bias-adjusted using this algorithm with minor modifications described below. As discussed in Chapter 6, careful consideration is warranted prior to bias-adjusting for disease misclassification in a case-control study, however.

Before we continue, we note that in the previous edition of this textbook we applied a slightly different algorithm for probabilistic bias analysis adjustments. In that algorithm, step 4b above did not entail any probabilistic sampling of the true cell counts or the probability of exposure among cases and controls. Instead, the previous algorithm used sensitivity and specificity parameters in step 4a, applied them to the observed data using equations in Chapter 6 to calculate a 2 × 2 table of case status by misclassification-adjusted exposure. Finally, ORs were calculated from each of these tables and random error was introduced as in step 4c. In many cases, that approach works quite well and is clearly superior to the conventional approach of completely ignoring misclassification. However, while that approach worked well, we found in simulation studies that the coverage probability could be somewhat lower than the nominal level, particularly if the bias parameter distributions were very precise.

Misclassification Implementation: Predictive Values

In this misclassification problem, instead of using sensitivity and specificity to inform the probabilistic bias analysis, we could have used positive and negative predictive values directly to adjust for the bias. This approach works well if we have good estimates of the positive and negative predictive values, which must usually be obtained from an internal validation study (see Chapter 3). An internal validation study would ordinarily be required to support use of predictive values directly because, as explained above and in Chapter 3, predictive values depend on the prevalence of the variable in the population. Estimates of predictive values from one population are therefore seldom validly transportable to a second population.

To conduct a probabilistic bias analysis using predictive values, we can adapt the algorithm above in a straightforward manner. In the previous algorithm, we sampled sensitivities and specificities and then converted them into positive and negative predictive values in order to draw samples of the misclassification-adjusted data. If positive and negative predictive values are available to the analyst, the algorithm is shortened substantially. We describe only the steps from Table 8.1 which differ if predictive values are used. In step 2, the analyst would decide to use positive and negative predictive values for the bias parameters. In step 3, the analyst would use internal validation data to specify distributions for the four predictive values. In step 4a the analyst would draw random samples from the predictive values to propagate uncertainty in the bias parameters. Step 4b would ignore the steps on estimating A^0, B^0, C^0, D^0 and sampling $P(E_1)$ and $P(E_0)$. Because samples from the distribution for the predictive values are immediately available, Step 4b would simplify to sampling the bias-adjusted cell counts among cases:

$$W_1 \sim \text{binomial}(a, ppv_1)$$
$$Y_1 \sim \text{binomial}(b, (1 - npv_1))$$
$$A = W_1 + Y_1$$
$$B = D_{1total} - A$$

and controls

$$W_0 \sim \text{binomial}(c, ppv_0)$$
$$Y_0 \sim \text{binomial}(d, (1 - npv_0))$$
$$C = W_0 + Y_0$$
$$D = D_{0total} - C$$

Step 4c would incorporate random error exactly as in the previous approach and these steps would be iterated a large number of times to produce a summary misclassification adjusted OR and its simulation interval. This approach to implementing bias adjustments for misclassification using positive and negative predictive values is very flexible and can be implemented for cohort or case-control studies with exposure or disease misclassification. We describe the small changes necessary for disease misclassification below.

If the distributions assigned to predictive values are derived from an internal validation study, then there is an additional important nuance. The subset of participants who were in the internal validation study have a gold standard measurement, which should be used in the computation of the effect estimate. These participants should be removed from the bias-adjustment procedure, and then added back into the computations to estimate the bias-adjusted effect. The simplest way to accomplish this is to separately estimate the effect of the gold-standard exposure on the outcome among participants in the validation study as well as the bias-adjusted effect (using methods described above) among those not in the validation study. The two effect estimates can be pooled by weighting them by the inverse of their respective variances. If we estimate an effect (on the ln scale of relative measures) θ_1 with a variance $V(\theta_1)$ among those in the validation study and a bias adjusted estimate θ_2 with a variance $V(\theta_2)$ among those not in the validation study, a pooled estimate of effect, θ_{pool}, and pooled variance, $V(\theta_{pool})$, can be estimated using:

$$\theta_{pool} = \frac{(\theta_1 wt_1 + \theta_2 wt_2)}{wt_1 + wt_2}$$

$$V(\theta_{pool}) = \frac{1}{wt_1 + wt_2}$$

$$wt_1 = 1/V(\theta_1)$$

$$wt_2 = 1/V(\theta_2)$$

Misclassification Implementation: Predictive Values – Alternative

The misclassification method we described above using predictive values is quite general and easy to implement. However, the statistical structure of misclassification in certain study designs lends itself to an alternative procedure. This algorithm is specific to exposure misclassification in case-control studies. It can be modified to work with disease misclassification in cohort studies. It is not suitable for use in other situations, namely it is not suited for disease misclassification in case-control studies or exposure misclassification in cohort studies. However, in the two situations where it can be used, it is extremely fast.

When using predictive values, the bias model is much more straightforward than the structure imposed when using sensitivities and specificities. In particular, it is easy to specify a statistical model for misclassification with positive and negative predictive values that maintains key independencies among random variables, allowing for a very easy algorithm to update the parameters [8]. Further, this algorithm appears to perform slightly better (closer to nominal coverage) in preliminary simulations than modifying the algorithm designed for misclassification adjustment using sensitivities and specificities. Specifically, we proceed by modifying steps 4a–4c in Table 8.1 and rearranging the order to be 4a, 4c, and then 4b. First, we implement step 4a and propagate uncertainty by sampling positive and negative predictive values from their respective bias parameter distributions. Second, we

Table 8.9 Validation data comparing birth certificate (misclassified) BMI category to medical record review (gold standard) by early preterm birth from Lash et al., 2009 [9].

	Term		Preterm	
	Medical record			
Birth certificate	Underweight	Normal weight	Underweight	Normal weight
Underweight	132	47	50	27
Normal weight	2	115	0	120

implement step 4c and propagate random error by sampling from the probability of *observed* exposure among cases and controls, $P(E_1^*)$ and $P(E_0^*)$. Recall that above we sampled from the probability of modeled true exposure [$P(E_1)$ and $P(E_0)$]. Third, we implement step 4b and adjust the observed data for the bias by using equations in Chapter 6 to estimate a misclassification-adjusted probability of true exposure: $P(E_i) = ppv_i P(E_i^*) + (1 - npv_i)(1 - P(E_i^*))$. These misclassification-adjusted estimates of the true exposure prevalence among cases($i = 1$) and controls($i = 0$) can be used to estimate the misclassification-adjusted effect of interest. We show in Chapter 11 that this procedure mimics a Bayesian estimation procedure with a prior distribution designed to maximize the influence of the observed data.

We demonstrate this method using an example introduced in Chapter 6 on a study of pre-pregnancy maternal weight category and risk of preterm delivery. While the data actually arise from a cohort study, we pretend for purposes of this example that this is a case-control study. Following the outline in Table 8.1, in step 1 we identify misclassification of self-report of pre-pregnancy BMI as the bias of particular interest since it is only ascertained by self-report on birth certificates following delivery. In step 2, we identify positive and negative predictive values as the relevant bias parameters. This choice of bias parameter follows immediately from the study by Bodnar and colleagues who conducted a validation study using pre-pregnancy clinic data to estimate positive and negative predictive values for women who did and did not deliver preterm. In step 3, we assign bias parameter distributions to these parameters. Validation data for this study are shown in Table 8.9 and we use them to directly inform the parameters of our bias distributions. For example, the PPV among preterm births is the fraction of mothers whose infants were underweight via medical record given they reported being underweight on the birth certificate ($50/(50 + 27)$). Analyzing the validation data, we obtain a maximum likelihood estimate of $ppv_1 \approx 65\%$ with exact 95% CI: 53%, 75%. We choose a bias distribution for ppv_1 that reflects these values by assuming $ppv_1 \sim beta(\alpha = 50, \beta = 27)$. This gives values of 54%, 65% and 75% for the 2.5th, 50th, and 97.5th percentiles of the distribution, closely matching the maximum likelihood point estimate and the 95% confidence interval. We use similar logic and data from Table 8.9 to inform the remaining parameters and note that we add 0.5 to the zero cell in the table, otherwise the bias distribution of npv_1 would be a point mass at 1, with no uncertainty:

$$ppv_1 \sim \text{beta}(\alpha = 50, \beta = 27)$$
$$ppv_0 \sim \text{beta}(\alpha = 132, \beta = 47)$$
$$npv_1 \sim \text{beta}(\alpha = 120, \beta = 0.5)$$
$$npv_0 \sim \text{beta}(\alpha = 115, \beta = 2)$$

Having specified the bias distributions, we can begin the steps to incorporate uncertainty in the bias parameters and random error in the bias-adjusted effect. The three steps, outlined above, proceed by first sampling positive and negative predictive values from their bias parameter distributions. In our first random sample, we find $ppv_1 = 0.704$, $ppv_0 = 0.779$, $npv_1 = 0.998$, $npv_0 = 0.984$. We next incorporate random error by sampling the observed (potentially misclassified) exposure probability for cases and controls. This step parameterizes the distribution of observed exposure using beta distributions with parameters equal to the observed cell counts, which were presented in Chapter 6. Since there were 599 exposed cases and 4978 unexposed cases, we have $\Pr(E_1^*) \sim \text{beta}(599, 4978)$. This specification gives a mean for $\Pr(E_1^*) = \frac{599}{599+4978} = 0.107$ with 2.5th and 97.5th percentiles of 0.099 and 0.116. These values are identical (with rounding) to the maximum likelihood estimate for this parameter and its exact 95% confidence intervals. Thus, sampling from this beta distribution will reproduce the random error in the observed data. Similarly, we parameterized the prevalence of exposure among controls based on the 31,175 term births whose mothers reported underweight and 391,851 term births whose mothers reported normal weight. These crude numbers are used to parameterize $\Pr(E_0^*) \sim Beta(31175, 391851)$. In our first random sample, we find $\Pr(E_1^*) = 0.108$ and $\Pr(E_0^*) = 0.074$. Finally, we combined the uncertainty and random error while estimating the misclassification-adjusted prevalence of exposure among cases and controls:

$$
\begin{aligned}
\Pr(E_1) &= ppv_1\Pr(E_1^*) + (1 - npv_1)(1 - \Pr(E_1^*)) \\
&= 0.704(0.108) + (1 - 0.998)(1 - 0.108) \\
&= 0.078
\end{aligned}
$$

$$
\begin{aligned}
\Pr(E_0) &= ppv_0\Pr(E_0^*) + (1 - npv_0)(1 - \Pr(E_0^*)) \\
&= 0.779(0.074) + (1 - 0.984)(1 - 0.074) \\
&= 0.072
\end{aligned}
$$

A misclassification-adjusted OR is estimated as $\left(\frac{0.078}{1-0.078}\right) / \left(\frac{0.072}{1-0.072}\right) = 1.09$. Iterating these 3 steps 1,000,000 times gives an estimate of the distribution of the bias-adjusted effect. The median, 2.5th and 97.5th percentiles of the distribution gives us final estimates of OR $= 1.07$; 95% SI: 0.68, 1.51. As we stated at the beginning of this section, this algorithm can be implemented for disease misclassification in cohort studies, with the adjustment noted below.

Misclassification of Outcomes and Confounders

The steps described above for the main approach to bias modeling of exposure misclassification are the same procedures we would use for adjustment of a misclassified outcome or confounder. No differences need to be made except to choose which variable to bias-adjust using the corresponding simple bias analysis models provided in Chapter 6. This necessitates minor changes to the algorithm above. For disease misclassification, in step 4b, rather than sampling $P(E_1)$ and $P(E_0)$ from beta distributions, we would sample $P(D_1) \sim beta(\alpha = A_0, \beta = C_0)$ and $P(D_0) \sim beta(\alpha = B_0, \beta = D_0)$. Subsequent positive and negative predictive formula would be for disease status, rather than exposure:

$$P(D_1) = ppv_1 P(D_1^*) + (1 - npv_1)(1 - \Pr(D_1^*))$$
$$P(D_0) = ppv_0 P(D_0^*) + (1 - npv_0)(1 - \Pr(D_0^*))$$

The bias adjusted estimates of disease probability among the exposed and unexposed can be used to estimate the desired measure of effect.

Uncontrolled Confounding Implementation

We can apply the same steps outlined in Table 8.1 to conduct a probabilistic bias analysis that models error from uncontrolled confounding. We present two ways of conducting a probabilistic bias analysis for uncontrolled confounding. The first way is similar in spirit to the methods used to adjust for misclassification with sensitivities and specificities. In this method we randomly sample from the bias parameters and use those samples to compute the probability of the unobserved exposure-disease-confounder specific cell counts. We randomly sample the cell counts, compute a confounding-adjusted effect, and then incorporate random error. The second method directly models the confounding relative risk that was discussed in Chapter 5. In this method, we use samples from the bias parameter distributions to compute a confounding relative risk, which is then used to directly bias-adjust the observed risk ratio before incorporating random error. These two methods accomplish the same goal of adjusting for an uncontrolled confounder and will often produce similar results. McCandless and Gustafson examined the performance of probabilistic bias analysis methods similar to those we present here, relative to formal Bayesian methods [2]. Their results indicate that, in general, the methods presented here perform approximately comparably to Bayesian methods and may perform slightly better when the bias parameter distribution is misspecified [10].

To demonstrate these methods, we will continue an example that first appeared in Chapter 5 and pertained to a study of the association between male circumcision and infection with HIV [11]. In the observed data, men who were circumcised had approximately one third the risk of HIV infection as men who were not circumcised

Table 8.10 Observed data for the association between male circumcision (E) and HIV (D).

	Total	
	E_1	E_0
D_1	105	85
D_0	527	93
Total	632	178
RR	0.35	

Crude data from Tyndall et al., 1996 [11].

(conventional RR 0.35; 95% CI 0.28, 0.44). The observed data from this study are shown in Table 8.10. We note that we will use an example in which the risk ratio is the parameter of interest, but similar results could be generated for the risk difference or odds ratio. These would require amending some of the calculations described below using equations from Chapter 5 for the measure of interest.

Step 1: Identify the Source of Bias

In the study of the association between male circumcision and acquisition of HIV, the threat to validity was the possibility that the results were confounded by religious category (Muslim versus other religion), a variable for which data were not collected. We will use probabilistic bias analysis to model the bias that the unmeasured confounder might have introduced.

Step 2: Identify the Bias Parameters

In Chapter 5 we introduced a bias model with three bias parameters to conduct a simple bias analysis for an unmeasured confounder: p_1, p_0, and RR_{CD}. For the simple bias analysis, we specified that the prevalence of being Muslim versus any other religion was 80% among circumcised men (p_1) and 5% among uncircumcised men (p_0). Finally, we specified that Muslim men had 0.63 times the risk of acquiring HIV as men who were not (RR_{CD}) [12].

Step 3: Assign Probability Distributions to Each Bias Parameter

In the simple bias analysis, we implicitly assumed that the values assigned to the bias parameters were known with certainty. Now, as with the misclassification problem above, we will assign probability distributions to each of the three bias parameters (p_1, p_0, and RR_{CD}). For this example, we assign a trapezoidal distribution to each of

Table 8.11 Bias parameter distributions for a probabilistic bias analysis applied to the relation between male circumcision and HIV stratified by an unmeasured confounder (religion).

Bias parameter	Description	Min	Mod$_{low}$	Mod$_{up}$	Max
p_1 (%)	Prevalence of being Muslim among circumcised	70	75	85	90
p_0 (%)	Prevalence of being Muslim among uncircumcised	3	4	7	10
RR$_{CD}$	Association between being Muslim and HIV acquisition	0.5	0.6	0.7	0.8

the three bias parameters. For each bias parameter, we center the midpoint between the upper and lower modes approximately on the value used in the simple sensitivity analysis in Chapter 5, choose a range for the mode that seems reasonable, and then extend the trapezoidal distribution to lower and upper bounds such that the width of the trapezoid is approximately twice the width of the range between modes. The minimum, modes and maximum for these distributions are shown in Table 8.11.

Generally, in implementation of confounding bias-adjustment, it is reasonable to expect that p_1 and p_0 would be correlated. Our specification of the prevalence of Muslim religion among the circumcised is roughly 70 percentage points higher than the prevalence among the uncircumcised. If we sampled a very low prevalence for p_0 we would expect a lower prevalence for p_1. In this first example of a probabilistic bias analysis for confounding, we do not include a correlation between these parameters. In Chapter 12, however, we include correlations between these parameters. There are two ways to incorporate a correlation. The first is to use the NORTA algorithm from Chapter 7 and directly specify a correlation. The difficulty with that approach is that little information might exist on the extent of the correlation. An alternative approach, shown in Chapter 12, is to specify a prior distribution on p_0 as well as a normal distribution on $\ln(\mathrm{RR}_{ec}) = \ln\left(\frac{p_1}{p_0}\right) \sim N(\mu, \sigma)$. We can sample p_0 and then $\ln(RR_{ec})$ and then generate $p_1 = p_0\, RR_{ec}$. This will generate a natural correlation between the two parameters.

Step 4: Use Simple Bias Analysis Methods to Incorporate Uncertainty in the Bias Parameters and Random Error

Step 4a: Sample from the Bias Parameter Distributions

In this analysis we have chosen to use trapezoidal distributions rather than beta distributions, which means that we cannot simply use existing procedures within each of the major software packages to sample from the distributions but will have to use the approach we described in Chapter 7. We assume in this example that the three bias parameters are independent and have no correlation with one another, though analysts may want to allow a correlation in some scenarios. With these trapezoidal distributions assigned to each of the three bias parameters, we can sample from each to choose initial values for p_1, p_0, and RR_{CD}. In the initial samplings, we obtained values of 75.1% for p_1, 7.2% for p_0, and 0.567 for RR_{CD}.

Step 4b: Generate Bias-Adjusted Data Using Simple Bias Analysis Methods and the Sampled Bias Parameters

The bias model used in Chapter 5 to model data stratified by the unmeasured confounder can now be applied to the observed data using the values sampled for the three bias parameters. Given the values selected for the bias parameters, the simulated data would be as depicted in Table 8.12. This simple bias analysis gives a bias-adjusted RR associating circumcision with acquisition of HIV of 0.50, nearer to the null than the conventional RR (0.35) without adjusting for religion category.

From these data we can calculate the stratum specific RR and standardized morbidity ratio (SMR) of 0.50, which is substantially closer to the null than the crude RR = 0.35. However, this estimate only uses one set of bias parameter samples, computes an expected number of people with and without the confounder and does not include any random error.

We now impute the number of people in each stratum, rather than using the expected value. To do this we estimate the probability of any cell in the observed 2×2 table having the uncontrolled confounder, using information in Table 8.12. For instance, for the "a" cell of exposed people with the outcome, $P(C = 1|E = 1, D = 1) = 66.3/105$. Similar probabilities can be found for the remaining four cells:

$$P(C = 1|E = 1, D = 1) = \frac{66.3}{105} \cong 63.1\%$$

$$P(C = 1|E = 0, D = 1) = \frac{3.6}{85} \cong 4.2\%$$

$$P(C = 1|E = 1, D = 0) = \frac{408.4}{527} \cong 77.5\%$$

$$P(C = 1|E = 0, D = 0) = \frac{9.2}{93} \cong 9.9\%$$

Table 8.12 Bias-adjusted data for the association between male circumcision (E) and HIV (D) stratified by an unmeasured confounder, religious category (C; C_1 – Muslim, C_0 – any other religion).

	Crude		C_1		C_0	
	E_1	E_0	E_1	E_0	E_1	E_0
D_1	105	85	66.3	3.6	38.7	81.4
D_0	527	93	408.4	9.2	118.6	83.8
Total	632	178	474.6	12.8	157.4	165.2

Crude data from Tyndall et al., 1996 [11].

These probabilities can be used to propagate the random error associated with not knowing the confounder status. We can randomly impute the confounder status by sampling the missing cells (the four exposure-disease counts among those who have the confounder: A_1, B_1, C_1, D_1) from binomial distributions using these four probabilities. This imputation is similar to using the predictive values to sample the unknown exposure value in the methods for exposure misclassification.

$$A_1 \sim \text{binomial}(a, P(C = 1|E = 1, D = 1))$$
$$B_1 \sim \text{binomial}(b, P(C = 1|E = 0, D = 1))$$
$$C_1 \sim \text{binomial}(c, P(C = 1|E = 1, D = 0))$$
$$D_1 \sim \text{binomial}(d, P(C = 1|E = 0, D = 0))$$

In this iteration of the algorithm, these equations would be implemented as:

$$A_1 \sim \text{binomial}(105, 0.631)$$
$$B_1 \sim \text{binomial}(85, 0.042)$$
$$C_1 \sim \text{binomial}(527, 0.775)$$
$$D_1 \sim \text{binomial}(93, 0.099)$$

Drawing a random sample from these distributions yields values of $A_1 = 64$, $B_1 = 4$, $C_1 = 436$, $D_1 = 12$. Cell totals for individuals without the confounder can be calculated by subtraction: $A_0 = a - A_1 = 105 - 64 = 41$. Similar calculations give $B_0 = 81$, $C_0 = 91$, and $D_0 = 81$. Over many repetitions, the sampling in this step can lead to instances in which sampled cell counts are zero. In this case, an analyst can either add 0.5 to each cell in the table or exclude those iterations. We have chosen the latter approach and will report the total number of excluded iterations.

Step 4c: Sample the Bias-Adjusted Effect Estimate

After bias-adjusting the cell counts for the sampled values of p_1, p_0, and RR_{CD}, we now incorporate random error due to the implied sampling model (RE_1, above). There are multiple ways to do this, and the analyst must take care. All of the choices will include sampling a random variable z from a standard normal distribution and including the random error as we did in the misclassification example. The distinction is how we estimate the standard error, which will be guided by choice of estimate and practicalities. Often in this book we have used SMRs to summarize estimates of ratio measures of effect across strata. If one wishes to use an SMR here, the proper variance for this weighted measure should be used. In this example, given the sparse data in some cells, the standard error becomes extremely large. An alternative approach that we use is to estimate the Mantel-Haenszel risk ratio and its associated standard error [13]. Applying the Mantel-Haenszel equation to the data we sampled in step 4b, we find RR $= 0.61$ and $SE_1^{adj} = 0.14$. In the first iteration, we sampled z_1 from a normal($\mu = 0, \sigma = 1$) distribution, which returned $z_1 = 0.498$. Using this standard normal deviate in Equation 8.5 with the bias-adjusted estimate of the RR and the standard error of the Mantel-Haenszel risk ratio, we obtain:

$$RR_1^{total} = e^{[\ln(0.61) - 0.498 \cdot 0.14]} = 0.57$$

This estimate is bias-adjusted for one randomly sampled set of bias parameters and also includes one randomly sampled random error term.

Steps 5 and 6: Resample, Save and Summarize

In the final two steps we repeat steps 4a–c, each time saving the RR that was bias-adjusted for religious category, the uncontrolled confounder. We repeated this process 100,000 times in R and summarized the results in Table 8.13. The median bias-adjusted RR, adjusting for uncontrolled confounding, is approximately 0.47 (for both systematic only and total adjustments), closer to the null than the conventional point estimate of 0.35, and similar to the result from the simple bias analysis in Chapter 5. Note that this is one reason we strongly recommend conducting simple bias analysis first: discrepancies between the probabilistic bias analysis results and the simple bias analysis results may indicate errors in coding or in the specification of the distributions. Unlike with the simple bias analysis, however, we can now evaluate uncertainty using the simulation intervals. In addition to summarizing the distribution of the bias-adjusted RRs, we also compute and store RRs that are adjusted only for systematic error. The results are bias adjusted and account for the uncertainty in the bias parameters, but they do not include other sources of random error. At step 4b, we do not resample cell counts, and instead, use the expected cell count because here we are only interested in the error that does not decrease as sample size increases. At step 4c, we do not include the additional source of random error and simply compute the Mantel-Haenszel RR.

The bias-adjusted estimate and its interval are shifted towards the null, suggesting the bias due to unmeasured confounding by religious category had been away from

Table 8.13 Results of a probabilistic bias analysis of the relation between male circumcision and HIV acquisition bias-adjusting for religion as an unmeasured confounder using 100,000 iterations in R.

Analysis	Results from R	
	RR	95% Interval
Conventional result, assuming no bias	0.35	(0.28, 0.44)[a]
Systematic error only[b]	0.47	(0.42, 0.55)
Random and systematic error	0.47	(0.29, 0.72)

[a]Computed using Wald confidence intervals.
[b]Systematic error only analysis incorporates uncertainty in bias parameters and adjusts for bias.

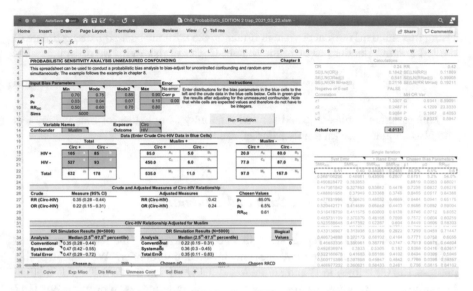

Figure 8.3 Screenshot of Excel spreadsheet to conduct a probabilistic bias analysis for an unmeasured confounder.

the null. When accounting for bias due to the uncontrolled confounding and only systematic error, the probabilistic bias analysis yielded a median RR of 0.47 with a 95% simulation interval ranging from 0.42 to 0.55 (Table 8.13). The next row of Table 8.13 shows the estimate and 95% simulation interval for the RR adjusting for systematic error and propagating uncertainty in the bias parameter as well as random error. This analysis yielded a median RR of 0.47 and a 95% simulation interval ranging from 0.29 to 0.72. We can evaluate the importance of each source of uncertainty by computing and comparing the widths of their respective intervals. Using the results in Table 8.13, the width of the 95% confidence interval for the conventional estimate assuming no uncontrolled confounding is 0.44/0.28 = 1.57. The interval accounting only for the uncertainty in the distributions for the bias parameters is 0.55/0.42 = 1.31. The simulation interval that adjusts for systematic error, incorporates uncertainty in the bias parameter, and incorporates random error has width 0.72/0.29 = 2.48. Random error seems to be the largest driver of the simulation interval width; however, the contribution of uncertainty is not negligible. None of the simulation intervals are wide enough to cast doubt on the protective effect of circumcision on HIV acquisition. Of the 100,000 simulations, 4844 resulted in a zero cell. We excluded these draws, however, we could have included a small constant (0.5) to each cell and retained them. Doing so made little practical difference. As always, these results and inferences assume that the specifications of the bias distributions are valid and there are no other sources of error.

Figure 8.3 shows a screenshot of an Excel spreadsheet to conduct a probabilistic bias analysis for an unmeasured confounder.

Confounding Implementation Alternative: Relative Risk Due to Confounding

An alternative probabilistic bias analysis approach for unmeasured confounding would be to either directly or indirectly specify a distribution for the relative risk due to confounding (RR_C) [14]. After sampling from this distribution, divide the observed crude risk ratio (or a risk ratio adjusted for a set of measured confounders) by the relative risk due to confounding to obtain a distribution of estimates bias-adjusted for the unmeasured confounder. This approach has the advantage of being relatively simple to program in statistical analysis software and is simpler to explain in a methods section of a manuscript. However, the relative risk due to confounding may be difficult to specify directly because it is less intuitive than specifying the individual parameters that contribute to the RR_C—and it may be difficult to obtain data from the literature to specify a distribution directly for RR_C. Analysts may inadvertently specify distributions for RR_C that portray less uncertainty in the bias-adjusted estimates of effect (*i.e.,* a narrower simulation interval) than they would have obtained by specifying the individual parameters that influence the RR_C, even accounting for just the first source of uncertainty.

Bodnar et al. used the approach of directly specifying a bias parameter distribution for the confounding relative risk in a report on the association between periconceptual vitamin use and risk of preeclampsia [15]. Because data were not collected on fruit and vegetable consumption, a variable considered an important, but unmeasured, confounder, the authors assigned a trapezoidal distribution to the relative risk due to confounding due to fruit and vegetable intake and simulated the impact of the unmeasured confounding on the results.

The difficulty in directly specifying a bias parameter distribution for the confounding relative risk can be avoided by specifying distributions for p_1, p_0, and RR_{cd} as above. By randomly sampling from these three distributions, we can use Monte Carlo methods to determine the distribution of the confounding relative risk and, in turn, directly adjust the observed RR. We demonstrate this approach using the circumcision and HIV risk example from the previous section. Steps 1–3 proceed as before. We identify uncontrolled confounding as the bias of interests, choose p_1, p_0, and RR_{cd} as the bias parameters, and the use the same trapezoidal distributions that we did above.

Step 4: Use Simple Bias Analysis Methods to Incorporate Uncertainty in the Bias Parameters and Random Error

Step 4a: Sample from the Bias Parameter Distributions

To proceed with the confounding relative risk approach, we must identify the distribution of the confounding relative risk. This can be done by iteratively sampling from p_1, p_0, and RR_{cd} and using them in the equation for the confounding relative risk. However, care is required at this step. As we discussed in Chapter 5, a different equation is necessary depending on the effect estimate of interest (RR, RD or OR) as well as the target population of interest and whether there is effect modification. A comprehensive list of these choices is found in Arah et al. [16]. In our example, we are interested in bias-adjusting the relative risk and have assumed there is no effect measure modification, so the target population does not matter. The equation for the confounding relative risk is:

$$RR_c = \frac{RR_{cd} \cdot p_1 + (1 - p_1)}{RR_{cd} \cdot p_0 + (1 - p_0)}$$

We use the same initial samples as in the previous example where we obtained values of 88.1% for p_1, 9.9% for p_0, and 0.63 for RR_{CD}. These values are inserted in the equation above to obtain a $RR_C = 0.699$.

Step 4b: Generate Bias-Adjusted Results Using Simple Bias Analysis Methods and the Sampled Bias Parameters

Instead of imputing the bias-adjusted contingency table, we can use the RR_C to adjust the observed RR for uncontrolled confounding using the formula:

$$RR_{adj} = RR_{obs}/RR_C$$

Because $RR_{obs} = 0.35$, we obtain a confounding-adjusted $RR_{adj} = 0.35/0.699 = 0.50$ for this iteration of the algorithm. It is not a coincidence that the adjusted RR in this step is the same as the adjusted RR obtained from step 4b in the last algorithm (Table 8.12). The two methods are equivalent, and the same sampled values have been used to bias-adjust the observed estimate.

Step 4c: Sample the Bias-Adjusted Effect Estimate

After obtaining a bias-adjusted estimate for the sampled values of p_1, p_0, and RR_{CD}, we now incorporate random error. We follow the same procedure as above and propagate random error by adding a normal deviate (with standard deviation equal to the standard error of the observed effect estimate). In this example, we use the standard error based on the crude cell counts rather than the bias-adjusted cell counts, because we have not modeled the cells counts of the stratified data. A procedure similar to this has been examined in simulation studies and has performed well [2, 10]. The standard error of the crude effect estimate is computed as:

$$SE_{crude} = \sqrt{\frac{1}{105} + \frac{1}{85} - \frac{1}{105 + 527} - \frac{1}{85 + 93}} = 0.12$$

In the first iteration, we sampled z_1 from a normal($\mu = 0, \sigma = 1$) distribution, which returned $z_1 = 0.498$. Using this standard normal deviate in Equation 8.5 with the bias-adjusted estimate of the SMR and the standard error of the crude risk ratio, we obtain:

$$RR^{total} = e^{[\ln(0.50) - 0.498 \cdot 0.12]} = 0.47$$

Steps 5 and 6: Resample, Save and Summarize

We repeat steps 4a–4c and save the bias-adjusted result after each step. We programmed this algorithm in R and repeated it for 10^6 iterations, which took only a few seconds to complete. The median of the iterations was used as an estimate of the bias adjusted risk ratio and the 2.5th and 97.5th percentiles were used to estimate the uncertainty in the bias-adjusted effect: $RR_{adj} = 0.47$ with a 95% simulation interval of 0.36, 0.62. This result is similar to the one we obtained using the previous method. The point estimates are identical, but the simulation interval is narrower when using the confounding relative risk method. This narrowing of the interval is likely due to either 1) the fact that the confounding relative risk implementation of bias adjustment does not incorporate the additional random error from imputing the missing confounder or 2) the confounding relative risk method is not using the most appropriate variance estimate to incorporate traditional random error. In preliminary simulations, the confounding relative risk method seems to have narrower confidence intervals on average; however, the difference depends on the observed data and bias parameter distributions. In this example, we specified confounder prevalences that were quite different from one another (a very large exposure-

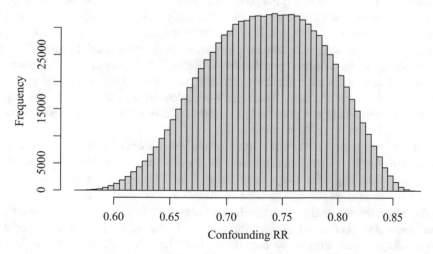

Figure 8.4 Histogram of the estimated distribution of the confounding relative risk (RR_C) obtained by separately sampling p_1, p_0, and RR_{cd}.

confounder association). This resulted, as can be seen above in simulated datasets (Table 8.12), in small sample sizes in some strata. In this situation, we see a modest difference in the width of the simulation intervals between the two approaches. When sparse data is not an issue, the discrepancy between these two methods diminishes considerably.

The distribution of RR_C, obtained by sampling from p_1, p_0, and RR_{cd} is shown in Figure 8.4. Note that the distribution of RR_C in this case is non-normal, which emphasizes the point that using prior information to directly specify the distribution of RR_C (as opposed to specifying distributions for p_1, p_0, and RR_{cd} and using Monte Carlo simulation to derive the distribution of RR_C) could be difficult. In either case, conducting a bias analysis using the confounding relative risk may be easier in some situations but requires the user to carefully select the proper equation for the confounding RR (or confounding RD or confounding OR). This approach works for bias analyses of an unmeasured confounder, and would work for baseline selection bias, because the relation between the bias-adjusted estimate and the conventional estimate are multiplicative functions of the relative risk due to confounding or the selection odds, respectively (see Chapters 4 and 5). There is no analogous factorized solution for most misclassification problems, with rare exceptions [14], because the bias from misclassification does not ordinarily factorize.

Selection Bias Implementation

Following the outline in Table 8.1, we can use probabilistic bias analysis to address selection bias. As we have discussed above, the implementation of any probabilistic bias analysis will depend on the context of the research and selection bias is no exception. There are multiple methods for implementing a probabilistic bias analysis, which we will work through in detail. The methods generally differ based on the extent of information on the selection process. In the first example, we work through an approach to dealing with selection bias when there is substantial information in the study about the selection process. In particular, the number of non-participating cases and controls is known and we can estimate the exposure prevalence in those two groups from study data. After that we implement a method for the, perhaps, more common setting where an analyst suspects selection bias but may not have data to directly inform the bias parameters.

Before we proceed, we note that for selection bias problems, we recommend almost always using the conventional estimate of the standard error when incorporating the study sampling error (Step 4c in Table 8.1). That is, we set $SE_i^{adj} = SE_{crude}$. For misclassification and uncontrolled confounding examples, we sometimes recommend using the standard error for the bias-adjusted data (SE_i^{adj}), but for selection bias problems we recommend against this choice for most bias-adjustment models. We make this recommendation because we hold the view that bias analysis should almost always increase our uncertainty in the study results, implying that the analysis should yield a wider simulation interval than the conventional confidence interval.

Bias analysis requires that we make assumptions to model the various sources of bias, and these assumptions increase uncertainty beyond that of the conventional analysis. With selection bias, the bias model imputes data for people or person-time that was not included in the conventional result. These imputed data will often have the effect of decreasing the standard error of the bias-adjusted estimate compared with the standard error of the conventional estimate ($SE_i^{adj} < SE_{crude}$) and therefore potentially decreasing the uncertainty rather than increasing it (depending on the balance of the reduced uncertainty from Step 4c and the additional uncertainty from Steps 4a and 4b in Table 8.1). We therefore recommend using the conventional standard error (SE_{crude}) in Step 4c of Table 8.1 for selection bias problems to assure that the intervals appropriately grow wider. It may be defensible to use the standard error from the bias-adjusted estimate in Step 4c for some selection bias problems, such as in an example below, but we expect these occasions will be rare.

An Example of Probabilistic Bias Analysis in the Presence of Substantial Source Population Data

In Chapter 4, we described an example in which there was concern about selection bias in a case-control study of the association between mobile phone use and the occurrence of uveal melanoma [17]. Table 8.14 shows the data from the study and the allocation of nonparticipants, informed by the numbers of nonparticipants who filled out only a shorter questionnaire indicating their exposure status. Recall that the crude odds ratio associating regular mobile phone use with uveal melanoma occurrence was:

$$OR_{crude,participants} = \frac{136/297}{107/165} = 0.71$$

Among nonparticipants who answered the short questionnaire, the crude odds ratio was:

$$OR_{crude,nonparticipants} = \frac{3/72}{7/212} = 1.26$$

Table 8.14 Depiction of participation and mobile phone use in a study of the relation between mobile phone use and the occurrence of uveal melanoma.

	Participants		Nonparticipants	
	Regular use	No use	Regular use	No use
Cases	136	107	$3 + 0.3{\cdot}17 = 8.1$	$7 + (1-0.3){\cdot}17 = 18.9$
Controls	297	165	$72 + 0.254{\cdot}379 = 168.1$	$212 + (1-0.254){\cdot}379 = 494.9$

Original data from Stang et al., 2009 [17].

Step 1: Identify the Source of Bias

In addition to the nonparticipants who completed only the short questionnaire, there were 17 cases and 379 controls who did not participate at all. These crude data can be bias-adjusted for possible selection bias in a simple bias analysis by using an estimate of the exposure proportion for cases and controls among those who participated in the short survey. Those who did not participate at all will be allocated to exposure groups according to the exposure prevalences observed in the cases and controls who answered only the short questionnaire. This approach to selection bias is similar in spirit to multiple imputation of missing exposure status, which we discuss in Chapter 10. The simple bias-adjusted odds ratio, as shown also in Equation (4.1), is:

$$OR^s = \frac{136 + 3 + \frac{3}{10}17}{297 + 72 + \frac{72}{284}379} \Bigg/ \frac{107 + 7 + \frac{7}{10}17}{165 + 212 + \frac{212}{284}379} = 1.62 \qquad (8.8)$$

We will use probabilistic bias analysis to model the bias from non-participation. Note that we assume the exposure status of those who answered the complete or the short questionnaire is accurate, so the only source of uncertainty is in application of the exposure prevalence in those who answered the short questionnaire to those who did not participate at all. One could simultaneously model information bias by assuming inaccuracies in measuring mobile phone use in the complete participants and in those who completed only the short questionnaire (see Chapter 12 for methods of multiple bias analysis).

Step 2: Identify the Bias Parameters

The two bias parameters are the prevalence of mobile phone use among cases who were invited to participate but refused to answer either questionnaire (p_1) and among controls who were invited to participate but refused to answer either questionnaire (p_0). Rewriting Equation 8.8 with these bias parameters:

$$OR^s = \frac{136 + 3 + p_1 17}{297 + 72 + p_0 379} \Bigg/ \frac{107 + 7 + (1 - p_1)17}{165 + 212 + (1 - p_0)379} \qquad (8.9)$$

Step 3: Assign Probability Distributions to Each Bias Parameter

In the simple bias analysis, we implicitly assumed that the values assigned to the bias parameters were known with certainty ($p_1 = 3/10$ and $p_0 = 72/284$). For the probabilistic bias analysis, we will assign probability distributions to them, informed directly by the data from the cases and controls who answered only the short questionnaire. Without this type of information, one would most likely have to assign distributions based on an educated guess, as discussed in Chapters 3 and 4. The exposure prevalence is a proportion, so the data from the cases and controls who answered only the short questionnaire can be used to directly assign beta distributions to the bias parameters: $p_1 \sim$ beta($\alpha = 3$, $\beta = 7$) and $p_0 \sim$ beta($\alpha = 72$, $\beta = 212$). We will use these assignments of distributions to bias parameters in a probabilistic bias analysis with 100,000 iterations, implemented in R.

Step 4: Use Simple Bias Analysis Methods to Incorporate Uncertainty in the Bias Parameters and Random Error

There are three sources of uncertainty or random error in this probabilistic bias analysis. The first is from the uncertainty in the bias parameters. The second is from the application of the bias parameters to the individual participants' missing data in the study; that is, when we impute missing exposure status, we do so randomly allowing the missing number of exposed to be different at each iteration. The third is the random error associated with the specified statistical outcome model: that is that the number of exposed people follows a binomial distribution among cases and controls.

Step 4a: Sample from the Bias Parameter Distributions

The first step is to sample from the bias parameter distributions. With the beta distributions we assigned to the bias parameters, we sample from each to choose initial values for p_1 and p_0. In our initial samplings we obtained values of 43.1% for p_1 and 27.2% for p_0. Given the values selected for the bias parameters in the first iteration, the simulated data would be as depicted in Table 8.15. This iteration of the bias analysis gives a bias-adjusted OR associating regular mobile phone use with uveal melanoma of:

$$OR^s = \frac{136 + 3 + 0.431 \cdot 17}{297 + 72 + 0.272 \cdot 379} \Big/ \frac{107 + 7 + (1 - 0.431)17}{165 + 212 + (1 - 0.272)379} = 1.64$$

Table 8.15 Depiction of participation and mobile phone use in a study of the relation between mobile phone use and the occurrence of uveal melanoma. Results from first iteration of the bias analysis, first source of uncertainty.

	Participants		Nonparticipants	
	Regular use	No use	Regular use	No use
Cases	136	107	3 + 0.431·17=10.3	7 + (1–0.431)·17=16.7
Controls	297	165	72 + 0.272·379=175.1	212 + (1–0.272)·379=487.9

Original data from Stang et al., 2009 [17].

The difference between the bias-adjusted estimate from the first iteration (1.64) and the simple bias-adjusted estimate of 1.62 is due to the first source of uncertainty: assignment of probability distributions to the bias parameters.

Step 4b: Generate Bias-Adjusted Data Using Simple Bias Analysis Methods and the Sampled Bias Parameters

The second source of uncertainty arises from application of the values of the bias parameters to the individual participants in the study (or, for this selection bias problem, application to the individual nonparticipants in the study). To model this uncertainty, we apply the sampled exposure prevalences to the individual records and impute the number of exposed cases from the 17 cases who did not participate in any survey. We simulate this result using draws from a binomial distribution. For example, there were 17 cases who answered neither questionnaire and the initial draw from the beta distribution yielded 43.1% for the prevalence of regular mobile phone use among them. The expectation for the number of regular mobile phone users would be $17 \cdot 0.431 = 7.3$. Modeling the second source of uncertainty, however, requires that we draw from a binomial distribution, with probability of being a regular mobile phone user of 0.431.

$$N_{E+,D+} = \text{binomial}(17, 0.431)$$
$$N_{E-,D+} = N_{D+} - N_{E+,D+}$$

In the first iteration, this simulation returned 4 regular mobile phone users among the cases who answered neither questionnaire, which is a bit different than the 7.3 expected (note that the binomial draw always returns an integer, as described in Chapter 7). The number of unexposed cases among those who answered neither questionnaire is computed by difference ($17-4 = 13$).

Similarly, there were 379 controls who answered neither questionnaire and the initial draw from the beta distribution yielded 27.2% for the prevalence of regular mobile phone use among them. The expectation for the number of regular mobile phone users would be $379 \cdot 0.272 = 103.1$, as shown in Table 8.16. Modeling the

Table 8.16 Depiction of participation and mobile phone use in a study of the relation between mobile phone use and the occurrence of uveal melanoma. Results from first iteration of the bias analysis, first and second sources of uncertainty.

	Participants		Nonparticipants		Selection Bias-adjusted	
	Regular use	No use	Regular use	No use		
Cases	136	107	3 + 4 = 7	7 + 13 = 20	143	127
Controls	297	165	72 + 99 = 171	212 + 280 = 492	468	657

Original data from Stang et al., 2009 [17].

second source of uncertainty, however, requires that we draw from a binomial distribution, with probability of being a regular mobile phone user of 0.272.

$$N_{E+,D-} = \text{binomial}(379, 0.272)$$

$$N_{E-,D-} = N_{D-} - N_{E+,D-}$$

In the first iteration, this simulation returned 99 regular mobile phone users among the controls who answered neither questionnaire, which is similar to the 103.1 expected. The number of unexposed controls among those who answered neither questionnaire is computed by difference ($379-99 = 280$). Table 8.16 shows the results of the bias analysis after incorporating the first and second sources of uncertainty. We can now compute the bias-adjusted estimate of the OR for this first iteration, accounting for the second source of uncertainty:

$$OR_1 = \frac{{}^{136+3+4}/_{297+72+99}}{{}^{107+7+13}/_{165+212+280}} = 1.58$$

The bias-adjusted estimate accounting for only the first source of uncertainty was 1.64. The difference between 1.58 and 1.64 is due to the second source of uncertainty: application of the values of the bias parameters to the individual nonparticipants in the study.

Steps 4c: Sample the Bias-Adjusted Effect Estimate

After bias-adjusting the estimate for the sampled values of p_1 and p_0 and for application of the values of the bias parameters to the individual participants in the study, we now incorporate conventional random error. The standard error of the bias-adjusted effect estimate is:

$$SE_1^{adj} = \sqrt{\frac{1}{143} + \frac{1}{468} + \frac{1}{127} + \frac{1}{657}} = 0.14$$

This approach, in which the standard error is based on the bias-adjusted cell counts, is defensible in this selection bias-adjustment because we know exactly how many cases and controls were invited to participate and we have good information about the exposure prevalence in those who did not fully participate. In general, this information will not be available, and we discuss alternate methods below. In the first iteration, we sampled z_1 from a normal($\mu = 0, \sigma = 1$) distribution, which returned $z_1 = -0.529$. Using this standard normal deviate in Equation 8.5 with the bias-adjusted estimate of the OR accounting for the first two sources of uncertainty and the standard error of the bias-adjusted estimate, we obtain:

$$OR_1^{total} = e^{[\ln(1.58)-0.529\cdot0.14]} = 1.70$$

These estimates are bias-adjusted for one randomly sampled set of bias parameters and also include one randomly sampled conventional random error term.

Steps 5 and 6: Resample, Save and Summarize

In the final two steps, we repeat steps 4a–c above, each time saving the OR bias-adjusted for non-participation. We repeated this process 1,000,000 times in R and summarized the results in Table 8.17. The median bias-adjusted OR, adjusting for nonparticipation, is about 1.62 (for both systematic only and total adjustments), in the direction suggesting a causal effect of regular mobile phone use—compared with no mobile phone use—whereas the conventional point estimate of 0.71 suggested a protective effect. The median bias-adjusted estimate is similar to the result from the simple bias analysis in Chapter 5. Unlike with the simple bias analysis, however, we can now evaluate uncertainty and sources of uncertainty using the simulation intervals.

When accounting only for bias due to the nonparticipation (and ignoring random error), the probabilistic bias analysis yielded a median OR of 1.62 with a 95% simulation interval ranging from 1.48 to 1.80 (Table 8.17). The last row of Table 8.17 shows the point estimate and 95% simulation interval adjusting for the nonparticipation and accounting for conventional random error using the standard error of the bias-adjusted estimate. The median is 1.62 and the simulation interval is wider still, with limits extending from 1.21 to 2.19. If we use the standard error of the conventional result (SE_{crude}) instead of the standard error of the bias-adjusted result (SE_i^{adj}), then the bias-adjusted estimate remains 1.62 and the 95% simulation interval widens slightly with limits of 1.15 and 2.29.

We can evaluate the importance of each source of error by computing and comparing the widths of their respective intervals, as we have above. Using the results generated by R and reported to three decimals in Table 8.17, the width of the interval corresponding to uncertainty in the bias parameters is, $1.797/1.477 = 1.217$.

Table 8.17 Results of a probabilistic bias analysis of the relation between regular mobile phone use and uveal melanoma bias-adjusting for non-participation using 1,000,000 iterations in R.

Analysis	Results from R	
	Median[a]	95% Interval[a]
Conventional result, assuming no bias	0.71	·(0.51, 0.97)
Systematic error only[b]	1.62	(1.48, 1.80)
Random and systematic error	1.62	(1.21, 2.19)

Original data from Stang et al., 2009 [17].
[a]Results presented to three decimals to facilitate comparisons.
[b]Systematic error only analysis incorporates uncertainty in bias parameters and adjusts for bias.

The width of the interval that includes random and systematic error (conventional random error plus distributions for the bias parameters) is $2.186/1.210 = 1.807$. The uncertainty added by the third source of uncertainty (conventional random error) beyond the first and second sources of uncertainty (distributions for the bias parameters and application to the sample size of the study) is $1.807/1.217 = 1.485$. The largest component of the uncertainty, therefore, arises from conventional random error, although it is less than the uncertainty from conventional random error portrayed in the conventional confidence interval because we are adding information by imputing the missing exposure status ($0.969/0.514 = 1.885$). If we instead compute the width based on the limits from the bias model that used the conventional standard error, we obtain a ratio of $2.29/1.15 \approx 1.98$, which is wider than the conventional confidence interval because we have enforced the recommendation that selection bias should only add to the uncertainty (see note at the beginning of this section).

These changes in interval widths are small compared with the change in the median estimates between the bias-adjusted result and the conventional result ($1.623/0.706 = 2.299$). Recall that the bias model assumed that the prevalence of regular mobile phone use in nonparticipant cases could be directly informed by what was observed in the case participants who answered only the short questionnaire and that the prevalence of regular mobile phone use in nonparticipant controls could be directly informed by what was observed in the control participants who answered only the short questionnaire. If we instead assume that these exposure prevalences correspond to what was observed in the case and control participants who answered the complete questionnaire, and assign the corresponding beta distributions, then the bias-adjusted odds ratio accounting for all three sources of uncertainty is 1.02 (95% simulation interval 0.77, 1.36; using SE_i^{adj} for Step 4c in Table 8.1).

To further explore this uncertainty, we can set bounds on p_1 with lower limit set to the 2.5th percentile of $p_1 \sim$ beta($\alpha = 3$, $\beta = 7$), which is a value of 0.075 (informed by the participants who answered only the short questionnaire) and upper limit set to the 97.5th percentile of $p_1 \sim$ beta($\alpha = 136$, $\beta = 107$), which is a value of 0.621 (informed by the participants who answered the full questionnaire). We likewise set bounds on p_0 with lower limit set to the 2.5th percentile of $p_0 \sim$ beta($\alpha = 72$, $\beta = 212$) which occurs at 0.205 (informed by the participants who answered only the short questionnaire) and upper limit set to the 97.5th percentile of $p_0 \sim$ beta($\alpha = 297$, $\beta = 165$) which occurs at 0.686 (informed by the participants who answered the full questionnaire). Setting $p_1 \sim$ uniform(0.075, 0.621) and $p_0 \sim$ uniform(0.205, 0.686), the bias-adjusted odds ratio accounting for all three sources of uncertainty is 1.26 (95% simulation interval 0.81, 1.99; using SE_i^{adj} for Step 4c in Table 8.1). This bias model allows the prevalence of regular mobile phone use to range from a lower limit of the beta distribution informed by participants who answered only the short questionnaire to an upper limit of the beta distribution informed by participants who answered the full questionnaire, with no most likely value assigned to the exposure prevalence and no correlation assigned between the exposure prevalence in cases and the exposure prevalence in controls. Setting $p_1 \sim$ uniform(0, 1) and $p_0 \sim$ uniform(0, 1), which effectively ignores information from either questionnaire when imputing the use of mobile phones in the nonparticipants, the bias-adjusted odds ratio accounting for all three sources of uncertainty is 1.22 (95% simulation interval 0.57, 2.63, using SE_i^{adj} for Step 4c in Table 8.1). Recall that the conventional odds ratio and its 95% confidence interval were 0.71 (95% CI: 0.51, 0.97). These latter three models are sensitivity analyses of the probabilistic bias analysis, evaluating the sensitivity of the results to changes in the assumptions. In this case, the sensitivity analysis assumptions were more extreme than the assumptions of the original analysis, in that they borrowed less and less information from the non-participants who answered only the short questionnaire. None of the bias models accounting for nonparticipation overlap substantially with the conventional result, suggesting that nonparticipation biased the estimate of association towards, and probably beyond, the null (conditional on the accuracy of the bias model).

Selection Bias Adjustment Using Selection Probabilities

In the previous example, we knew the total number of cases and controls who did not participate and we were able to use data from the additional sample who completed a short questionnaire to impute exposure status among those who did not participate. In many applied settings, this level of information about selection forces will not be available. More often, we will suspect that those with and without the disease and those with and without the exposure enrolled in the study differentially, but there will be less than definitive information on how many people did not enroll or what the exposure and disease specific probabilities of selection into the study were. As we stated in Chapter 4, selection is both the easiest bias to model and the most vexing; while the mathematics of adjusting for selection bias are simple, the information needed to do so credibly is generally unavailable. To illustrate a general approach to probabilistic bias analysis for selection bias, we revisit the previous case-control study of uveal melanoma and mobile phone use. However, to illustrate the general method, we will adjust for selection bias by specifying four selection probabilities (instead of the two exposure probabilities of the previous approach). This approach will be generally applicable even when analysts are lacking additional study data that were available to Stang et al. [17].

Step 1: Identify the Source of Bias

As described above, the source of bias is selection bias. By design, cases were more likely to be invited to participate than controls and may have been more motivated to enroll. Additionally, those who use mobile phones regularly may be more motivated to enroll in the study. That exposure and disease are both associated with participation in the study raises the possibility of selection bias. We will use probabilistic bias analysis to adjust the observed measure of effect for selection into the study.

Step 2: Identify the Bias Parameters

In our first approach, we were able to adjust for selection bias by specifying two bias parameters. More generally, however, bias analysis will proceed by specifying the four parameters presented in Chapter 4. These probabilities determine selection into the study conditional on exposure and disease status: the probability of selection into the study for cases who were exposed ($S_{case, 1}$), cases who were unexposed ($S_{case, 0}$), controls who were exposed ($S_{control, 1}$), and controls who were unexposed ($S_{controls, 0}$). With these bias parameters specified, we can multiply their inverse by the observed cell counts to adjust for selection bias.

Step 3: Assign Probability Distributions to Each Bias Parameter

To assign probability distributions to the four parameters we refer to Table 8.14, in which we used the exposure prevalence among cases and controls who answered the short survey to impute the expected number of non-participants who would be exposed and unexposed among cases and controls. To assign bias parameters in this example, we consider both regular participants and those who completed the short questionnaire to have been selected into the study. Only those who completely refused are non-participants. Thus, for exposed cases, we have 136 (regular questionnaire) + 3 (short questionnaire) =139 participants. The expected number of exposed case non-participants is $(3/10) \cdot 17 = 5.1$. Thus, $139/(139 + 5.1) = 96.4\%$ of all exposed cases enrolled in the study. We model $s_{case, 1} \sim beta(\alpha = 139, \beta = 5.1)$. All the bias parameter distribution probabilities are specified in a similar way:

$$S_{case,1} \sim beta(\alpha = 139, \beta = 5.1)$$
$$S_{case,0} \sim beta(\alpha = 114, \beta = 11.9)$$
$$S_{control,1} \sim beta(\alpha = 369, \beta = 96.1)$$
$$S_{control,0} \sim beta(\alpha = 377, \beta = 282.9)$$

In many settings, analysts will not be able to parameterize the bias parameter distribution as easily as we have in this example. They may need to rely on previous studies in the literature or expert knowledge to specify these four distributions.

Step 4: Use Simple Bias Analysis Methods to Incorporate Uncertainty and Random Error

There are two sources of uncertainty or random error in the way we are implementing this probabilistic bias analysis. The first is the uncertainty in the bias parameters. The second is the random error associated with the specified statistical model for the case-control study in which the number of exposed people among cases and controls is assumed to follow a binomial distribution.

Step 4a: Sample from the Bias Parameter Distributions

We first sample from each of the four bias parameter distributions. Our initial sample for the four bias parameters results in $s_{case,\,1}= 96.4\%$, $s_{case,\,0}=89.8\%$, $s_{control,\,1}= 81.2\%$, and $s_{control,\,0}= 56.2\%$. We then use these sampled values to bias-adjust the observed data.

Step 4b: Generate Bias-Adjusted Data Using Simple Bias Analysis Methods and the Sampled Bias Parameters

We use the simple bias adjustment techniques presented in Chapter 4 to adjust the observed data for possible selection bias. Specifically, we divide each of the four observed cell counts by the sampled selection probability for that cell to generate selection bias adjusted cell counts, as shown in Table 8.18. These bias-adjusted cell counts produce a bias-adjusted OR = 1.68 for this first iteration of the bias adjustment algorithm.

Steps 4c: Sample the Bias-Adjusted Effect Estimate

After bias-adjusting the estimate using the sample selection probabilities, we must propagate random error in the bias-adjusted effect estimate. We do this using the

Table 8.18 Depiction of participation and mobile phone use in a study of the relation between mobile phone use and the occurrence of uveal melanoma. Results from first iteration of the bias analysis, first source of uncertainty.

	Any participants[a]		Bias-Adjusted	
	Regular use	No use	Regular use	No use
Cases	139	114	139/0.964 = 144.2	114/0.898 = 126.9
Controls	369	377	369/0.812 = 454.6	377/0.562 = 670.6

Original data from Stang et al., 2009 [17].
[a]Full participants and those who only answered the short survey are grouped together as "Any Participant".

same procedure as above. We estimate the magnitude of the random error by calculating the standard error of the data and then propagating it by drawing a value from a standard normal distribution. However, there is one important distinction. In the previous algorithm, we used all the data (including the data on non-participants) to estimate the standard error (SE_i^{adj}). In this algorithm, we use only the data on those who fully participated in the study to calculate the conventional standard error (SE_{crude}). The reason for this change is that, in general, most analysts will have relatively little information on the source population. As such, basing the standard error on the size of the source population may be difficult. Further, even if the source population is known, it is easy to see that using this information could result in a poorly performing procedure. An analyst could make the random error as small as they like by modeling a large source population with low participation proportions. Perhaps, the source population is the entire state in which the study was carried out or the entire country. Including such a large sample would negate the random error component, regardless of the study size. For these reasons and the reasons described at the outset of the selection bias section, we recommend analysts modeling selection bias most often use the standard error of the observed sample:

$$SE_{crude} = \sqrt{\frac{1}{139} + \frac{1}{114} + \frac{1}{369} + \frac{1}{377}} = 0.15$$

To incorporate this error, we calculate the OR $= 1.68$ at this step from the adjusted data and sample from a standard normal distribution, find a value of $z = -0.02$, and generate a new bias-adjusted OR:

$$OR = e^{[\ln(1.68)+0.02 \cdot 0.15]} = 1.69$$

This is the first iteration of the algorithm and produces a single bias-adjusted OR that propagates the uncertainty in the bias parameter and random error. All that is left is to repeat this procedure many times.

Steps 5 and 6: Resample, Save and Summarize

We repeat steps 4a–4c 1,000,000 times, saving the bias adjusted OR after each iteration. It took less than a second to run the simulation in R and we evaluated the quantiles of the resulting distribution to summarize the data. Using this second method to adjust for selection bias, we find a bias adjusted OR $= 1.62$ with 95% simulation intervals (1.20, 2.20). Comparing these to the results in Table 8.17, we see that the point estimate is identical while the intervals are slightly wider in the second method, likely due to the fact that we did not base the standard error on non-participants as well as participants. The interval of this result is slightly narrower

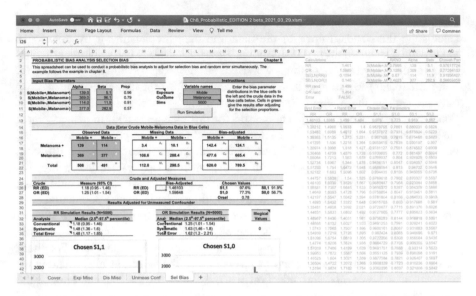

Figure 8.5 Screenshot of Excel spreadsheet to conduct probabilistic bias analysis for selection bias. Original data from Stang et al., 2009 [17].

than what we showed above using the standard error of the conventional result (OR = 1.62, 95% SI 1.15, 2.29) because we included individuals who completed the short form as study participants in this example, thus increasing the sample size and narrowing the simulation intervals. Figure 8.5 shows a screenshot of the Excel program for conducting probabilistic bias analysis for selection bias.

Computing Issues with Summary Level Probabilistic Bias Analysis

Unlike the computing issues we will discuss when we describe individual record-level probabilistic bias analysis, summary level probabilistic bias analysis typically can be run in seconds, even with a 1,000,000 or more iterations. Still, it is helpful to organize the dataset on which the analysis is run as efficiently as possible, containing only the variables necessary, and deleting any variables that are created in interim calculations. For summary level bias analyses there are typically few of these, but should they be identified, removing them can help reduce computing time.

Note that for the first source of uncertainty, what varies are the values assigned to the bias parameters. The observed dataset, which exists only at the level of a data summary (often just four cell counts) does not influence this distribution. Thus, to implement a summary level probabilistic bias analysis that accounts for only the first source of uncertainty, the analyst can sample as many values for the bias parameters as there are iterations of the analysis, without needing to recreate the dataset. It is usually much quicker (if software allows it) to randomly sample all these parameters at once, rather than iterating through a loop for each iteration. After all parameters have been sampled, the equations for bias-adjustment can be implemented. This analysis strategy is implemented in R using the vectorized features.

The analyst must also decide how many iterations of the analysis to run. The analyst does not require more iterations when the values that summarize the distribution (*i.e.,* the median and the 2.5th and 97.5th percentiles) cease to vary meaningfully with more iterations. Since large summary level probabilistic bias analyses generally can be run in a short time, for summary level analyses we recommend running a very large number of iterations, say 100,000, unless the available computing resources cannot complete the targeted number of iterations in a reasonable time. We will discuss the number of iterations and convergence again in Chapter 9 on individual-level probabilistic bias analysis and Chapter 11 on Bayesian bias analysis. These methods may require far greater computing resources, so take much longer to reach convergence on many desktop machines. An excellent way to determine whether enough samples have been drawn is to simply repeat the analysis (with different seeds) several times and determine whether the point estimates change after each run. If they have changed between runs, the analyst can increase the number of iterations by a factor of 10 and try again.

Bootstrapping

In the algorithms above, we modeled random error by sampling from the asymptotic distribution of the bias-adjusted effect estimate. An alternative approach is to use the bootstrap. The bootstrap, developed by Efron [18], uses resampling from the original dataset, with replacement, to estimate the variance of the effect of interest. The bootstrap proceeds by resampling the observed data numerous times and computing the parameter of interest in the resampled data. To implement the nonparametric bootstrap in a probabilistic bias analysis, we generally need record-level data for resampling. Although it is possible to use the parametric bootstrap for summary level bias adjustment, this has not been implemented in the literature.

Impossible Values for Bias Parameters and Model Diagnostic Plots

As noted at various points in the text, when using sensitivity and specificities to adjust for misclassification, an analyst may obtain bias-adjusted data that are impossible (*e.g.,* a negative cell frequency in a 2×2 table). Dealing with impossible values requires great care and we propose the following general approach, which includes substantial planning before analyzing any data.

First, in Table 8.1, step 3 of any probabilistic bias analysis is to specify the distribution for the bias parameter. As we describe below, one way in which impossible values can be obtained is by poorly specifying the bias parameter distribution. As such, a great deal of thought should be given to these distributions to ensure that the bias parameter distributions truly capture the analyst's prior knowledge. If external studies are used to inform the bias parameters, the analyst should be sure to find external studies that are transportable to the current study.

Second, some studies are more prone to generating impossible values than others, particularly those with low exposure prevalences or wide specificity distributions. Analysts are encouraged to consider the possibility of obtaining impossible values prior to implementing the probabilistic bias analysis. As we emphasize throughout the text, *a priori* specified sensitivity analyses of bias parameter distributions is an important way to test the sensitivity of results to necessary assumptions. Plausible alternative bias parameter distributions should be specified *a priori*, and special consideration should be given to alternate distributions for specificity as this is the parameter most likely (in our experience) to cause impossible values when the exposure prevalence is low (see Chapter 6). The analyst may wish to consider a series of specificity distributions with varying ranges. It is important to specify these alternative distributions prior to looking at the bias analysis results to avoid a "results-driven bias analysis."

Third, when bias adjustments have been conducted, the results of all analyses should be reported (the main analysis and any sensitivity analyses), including the proportion of all simulations that resulted in impossible values, and especially any changes to the distributions assigned to the bias parameters in reaction to results (such as a high proportion of results with impossible values).

Fourth, if the proportion of results with impossible values is substantial, the analyst should carefully consider and report why this has occurred. We do not recommend interpreting bias-adjusted effects if there is a substantial proportion of impossible values as the interval width may be far too narrow. We give much more detail on this below.

Fifth, if a substantial proportion of results have impossible values, the analyst should consider implementing a Bayesian approach (Chapter 11), which avoids this problem in a statistically appropriate way.

As an example, in the misclassification application above, we assigned probability distributions of sensitivity~beta($\alpha = 57$, $\beta = 16$) and specificity~beta($\alpha = 70$, $\beta = 1$) for exposure classification. Although unlikely, at some point in the simulation, we might have drawn a value of 90% for sensitivity and 87% for specificity. Table 8.19 depicts the simulated data implied by a bias-adjustment for misclassification with these values, incorporating only this first source of uncertainty.

Note that, after adjusting for the misclassification with these values assigned to the bias parameters, the bias-adjusted estimate of the frequency of breast cancer cases who were smokers equals -1.7. The combinations of values assigned to the

Table 8.19 Bias-adjustment for misclassification of smoking in a study of the effect of smoking during pregnancy on breast cancer risk assuming a nondifferential sensitivity of 90% and specificity of 87%, which produces a negative cell frequency.

	Observed		Bias-adjusted	
	Smokers	Nonsmokers	Smokers	Nonsmokers
Cases	215	1449	−1.7	1665.7
Controls	668	4296	29.5	4934.5

Observed data from Fink and Lash, 2003 [7].

sensitivity and specificity are impossible given the data. In this case, either the values assigned to the bias parameters are incorrect, or the data are incompatible with these assigned values (perhaps due to chance or to a systematic error that had not been considered, or both).

The values assigned to the bias parameters may be implausible for several reasons. As we have noted throughout, the exact values of the bias parameters are not identifiable. One can never know with certainty the sensitivity and specificity of exposure classification that operated in the single obtained result, for example. To know the exact values, one would need both the error-prone classification and the gold-standard measure for all participants but having the gold-standard measure on all participants would obviate the need for the bias-adjustment. Assignment of values to the bias parameters is always, therefore, uncertain. When the assigned probability distributions are informed by external validation studies, it may be that they are not transportable, resulting in at least part of the assigned probability distribution assigning values to the bias parameters that are not possible given the data. When the assigned probability distributions are informed by an internal validation study, it may be that the sample size allows too wide an interval, again resulting in at least part of the assigned probability distribution assigning values to the bias parameters that are not possible given the data. In the example above, the negative cell frequency arises primarily from an assigned specificity of 87%. A specificity of 87% for smoking status, as recorded by self-report of new mothers on their newborn's birth certificate, implies that 13% of true non-smokers will misreport that they smoked during pregnancy. This degree of misreporting seems implausible. Recall, for example, that the observed exposure prevalence among controls was $668/(668 + 4296) = 13.5\%$. Specificity of 87% suggests that almost all the observed cases were in fact misreported false-positive smokers. Nonetheless, specificities as low as 80% have been used for self-reported smoking status in probabilistic bias analyses on the basis of external validation studies [19]. If 80% were adopted as the lower limit of a trapezoidal distribution assigned to the specificity in this bias analysis, for example, then implausible values would often have been selected. In fact, we assigned a beta distribution to the specificity with $\alpha = 70$ and $\beta = 1$. A specificity of 87% is allowed under this distribution, but specificities of 87% or lower would only be expected in 5 or 6 of 100,000 iterations drawing from the assigned beta distribution. If the values of the beta distribution were assigned based on an internal validation study of 71 persons, with only 1 of 71 true non-smokers misreporting on the birth certificate that they were a smoker (observed specificity 70/71 = 98.6%), then implausible values would be quite rare. Imagine, though, that the validation study was much smaller, with none of 10 true nonsmokers misreporting. Given the small sample size, one might choose to add a prior for the assigned values of the beta distribution, with $\alpha = 10 + 1$ and $\beta = 0 + 1$. A specificity of 87% is much more likely under this distribution. Even though the point estimate of the observed specificity (10/10 = 100%) is higher than in the validation study with 71 participants (70/71 = 98.6%), specificities of 87% or lower would be common. They would be expected in >21,600 of 100,000 iterations drawing from the assigned beta distribution. One may therefore have to use subject matter knowledge to assign a narrower distribution to the specificity. See the sections in Chapters 6 and 7 for more on how to determine which values will produce negative expected values in a misclassification problem.

It is also possible that the values and distributions assigned to the bias parameters are valid, but the data are incompatible with these assigned values due to chance or to a systematic error that had not been considered in the bias analysis. We use a hypothetical example to illustrate. Imagine a case-control study with true odds ratio of:

$$OR_{true} = \frac{10/5}{60/80} \cong 2.7$$

If the sensitivity of exposure classification is 80% and the specificity is 90%, then the odds ratio we expect to observe would be:

$$OR_{expected} = \frac{14/12}{56/73} \cong 1.5$$

As we have noted above, however, the exposure classification would operate on each case and control individually. Anywhere between 0 and 70 exposed cases might be observed. Imagine that the following data were observed:

$$OR_{observed} = \frac{15/8}{55/77} \cong 2.6$$

Bias-adjusting the observed odds ratio would yield a result with a negative cell frequency:

$$OR_{bias-adjusted} \cong \frac{11.4/-0.7}{58.6/85.7} \cong -23.4$$

Of the five truly exposed controls, all five were correctly classified as exposed. While the true sensitivity of exposure classification was 80%, the realized sensitivity of exposure classification was 100%. Of the 80 truly unexposed controls, 77 were correctly classified as unexposed. While the true specificity of exposure classification was 90%, the realized specificity of exposure classification was 77/80 = 96%. Using the realized sensitivity and specificity instead of the true sensitivity and specificity would have returned five as the bias-adjusted number of exposed controls. However, one has no way to know the realized sensitivity and specificity outside of a simulation setting. Using the true sensitivity and specificity returns −0.7 for the bias-adjusted number of exposed controls.

The frequency with which a negative bias-adjusted number of exposed controls or cases might be returned in a probabilistic bias analysis depends on the sample size. As with any chance procedure, the probability of obtaining exactly what is expected is low, but not vanishingly small, for small sample sizes. For example, in the problem above, the expected result for the observed number of exposed controls is $0.8 \cdot 5 + (1 - 0.9) \cdot 80 = 12$, and bias-adjustment with the true sensitivity and specificity would return the correct number of truly exposed controls. The probability of observing exactly this result is ~6.03%. Unfortunately, with small sample size, the probability of observing results far from the expected result is also large. The probability of obtaining a negative bias-adjusted cell frequency in this example is ~10.23%. Increasing the sample size actually reduces the probability of obtaining exactly what is expected. For example, increasing the sample size in this example tenfold would mean that the expected result for the observed number of exposed controls is $0.8 \cdot 50 + (1 - 0.9) \cdot 800 = 120$, and bias-adjustment with the true sensitivity and specificity would return the correct number of truly exposed controls (50). The probability of obtaining exactly this result is ~0.657%, about a 9.2-fold reduction from the ~6.03% probability of observing exactly the expected result with the smaller sample size. Fortunately, with the larger sample size, the probability of observing results far from the expected result decreases even faster. The probability of obtaining a negative bias-adjusted cell frequency in the tenfold larger data set is only ~0.0015%. Increasing the sample size reduces the probability of obtaining a negative cell frequency by ~7000-fold from the ~10.23% in the smaller sample size. We are willing to trade the ~9.2-fold reduced probability of getting exactly what is expected for the much larger ~7000-fold reduction in the probability of getting results far from what is expected; that is the advantage of a larger sample size.

When conducting simple bias analysis, the analyst would immediately notice that a bias-adjustment returned a negative cell frequency. Diagnosing the explanation with certainty is impossible. If the sample size is large, then the issue may lie with the assignment of values or distributions for the bias parameters. If the sample size is small, then it may be more likely that the observed data deviated substantially from what was expected, and still possible that the values assigned to the bias parameters are different from the true values. There is, of course, a continuum for both explanations, with overlapping regions where both explanations may be acting. One could adjust the values assigned to the bias parameters to avoid the negative cell frequencies, but this solution would be suboptimal if the underlying problem arose from an unexpected result and not from assignment of incorrect values to the bias parameters. With probabilistic bias analysis, it can be even more difficult to note the problem and to know what to do about it. For example, one might obtain a negative cell frequency for both the bias-adjusted number of exposed cases and the bias-adjusted number of exposed controls in the same iteration, and the negative signs in each cell will cancel, resulting in a positive but implausibly-derived bias-adjusted estimate. Thus, the analyst ought to look beyond the bias-adjusted estimate at the bias-adjusted individual cell counts to ensure the problem is identified.

There are generally two diagnostics that can be used to identity impossible values in probabilistic bias analysis. First, the bias analysis can be coded to count the number and proportion of iterations that produce impossible bias-adjusted values. If impossible values are being produced, it suggests that the observed data and at least part of the values assigned to distributions of the bias parameters are incompatible, for one or both of the reasons described above. Second, the analyst can plot a histogram of the values of the bias parameters as they were realized in the analysis, excluding iterations that produced impossible values (note the spreadsheets developed for this book do this by default). These plots show where the data are incompatible with the input density distribution by comparing the realized histogram with the density expectation based on the user inputs.

The analyst has several options when impossible values are generated in a probabilistic bias analysis. It is important to recognize that these impossible values are suggesting that the bias parameter distributions and the observed data are not completely compatible. It also suggests that some values for bias parameters are producing results that are nearly incompatible, while remaining technically compatible. In a bias-adjusted 2×2 table, these nearly incompatible tables are those with very small bias-adjusted frequencies in at least one cell of the table. While these results are logically plausible, analysts should remember that bias-adjusted results can be extremely unstable when computed using these very small bias-adjusted frequencies. Bias-adjusted ratio estimates may, for example, be very large or near zero.

As described above, it is important that the analyst attempt to determine why the bias distribution and data are incompatible and should not simply alter the distributions assigned to the bias parameters to avoid the problem. In searching for reasons for the incompatibility, the analyst should examine both the bias distribution and the data. For instance, it is possible that the bias distribution was poorly specified. Perhaps the study population used to determine sensitivity and specificity of exposure was much older or had a different gender distribution than the population used to collect data for the analyst's study, and that these differences may change the sensitivity and specificity. Further, it is possible that the incompatibility has arisen due to other systematic errors in the data. If the data were subject to both selection bias and misclassification, only bias-adjusting for misclassification while ignoring selection bias could result in impossible bias-adjusted values. Finally, as noted above, small sample sizes in the original study may have yielded an original result that differs substantially from the expectation. The reasons for the incompatibility may vary, but it requires due diligence on the analyst's part to attempt to identify them.

If there is no suitable explanation for the incompatibility, one could simply leave the distributions assigned to the bias parameters as they are but remove all iterations that produce impossible values. If this is done, it is preferable to report the number of iterations that have been removed from the simulation so that consumers of the bias analysis can inspect for themselves whether or not the probability distributions were compatible with the data. This approach can help ensure that the analyst produces only plausible bias-adjusted values but has the disadvantage that the analyst may not realize that the analysis they are implementing is not the one they think they are implementing. In other words, the values for the bias parameters will be truncated at values that produce impossible results and as such the distributions for the bias parameters that are used in practice will be different from those desired. However, if the proportion of impossible iterations is relatively small (say, <1%), the impact may be negligible.

As foreshadowed above, a more technically appropriate way to deal with impossible values in the bias analysis is through a fully Bayesian analysis (see Chapter 11 for a full discussion). Bayesian approaches update the bias parameter distribution conditional on the observed data and eliminate the possibility of producing negative cell counts. Simulations suggest that while Bayesian methods and probabilistic bias analysis typically produce similar results, when data and bias parameters are incompatible, the two approaches may differ [20]. In particular, the probabilistic bias analysis intervals may be far narrower than Bayesian intervals. However, if a substantial portion of the samples result in impossible cell counts, it also implies a large number of valid cell counts will be very small (but greater than zero) and the final bias-adjusted estimate will potentially have a large interval estimate indicating a very imprecise effect.

It is tempting to adjust the bias parameter distribution such that it does not include (or rarely allows) bias parameter estimates that result in impossible bias-adjusted cell counts. There are two ways to view a procedure such as this. Less generously, this amounts to altering prior information using the observed data, a type of results driven bias analysis. In this light, it can be argued that this should be avoided in bias analysis. Alternatively, altering the bias parameter distribution could be considered an attempt to mimic a formal Bayesian approach (which would alter the prior distribution in light of the observed data). This ad-hoc solution is very unlikely to accomplish what is intended. Even with simple bias parameter distributions, the proper Bayesian adjustment in light of the bounds has a mathematically complicated form and it is very unlikely an analyst could mimic that distribution without writing out the math. In fact, we think it is quite likely that altering the bias parameter distribution to avoid impossible values would result in a bias distribution that inadvertently reduces the uncertainty in the final bias-adjusted effect estimate. For example, consider a distribution for specificity that has 10% of its probability density that is lower than specificity 0.8 and those values result in negative cell counts. A proper statistical solution to this problem modifies the bias distribution in light of the bound in such a way that the inadmissible 10% of the distribution is reallocated to values greater than specificity=0.8. However, that probability mass is reallocated to admissible values near 0.8, because those values are most similar to the inadmissible values the analyst specified in their bias parameter distribution. This makes intuitive sense: if an analyst says a value of 0.78 is likely based on prior knowledge but, after seeing the data, no value less than 0.8 is possible, the logical solution is to increase the probability of specificities that are near, but slightly larger than, 0.8. This can result in very odd, bimodal, bias parameter distributions, which are not what most analysts would choose if they were re-specifying their prior. Finally, and of direct relevance, the inadmissible probability density that is reallocated to values near specificity 0.8 will result in very small bias-adjusted cell counts and cause very large standard errors. Therefore, there is a strong concern that an analyst who re-specifies their bias distribution after seeing inadmissible values could seriously underestimate the uncertainty in their bias-adjusted effect. See Chapter 11 for a more complete description of these alternatives and their consequences for bias modeling.

We do not recommend altering the bias parameter distribution simply because impossible values have been realized, as this would result in a data-driven bias analysis. Rather, the analyst should report the results that are obtained. If there are serious concerns about incompatibility, the analyst can opt for a Bayesian solution to the problem. However, the final answer is likely to include very wide intervals. It may be wise to consider sources of the incompatibility and explore multidimensional bias analysis.

Conclusions

Probabilistic bias analysis has many advantages over simple bias analysis. First, simple bias analysis assumes that the bias parameters are known with certainty, a condition that rarely, if ever, occurs. Probabilistic bias analysis addresses this shortcoming by assigning probability distributions to the bias parameters, which more accurately reflects the uncertainty in the values assigned to those parameters. In the example above of smoking misclassification, data on the sensitivity of using birth certificate records as a measure of maternal smoking status were assessed using a validation study. The validation study found that birth certificates had a sensitivity of 78%, but this estimate of sensitivity was measured with random error that is not reflected in a simple bias analysis. The resulting simulation intervals from a probabilistic bias analysis reflect this uncertainty in the bias parameters and give a more accurate representation of the total uncertainty about a given result.

Another important advantage of the probabilistic bias analysis approach is ease and familiarity of presentation. The results of the probabilistic bias analysis can be summarized with a median estimate of bias-adjusted association and a 95% simulation interval. Tables such as those presented above can be used to easily compare the conventional estimates with the systematic error intervals and an interval adjusted for both systematic and random error, as well as sensitivity analyses that revise the bias-analysis to probe the sensitivity of its results to different bias-adjustment assumptions.

Bias analysis is sometimes criticized as being subjective because the analyst chooses what values or distributions to assign to the bias parameters. The bias parameters chosen should reflect a credible estimate of the actual biases at work in a study, but one never knows for sure if accurate distributions have been assigned to the bias parameters. Reporting estimates that reflect only random error, as if there were no systematic error in the data, can be a more misleading approach. By providing the distributions for the bias parameters and the rationale for the assignments, reviewers and readers can judge for themselves whether the distributions were appropriate and alternative distributions can be assigned to reflect the views of interested parties other than the original analysts. Making bias analysis code available can help interested parties to investigate the sensitivity of results to different assumption sets [3].

If simple bias analysis is more common than probabilistic bias analysis, it may be due, in part, to the perception that probabilistic bias analysis is difficult to implement and no inexpensive software exists to carry out probabilistic bias analysis. However, this chapter seeks to demonstrate that probabilistic bias analysis can be conducted by those who are able to use simple bias analysis with little additional work. We have been able to conduct probabilistic bias analysis using Microsoft Excel, SAS and R, and provide spreadsheets and code for users to implement their own on the website associated with this text (https://sites.google.com/site/biasanalysis/).

Appendix: Sampling Models for Exposure Misclassification

A common approach to solving a statistical problem is to write down a model that represents the random process that is giving rise to the data in question. As an example, if one had collected data from a case-control study in which there had been no bias of any kind, a common way of writing down a model for this study would be:

$$A \sim \text{binomial}(N_1, \pi_1)$$
$$C \sim \text{binomial}(N_0, \pi_0)$$

That is, among a sample of N_1 cases, with π_1 equal to the probability of being exposed among cases, the random variable A is the number of exposed cases we would see. We are assuming that the number of exposed cases arises randomly based on the number of cases and the probability of exposure among cases. Similarly, among N_0 controls, C of them were randomly found to be exposed with probability π_0. If we want to estimate the odds ratio from this case-control study, then we wish to find an estimate of $OR = \frac{(\pi_1)/(1-\pi_1)}{(\pi_0)/(1-\pi_0)}$. Writing down this model and following standard maximum likelihood statistical procedures allow us to find an estimator of this effect as well as its standard error.

In conducting a probabilistic bias analysis for exposure misclassification in a case-control study, we have not yet been explicit about writing down our inferential model. The model most often seen in the literature is:

$$p_i = \pi_i \cdot se_i + (1 - \pi_i)(1 - sp_i)$$
$$a \sim \text{binomial}(N_1, p_1)$$
$$c \sim \text{binomial}(N_0, p_o)$$

where π_i is, as above, the probability of exposure among cases (i = 1) and controls (i = 0). The sensitivity and specificity among cases and controls are given by se_i and sp_i, respectively, and p_i is the probability of observing a potentially misclassified exposure. Observed and potentially misclassified exposed people (a and c) arise randomly from the cases and controls dependent on p_i. That is, in contrast to the first model in which the number of truly exposed individuals arises randomly from cases and controls, we now have the number of potentially misclassified exposed individuals arising randomly from cases and controls. This model has often been used as the basis of methods to estimate $OR = \frac{(\pi_1)/(1-\pi_1)}{(\pi_0)/(1-\pi_0)}$, even when we cannot directly estimate π_i [6, 21].

It may be confusing to write the model as we have above rather than using the true but unknown number of exposed cases (A) and controls (C). We show here that we can capture the same statistical information using an equivalent model that includes the latent variables A and C.

$$A \sim \text{binomial}(N_1, \pi_1)$$
$$C \sim \text{binomial}(N_0, \pi_0)$$
$$W_1 \sim \text{binomial}(A, se_1)$$
$$X_1 \sim \text{binomial}(N_1 - A, 1 - sp_1)$$
$$W_0 \sim \text{binomial}(C, se_0)$$
$$X_0 \sim \text{binomial}(N_0 - C, 1 - sp_0)$$

In this model W_i and X_i are the number of people who are truly exposed and classified as exposed and the number of people who are truly unexposed but classified as exposed, respectively among cases (i = 1) and controls (i = 0). It follows that the observed number of apparently exposed cases $a = W_1 + X_1$ and the number of apparently exposed controls is $c = W_0 + X_0$. This model may adhere more closely to what we believe is occurring scientifically. First, individuals are sampled into the study and have a true and latent exposure status (A and C). We administer an imperfect test to them, resulting in W_1 of the A truly exposed individuals accurately classified and the remaining $A - W_1$ incorrectly classified as unexposed. Similarly, among the $N_1 - A$ people who were truly not exposed, X_1 of them are incorrectly classified as exposed while $N_1 - A - X_1$ of them were accurately classified as unexposed.

These two statistical models of misclassification convey the same information about π_i and, all else being equal, we would expect to estimate the same effect using either model. The first model is often used because it is more parsimonious and excludes information on the latent variables W_i and X_i since we cannot directly observe them and since they are not immediately necessary for inference on π_i. However, the second model can also be useful. For example, the justification for the probabilistic bias analysis we propose for exposure misclassification in this Chapter is made clearer by using the second model. We present a more formal justification in Chapter 11.

References

1. Dietz PM, Adams MM, Kendrick JS, Mathis MP. Completeness of ascertainment of prenatal smoking using birth certificates and confidential questionnaires: variations by maternal attributes and infant birth weight. PRAMS Working Group. Pregnancy Risk Assessment Monitoring System. Am J Epidemiol. 1998;148:1048–54.
2. McCandless LC, Gustafson P. A comparison of Bayesian and Monte Carlo sensitivity analysis for unmeasured confounding. Stat Med. 2017;36:2887–901.
3. Lash TL, Fox MP, MacLehose RF, Maldonado G, McCandless LC, Greenland S. Good practices for quantitative bias analysis. Int J Epidemiol. 2014;43:1969–85.
4. Buescher PA, Taylor KP, Davis MH, Bowling JM. The quality of the new birth certificate data: a validation study in North Carolina. Am J Public Health. 1993;83:1163–5.
5. Piper JM, Mitchel EF Jr, Snowden M, Hall C, Adams M, Taylor P. Validation of 1989 Tennessee birth certificates using maternal and newborn hospital records. Am J Epidemiol. 1993;137:758–68.
6. Gustafson P. Measurement error and misclassificaion in statistics and epidemiology impacts and Bayesian adjustments. Boca Raton, Fla: Chapman & Hall/CRC; 2004.
7. Fink AK, Lash TL. A null association between smoking during pregnancy and breast cancer using Massachusetts registry data (United States). Cancer Causes Control. 2003;14:497–503.
8. Gustafson P. Invited Commentary: Toward Better Bias Analysis. Am J Epidemiol. 2021;190: 1613–6.
9. Lash TL, Abrams B, Bodnar LM. Comparison of Bias Analysis Strategies Applied to a Large Data Set. Epidemiol. 2014;25:576.
10. Greenland S. A commentary on 'A comparison of Bayesian and Monte Carlo sensitivity analysis for unmeasured confounding'. Stat Med. 2017;36:3278–80.
11. Tyndall MW, Ronald AR, Agoki E, Malisa W, Bwayo JJ, Ndinya-Achola JO, et al. Increased risk of infection with human immunodeficiency virus type 1 among uncircumcised men presenting with genital ulcer disease in Kenya. Clin Infect Dis. 1996;23:449–53.
12. Cameron DW, Simonsen JN, D'Costa LJ, Ronald AR, Maitha GM, Gakinya MN, et al. Female to male transmission of human immunodeficiency virus type 1: risk factors for seroconversion in men. Lancet. 1989;2:403–7.
13. Haneuse S. Stratification and Standardization. In: Lash T, VanderWeele TJ, Haneuse S, Rothman KJ, editors. Modern Epidemiology. 4th ed. Philadelphia: Wolters Kluwer; 2021.
14. Lash TL, Schmidt M, Jensen AØ, Engebjerg MC. Methods to apply probabilistic bias analysis to summary estimates of association. Pharmacoepidemiol Drug Saf. 2010;19:638–44.
15. Bodnar LM, Tang G, Ness RB, Harger G, Roberts JM. Periconceptional multivitamin use reduces the risk of preeclampsia. Am J Epidemiol. 2006;164:470–7.
16. Arah OA, Chiba Y, Greenland S. Bias Formulas for External Adjustment and Sensitivity Analysis of Unmeasured Confounders. Ann Epidemiol. 2008;18:637–46.
17. Stang A, Schmidt-Pokrzywniak A, Lash TL, Lommatzsch P, Taubert G, Bornfeld N, et al. Mobile Phone Use and Risk of Uveal Melanoma: Results of the RIFA Case–Control Study. J Natl Cancer Inst. 2009;101:120–3.

18. Efron B, Tibshirani RJ. An introduction to the bootstrap. New York: Chapman and Hall; 1994.
19. Corbin M, Haslett S, Pearce N, Maule M, Greenland S. A comparison of sensitivity-specificity imputation, direct imputation and fully Bayesian analysis to adjust for exposure misclassification when validation data are unavailable. Int J Epidemiol. 2017;46:1063–72.
20. MacLehose RF, Gustafson P. Is probabilistic bias analysis approximately Bayesian? Epidemiology. 2012;23:151–8.
21. Chu H, Wang Z, Cole SR, Greenland S. Sensitivity analysis of misclassification: a graphical and a Bayesian approach. Ann Epidemiol. 2006 Nov;16:834–41.

Chapter 9
Probabilistic Bias Analysis for Simulation of Record-Level Data

Introduction

The probabilistic bias analyses shown in the previous chapter were conducted using summarized data. As noted, a major limitation of using the summarized data approach is that it is difficult to adjust for other measured confounders for which adjustment may have been made in the conventional analysis. However, probabilistic bias analysis can be conducted using record-level data (record-level bias-adjustment), which retains information on other covariates and allows for multiple adjustments to be made in the final analysis. The methods in this chapter follow the same logic and outline described in Chapter 8, with some changes for the record level scenario, so we strongly suggest working through that chapter before this one even if record level correction is the intended approach.

The main difference between summary-level and record-level adjustments is that with summary-level analysis, the dataset is already aggregated into a contingency table and therefore all the calculations are performed on either the observed cell counts or the summary effect estimate. For example, if we run 10,000 iterations, we will have a dataset with 10,000 records. With record-level bias-adjustments, we need to take the full record-level dataset and then sometimes summarize it into a contingency table, use that summarized data to run the same types of adjustments we would in the summary-level case, and then turn those calculations into probabilities used for bias-adjusting the record-level dataset. Then we need to merge those probabilities back into the record-level dataset and sample the unobserved data. This process can require additional coding and familiarity with how to aggregate datasets into summary-level data and familiarity with how to merge datasets. In addition, the number of records we need to work with will equal the number of iterations multiplied by the original dataset study size (unless some collapsing with reweighting is possible [1]). A dataset of 10,000 records requiring 10,000 simulations means a dataset of 100,000,000 records. Figure 9.1 shows a flowchart of the record-level approach.

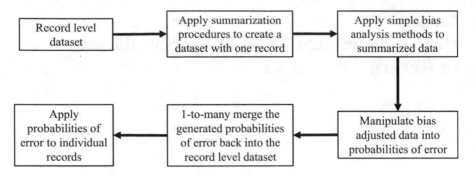

Figure 9.1 Flowchart for record-level bias analysis.

Table 9.1 General steps for a record-level probabilistic bias analysis.

Step	Explanation
1	Identify the likely sources of important bias in the data
2	Identify the bias parameters needed to bias-adjust
3	Assign probability distributions to each bias parameter
4	Use simple bias analysis methods to incorporate uncertainty in the bias parameters and random error
4a	Sample from each bias parameter distribution
4b	Use simple bias analysis methods to generate bias-adjusted data to inform the bias analysis and apply bias parameters probabilistically
4c	Incorporate additional sources of random error
5	Save the bias-adjusted estimate and repeat steps 4a–c
6	Summarize the bias-adjusted estimates with a frequency distribution that yields a central tendency and simulation interval

Although record-level bias-adjustment follows the same approach as the summary-level bias-adjustment (Table 9.1), it often requires additional programming and may substantially increase computing time due to the larger size of the datasets involved. In addition, while one can skip some of the summarizing and merging with additional programming (*i.e.* write lines of code that calculate the cell counts for the summarized dataset rather than using a summarization procedure for a frequency table and outputting the data into a dataset that can be merged back into the record-level dataset), we find that using summarization procedures and merges allows for more flexible coding that can be reused for other analyses. For general recommendations on conducting record-level analyses, see the section at the end of this chapter on the complexities of coding record-level bias analyses.

There are multiple ways to estimate the variance of bias-adjusted estimates, particularly with record-level data, and we discuss alternatives at the end of the chapter. Our principal approach, as with Chapter 8, is to combine uncertainty in the bias parameter with random error from the data in our final variance estimate through a Monte Carlo procedure. In Chapter 8, this process involves resampling the bias-adjusted effect in step 4c using SE^{adj} (for misclassification and confounding adjust-ments) or SE^{crude} (for selection bias adjustment). One of the major benefits of a record-level analysis is that the analyst can run a regression model to adjust for measured confounders. We will use standard errors from the regression models, SE^{reg}, in this chapter when we implement step 4c. We will discuss the specifics of the implementation below.

Exposure Misclassification Implementation

The steps for a record-level probabilistic bias analysis are largely the same as for summary-level probabilistic bias analysis, but we note where additional consider-ations are needed to account for the use of record-level data.

Step 1: Identify the Source of Bias

To demonstrate record-level data adjustments, we will again use the example of misclassification of smoking on the birth certificate. The goal in the example is to estimate the association between smoking during pregnancy and incidence of breast cancer, with observed data presented in the left panel of Table 9.2 and the notation for the misclassification-adjusted data in the right panel.

Step 2: Select the Bias Parameters

For record-level bias-adjustment, like in the summary-level case, we need to know the probability that, given a woman's classification as a smoker or a non-smoker, the classification was correct (*i.e.*, the predictive values). But unlike in the summary-level case, where we applied those probabilities to the summarized cell counts, we can simulate, for each woman in the dataset, a new version of her smoking status. Namely, rather than using summary-level draws from binomial distributions to generate bias-adjusted cell counts as we did in Chapter 8, we will sample from

Table 9.2 Observed data in a study of the effect of smoking during pregnancy on breast cancer risk.

	Observed		Misclassification-adjusted	
	Smokers	Nonsmokers		
Cases	215	1449	A	B
Controls	668	4296	C	D
Odds ratio	0.95			

Data from Fink and Lash, 2003 [2].

Bernoulli distributions and generate a bias-adjusted exposure value for each woman in the dataset. As we have noted in previous chapters, for some internal validation studies, we can estimate the predictive values directly. If we have done so, then our job is less computationally intensive; we can assign a probability distribution to the positive and negative predictive values and ignore the steps described below that lead up to the estimation of the positive and negative predictive values. However, unless an internal validation study was conducted and predictive values were estimated, we cannot apply the predictive values from another study to our results as positive and negative predictive value are functions of both sensitivity and specificity and also prevalence. Since it is most common to have sensitivity and specificity and not positive and negative predictive values, we will proceed assuming those are the required bias parameters and as in Chapter 8, we assume the misclassification is nondifferential.

Step 3: Assign Probability Distributions to Each Bias Parameter

For this example, we will use the same distributions for the bias parameters that we used in the previous chapter, allowing a direct comparison of the results of a summary-level and record-level probabilistic bias analysis. We specify a distribution for sensitivity of exposure classification of beta($\alpha = 50.6$, $\beta = 14.3$) and will set a distribution for specificity of exposure classification of beta($\alpha = 70$, $\beta = 1$).

If predictive values are available to the analyst, one can specify distributions for the predictive values, sample from them, and then skip directly to the middle of step 4b where the predictive values are applied.

Step 4: Use Simple Bias Analysis Methods to Incorporate Uncertainty in the Bias Parameters and Random Error

Step 4a: Randomly Sample from the Bias Parameter Distributions

Step 4a proceeds exactly as it did for the summary-level probabilistic bias analysis: we randomly sample from the sensitivity and specificity distributions (possibly with correlated sampling, for example when sampling from the sensitivity distributions assigned to cases and controls under differential misclassification). To ease comparisons with the summary-level approach in Chapter 8, we use the same sampled sensitivity of 71.22% and a specificity of 95.78% from that chapter. Code for sampling from distributions is included in Chapter 8 and we do not repeat it in this chapter.

Step 4b: Use Simple Bias Analysis Methods and Incorporate Uncertainty and Conventional Random Error

As noted above, the approach to record-level analyses will parallel the summary-level approach, accounting for the same uncertainty and applying many of the same equations. The main difference is that we need to work with both the record-level dataset and a contingency table summarizing that dataset. In addition, while we will use the equations from Chapter 8 to estimate positive and negative predictive values from the sampled sensitivity and specificity, there is a key distinction at the end of step 4. In Chapter 8, we used the predicted probabilities to sample the aggregate bias-adjusted cell counts from a binomial distribution. In this chapter, we will apply the probabilities to each individual record and sample their bias-adjusted exposure value from Bernoulli distributions. Because each individual record will have a bias-adjusted exposure, we can use standard regression techniques to adjust for confounding.

Step 4b begins with a record-level dataset, as shown in Figure 9.2 (first 10 observations only). Here we have one record for each subject in the dataset, for a total of 6628 records. We code the observed exposure (with a variable name of "e_obs") as 1 if exposed and 0 if unexposed and the outcome (variable name = "d") as 1 if a case and 0 if a control. We then summarize the dataset using any procedure that generates a contingency table and can output the summarized information into a dataset (e.g. using PROC FREQ in SAS with an output statement, the "postfile" command in Stata, or "table" in R). We do this to turn the record-level dataset into an aggregate dataset with a single record that contains one variable for each of the four cells in the summarized contingency table, a, b, c and d (continuing our notational scheme in which lower case letters indicate observed cell counts and capital letters indicate bias-adjusted cell counts). In SAS this requires some manipulation of the output dataset to get the data into a single record. One could skip this step by writing code that adds the number of subjects in each cell into new variables in the record-level dataset, but our experience is that mixing the aggregate and record-level data at this step is prone to errors. Figure 9.3 shows the summarized dataset.

Figure 9.2 First 10 observations in a record-level dataset in a study of the effect of smoking during pregnancy on breast cancer risk for use in a record-level bias analysis for exposure misclassification. Observed data from Fink and Lash, 2003 [2]; "e_obs" is the observed exposure, "d" is the outcome).

id	e_obs	d
1	1	1
2	0	0
3	0	0
4	0	0
5	1	0
6	0	1
7	0	0
8	1	1
9	1	0
10	0	0

a	b	c	d
215	1449	668	4296

Figure 9.3 Aggregated dataset at the summary level in a study of the effect of smoking during pregnancy on breast cancer risk for use in a record-level bias analysis for exposure misclassification. (Observed data from Fink and Lash, 2003 [2]; a = observed exposed with the outcome, b = observed unexposed with the outcome, c = observed exposed without the outcome, d = observed unexposed without the outcome).

Table 9.3 Process for calculating simple misclassification-adjusted data given the observed data with exposure misclassification.

	Observed		Misclassification-adjusted	
	E+	E−	E+	E−
D+	a	b	$A^O = [a - D_{1\ \text{Total}} (1 - SP_{D1})] / (SE_{D1}-1 + SP_{D1})$	$B^O = D_{1\ \text{Total}} - A$
D-	c	d	$C^O = [c - D_{0\ \text{Total}} (1 - SP_{D0})] / (SE_{D0}-1 + SP_{D0})$	$D^O = D_{0\ \text{Total}} - C$
Total	a + c	b + d	$A^O + C^O$	$B^O + D^O$

Table 9.4 Simple bias-adjustment for misclassification of smoking in a study of the effect of smoking during pregnancy on breast cancer risk assuming a nondifferential sensitivity of 71.22% and nondifferential specificity 95.78%.

	Observed		Misclassification adjusted data given bias parameters	
	Smokers	Nonsmokers	Smokers	Nonsmokers
Cases	215	1449	216.1	1447.9
Controls	668	4296	684.4	4279.6
OR	0.954		0.933	

Observed data from Fink and Lash, 2003 [2].

Next, using the sampled values of sensitivity and specificity from step 4a and the summarized data in Figure 9.3, we apply the same simple bias analysis methods that we did in Chapter 8 to obtain initial estimates of the misclassification-adjusted cell counts. Table 9.3 repeats the equations from Chapters 6 and 8 to accomplish this. Table 9.4 gives the misclassification-adjusted data that are calculated using the first draw of sensitivity = 71.22% and a specificity = 95.78%. The numbers obtained at this step are identical to those in Chapter 8 because we have used the same sampled values of sensitivity and specificity and applied them to the same data using the same equations.

Although we show the bias-adjusted cell counts in a contingency table, when conducting a record-level analysis we would implement this bias-adjustment within the summarized, single record dataset. The dataset would have one variable (column) for each of the observed cell counts (a, b, c and d), one variable for each of the bias parameters (sensitivities and specificities), and one variable for each of the bias-adjusted cell counts (A, B, C and D). We note that although we have assumed nondifferential misclassification, we include sensitivity (specificity) among cases and sensitivity (specificity) among controls in the dataset to allow easier adaptation

iter	se_D1	se_D0	sp_D1	sp_D0	A0	B0	C0	D0
1	0.7122	0.7122	0.9578	0.9578	216.1	1447.9	684.4	4279.6
2	0.7800	0.7800	0.9732	0.9732	226.3	1437.7	710.4	4253.6
3	0.8019	0.8019	0.9629	0.9629	200.3	1463.7	632.4	4331.6
4	0.8237	0.8237	0.9926	0.9926	248.3	1415.7	773.4	4190.6
5	0.8349	0.8349	0.9874	0.9874	235.9	1428.1	736.2	4227.8
6	0.7526	0.7526	0.9641	0.9641	216.6	1447.4	683.3	4280.7
7	0.7718	0.7718	0.9991	0.9991	277.0	1387.0	860.9	4103.1
8	0.8466	0.8466	0.9562	0.9562	177.0	1487.0	561.3	4402.7
9	0.7646	0.7646	0.9771	0.9771	238.5	1425.5	747.3	4216.7
10	0.7382	0.7382	0.9811	0.9811	255.3	1408.7	798.5	4165.5

Figure 9.4 The first 10 iterations of a summarized level dataset with the misclassification-adjusted data in a study of the effect of smoking during pregnancy on breast cancer risk for use in a record-level bias analysis for exposure misclassification. (Observed data from Fink and Lash, 2003 [2]; iter = iteration number, se_D1 = chosen sensitivity among those with the outcome, sp_D1 = chosen specificity among those with the outcome, se_D0 = chosen sensitivity among those without the outcome, sp_D0 = chosen specificity among those without the outcome, A0 = misclassification-adjusted exposed with the outcome, B0 = misclassification-adjusted unexposed with the outcome, C0 = misclassification-adjusted exposed without the outcome, D0 – misclassification-adjusted unexposed without the outcome).

of the example to problems with differential misclassification. While we are only discussing the first iteration of this algorithm, with the first sampled sensitivity and specificity, in practice it is more efficient to sample sensitivities and specificities for all iterations of the algorithm at this stage. If we intend to repeat the algorithm for 10,000 iterations, we sample 10,000 sensitivities and 10,000 specificities and compute 40,000 bias-adjusted cell counts. Each of the 10,000 rows in the dataset would correspond to different samples of sensitivities and specificities and the resulting values of the bias-adjusted cell counts. The first 10 rows of this dataset are shown in Figure 9.4 (with the observed data removed, to allow the remaining variables to be displayed on one page).

Now that we have initial estimates of the misclassification-adjusted data, we can follow the same approach as in Chapter 8 for summary-level analysis. To account for the fact that we do not know the true prevalence of exposure, we estimate it by sampling from the distribution of expected prevalence. We do this by sampling from beta distributions using the misclassification adjusted data as inputs:

$$P(E_{D1}) \sim \text{beta}(\alpha = A^0, \beta = B^0) = \text{beta}(\alpha = 216.09, \beta = 1447.91)$$

$$P(E_{D0}) \sim \text{beta}(\alpha = C^0, \beta = D^0) = \text{beta}(\alpha = 684.36, \beta = 4279.64)$$

To make the comparison with Chapter 8 clear, we use the same sampled values from this step in Chapter 8, and have $P(E_{D1}) = 13.28\%$ and $P(E_{D0}) = 13.99\%$. We use these values to estimate the positive and negative predictive values using the equations:

$$PPV_{D1} = \frac{SE_{D1}P(E_{D1})}{SE_{D1}P(E_{D1}) + (1 - SP_{D1})(1 - P(E_{D1}))}$$

$$NPV_{D1} = \frac{SP_{D1}(1 - P(E_{D1}))}{(1 - SE_{D1})P(E_{D1}) + SP_{D1}(1 - P(E_{D1}))}$$

$$PPV_{D0} = \frac{SE_{D0}P(E_{D0})}{SE_{D0}P(E_{D0}) + (1 - SP_{D0})(1 - P(E_{D0}))}$$

$$NPV_{D0} = \frac{SP_{D0}(1 - P(E_{D0}))}{(1 - SE_{D0})P(E_{D0}) + SP_{D0}(1 - P(E_{D0}))}$$

Using the sampled values, we find $PPV_{D1} = 72.10\%$, $NPV_{D1} = 95.60\%$, $PPV_{D0} = 73.30\%$ and $NPV_{D0} = 95.34\%$. For a more detailed explanation of step 4b, see Chapter 8 on exposure misclassification for summary-level adjustment. Again, we note that even if we specified non-differential misclassification of exposure with respect to the outcome, we would expect in most cases that the predictive values will differ within levels of the outcome because they also depend on the true prevalence of the exposure, which will differ within levels outcome unless there is no association between exposure and outcome. Figure 9.5 shows the dataset of the sampled predictive values (in this case we see 10 iterations with different values for the exposure prevalence and predictive values due to different values drawn for sensitivity and specificity).

Before we can use these predictive values, we need to merge them back into the record-level dataset. We take the first set of four predictive values and combine them with the original data using a 1-to-many merge such that each set of the four predictive values is repeated for each observation in the original dataset. Figure 9.6

iter	se_D1	se_D0	sp_D1	sp_D0	A0	B0	C0	D0	ped1	ped0	PPV_d1	PPV_d0	NPV_d1	NPV_d0
1	0.7122	0.7122	0.9578	0.9578	216.1	1447.9	684.4	4279.6	0.1328	0.1399	0.7210	0.7330	0.9560	0.9534
2	0.7757	0.7757	0.9887	0.9887	256.7	1407.3	800.6	4163.4	0.1443	0.1543	0.9206	0.9261	0.9632	0.9603
3	0.7505	0.7505	0.9847	0.9847	257.8	1406.2	805.2	4158.8	0.1519	0.1597	0.8976	0.9030	0.9566	0.9541
4	0.7980	0.7980	0.9922	0.9922	255.7	1408.3	796.5	4167.5	0.1475	0.1604	0.9466	0.9514	0.9660	0.9626
5	0.7146	0.7146	0.9823	0.9823	266.2	1397.8	832.3	4131.7	0.1640	0.1818	0.8877	0.8995	0.9461	0.9393
6	0.8174	0.8174	0.9956	0.9956	255.5	1408.5	794.8	4169.2	0.1432	0.1575	0.9688	0.9720	0.9703	0.9668
7	0.7706	0.7706	0.9829	0.9829	247.7	1416.3	774.2	4189.8	0.1394	0.1557	0.8798	0.8929	0.9636	0.9587
8	0.7214	0.7214	0.9858	0.9858	270.7	1393.3	845.2	4118.8	0.1588	0.1772	0.9059	0.9165	0.9493	0.9426
9	0.7904	0.7904	0.9704	0.9704	217.8	1446.2	684.9	4279.1	0.1262	0.1344	0.7941	0.8057	0.9697	0.9675
10	0.7987	0.7987	0.9343	0.9343	144.2	1519.8	466.4	4497.6	0.0863	0.0900	0.5346	0.5459	0.9801	0.9791

Figure 9.5 10 iterations of a summarized level dataset with the sampled prevalences of exposure and estimated predictive values in a study of the effect of smoking during pregnancy on breast cancer risk for use in a record-level bias analysis for exposure misclassification. (Observed data from Fink and Lash, 2003 [2]; Variables as defined in previous figures and ped1 = $P(E_{D1})$, ped0 = $P(E_{D0})$, PPV_d1 = positive predictive value among cases, PPV_d0 = positive predictive value among controls, NPV_d1 = negative predictive value among cases, NPV_d0 = negative predictive value among controls).

id	e_obs	d	PPV_d1	NPV_d1	PPV_d0	NPV_d0	p	iter
1	1	1	0.721	0.956	0.733	0.9534	0.7210	1
2	0	0	0.721	0.956	0.733	0.9534	0.0466	1
3	0	0	0.721	0.956	0.733	0.9534	0.0466	1
4	0	0	0.721	0.956	0.733	0.9534	0.0466	1
5	1	0	0.721	0.956	0.733	0.9534	0.7330	1
6	0	1	0.721	0.956	0.733	0.9534	0.0440	1
7	0	0	0.721	0.956	0.733	0.9534	0.0466	1
8	1	1	0.721	0.956	0.733	0.9534	0.7210	1
9	1	0	0.721	0.956	0.733	0.9534	0.7330	1
10	0	0	0.721	0.956	0.733	0.9534	0.0466	1

Figure 9.6 First 10 observations in a record-level dataset in a study of the effect of smoking during pregnancy on breast cancer risk for use in a record-level bias analysis for exposure misclassification with predictive values merged. (Observed data from Fink and Lash, 2003 [2]; Variables as defined in previous figures, iter = iteration number, p = probability of exposure for the record in that row, selected from the four predictive values, given observed exposure and case or control status.).

shows the new dataset (with some variables removed from the figure to save space) showing how we have merged the four predictive values into each record of the main record-level dataset.

We can now use Bernoulli trials to simulate, at the record-level, a new exposure variable. As described in Chapter 7, a Bernoulli trial is essentially a coin flip that returns a 1 or a 0 with a probability decided by the user. To implement the Bernoulli trials, we need to ensure that: (a) we apply the correct probability to each person; and (b) that we apply the probability in the correct way. There are four types of people in our dataset based on the way they were originally classified in the dataset, those who were classified as: (1) exposure = 1, outcome = 1; (2) exposure = 1, outcome = 0; (3) exposure = 0, outcome = 1; and (4) exposure = 0, outcome = 0. For the first two groups, the modeled probability of being correctly classified as exposed is equal to the positive predictive value among the cases $(PPV_{D1} = 72.10\%)$ and controls $(PPV_{D0} = 73.30\%)$, respectively. For the second two groups, the modeled probability of being truly unexposed is equal to the negative predictive value among the cases $(NPV_{D1} = 95.60\%)$ and controls $(NPV_{D0} = 95.34\%)$, respectively (Table 9.5).

To simulate a misclassification adjusted exposure variable, we create a new exposure variable and write code to apply the appropriate positive and negative predictive value for each subject. We need to calculate the modeled probability, p, that an individual is truly exposed given their observed exposure and disease status. This calculation can be accomplished using a series of if-then statements to find the right predictive value. For example, if a subject is observed to be a case and exposed, the modeled probability they were truly exposed is p = 72.10%, or PPV_{D1}. If a subject is observed to be an exposed control, the modeled probability they were truly exposed is p = 73.30%, or PPV_{D0}. If a subject is observed to be an unexposed case, the modeled probability they were truly exposed was $1 - NPV_{D1}$ (p = 100%–

Table 9.5 Estimated positive and negative predictive values used in the analysis of data from Fink and Lash, 2003 [2] in a study of the effect of smoking during pregnancy on breast cancer risk.

Classified Group	Observed Exposure	Outcome	Parameter	Estimate
Cases-exposed	1	1	PPV_{D1}	72.10%
Controls-exposed	1	0	PPV_{D0}	73.30%
Cases-unexposed	0	1	NPV_{D1}	95.60%
Controls-unexposed	0	0	NPV_{D0}	95.34%

Table 9.6 Examples of how to conduct Bernoulli trials to model true exposure status with a positive predictive value of 72.10% or a negative predictive value of 95.6%.

Program	Syntax for positive predictive value	Syntax for negative predictive value
R	*rbinom(1,1,0.721)*	*rbinom(1,1,1−0.956)*
Stata	*gen x=rbinomial(1,0.721)*	*gen x=rbinomial(1, 1−0.956)*
SAS	*x=rand('bernoulli', 0.721)*	*x=rand('bernoulli', 1−0.956)*
Excel	*=if(RAND()<0.721,1,0)*	*=if(RAND()>0.956,1,0)*

$95.60\% = 4.40\%$). If a subject is observed to be an unexposed control, the modeled probability they were truly exposed was $1 - NPV_{D0}$ (p = 100%–$95.34\% = 4.66\%$). Alternatively, because if-then statements can be time-consuming for software to evaluate, a quicker alternative is to assign p (shown in Figure 9.6) using the following equation:

$$p = e_{obs} \cdot d \cdot PPV_{D1}$$
$$+ e_{obs} \cdot (1 - d) \cdot PPV_{D0}$$
$$+ (1 - e_{obs}) \cdot d \cdot (1 - NPV_{D1})$$
$$+ (1 - e_{obs}) \cdot (1 - d) \cdot (1 - NPV_{D0})$$

Using either approach, we have assigned each observation a modeled probability of truly being exposed and we impute the bias-adjusted exposure status by sampling from a Bernoulli(p) distribution where, again, p is the probability of that individual being truly exposed. Note that we do this by creating a new variable rather than overwriting the original exposure variable so that we can compare the classified result to the bias-adjusted result. A Bernoulli trial can be implemented in standard software packages as shown in Table 9.6. Figure 9.7 shows the dataset with the newly created exposure variable.

The record-level data are summarized in Table 9.7. Recall that the input $PPV_{D1} = 72.10\%$. However, when applied to the 215 exposed cases, we found that 75.81% of the classified exposed cases remained classified as exposed cases and 24.19% were reclassified as unexposed cases. Similarly, recall that the input $NPV_{D0} = 95.34\%$. When applied to the 4296 unexposed controls, we found that 95.07% of them remained classified as unexposed controls while 4.93% were reclassified as exposed controls. This mimics the uncertainty we simulated in the summary-level version using the positive and negative predictive values and

id	e_obs	d	PPV_d1	NPV_d1	PPV_d0	NPV_d0	p	e_adj	iter
1	1	1	0.721	0.956	0.733	0.9534	0.7210	1	1
2	0	0	0.721	0.956	0.733	0.9534	0.0466	0	1
3	0	0	0.721	0.956	0.733	0.9534	0.0466	0	1
4	0	0	0.721	0.956	0.733	0.9534	0.0466	0	1
5	1	0	0.721	0.956	0.733	0.9534	0.7330	1	1
6	0	1	0.721	0.956	0.733	0.9534	0.0440	0	1
7	0	0	0.721	0.956	0.733	0.9534	0.0466	0	1
8	1	1	0.721	0.956	0.733	0.9534	0.7210	1	1
9	1	0	0.721	0.956	0.733	0.9534	0.7330	1	1
10	0	0	0.721	0.956	0.733	0.9534	0.0466	0	1

Figure 9.7 First 10 observations in a record-level dataset after reclassifying the exposure in a study of the effect of smoking during pregnancy on breast cancer risk in a record-level bias analysis for exposure misclassification. (Observed data from Fink and Lash, 2003 [2]; Variables as defined in previous tables and e_adj = bias-adjusted exposure value, sampled for each individual).

Table 9.7 Exposure misclassification-adjusted data created by applying positive and negative predictive values probabilistically to the observed data in a study of the effect of smoking during pregnancy on breast cancer risk using the positive and negative predictive values from Table 9.5.

Classified group	N observed	Retained original exposure category	Reclassified	% retained original exposure category
Cases-exposed	215	163	52	75.81%
Cases-unexposed	1449	1389	60	95.86%
Controls-exposed	668	481	187	72.01%
Controls-unexposed	4296	4084	212	95.07%

Observed data from Lash and Fink (2003) [2].

Table 9.8 A single exposure misclassification adjusted dataset created by applying positive and negative predictive values probabilistically to the observed data in a study of the effect of smoking during pregnancy on breast cancer risk using the positive and negative predictive values from Table 9.5. Observed data from Fink and Lash 2003.

	Observed		Single exposure-misclassification adjusted data	
	Smokers	Nonsmokers	Smokers	Nonsmokers
Cases	215	1449	223 = (163 + 60)	1441 = (1389 + 52)
Controls	668	4296	693 = (481 + 212)	4271 = (4084 + 187)
Odds ratio	0.9542		0.9538	

binomial draws. These two approaches accomplish exactly the same goal since, as we noted in Chapter 7, a series of N Bernoulli(p) trials (the Chapter 9 approach) is mathematically identical to one Binomial(N,p) trial (the Chapter 8 approach).

After completing this step, we now have a dataset that contains both the classified exposure and the newly simulated adjusted exposure along with any other covariates we had in the dataset as shown in Table 9.8. Note that, at this step, we are deviating from the results in Chapter 8. This is only due to random sampling error; the results are the same in expectation. We can analyze the resulting dataset using standard techniques for record-level data. As with simple bias analysis and summary-level probabilistic bias analysis, we could use a risk ratio, odds ratio, risk difference or any other measure of interest that is relevant, calculated directly from the data. In this case, because the data came from a case-control study, only the odds ratio is relevant.

Importantly, one could use the records with reclassified exposure status in regression analyses controlling for other confounders, in a mediation analysis, or in any other analysis that would have been completed with the original dataset, but now using the newly created exposure variable rather than the originally classified exposure variable. Unlike in the summary-level approach, we can now easily adjust for multiple measured confounders by including them in a regression model at this step. However, to show the equivalence with the approach in Chapter 8, we omit confounder adjustment at this step. As in the summary level approach, we need to ensure that we can save the resulting beta coefficient and its standard error from the regression model in another dataset so that we can save the bias-adjusted result. In SAS this can be done in some procedures (*e.g.*, proc. logistic) with an output statement, which saves the resulting beta coefficients in a new dataset. For other procedures (*e.g.* proc genmod), this is not possible and ODS output is needed, which requires more programming. In Stata, regression coefficients are stored in a matrix named *e(b)* and their variances are stored in the matrix e(V). In R, regression results are easily returned from the summary table.

We fit a logistic regression using the case-control data with the misclassified exposure data and it returned an odds ratio of $OR^{reg}=0.9542$ ($SE^{reg} = 0.083$) compared to the misclassification-adjusted estimate (with "adjusted" here referring to the exposure bias-adjustment) from a logistic model that returned an odds ratio of 0.9538, exactly what we calculated in Table 9.8. The "reg" superscript serves as a reminder that these parameter estimates came from a regression model of the bias-adjusted data.

Step 4c: Sample the Bias-Adjusted Effect Estimate

The final step in accounting for sources of random error is to simulate the conventional random error. We repeat the process from the summary-level analysis where we choose a standard normal deviate (z_i) for each iteration (i) and multiply it by the standard error of the bias-adjusted association (SE_i^{reg}) and combine the systematic error and random error as:

$$\text{estimate}_i^{total} = \text{estimate}_i^{reg} - z_i \cdot SE_i^{reg}$$

where $\text{estimate}_i^{total}$ is a single simulated estimate of association that incorporates all modeled sources of uncertainty, *estimate*$_i^{reg}$ is a single simulated estimate of association bias-adjusted from step 4b, and SE_i^{reg} is the standard error of the bias-adjusted estimate. The systematic adjusted estimates are typically generated from the output beta coefficient from a regression model. In this case, a logistic model fit to these bias-adjusted data gives OR = 0.9538, the same answer we obtained in the previous step since we are not adjusting for any covariates. Generally, however, the OR from the regression model will differ from the crude if we are adjusting for possible confounders. The standard error estimate of the log odds ratio is obtained from the same regression model, obtaining a value of $SE_i^{reg} = 0.083$. Since we are using the odds ratio and the estimate is on the log scale, we modify the equation above to reintroduce random error as:

$$OR_i^{total} = e^{[\ln(OR_i^{reg}) - z_i \cdot SE_i^{reg}]}$$

The same code can be used from Chapter 8 to sample from these distributions. We sample z_1 from a standard normal($\mu = 0$, $\sigma = 1$) distribution and obtained $z_1 = 0.291$. Using this standard normal deviate with the bias-adjusted estimate we obtain:

$$OR_1^{total} = e^{[\ln(0.9538) - 0.291 \cdot 0.083]} = 0.9770$$

Step 5: Save the Bias-Adjusted Estimate and Repeat Steps 4a–c

Just as with the summary-level bias analysis, the single adjusted estimate is the result of a single simulation to bias-adjust for smoking misclassification. We would not want to draw conclusions based on this single adjustment and would want to realize the entirety of the possible results that could have occurred given the distributions we assigned for sensitivity and specificity and the random nature of the reclassification. Including this additional source of uncertainty means that, in addition to the computing power needed to run simulations on the full dataset, we also need to run more simulations to realize the entirety of the possible results. The solution to this problem is the same as with summary-level probabilistic bias analysis, to repeat steps 4a-c over and over, each time saving the bias-adjusted estimate of association and then summarize the resulting distribution. Figure 9.8 shows the results of 10 iterations.

Unlike with summary-level probabilistic bias analysis, record-level probabilistic bias analysis cannot be accomplished with a few lines of code and often needs to be tailored to the specific dataset. As such, we cannot easily provide code that runs the analyses. Still an example template using SAS, Stata, and R code for the smoking during pregnancy and breast cancer analysis that can be followed is provided in this text's website and should not be too difficult to adapt to the user's specific situation.

iter	se_D1	se_D0	sp_D1	sp_D0	or_reg	se_reg	or_tot
1	0.7122	0.7122	0.9578	0.9578	0.9538	0.0828	0.9770
2	0.8298	0.8298	0.9891	0.9891	0.9143	0.0813	0.9442
3	0.7412	0.7412	0.9831	0.9831	0.8736	0.0787	0.8453
4	0.6645	0.6645	0.9953	0.9953	1.0135	0.0713	1.1546
5	0.8406	0.8406	0.9528	0.9528	0.8814	0.0929	0.8015
6	0.8793	0.8793	0.9987	0.9987	0.9263	0.0802	1.0440
7	0.7922	0.7922	0.9966	0.9966	0.9876	0.0756	0.9166
8	0.8821	0.8821	0.9956	0.9956	0.9840	0.0798	0.7862
9	0.8280	0.8280	0.9867	0.9867	0.9000	0.0819	0.9535
10	0.7666	0.7666	0.9924	0.9924	0.9962	0.0764	1.0272

Figure 9.8 First 10 regression coefficients for the systematic error (exposure misclassification) adjusted and total error adjusted association of smoking during pregnancy on breast cancer risk in a record-level probabilistic bias analysis. Observed data from Fink and Lash, 2003 [2]; Variables are as defined in previous figures and or_reg = the odds ratio estimated from the regression model, se_reg = the bias adjusted standard error, or_tot = the odds ratio accounting for systematic and random error.

Step 6: Summarize the Distribution of Bias-Adjusted Estimates in a Simulation Interval

The last step in the process is the same as for summary-level probabilistic bias analysis, we summarize the resulting output distribution of adjusted estimates, add in random error using any of the methods described in Chapter 8, and make inferences on the full set of results. After repeating the simulation described above in R for 100,000 iterations using the code provided on the text's website, the results are as shown in Table 9.9. As opposed to summary level results that were typically obtained in seconds, the record-level results took 22 minutes to run; however, this run-time could be reduced by optimizing the code further. The median bias-adjusted estimate was 0.95 and the 2.5th and 97.5th percentiles of the distributions were 0.78 and 1.14. Compare this to the conventional analysis (OR 0.95, 95% CI: 0.81, 1.13), which only accounts for random error and we get little change in the point estimate, just as we did with the summary-level probabilistic bias analysis. Comparing our individual record-level probabilistic bias analysis to our summary-level probabilistic bias analysis (also shown in Table 9.9), we see that we have identical results. This is to be expected since we adjusted for no confounders in the record level approach. Mathematically, the two approaches are identical and with enough iterations of the algorithm, they should match perfectly. The key benefit to the record-level approach is the ease with which it can flexibly model the data and the ability to adjust for covariates. This is offset by the increased time it takes to run the model, which is many orders of magnitude higher than summary level analyses. We also estimated the bias-adjusted effect estimate that adjusts for systematic bias and accounts only

Table 9.9 Results of a record-level and summary-level probabilistic bias analysis of the relationship between smoking and breast cancer bias-adjusting for nondifferential misclassification of smoking.

Analysis	Record-level analysis		Summary-level analysis[a]	
	Median	95% Interval	Median	95% Interval
Conventional result, assuming no bias	0.954	$(0.806, 1.129)^b$	0.954	$(0.806, 1.129)^b$
Systematic error only	0.949	$(0.928, 0.953)^c$	0.949	$(0.928, 0.953)^c$
Systematic and random error	0.948	$(0.780, 1.141)^c$	0.947	$(0.780, 1.140)^c$

Observed data from Fink and Lash, 2003 [2].
[a]See Chapter 8.
[b]Computed using exact confidence intervals.
[c]Computed using simulation.

for uncertainty in the bias parameters. To do this, we sampled sensitivity and specificity values, but at other steps that involve sampling, we set parameters equal to their mean value. A final systematic OR was calculated by not sampling the exposure probability in step 4b and by setting each individual exposure equal to its predicted probability (rather than sampling from a Bernoulli), and finally aggregating record-level bias-adjusted exposure probabilities and disease values to the summary level and computing an OR.

Computing Issues with Record-Level Probabilistic Bias Analysis

Although we describe the steps above as a series of sequential steps repeated thousands or even millions of times, in reality, we often implement probabilistic bias analysis with multiple steps in parallel. This is because, at least in some software packages such as SAS, it is computationally more efficient to run a series of regression models on a single combined dataset with individual copies of the data contained within the dataset and indexed by an iteration identifier. Specifically, it is often computationally more efficient to create a single combined dataset that contains within it as many copies of the data as iterations of the probabilistic bias analysis that the user wishes to run. If a probabilistic bias analysis will be repeated for 10,000 iterations, there will be 10,000 copies of the original record-level data within the large combined dataset, with indices ranging from 1–10,000. The calculations described in steps 4a and 4b can then be conducted within the combined dataset (*i.e.,* for each dataset, sample a set of bias parameters, use these values to generate the positive and negative predictive values from the summary adjusted dataset, then apply the predictive values to individuals within each of the copies of the dataset within the combined dataset). Once this has been completed, the user can run a separate regression model on the each of the 10,000 copies of the original data using a "by" statement in SAS or one of the "for" commands in Stata, which serve to run regressions on each of the indexed copies of the dataset and the user can output each of the adjusted estimates into a single dataset. Although this means that each of the steps needs to be run on a single massive dataset (*e.g.,* a dataset of 500 individuals, with 100,000 iterations, creates a dataset of 50,000,000 observations), this is often much faster to run than processing each iteration of the dataset one at a time and appending each new result into a results dataset. In R, the family of "apply()" functions can be used to avoid loops and improve computing time substantially.

Such an approach may be computationally more efficient, but it will still be computationally intensive and can take hours to days depending on dataset size, number of iterations, and computing power. Because of this, it is critical to ensure the code is working correctly before running the full set of iterations. This includes checking that the analysis is not creating predictive values outside the range of 0–1, there are no 0 cells or negative values in the summarized or summarized adjusted data, and ensuring the code has no other errors in it. We recommend always running the analysis for a single iteration and calculating the predictive values, bias-adjusted cell counts, and bias-adjusted effects by hand (or using a regression model) to catch programming errors. Next, we suggest running the program for 100 iterations. Even with this small number of iterations, the estimated median results should be near to what would be expected based on a simple bias analysis in which the median of the distributions for the bias parameters is used as input values. If the median of the simulated probabilistic bias analysis results differs substantially from the simple bias analysis results (more than what can be attributed to the random variation from only running 100 iterations) the code should be debugged. Further, the analyst should examine the distribution of the 100 bias-adjusted effect estimates to ensure there are no values that are impossible or implausible. Once this is complete, the analyst can proceed to the full analysis.

Another way to reduce the computation time is to reduce the size of the dataset as much as possible before running. Because the analyst will create thousands to millions of copies of the analytic dataset, it is important to limit the analytic dataset to only the analytic variables needed. This is typically only the exposure (reclassified and as observed), outcome, confounders, and modifiers (or potentially a measure of person time or censoring for time to event models) and any variables necessary for summarization of the dataset (*i.e.,* the chosen values of sensitivity and specificity). Removing from the dataset all other variables will reduce the total computation time.

A second strategy, alluded to above, is to collapse records with identical values for all original variables to be used in the analysis and assign the record a frequency weight equal to the number of times the pattern of values appears in the dataset. At the extreme limit, this amounts to summarizing the dataset with a 2×2 table and using the summary bias analysis methods described in Chapter 8. In that setting, the data records are summarized with the number of exposed cases (a), unexposed cases (b), exposed non-cases or controls (c), and unexposed non-cases or controls (d), using the originally observed exposure status. Were the 2×2 table expanded to a record-level dataset, the 2×2 table could be represented by individual records where the number of records equals N=a+b+c+d, or it could be represented by four records where the record for exposed cases has the value "a" assigned to a weighting variable, and so on for b, c, and d. So long as the frequency weight variable is properly used in the analyses, the two datasets will give the same results, but the weighted dataset will be computationally more efficient because only four records need to be read instead of N records. When there are more variables than just exposure and outcome, the same data reduction strategy can still be applied [1]. For example, in a study that involved 773,625 birth records [1], the analysis included 12 variables with 39,120 possible combinations of values assigned to the variables, of which only 22,417 were observed. The frequency with which each observed combination appeared ranged from 1 to 20,598. About 39% of the records had frequency of 1 or 2; for these records, there was little computational advantage to using the weighted coalesced records over the individual records. Only ~1% of possible combinations had frequency of 500 or more; but these accounted for 48% of the individual records. We could replace their >370,000 records with just 211 records, so it is these records for which the weighted analysis resulted in substantial gains in computational efficiency compared with the full cohort analysis. The conventional analysis required 4.1 seconds to run on the dataset with 773,625 individual records and 0.1 seconds to run on the dataset with weighted aggregated records. The probabilistic bias analysis to address exposure misclassification required 7.75 days to run when applied to the full dataset and only 8.5 hours when applied to the weighted aggregated records. The investment of time to write code to aggregate records might not be worthwhile for the conventional analysis but would pay important dividends for the probabilistic bias analysis. Further, this approach is only possible if all variables are categorical and none are continuous. Attention needs to be paid to using the weights in different statistical packages. For instance, in Stata, these weights need to be specified as "frequency weights" rather than "probability weights."

The analyst will also need to decide how many iterations to run. Although there are approaches to determining when the results have "converged" (see Chapter 11), there is no consistently agreed standard for convergence and no accepted number of iterations. The key is to run enough iterations that the values estimated for the median, 2.5th and 97.5th percentile of the distribution are stable [3], assuming those are the results that will most influence inferences. One simple way to evaluate this stability is to run the analysis with a reasonable number of iterations and then run it again (with a different value for the random number generating seed if one was used). The key percentile results can then be compared to see how different they are. If there is no difference at the desired level of significant digits that will be reported (typically one or two decimal places for ratio measures, but this will depend on context), adding more iterations may not be worth the additional computing time, whereas if the percentile results are unstable, more runs will be needed. In our experience, the medians tend to stabilize more quickly than the limits of the simulation interval, but this can be analysis specific.

Another factor that will determine the number of iterations that need to be run is the number of bias parameters being used and the spread of the distributions used to describe the uncertainty in the bias parameters. For non-differential misclassification problems, two parameters are typically needed (one sensitivity and one specificity) whereas differential misclassification problems require four (two sensitivities and two specificities); it is reasonable to expect that with more bias parameters or random variables, more iterations will be required to achieve stable estimates. If we assign wide distributions representing the uncertainty in the sensitivities and specificities, more iterations will be required to realize the full distribution of bias-adjusted estimates than with narrower distributions. All of these considerations will also interact with the size of the observed data set. Therefore, the number of iterations needed will depend on a number of factors and cannot be identified *a priori*.

Although computers have increased in power over time, record-level probabilistic bias analyses can still take a long time to complete and it can be concerning to the analyst who may worry that the analysis is not functioning correctly (due to a coding error, such as an infinite loop, or computing problem). One way to avoid this is to run a reasonably large number of iterations, but still only some fraction of the total number of runs desired. For example, if 100,000 iterations are needed, perhaps the analyst only runs 20,000 iterations. Once this is completed, the entire approach can be run an additional four times and at the end the results can be combined to create a total of 100,000 runs. This approach has the advantage of ensuring that errors are identified before the full 100,000 iterations are completed and allows the user to create a summary of the results each time the new iterations are run. If the median and 95% simulation limits are reaching stability, the user may be able to stop early. If all five groups of iterations are run and the estimates are not stable, the user knows to run even more than the planned 100,000.

Diagnostic Plots

As with summary-level probabilistic bias analysis, it is critical to identify illogical values when conducting a record-level probabilistic bias analysis. Record-level analyses are also susceptible to impossible values at the same steps, particularly step 4b where adjusted cell counts are being used to inform the beta distribution for the exposure prevalence. Further, with record-level probabilistic bias analysis, the problem may arise in the calculation of positive and negative predictive values. When these result in values outside the range of 0–1, such as when sampled from a normal distribution, one cannot simulate a bias-adjusted exposure. Thus, it is important to write code that will identify when impossible values for positive or negative predictive values have been generated. As with summary-level probabilistic bias analysis, one can also display a histogram of the sensitivity and specificity values that arise when removing iterations that produce illogical values of positive and negative predictive value to identify areas of the distribution that are not compatible with the data and then readjust the input distributions to ensure compatibility with the data if justified (see Chapter 11).

Misclassification Implementation Alternative: Predictive Values

In this misclassification problem, instead of using sensitivity and specificity to inform the probabilistic bias analysis, we could have used positive and negative predictive values as inputs to adjust for misclassification bias. This approach would work well in the example above if we had good estimates of the positive and negative predictive values of using birth certificate information to measure actual smoking status. As discussed in Chapter 6, predictive values depend on the prevalence of the variable in the population, so estimates of predictive values in one population may not be transportable to other populations.

To conduct this analysis, we could assign distributions to both the positive and negative predictive values in the same way as above for sensitivity and specificity. The user can modify Step 2 above by using positive and negative predictive values and Step 3 can be modified to put distributions around the predictive values. In Step 4 the user randomly samples from those distributions, and then skips the calculations to generate the predictive values. The remaining steps procced as shown.

Misclassification of Outcomes and Confounders

The steps above for exposure misclassification are the same procedures we would use for adjustment of a misclassified outcome or confounder. The key difference (as described in Chapter 6) is the variable to be bias-adjusted and the levels within which the misclassification is bias-adjusted. For outcome misclassification, positive and negative predictive values are calculated within levels of the exposure. This necessitates minor changes to the algorithm above. In step 4b, rather than sampling $P(E_1)$ and $P(E_0)$ from beta distributions, we would sample $P(D_1) \sim$ beta $(\alpha = A_0, \beta = C_0)$ and $P(D_0) \sim beta(\alpha = B_0, \beta = D_0)$. Subsequent positive and negative predictive formula would be for disease status, rather than exposure:

$$P(D_1) = ppv_1 \cdot P(D_1^*) + (1 - npv_1)(1 - \Pr(D_1^*))$$
$$P(D_0) = ppv_0 \cdot P(D_0^*) + (1 - npv_0)(1 - \Pr(D_0^*))$$

For misclassified categorical confounders, the bias-adjustments are typically done within levels of both the exposure and the outcome. See Chapter 10 for bias analyses for mismeasured continuous confounders.

Table 9.10 Observed data on the association between male circumcision (E) and HIV (D) stratified by an unmeasured confounder, religious category (C; C_1 – Muslim, C_0 – any other religion).

	Total		C_1		C_0	
	E_1	E_0	E_1	E_0	E_1	E_0
D_1	105	85	A_1	B_1	A_0	B_0
D_0	527	93	C_1	D_1	C_0	D_0
Total	632	178	M_1	N_1	M_0	N_0

Crude data from Tyndall et al., 1996 [4].

Unmeasured Confounding Implementation

Record-level probabilistic bias analysis for unmeasured confounders follows a similar logic as the summary-level approach, but like with the misclassification problem, we need to make some adjustments to deal with the record-level analysis. Here, rather than identifying positive and negative predictive values as we did with a misclassification problem, we need to identify the probability of having the unmeasured confounder within levels of the exposure and outcome, just as we did for summary-level bias analysis. For exposition, we will assume a dichotomous confounder, but the same logic as used for extension to polytomous confounders in Chapter 5 can be applied. We will use the same example from Chapter 5 that we used with summary-level probabilistic bias analysis to describe the steps necessary for record-level probabilistic bias analysis for an unmeasured confounder, so that we can compare the two types of analyses. As we saw in Chapters 5 and 8, this example uses the association between male circumcision and acquisition of HIV where male circumcision reduced the risk of acquisition of HIV (RR 0.35; 95% CI 0.28, 0.50) but we had concerns about unmeasured confounding by religious category. The data and the set up for stratifying by an uncontrolled confounder are summarized in Table 9.10.

Step 1: Identify the Source of Bias

This step is exactly as for any bias analysis. Here we are concerned about the impact of confounding by religious category, which was not measured in the main analysis.

Step 2: Identify the Bias Parameters

We know from Chapters 5 and 8 that the three bias parameters required to conduct a bias analysis for a dichotomous unmeasured confounder are: p_1, p_0, and RR_{CD}.

Table 9.11 Bias parameter distributions (trapezoidal) for a probabilistic bias analysis of the relationship between male circumcision and HIV stratified by an unmeasured confounder (religious category).

Bias parameter	Description	Min	Mod$_{low}$	Mod$_{up}$	Max
p_1 (%)	Prevalence of being Muslim among circumcised	70	75	85	90
p_0 (%)	Prevalence of being Muslim among uncircumcised	3	4	7	10
RR$_{CD}$	Association between being Muslim and male HIV acquisition	0.5	0.6	0.7	0.8

Step 3: Assign Probability Distributions to Each of the Bias Parameters

In order to compare to the summary-level probabilistic bias analysis, we will use the same trapezoidal distribution for each of the three bias parameters in this record-level bias analysis that we used in the summary-level bias analysis. The minimum, modes and maximum for these distributions are depicted in Table 9.11.

Step 4: Use Simple Bias Analysis Methods to Incorporate Uncertainty in the Bias Parameters and Random Error

Step 4a: Randomly Sample from the Bias Parameter Distributions

This step proceeds exactly as for a summary-level probabilistic bias analysis, by sampling from each of the three trapezoidal distributions for p_1, p_0, and RR$_{CD}$. Our initial draw gives us values of 0.567 for RR$_{CD}$, 75.1% for p_1, and 7.2% for p_0.

Step 4b: Use Simple Bias Analysis Methods and Incorporate Uncertainty and Conventional Random Error

The main difference between the summary-level approach and the record-level approach is that we start with a dataset that contains one record for each individual (Figure 9.9). We will need to summarize the dataset into a single record that contains the cell count frequencies from the summarized contingency table. Here we have 810 records, one for each subject in the dataset. Following the same conventions used in the misclassification example, we code the exposure as 1 if exposed and 0 if unexposed and the outcome as 1 if a person has the outcome and 0 if not (recognizing this as a cohort study, not a case-control study). The dataset must then be summarized into a contingency table (*e.g.,* using PROC FREQ in SAS with an output statement) creating a dataset that has a single record with one variable for

Figure 9.9 First ten observations for a record-level dataset for use in a record-level bias analysis for an uncontrolled confounder of the association between male circumcision (e) and HIV (d). Original data from Tyndall et al., 1996 [4].

id	e	d
1	1	0
2	1	0
3	0	0
4	1	0
5	1	0
6	1	0
7	1	0
8	0	1
9	1	1
10	0	0

id	a	b	c	d
1	105	85	527	93

Figure 9.10 Summarized dataset for use in a record-level bias analysis for an uncontrolled confounder of the association between male circumcision (E) and HIV (D). Original data from Tyndall et al., 1996 [4].

each of the four cells in the summarized contingency table: A = 105, B = 85, C = 527 and D = 93. Figure 9.10 shows the summarized dataset.

Now that we have a summarized dataset, we can use the sampled values of the three bias parameters and apply the same simple bias analysis bias-adjustment that we did in Chapters 5 and 8. Then, as we did in Chapter 8, we will use the bias-adjusted data to generate the probability of having the confounder within levels of the exposure and outcome using the simple bias analysis adjusted data. Our approach will mirror the main approach in Chapter 8.

Given the values selected of 0.567 for RR_{CD}, 75.1% for p_1, and 7.2% for p_0, and the total exposed and unexposed, we can use the equations from Chapter 5 for simple bias analysis:

$$M_1 = mp_1 = 623 \cdot 0.751 = 474.6$$

$$M_0 = m - M_1 = 623 - 474.7 = 157.4$$

$$N_1 = np_0 = 178 \cdot 0.072 = 12.8$$

$$N_0 = n - N_1 = 178 - 12.8 = 165.2$$

$$A_1 = (RR_{CD}M_1a)/(RR_{CD}M_1 + m - M_1) = 66.3$$

$$C_1 = M_1 - A_1 = 474.7 - 66.3 = 408.4$$

$$B_1 = (RR_{CD}N_1b)/(RR_{CD}N_1 + n - N_1) = 3.6$$

$$D_1 = N_1 - B_1 = 12.8 - 3.6 = 9.2$$

$$A_0 = A - A_1 = 105 - 66.3 = 38.7$$

$$B_0 = B - B_1 = 85 - 3.6 = 81.4$$

$$C_0 = C - C_1 = 527 - 408.4 = 118.6$$

$$D_0 = D - D_1 = 93 - 9.2 = 83.8$$

Table 9.12 shows the expected data within levels of the unmeasured confounder given the values assigned to the three bias parameters in the first draw from their distributions. While we show the contingency table, in record-level analysis we would conduct this bias-adjustment within the summarized, one record dataset and have one variable for each of the variables we will need, A_1, B_1, C_1 and D_1, the interior cells of the contingency table for those with the confounder ($C = 1$). Note that if we had assumed there was an interaction between C and E, we would need to store all eight interior cells. However, in this example the first 10 samples and respective bias-adjusted cell frequencies of the summarized dataset would look as shown in Figure 9.11.

Table 9.12 Data on the association between male circumcision (E) and HIV (D) stratified by an unmeasured confounder, religious category when $RR_{CD} = 0.567$, $p_1 = 75.1\%$, and $p_0 = 7.2\%$ (C; C_1 – Muslim, C_0 – any other religion).

	Total		C_1		C_0	
	E_1	E_0	E_1	E_0	E_1	E_0
D_1	105	85	66.3	3.6	38.7	81.4
D_0	527	93	408.4	9.2	118.6	83.8
Total	632	178	474.6	12.8	157.4	165.2

Crude data from Tyndall et al., 1996 [4].

iter	rr.cd	p1	p0	M1	M0	N1	N0	A1	B1	C1	D1	A0	B0	C0	D0
1	0.567	0.751	0.072	474.6	157.4	12.8	165.2	66.3	3.6	408.4	9.2	38.7	81.4	118.6	83.8
2	0.587	0.783	0.039	494.9	137.1	6.9	171.1	71.3	2.0	423.6	4.9	33.7	83.0	103.4	88.1
3	0.623	0.779	0.058	492.0	140.0	10.3	167.7	72.1	3.1	420.0	7.2	32.9	81.9	107.0	85.8
4	0.533	0.875	0.078	553.3	78.7	13.8	164.2	82.9	3.7	470.4	10.2	22.1	81.3	56.6	82.8
5	0.627	0.783	0.045	494.6	137.4	7.9	170.1	72.8	2.4	421.8	5.5	32.2	82.6	105.2	87.5
6	0.696	0.807	0.036	510.0	122.0	6.4	171.6	78.1	2.1	431.8	4.2	26.9	82.9	95.2	88.8
7	0.571	0.867	0.061	547.6	84.4	10.8	167.2	82.7	3.0	464.9	7.8	22.3	82.0	62.1	85.2
8	0.732	0.865	0.075	546.5	85.5	13.3	164.7	86.5	4.8	460.0	8.6	18.5	80.2	67.0	84.4
9	0.650	0.761	0.075	481.2	150.8	13.3	164.7	70.8	4.3	410.4	9.1	34.2	80.7	116.6	83.9
10	0.777	0.833	0.077	526.2	105.8	13.7	164.3	83.4	5.2	442.8	8.5	21.6	79.8	84.2	84.5

Figure 9.11 Summarized dataset for use in a record-level bias analysis for the association between male circumcision (E) and HIV (D) stratified by an unmeasured confounder, religious category with variables for each key cell of the contingency table. Original data from Tyndall et al., 1996 [4]. Cell counts are as described in previous figures. Iter = algorithm iteration, rr.cd = the sampled confounder-outcome association, p1 = the prevalence of the confounder among the exposed, p0 = the prevalence of the confounder among the unexposed.

Table 9.13 Estimated probability of having the confounder, religious category, within levels of the exposure and outcome in a record-level bias analysis of the relationship between circumcision and acquisition of HIV.

Parameter	Estimate
pr(C + IE + D+)	63.1%
pr(C + IE-D+)	4.2%
pr(C + IE + D-)	77.5%
pr(C + IE-D-)	9.9%

Using data from Tyndall et al., 1996 [4].

iter	rr.cd	p1	p0	M1	N1	A1	B1	C1	D1	prc.e1d1	prc.e0d1	prc.e1d0	prc.e0d0
1	0.567	0.751	0.072	474.6	12.8	66.3	3.6	408.4	9.2	0.631	0.042	0.775	0.099
2	0.587	0.783	0.039	494.9	6.9	71.3	2.0	423.6	4.9	0.679	0.023	0.804	0.053
3	0.623	0.779	0.058	492.0	10.3	72.1	3.1	420.0	7.2	0.686	0.037	0.797	0.077
4	0.533	0.875	0.078	553.3	13.8	82.9	3.7	470.4	10.2	0.789	0.043	0.893	0.109
5	0.627	0.783	0.045	494.6	7.9	72.8	2.4	421.8	5.5	0.693	0.028	0.800	0.059
6	0.696	0.807	0.036	510.0	6.4	78.1	2.1	431.8	4.2	0.744	0.025	0.819	0.045
7	0.571	0.867	0.061	547.6	10.8	82.7	3.0	464.9	7.8	0.788	0.036	0.882	0.084
8	0.732	0.865	0.075	546.5	13.3	86.5	4.8	460.0	8.6	0.824	0.056	0.873	0.092
9	0.650	0.761	0.075	481.2	13.3	70.8	4.3	410.4	9.1	0.675	0.050	0.779	0.098
10	0.777	0.833	0.077	526.2	13.7	83.4	5.2	442.8	8.5	0.794	0.061	0.840	0.092

Figure 9.12 Summarized dataset for the first 10 sampled iterations in a record-level bias analysis for the association between male circumcision (E) and HIV (D) stratified by an unmeasured confounder, religious category, after estimating the probability of having the confounder, religious category. Original data from Tyndall et al., 1996 [4]; Variables are as defined in previous figures, prc.e1d1 = the probability of having the confounder among the exposed with the outcome, prc. e0d1 = the probability of having the confounder among the unexposed with the outcome, prc. e1d0 = the probability of having the confounder among the exposed without the outcome and prc. e0d0 = the probability of having the confounder among the unexposed without the outcome.

Calculation of the probability of having the confounder within levels of the exposure and outcome is now straightforward given the simulated data of the exposure and outcome within levels of the unmeasured confounder in Table 9.13. For example, the probability of having the unmeasured confounder for those with the exposure and the outcome is equal to the cell count of those with the exposure, disease, and confounder divided by the total with the exposure and disease, or 66.3/ 105 = 63.1%. This calculation can be done for all four cells, as shown in Table 9.13, to obtain pr(C + IE-D+) = 3.6/85 (4.2%), pr(C + IE + D-) = 408.4/527 (77.5%) and pr(C + IE-D-) = 9.2/93 (9.9%). Figure 9.12 shows the same dataset shown in Figure 9.11 but with some variables removed for space considerations.

We can now simulate the unmeasured confounder for each person in the dataset, but to do so we need to merge the summarized dataset back into the record-level dataset with a 1-to-many merge so that we have all four confounder probabilities in each record of the main record-level dataset. To simulate the unmeasured confounder, we conduct a Bernoulli trial (as described above for exposure misclassification problems) with a probability equal to the relevant one of the four probabilities calculated in the previous step. This can be accomplished using a series

id	e	d	prc.e1d1	prc.e0d1	prc.e1d0	prc.e0d0	p	c
1	1	0	0.631	0.042	0.775	0.099	0.775	1
2	1	0	0.631	0.042	0.775	0.099	0.775	1
3	0	0	0.631	0.042	0.775	0.099	0.099	0
4	1	0	0.631	0.042	0.775	0.099	0.775	0
5	1	0	0.631	0.042	0.775	0.099	0.775	1
6	1	0	0.631	0.042	0.775	0.099	0.775	0
7	1	0	0.631	0.042	0.775	0.099	0.775	0
8	0	1	0.631	0.042	0.775	0.099	0.042	0
9	1	1	0.631	0.042	0.775	0.099	0.631	1
10	0	0	0.631	0.042	0.775	0.099	0.099	0

Figure 9.13 First 10 observations from a record-level dataset for use in a record-level bias analysis for the association between male circumcision (E) and HIV (D), after simulating the unmeasured confounder. Original data from Tyndall et al., 1996 [4]. Variables are as defined in previous figures. p = the probability of the confounder for that individual record's exposure and disease status. c = the imputed confounder value).

of if-then statements to find the right probability. So, for those who are unexposed and do not have the outcome, we would conduct a Bernoulli trial with a probability equal to 63.1% while for a person who is both unexposed and does not have the disease, we would conduct a Bernoulli trial with a probability equal to 9.9%. Alternatively, we can avoid slow if-then statements by using a similar formula to the one above:

$$p = e \cdot d \cdot \text{pr}(C + |E + D+)$$
$$+ e \cdot (1 - d) \cdot \text{pr}(C + |E + D-)$$
$$+ (1 - e) \cdot d \cdot \text{pr}(C + |E - D+)$$
$$+ (1 - e) \cdot (1 - d)\text{pr}(C + |E - D-)$$

This equation allows us to use one variable to simulate all Bernoulli trials. If the trial returns a 1, the subject is considered to have the confounder, whereas if it returns a 0, the subject is considered not to have the confounder. Figure 9.13 shows the dataset with the newly simulated confounder variable.

We now have a dataset with the simulated confounder that can be summarized as in Table 9.14 and we can calculate an adjusted measure of association using any measure we wish. Here we can use the standardized morbidity risk ratio (SMR).

Since we have record level data with an imputed confounder variable, we can easily fit a regression model to estimate the effect of interest. In keeping with the previous example, we wish to estimate a confounding-adjusted risk ratio and fit the data in Figure 9.13 using a Poisson regression with log-link and robust standard errors to obtain a coefficient that, when exponentiated, can be interpreted as a risk ratio [5]. Other modeling choices are possible as well, since we have individual level

Table 9.14 A single dataset created by applying the confounder probabilities to the observed data in a study of male circumcision (E) and HIV (D) stratified by an unmeasured confounder, religious category, when $\mathrm{RR_{CD}} = 0.567$, $p_1 = 75.1\%$, and $p_0 = 7.2\%$ (C; C_1 – Muslim, C_0 – any other religion).

	Total		C_1		C_0	
	E_1	E_0	E_1	E_0	E_1	E_0
D_1	105	85	65	2	40	83
D_0	527	93	409	13	118	80
Total	632	178	474	15	158	163

Crude data from Tyndall et al., 1996 [4].

data [6, 7]. The Poisson model returns an estimated risk ratio $RR_1^{\mathrm{reg}} = 0.53$ and $SE_1^{\mathrm{reg}} = 0.14$ where the subscript indicates this is the first iteration of the algorithm and the standard error is for the natural log of the risk ratio.

Step 4c: Sample the Bias-Adjusted Effect Estimate

We can now incorporate the conventional random error as we have done for the misclassification example, namely choose a standard normal deviate (z_i) for each iteration (i) and multiply it by the standard error of the bias-adjusted association (SE_i^{reg}) and combine the systematic error and random error as:

$$\mathrm{estimate}_i^{\mathrm{total}} = \mathrm{estimate}_i^{\mathrm{reg}} - z_i SE_i^{\mathrm{reg}}$$

where $\mathrm{estimate}_i^{\mathrm{total}}$ is a single simulated estimate of association that incorporates all sources of uncertainty, $\mathrm{estimate}_i^{\mathrm{adj}}$ is a single simulated estimate of association bias-adjusted (*i.e.*, the result of our probabilistic simulation for that iteration), and SE_i is the standard error of the bias-adjusted estimate. Since in this case we are using the risk ratio and the estimate is on the log scale, then the equation would be modified as:

$$RR_i^{\mathrm{total}} = e^{\left[\ln\left(RR_i^{\mathrm{reg}}\right) - z_i SE_i^{\mathrm{reg}} \right]}$$

In the first iteration, we sampled z_1 from a normal($\mu = 0, \sigma = 1$) distribution, which returned $z_1 = -0.6265$. Using this standard normal deviate with the bias-adjusted estimate of the SMR accounting for the first two sources of uncertainty and the standard error of the bias-adjusted risk ratio, we obtain:

$$RR_1^{\mathrm{total}} = e^{[\ln(0.53) + 0.6265 \cdot 0.14]} = 0.58$$

Steps 5 and 6: Resample, Save, Summarize

Finally, we repeat steps 4a–c above, for 100,000 iterations in R and save the output at the end of each iteration. Following this, we have 100,000 bias-adjusted estimates of the effect of circumcision on HIV adjusted for religious category. We show the results of the first 10 iterations in Figure 9.14. Full results from all iterations of the algorithm are shown in Table 9.15.

In Table 9.15 we see that the record-level analysis almost perfectly matches the summary-level analysis that imputes the confounder. As with the misclassification example, this is to be expected. The only difference in the summary and record level approaches is whether we sample from a binomial distribution (summary level) or sample from a series of Bernoulli distributions (record level). These approaches are equivalent. As we stated above, the benefit of the summary-level approach is the speed with which it can be conducted. The benefit of the record-level approach is

iter	rr.cd	p1	p0	rr_reg	se_reg	rr_tot
1	0.567	0.751	0.072	0.531	0.142	0.580
2	0.583	0.826	0.055	0.551	0.163	0.585
3	0.582	0.792	0.066	0.533	0.152	0.482
4	0.711	0.884	0.059	0.335	0.225	0.282
5	0.506	0.753	0.077	0.555	0.142	0.566
6	0.615	0.830	0.041	0.473	0.165	0.535
7	0.701	0.766	0.056	0.448	0.163	0.466
8	0.520	0.863	0.052	0.557	0.177	0.561
9	0.543	0.804	0.077	0.528	0.161	0.472
10	0.588	0.777	0.089	0.506	0.153	0.502

Figure 9.14 First 10 regression coefficients (exponentiated) for the uncontrolled confounder adjusted and total error adjusted association between male circumcision (E) and HIV (D) with bias-adjustment for unmeasured confounding by religious category. Original data from Tyndall et al., 1996 [4]. Variables as defined in previous figures. rr_reg = confounder-adjusted risk ratio obtained from regression model, se_adj = standard error of the ln(rr_reg), rr_tot = the final confounder-adjusted risk ratio that incorporates uncertainty in the bias parameter and all random error.

Table 9.15 Results of a record-level probabilistic bias analysis of the relationship between male circumcision and HIV adjusting for an unmeasured confounder, religious category.

Analysis	Record-level analysis		Summary-level analysis[a]	
	Median	95% Interval	Median	95% Interval
Conventional result, assuming no bias	0.348	(0.276, 0.439)	0.348	(0.276, 0.439)
Systematic error only[a]	0.472	(0.418, 0.550)	0.472	(0.419, 0.549)
Random and systematic error	0.471	(0.300, 0.722)	0.474	(0.293, 0.726)

[a]See Chapter 8

that the analyst has more flexibility to run regression models and control for additional measured confounders.

As we stated in Chapter 8, analysts should be careful to explore the data for zero cells that could result in unstable estimation. Removing iterations that result in zero cells (and reporting the total number) is one option. Using regression techniques that allow for zero cells is also possible. In this simulation, we excluded iterations that resulted in zero cell frequencies. Before leaving this section, we note that it is possible to use confounding relative risks to adjust for uncontrolled confounding with record-level data. The procedure would look similar to the one presented in Chapter 8 so we do not go into more detail here. Briefly, an analyst would use record-level data to run their preferred regression and obtain an effect estimate that is adjusted for everything but the uncontrolled confounder. That estimate and its standard error would be stored to use in combination with a confounding relative risk. The procedure would entail sampling from the bias parameter distributions and using them to generate a distribution of confounding relative risks (as in Chapter 8). The confounding relative risk (at each iteration) would be combined with the observed risk ratio as in Chapter 8 and the observed standard error would be used to propagate random error (again, identical to Chapter 8).

Selection Bias Implementation

We implement two approaches for record-level selection bias, as we did in Chapter 8. The first is for situations in which a greater degree of information is known about those who did not participate. The second is for the common case in which very little is known and the user must specify four selection probabilities. The first approach in this Chapter differs slightly from the one in Chapter 8. Here, we approach the problem using a weighting approach that allows the user to easily implement a regression model to control for confounders at the record level. The following steps implement a record-level probabilistic bias analysis for selection bias when there is substantial information on the non-participating individuals in the population.

Step 1: Identify the Source of Bias

To demonstrate record-level bias-adjustments to adjust for selection bias we will use the same example we used in Chapters 4 and 8, in which there was potential for selection bias in a study of the association between mobile phone use and uveal melanoma occurrence. Table 9.16 presents the study data for those who did complete the study and for those who completed a shorter questionnaire indicating only their exposure and disease status.

Table 9.16 Depiction of participation and mobile phone use in a study of the relation between mobile phone use and the occurrence of uveal melanoma. Original data from Stang et al., 2009 [8].

	Participants		Nonparticipants/short questionnaire		Nonparticipants
	Regular use	No use	Regular use	No use	Cannot categorize
Cases	136	107	3	7	17
Controls	297	165	72	212	379

Step 2: Identify the Bias Parameters

Step 2 proceeds just as for a summary-level bias analysis for selection bias. In Chapter 8 we described three populations, those who completed the full questionnaire, those who did not complete the full questionnaire but did complete the short questionnaire with exposure and outcome information, and those who did not complete any questionnaire. We used the data from those who completed only the exposure and outcome questionnaire to estimate values for bias parameters for those who completed neither questionnaire.

As in Chapter 8, the two bias parameters are the prevalence of mobile phone use among cases who were invited to participate but refused to answer either questionnaire (p_1) and among controls who were invited to participate but refused to answer either questionnaire (p_0).

Step 3: Assign Probability Distributions to Each Bias Parameter

For comparability we will use the same distributions for the bias parameters that we used in Chapter 8. For this analysis we assume that those who did not complete either questionnaire are like those who completed only the exposure outcome questionnaire where $p_0 = 0.300$ (3/10) and $p_1 = 0.253$ (72/284). We put distributions on the bias parameters informed by the sample sizes in each group, so we assign beta distributions to $p_1 \sim$ beta($\alpha = 3$, $\beta = 7$) and $p_0 \sim$ beta($\alpha = 72$, $\beta = 212$).

Step 4: Use Simple Bias Analysis Methods to Incorporate Uncertainty in the Bias Parameters and Random Error

Step 4a: Randomly Sample from the Bias Parameter Distributions

As with the other record-level bias analyses above, we now choose randomly sampled values from each of the two distributions. In this case, we chose $p_0 = 0.253$ and $p_1 = 0.208$. Note that these values are different from the values we chose in the corresponding summary level example in Chapter 8. This is because the method here will differ slightly in the way we use weighting and so we use a new draw to illustrate the method.

Step 4b: Use Simple Bias Analysis Methods and Incorporate Uncertainty and Conventional Random Error

Unlike with the summary-level approach, we cannot simply use the data that we have and make bias-adjustments (as with a misclassification problem) or simulate a new variable (as with an uncontrolled confounder problem) because we do not have records for those who did not participate. We could create blank records for each known missing subject, which would include their known case or control status, but then even after assigning mobile phone status using bias-adjustment, we would still lack data on confounders and any adjusted model would drop these subjects from the analysis. The easiest solution to this problem is to use the inverse probability of participation weighting we described in Chapter 4 to upweight those who were included to account for those who were not. We can do this only for those who completed no questionnaire or for both those who completed no questionnaire and those who completed the short questionnaire to allow for more use of the data.

To generate the weights, we first calculate the probability of participating in the study within each level of the exposure and outcome. To do so, we conduct a simple bias analysis (see Table 9.17). We then collapse the table into those who were included in the study (completed the full questionnaire) and those who were not (everyone else). Using these two strata (included vs. excluded) we can calculate the probability of participating in the study within levels of the exposure and the outcome. In our example, with the bias parameters we sampled, we obtain $136/(136 + 8.1) = 94.4\%$ of those who were exposed and had the outcome participating, $297/(297 + 168.1) = 63.9\%$ of those who were exposed and did not have the outcome participating, etc. We then calculate the weights as the inverse of the probability of participating in the study as shown in Table 9.18, so $1/0.944$ for those who were exposed and had the outcome $= 1.06$ while for those who were exposed and did not have the outcome the weight would be $1/0.639 = 1.57$.

Table 9.17 Depiction of participation and mobile phone use in a study of the relation between mobile phone use and the occurrence of uveal melanoma when $p_0 = 0.253$ and $p_1 = 0.208$.

	Participants		Nonparticipants short questionnaire		Nonparticipants		Total nonparticipants	
	Regular use	No use	Regular use	No use	Regular use	No use	Regular use	No use
Cases	136	107	3	7	$3.54 = (17 \cdot 0.208)$	$13.46 = (17 \cdot (1-0.208))$	6.54	20.46
Controls	297	165	72	212	$95.89 = (379 \cdot 0.253)$	$283.10 = (379 \cdot (1-0.253))$	167.89	495.10

Original data from Stang et al., 2009 [8].

Table 9.18 Inverse probability of participation weights for a study of mobile phone use in a study of the relation between mobile phone use and the occurrence of uveal melanoma when $p_0 = 0.253$ and $p_1 = 0.208$.

	Participants		Total nonparticipants		Participation probability		Inverse probability of participation weights	
	Regular use	No use	Regular use	No use	Regular use	No use	Regular use	No use
Cases	136	107	6.54	20.46	$=136/(136 + 8.1)$ $= 0.944$	$107/(107 + 18.9)$ $= 0.850$	1.06	1.18
Controls	297	165	167.89	495.10	$=297/(297 + 168.1)$ $= 0.639$	$165/(165 + 494.9)$ $= 0.250$	1.57	4.00

Original data from Stang et al., 2009 [8].

Table 9.19 Inverse probability weighted data in a study of the relation between mobile phone use and the occurrence of uveal melanoma when $p_0 = 0.253$ and $p_1 = 0.208$.

	Participants		Weights		Weighted data	
	Regular use	No use	Regular use	No use	Regular use	No use
Cases	136	107	1.05	1.19	$=136 \cdot 1.06 = 144.16$	$107 \cdot 1.18 = 126.26$
Controls	297	165	1.57	4.00	$=297 \cdot 1.57 = 466.29$	$165 \cdot 4.00 = 660.00$
Selection bias-adjusted odds ratio				1.62		

Original data from Stang et al., 2009 [8].

With the four weights calculated, we can apply the weights to each individual in the dataset who completed the full questionnaire based on their exposure and outcome combination. Although the crude selection bias-adjusted result would be 1.62 as shown in Table 9.19, the main reason to conduct a record-level probabilistic bias analysis is to be able to adjust for additional measured covariates, which requires a weighted regression analysis.

Step 4c: Sample the Bias-Adjusted Effect Estimate

From here we can implement an analysis using only those who were included in the study along with the weights and we can adjust for any variables selected for analytic control. For example, a regression analysis could be completed in SAS using PROC LOGISTIC or PROC GENMOD and a weight statement. In Stata, the command "logistic" can be implemented using the "[pweight=]" option. In R, the "survey" package can be used to include probability weights and estimate a wide range of regression models. We then output the beta coefficient and standard error from the model and use this as the estimate adjusted for the selection bias in a single iteration. The regression based standard error is in keeping with our advice from Chapter 8 that selection bias implementation should not be calculated from a dataset larger than the original. Using survey weighting procedures for regression ensures the proper standard error is used.

Steps 5 and 6: Resample, Save, Summarize and Include Random Error

The final steps are the same as for any record-level bias analysis, we repeat steps 4a–c a large number of times (here 50,000), each time saving the adjusted beta coefficient for the exposure outcome association from the regression. We then summarize the distribution of betas (or in this case, the exponentiated betas) to estimate an odds ratio, use the median as the point estimate, and use the 2.5th and 97.5th percentiles of the distribution as a simulation interval. We also obtain a median and interval that accounts for both systematic and random error using the regression-based random error from the weighted regression. As we discussed in Chapter 8, we usually recommend using the conventional standard error, which may need to be calculated as the regression model adjusted for the selection bias will be inappropriate.

Results for the example are shown in Table 9.20. We found a median of the bias-adjusted estimates of 1.62 with a 95% simulation interval from 1.48 to 1.80, very similar to the summary-level approach results shown in Chapter 8 (1.62; 95%SI: 1.44, 1.85). Our total error median and interval (1.62; 95%SI: 1.24, 2.14) is also very similar to the results from the summary-level approach. Note that this congruence is expected. If we had access to the complete dataset, adjustment for other variables in the regression might have yielded a different result.

Table 9.20 Results of a record-level probabilistic bias analysis of the relation between mobile phone use and uveal cancer adjusting for selection bias.

Analysis	Record-level analysis		Summary-level analysis[a]	
	Median	95% Interval	Median	95% Interval
Conventional result, assuming no bias	0.706	(0.515, 0.967)	0.706	(0.514, 0.970)
Systematic error only[a]	1.622	(1.477, 1.797)	1.622	(1.437, 1.849)
Random and systematic error	1.623	(1.236, 2.141)	1.624	(1.208, 2.184)

Original data from Stang et al., 2009 [8].
[a]See Chapter 8.

Selection Bias Implementation: Individual Selection Probabilities

In many instances, as noted in Chapter 8, we may be concerned about selection probability but know little or nothing about the number of people not participating in the study. In this case, we can proceed as we did in Chapter 8 and specify unique selection probabilities for each combination of exposure and outcome. We use the same example as above but now assume we have no short survey to inform the selection probabilities and no sense of how many people could have participated but did not.

Step 1: Identify the Source of Bias

As described before, the source of bias we are concerned with is selection bias.

Step 2: Identify the Bias Parameters

We proceed by specifying the four parameters presented in Chapters 4 and 8. These probabilities determine selection into the study conditional on exposure and disease status: the probability of selection into the study for cases who were exposed ($S_{\text{case}, 1}$), cases who were unexposed ($S_{\text{case}, 0}$), controls who were exposed ($S_{\text{control}, 1}$), and controls who were unexposed ($S_{\text{controls}, 0}$).

Step 3: Assign Probability Distributions to Each Bias Parameter

For consistency with Chapter 8, we use the same distributions for the bias parameters that we have already explained in the previous chapter.

$$s_{\text{case},1} \sim \text{beta}(\alpha = 139, \beta = 5.1)$$

$$s_{\text{case},0} \sim \text{beta}(\alpha = 114, \beta = 11.9)$$

$$s_{\text{control},1} \sim \text{beta}(\alpha = 369, \beta = 96.1)$$

$$s_{\text{control},0} \sim \text{beta}(\alpha = 377, \beta = 282.9)$$

Step 4: Use Simple Bias Analysis Methods to Incorporate Uncertainty in the Bias Parameters and Random Error

Step 4a: Sample from the Bias Parameter Distributions

We first sample from each of the four bias parameter distributions. For sake of comparability with Chapter 8, we retain the same initial samples that we did in that chapter resulting in $s_{case,\ 1} = 96.4\%$, $s_{case,\ 0} = 89.8\%$, $s_{control,\ 1} = 81.2\%$, and $s_{control,\ 0} = 56.2\%$. We then use these sampled values to bias-adjust the observed data.

Step 4b: Generate Bias-Adjusted Data Using Simple Bias Analysis Methods and the Sampled Bias Parameters

In Chapter 8, we divided the individual cell counts by the selection probabilities. In this chapter, we keep observations at the record-level and follow the approach in the previous example. We first compute weights, as the inverse of these participation probabilities: $w_{case,1} = \frac{1}{s_{case,1}} = 1.04$, $w_{case,0} = \frac{1}{s_{case,0}} = 1.11$, $w_{control,1} = \frac{1}{s_{control,1}} = 1.23$, and $w_{control,0} = \frac{1}{s_{control,0}} = 1.78$. These weights are merged with the observed data and every individual has a single weight depending on their outcome and exposure level. That is, an exposed case has $wt = w_{case,\ 1}$ and an unexposed case as $wt = w_{case,\ 0}$. The variable wt will be used as a probability weight in the final model. We fit a regression model that incorporates these probability weights to estimate a selection bias-adjusted effect estimate, OR^{adj} and standard error, SE^{reg}.

Steps 4c: Sample the Bias-Adjusted Effect Estimate

Using the stored effect estimate and standard error from step 4b, we propagate random error into the final estimate using the standard approach of sampling $z \sim N(0, 1)$ and computing:

$$OR_i^{tot} = e^{\left[\ln\left(OR_i^{reg}\right) - z \cdot SE_i^{reg}\right]}$$

where the i subscript represents the ith iteration of the algorithm.

Steps 5 and 6: Resample, Save and Summarize

The algorithm is repeated for a large number of iterations and the parameter estimates are stored at each iteration. In R, we implemented this procedure for 100,000 iterations to estimate a bias-adjusted effect. Examining the median, 2.5th and 97.5th percentiles we find a bias adjusted OR = 1.62 with 95% simulation intervals (1.19, 2.20). These are essentially the same results we obtained from the summary level analysis of the same data with the same bias parameter distributions (see Chapter 8). The advantage of this approach is that we are able to adjust for confounders in the regression model.

Alternative Methods for Incorporating Random Error in Bias-Adjusted Estimates

We present, in both Chapters 8 and 9, relatively straightforward ways to propagate random error in the final bias-adjusted effect estimate. The methods we present are not the only possibilities, and an analyst could, instead, turn to resampling methods to account for random error. The nonparametric bootstrap, developed by Efron [9], uses resampling from the original dataset, with replacement, to estimate the variance of the effect of interest. The bootstrap proceeds by resampling the observed data numerous times and computing the parameter of interest in the resampled data.

For example, for a non-differential exposure misclassification problem, one way to bootstrap would be to first specify the bias parameter distributions for the sensitivity and specificity. Next, we would draw sensitivity and specificity values from the bias parameter distributions. Third, for each of the sampled sensitivity and specificity values, we would draw bootstrap samples of the original data (that is for a dataset of size N, we would draw a sample of size N from the original data with replacement). Fourth, we would use the sensitivity and specificity values in conjunction with the bootstrap samples of data to produce a bias-adjusted estimate using the simple tabular approaches presented in Chapter 6. We would repeat these steps many times and summarize the results as described above. Just as with the error interval above, we can summarize the distribution using the median as the point estimate and the 2.5th to 97.5th percentile of the distribution to create a 95% simulation interval. Bootstrap confidence intervals can perform poorly in small sample sizes, contrary to popular misconceptions, and the bootstrap should not be viewed as a panacea [10]. Various improvements on the simple bootstrap, such as bias corrected and accelerated intervals have been proposed and can outperform the simple interval we describe here [9].

Because bootstrapping requires creating a new dataset with the same sample size as the original for each bias-adjusted dataset, it increases the computational workload and so increases the amount of time required to complete the analysis. It is important, therefore, to ensure that the code is as efficient as possible, that extraneous variables have been removed from the dataset, that there are no bugs in the code running many iterations, and that there are sufficient computing resources to complete the bootstrap analysis. Even with these measures, bootstrapping can add substantial time to the overall bias analysis.

Bootstrapping procedures have been built into many software packages and therefore do not require as much coding as they would if this had to be done directly by the analyst. For example, in SAS, bootstrapping can be done using the PROC SURVEYSELECT procedure. In STATA, the bsample command can be used and in R one can use the boot package. For each, consult the help files for full details on how to implement.

An alternative resampling procedure known as the jackknife estimate of the standard error may have somewhat better small sample properties and has been implemented in probabilistic bias analyses [11, 12]. The jackknife estimate is known as a 'leave-one-out' procedure because it proceeds iteratively through the dataset computing the statistic of interest using the entire dataset, lest the ith observation. The procedure moves through each observation in the dataset, removing that observation, estimating the statistic of interest $\left(\hat{\theta}_{-i}\right)$ on the remaining data and storing the estimate. The standard error is calculated as $se_{\text{jack}} = \frac{n}{n-1} \sum_i^n \left(\hat{\theta}_{-i} - \hat{\theta}^*\right)^2$, where $\hat{\theta}_{-i}$ is the parameter estimate computed on all but the ith sample, n is the total sample size and $\hat{\theta}^* = \frac{\sum \hat{\theta}_{-i}}{n}$. Lyles et al. noted the jackknife procedure performed quite well and experienced fewer technical difficulties than using the bootstrap [12]. Their procedure was to sample from the bias distribution and use each set of bias parameters to bias-adjust the observed data and estimate a jackknife standard error. For example, if 10,000 bias parameters were chosen, the data would be bias adjusted 10,000 times and 10,000 effect estimates and jackknife standard errors would be retained. To incorporate random error, 500 random draws were taken for each of the stored estimates and their standard errors for a total of $10,000 \cdot 500 = 5,000,000$ samples.

Conclusions

Record-level probabilistic bias analysis is often an improvement over summary-level probabilistic bias analysis in that it allows for more flexible modelling and confounder control. However, these improvements come at the expense of increased needs for computing power and time, as well as coding complexity, and cannot be conducted on datasets for which the analyst does not have access to the individual records. When possible, record-level probabilistic bias analysis may allow for a more flexible analysis plan while also modeling the impact of the bias.

References

1. Lash TL, Abrams B, Bodnar LM. Comparison of bias analysis strategies applied to a large data set. Epidemiol. 2014;25:576–82.

2. Fink AK, Lash TL. A null association between smoking during pregnancy and breast cancer using Massachusetts registry data (United States). Cancer Causes Control. 2003;14:497–503.
3. Lash T, Fink AK. Semi-automated sensitivity analysis to assess systematic errors in observational data. Epidemiol. 2003;14:451–8.
4. Tyndall MW, Ronald AR, Agoki E, Malisa W, Bwayo JJ, Ndinya-Achola JO, et al. Increased risk of infection with human immunodeficiency virus type 1 among uncircumcised men presenting with genital ulcer disease in Kenya. Clin Infect Dis. 1996;23:449–53.
5. Spiegelman D, Hertzmark E. Easy SAS calculations for risk or prevalence ratios and differences. Am J Epidemiol. 2005;162:199–200.
6. Muller CJ, MacLehose RF. Estimating predicted probabilities from logistic regression: different methods correspond to different target populations. Int J Epidemiol. 2014;43:962–70.
7. Richardson DB, Kinlaw AC, MacLehose RF, Cole SR. Standardized binomial models for risk or prevalence ratios and differences. Int J Epidemiol. 2015;44:1660–72.
8. Stang A, Schmidt-Pokrzywniak A, Lash TL, Lommatzsch P, Taubert G, Bornfeld N, et al. Mobile Phone Use and Risk of Uveal Melanoma: Results of the RIFA Case–Control Study. J Natl Cancer Inst. 2009;101:120–3.
9. Efron B, Tibshirani RJ. An introduction to the bootstrap. New York: Chapman and Hall; 1994.
10. Hesterberg TC. What teachers should know about the bootstrap: Resampling in the undergraduate statistics curriculum. Am Stat. 2015;69:371–86.
11. Corbin M, Haslett S, Pearce N, Maule M, Greenland S. A comparison of sensitivity-specificity imputation, direct imputation and fully Bayesian analysis to adjust for exposure misclassification when validation data are unavailable. Int J Epidemiol. 2017;46:1063–72.
12. Lyles RH, Lin J. Sensitivity analysis for misclassification in logistic regression via likelihood methods and predictive value weighting. Stat Med. 2010;29:2297–309.

Chapter 10
Direct Bias Modeling and Missing Data Methods for Bias Analysis

Introduction

Earlier chapters in this text presented methods for simple bias-adjustments (Chapters 4, 5 and 6). These methods yield bias-adjusted point estimates, but without any accompanying quantitative description of the uncertainty accompanying that estimate. We showed how the impact of uncertainty in the bias parameter could be described by assigning different values or sets of values to the bias parameters of the model. These were presented in a multidimensional or tabular bias analysis. Although this multidimensional bias analysis approach is useful to understand what value the adjusted effect estimate would take under various bias parameter values, it provides no single summary effect estimate and no sense of the relative likelihood of the various bias-adjusted estimates. Each estimate appears to be equally likely. Probabilistic bias analysis (Chapters 7, 8 and 9) uses simulation methods to model the uncertainty in the values assigned to the bias parameters, along with sampling error, and yields a point estimate of the central tendency and a sense of the relative frequency of different bias-adjusted estimates. These relative frequencies can be summarized in a simulation interval.

In this chapter we present two general classes of methods to implement bias analysis: direct bias modeling and missing data methods. In the first section of this chapter, we present three methods that directly incorporate bias parameters for misclassification (and their distributions) into the estimation process. One of these methods requires the analyst to include estimates of sensitivity and specificity. Two other methods use predictive values. These three methods are relatively quick to implement and may be preferable to the probabilistic methods presented earlier; however, they are somewhat less flexible when incorporating other types of bias at the same time (for example uncontrolled confounding bias and selection bias). In the second section we present missing data methods. We consider regression calibration as a type of missing data method as it uses validation study data to impute the expected value of the mismeasured variable. That expected value is used in a

M. P. Fox et al., *Applying Quantitative Bias Analysis to Epidemiologic Data*,
Statistics for Biology and Health, https://doi.org/10.1007/978-3-030-82673-4_10

regression model in place of the mismeasured version. Regression calibration is easily implemented but is generally only used for measurement error, so remains a somewhat specific method. Finally, we present missing data techniques as a general solution to the three main biases we have outlined in this book. Missing data techniques presented at the end of this chapter envision the source of bias as a missing data problem and use standard missing data techniques to estimate a bias-adjusted effect and interval.

Directly Incorporating Bias into Effect Estimation

We begin by describing methods to address misclassification of a dichotomous exposure using data displayed in a 2×2 contingency table. These methods yield the same bias-adjusted estimate as presented in Chapter 6. The extension is that they also provide a relatively simple method to estimate a confidence interval for the bias-adjusted estimate. This confidence interval can also incorporate the uncertainty in the bias parameter (e.g., the sensitivity and specificity). Unfortunately, these methods are somewhat limited in their scope, and apply to specific types of studies and measures of effect. They also are generally difficult to implement in the presence of measured confounders for which adjustment is necessary.

Standard Errors for Misclassification When Sensitivity and Specificity Are Used

If a summary level bias analysis for misclassification is conducted using estimates of sensitivity and specificity, then equations described by Greenland (1988) can be used to estimate the variance of the misclassification-adjusted odds ratio [1]. This approach is less computationally intensive than the probabilistic bias analysis approach presented in Chapter 8 and its 95% confidence intervals will have better asymptotic coverage, assuming that the classification parameters are properly specified. The method can be used for either case-control or cohort studies; we present the method for case-control studies here. This method assumes there is uncertainty in the sensitivity and specificity of exposure classification. In previous chapters in this book, this uncertainty was propagated to the final uncertainty intervals using probabilistic bias analysis and repeatedly drawing random samples. In Greenland's approach, the equation allows direct computation of the standard error of the bias-adjusted odds ratio (OR_b) that incorporates the uncertainty in the sensitivity and specificity parameters. The analyst must first specify the variance of the sensitivities, $Var(SE_i)$, and variance of the specificities, $Var(SP_i)$, where i = 1 for cases and i = 0 for controls. These variances may be available directly from an external study or could be estimated by the analyst using the equation for the variance of a binomial proportion: $\frac{p(1-p)}{n}$ with p being the sensitivity or specificity and n being the sample size used to estimate the sensitivity or specificity. If the exposure misclassification is expected by design to be <u>differential,</u> and sensitivity and specificity were estimated using external validation data, the variance for the bias-adjusted log odds ratio would be:

$$\text{Var}[\ln(OR_b)] = \sum_{i=0}^{i=1} \frac{\left(\frac{\text{Var}(SE_i)}{[1-P(E_i)]^2} + \frac{\text{Var}(SP_i)}{P(E_i)^2} + \frac{P(e_i)[1-P(e_i)]}{D_iP(E_i)^2[1-P(E_i)]^2}\right)}{(SE_i + SP_i - 1)^2}$$

where $P(E_i)$ is the proportion of cases ($i = 1$) or controls ($i = 0$) that are *truly* exposed. Because this proportion is unknown, we estimate it by reclassifying the observed cell frequencies using methods from Chapter 6. The proportion $P(e_i)$ is the proportion of cases and controls that were *observed* to be exposed (potentially misclassified). Finally, D_i is the total number of persons within the given strata.

If the misclassification is expected by design to be <u>nondifferential</u>, a different equation needs to be used to account for identical sensitivities and specificities among the cases and controls:

$$\text{Var}[\ln(OR_b)] = \frac{\text{Var}(SE)\left(\frac{1}{1-P(E_1)} - \frac{1}{1-P(F_0)}\right)^2 + \text{Var}(SP)\left(\frac{1}{P(E_1)} - \frac{1}{P(E_0)}\right)^2}{(SE + SP - 1)^2}$$

$$+ \frac{\sum_i \frac{P(e_i)[1 - P(e_i)]}{D_iP(E_i)^2[1 - P(E_i)]^2}}{(SE + SP - 1)^2}$$

The variance estimated using either of these equations is used in conjunction with a misclassification-adjusted odds ratio to produce 95% uncertainty intervals. The misclassification-adjusted odds ratio is obtained using the equations in Chapter 6.

We illustrate this method using the data from Fink and Lash, which estimated the effect of smoking during pregnancy on breast cancer risk using a case-control design [2]. As in Chapter 6, we assume the misclassification is nondifferential, an assumption supported by its prospective design, so we use the second variance formula. In Chapter 6, we used data from a validation study by Piper et al. (1993) to estimate a sensitivity of 78% and a specificity of 99% [3]. We estimate the variances using data from the validation study (see Table 6.2):

$$\text{Var}(SE) = \frac{SE(1 - SE)}{N} = \frac{0.78(1 - 0.78)}{164} = 0.001$$

$$\text{Var}(SP) = \frac{SP(1 - SP)}{N} = \frac{0.99(1 - 0.99)}{470} = 0.00002$$

Although these variances are quite small, they do admit reasonable uncertainty in the sensitivity and specificity estimates. For instance, the 95% prior interval for sensitivity is 72–84% and for specificity is 98–99.9%. Using data from the case-control study of Fink and Lash (see Table 6.9), we estimate the observed exposure probabilities among the cases as $P(e_1) = 0.129$ and among the controls as $P(e_0) = 0.135$. Finally, we need to estimate the expected number of exposed cases and controls after adjusting for misclassification with a sensitivity of 78% and a specificity of 99%. We have already presented these calculations in Chapter 6 (see Table 6.9). Inserting the observed cell counts and the estimates of sensitivity and specificity into equations from Chapter 6:

$$A = \frac{a - D_1(1 - SP_1)}{(SE_1 - 1 + SP_1)}$$

$$C = \frac{c - D_0(1 - SP_0)}{(SE_0 - 1 + SP_0)}$$

This returns estimates of A = 256.3 and C = 799.1 allowing us to estimate the bias-adjusted exposure prevalence as $P(E_1) = 0.154$ among the cases and $P(E_0) = 0.161$ among the controls (see Table 6.9). Substituting these values in the equation for the variance of the ln-odds ratio under non-differential misclassification above gives

$$\mathrm{Var}\left[\ln\left(OR_b\right)\right] = \frac{0.001\left(\frac{1}{1-0.154} - \frac{1}{1-0.161}\right)^2 + 2 \times 10^{-5}\left(\frac{1}{0.154} - \frac{1}{0.161}\right)^2}{(0.78 + 0.99 - 1)^2}$$

$$+ \frac{\dfrac{0.129(1 - 0.129)}{(215 + 1449)0.154^2(1 - 0.154)^2} + \dfrac{0.135(1 - 0.135)}{(668 + 4296)0.161^2(1 - 0.161)^2}}{(0.78 + 0.99 - 1)^2}$$

$$= 0.0089$$

$$(10.1)$$

To compute the final interval estimate, we first compute the misclassification-adjusted odds ratio:

$$OR_b = \frac{\frac{P(E_1)}{1-P(E_1)}}{\frac{P(E_0)}{1-P(E_0)}} = \frac{\frac{0.154}{1-0.154}}{\frac{0.161}{1-0.161}} = 0.949 \qquad (10.2)$$

This is the same bias-adjusted estimate we computed in Chapter 6 (see Table 6.8). Finally, interval estimates are computed as:

$$e^{\ln(OR_b) \pm 1.96\sqrt{\mathrm{var}[\ln(OR_b)]}} = e^{\ln(0.949) \pm 1.96\sqrt{0.0089}} = (0.79, 1.14) \qquad (10.3)$$

We note that this result is nearly equivalent to what we obtained using a probabilistic bias analysis in Chapter 8, where we found $OR_b = 0.95$ (0.78, 1.14).

The result in Chapter 8 was obtained by assigning beta distributions to the sensitivity and specificity with mode and (95% CI) of 0.78 (0.67, 0.87) and 0.99 (0.95, 0.99), respectively. Recall that the mean and interval for sensitivity and specificity used in the analysis above were 0.78 (0.72, 0.84) and 0.99 (0.98, 0.998), respectively. The wider intervals used in the probabilistic bias analysis were meant to incorporate additional uncertainty in the estimates of the sensitivity and specificity beyond the sampling error in the Piper et al. (1993) validation study [3]. It appears that this additional uncertainty only slightly increased the width of the interval of the ultimate result. In examining this method relative to the bias analysis method in Chapter 8, in other datasets we have found that they generally align well, though not always as well as in this example. In this example, the variance of the conventional $\ln OR = \left(\frac{1}{215} + \frac{1}{1449} + \frac{1}{668} + \frac{1}{4296} \right) = 0.0071$ accounts for 79% of the variance of $\ln OR_b$ (0.0089) computed in Equation 10.1 and 76% of the variance implied by the limits of the probabilistic bias analysis $\left(\frac{\ln(1.14) - \ln(0.78)}{3.92} \right)^2 = 0.0094$. Most of the uncertainty arises from conventional sampling error, so the slight differences in the bias model have little affect on the results. The choice of method has little effect on the point estimates: conventional $\widehat{OR} = 0.954$, $\widehat{OR_b}$ from Equation (10.2) $= 0.949$, $\widehat{OR_b}$ from Chapter 8 $= 0.95$. The small difference between the conventional and bias-adjusted estimates could arise because the exposure prevalence is low and the specificity is quite good (see Chapter 6), resulting in little bias-adjustment, and therefore little additional uncertainty.

Bias analysis using the equations provided in Greenland (1988) will be useful when bias-adjusted estimates are needed for ORs based on summary data. The equations may be altered for use with disease misclassification as well [1]. However, this approach is not easily adapted to settings in which record level adjustments are needed or in multiple bias modeling. Further, we note that the equations do not readily allow the user to specify a correlation between sensitivities or specificities among case and controls in the case of differential misclassification (see Chapter 8), nor are there existing equations for measures of effect other than the OR.

Standard Errors for Misclassification When Predictive Values Are Used

When predictive values are available to bias-adjust for misclassification, the equations derived by Marshall (1990) can be used to calculate the variance [4]. This approach can only be used for the odds ratio estimates of effect. As above, we assume that classification parameters are estimated from either an internal or external validation study. The uncertainty in the positive and negative predictive values stem from random sampling error. That is, there is some estimated positive predictive value (PPV$_i$) and negative predictive value (NPV$_i$) among cases (i $= 1$) and controls (i $= 0$) with corresponding variances V(PPV$_i$) and V(NPV$_i$). As above, the variances can be estimated by $\frac{p(1-p)}{n}$ with p being the PPV$_i$ or NPV$_i$ and n being the sample size used to estimate the predictive value of interest. In keeping with the previous section, assume E represents the true exposure status and e represents the observed exposure status, which has been potentially misclassified. The first step to computing the variance of the natural log of the odds ratio using predictive values is to compute the estimated proportion of true exposure, $P(E_i)$, among the cases and controls:

$$P(E_i) = PPV_i \, P(e_i) + (1 - NPV_i)(1 - P(e_i)) \tag{10.4}$$

A similar equation, for cell counts rather than proportions, was presented in Chapter 6. With these quantities, we can compute the variance of the misclassification adjusted odds ratio:

$$\text{Var}[\ln(OR_b)] = \frac{a_0}{[P(E_0)(1 - P(E_0))]^2} + \frac{a_1}{[P(E_1)(1 - P(E_1))]^2} \tag{10.5}$$

where

$$a_i = [PPV_i - (1 - NPV_i)]^2 \text{Var}[P(e_i)] + \text{Var}(PPV_i)P(e_i)^2 + \text{Var}(NPV_i)[1 - P(e_i)]^2 \tag{10.6}$$

OR_b and its confidence interval can then be calculated using Equations 10.2 and 10.3.

To demonstrate computing the variance of a bias-adjusted odds ratio when using predictive values, we present the example shown in Marshall (1990). Marshall used results from Greenland of a case-control study of the association between maternal antibiotic exposure and sudden infant death syndrome [5]. The exposure, maternal antibiotic use, was ascertained by self-report and potentially misclassified. The conventional crude odds ratio can be computed from data presented in Table 10.1, resulting in OR = 1.31 (95% exact CI: 0.97, 1.78). In this example, we use only the data from the portion of the study that was not included in the validation study. Because of concern about the accuracy of maternal self-report, an internal validation was conducted comparing maternal self-report to medical records. Table 10.2 summarizes the predictive values and their variances.

To implement the method of Marshall et al. [4], we calculate the prevalence of self-reported exposure among cases ($P(e_1) = \frac{122}{(122+442)} = 0.216$) and controls

Table 10.1 Data from a case-control study of the association between maternal antibiotic use and sudden infant death syndrome. Observed data from Greenland, 1988 [5]. Exposure data are self-reported and potentially misclassified.

	Reported exposure to antibiotics	No reported exposure
Case	122	442
Control	101	479

Table 10.2 Positive and negative predictive values of self-reported maternal antibiotic use among cases and controls. Observed data from Greenland, 1988 [5].

	Truly exposed/total reporting exposure (PPV)	Var(PPV)	Truly unexposed/total reporting no exposure (NPV)	Var(NPV)
Case	29/51 (56.9%)	29/51·(1−29/ 51)/51 = 0.0048	143/160 (89.4%)	143/160·(1−143/ 160)/160 = 0.00059
Control	21/33 (63.6%)	21/33·(1−21/ 33)/33 = 0.007	168/184 (91.3%)	168/184·(1−168/ 184)/184 = 0.00043

$(P(e_0) = \frac{101}{(101+479)} = 0.174)$, and use these to compute estimates of the prevalence of true exposure with Equation 10.4:

$$P(E_1) = 0.569 \cdot 0.216 + (1 - 0.894)(1 - 0.216) = 0.206$$
$$P(E_0) = 0.636 \cdot 0.174 + (1 - 0.913)(1 - 0.174) = 0.183.$$

The values can be inserted into Equations 10.5 and 10.6 to calculate the variance of the bias-adjusted $\ln OR_b$:

$$a_1 = [0.569 - (1 - 0.894)]^2 \cdot \frac{0.216(1-0.216)}{(122+442)} + 0.0048 \cdot 0.216^2$$
$$+ 0.00059(1 - 0.216)^2$$
$$= 0.00065$$
$$a_0 = [0.636 - (1 - 0.913)]^2 \cdot \frac{0.174\ (1-0.174)}{(101+479)} + 0.007 \cdot 0.174^2$$
$$+ 0.00043(1 - 0.174)^2$$
$$= 0.00058$$

$$\text{Var}[\ln(OR_b)] = \frac{0.00058}{[0.183(1-0.183)]^2} + \frac{0.00065}{[0.206(1-0.206)]^2}$$
$$= 0.051$$

Finally, the bias-adjusted $OR_b = \frac{\frac{0.206}{1-0.206}}{\frac{0.183}{1-0.183}} = 1.16$ with 95% confidence interval (0.75, 1.81).

This method has similar strengths and limitations to the method using sensitivities and specificities. The method cannot easily be extrapolated to record level bias analysis or multiple bias analysis. Further, it can only be used in conjunction with odds ratios and does not admit correlation between predictive probabilities. Furthermore, it assumes that sampling error from a validation study is the source of uncertainty about the predictive values measured in the validation study. When

Table 10.3 Example of bias-adjustment for misclassification of antibiotic use in a study of the effect of antibiotic use during pregnancy (E = 1 self-reported use, E = 0 no self-reported use) on Sudden Infant Death Syndrome (D = 1 case, D = 0 control) using the weighting approach described by Lyles et al., 2010 [6].

Measured		Reclassified					Weighted
E	D	E	N	Weight	Weight		N
1	1	1	122	PPV_{D+}	0.569		69.4
1	1	0	122	$1\text{-}PPV_{D+}$	0.431		52.6
1	0	1	101	PPV_{D-}	0.636		64.3
1	0	0	101	$1\text{-}PPV_{D-}$	0.364		36.7
0	1	1	442	$1\text{-}NPV_{D+}$	0.106		47.0
0	1	0	442	NPV_{D+}	0.894		395.0
0	0	1	479	$1\text{-}NPV_{D-}$	0.087		41.7
0	0	0	479	NPV_{D-}	0.913		437.3
Bias-adjusted data after weighting							
	E = 1	E = 0					
D = 1	116.34	447.67					
D = 0	105.92	474.08					
OR	1.16						

Observed data from Greenland, 1988 [5].

these conditions are plausible, the method by Marshall provides a straightforward approach to generating uncertainty intervals for bias-adjusted estimates.

Predictive Value Weighting

Another set of approaches to address bias problems uses weighted maximum likelihood methods in which the bias model is directly incorporated into the likelihood estimator. Below we describe one such method by way of application to an earlier exposure misclassification example, and then note that similar approaches are available for outcome misclassification problems. Lyles et al. described a weighting-based approach to quantitative bias analysis that can be used when positive and negative predictive values are available [6]. To illustrate this method, we will apply it to the example of the relation between antibiotic use and SIDs [5], which was presented in Table 10.1. Table 10.2 shows the positive and negative predictive values for cases and controls that we will use in the weighting-based approach to bias-adjust the data for misclassification. To do so, we create a copy of each subject in the dataset and in one copy we assign the exposure and in the second copy we assign lack of exposure. The original data could consist of record-level observations or summary-level frequencies (Table 10.3). Each copy receives a weight such that the sum of the weighted number of people in the duplicate data is the same as the overall number of people. For people who were classified as exposed, we weight their data using the relevant positive predictive value (*i.e.*, among the cases or controls). We weight by the positive predictive value in the copies where we continued to classify the person as exposed and we weight by 1 minus the positive predictive value in the scenario where we reclassified the person as unexposed. Note that this weighting guarantees that the information from each individual sums to one full unit of information. For people who were originally classified as unexposed, we weight by 1 minus the relevant negative predictive value for the copy where we reclassified the person as exposed and we weight by the negative predictive value for the copy where we continued to classify the person as unexposed. Table 10.3 shows the data with weights and how using these weighted data we recover the bias-adjusted point estimate of $OR_b = 1.16$ shown above. No interval is shown because, as described next, the method is best implemented by regression, and then intervals may be obtained by bootstrapping or jackknife procedures [6]. The predictive values may be estimated directly from validation data, or estimated from measures of sensitivity or specificity obtained from internal or external validation sources [6, 7].

This approach is similar to quantitative bias analysis methods presented in Chapter 6 and elsewhere. It uses the same inputs (the original data and assumptions about the predictive values of exposure classification) and instead of using it to directly recalculate the interior cells of the 2×2 table, it uses weighting to accomplish the same results. One advantage of this approach is that it can more easily be integrated into conventional regression-based approaches that would allow for control of multiple confounders and use more complex models. One would create two records for each record in the original data set and assign the relevant predictive value or its complement to the two records. Predictive values can be (as noted by Lyles et al) specified such that they vary across confounders, if a study is designed to collect these predictive values. Confounder control (or adjustment for loss-to-follow-up) could also be achieved using weighting-based approaches, such as inverse probability weighting. These weights could be combined with the predictive values to generate a single final weight. Thus, despite its simplicity and ease of implementation, such an approach has advantages over conventional quantitative bias analysis approaches presented earlier in the text because it can be relatively easily combined with other analysis techniques. Conceptually similar approaches have been developed for bias problems including misclassification of a dichotomous outcome [8, 9].

Thus far we have only demonstrated how to obtain a bias-adjusted effect estimate using predictive weighting. Incorporating random error and uncertainty in the bias parameter requires careful consideration. First, note that the data presented in Table 10.3 can be entered into a statistical package and used to estimate an effect with standard regression techniques, as long as the software allows for sampling weights. While these software packages will return standard error estimates (often robust standard errors), Lyles and Lin suggest using a jackknife approach for variance estimation. Jackknife standard errors are relatively simple to compute, but some programming is required. We provide R code to facilitate use of this method. The jackknife procedure to estimate standard errors is a resampling procedure, similar to the bootstrap, as discussed in Chapter 9. With the jackknife procedure, we iteratively eliminate one observation from the data, estimate the statistic of interest, and then aggregate those leave-one-out estimates to generate a standard error. For the predictive weighting approach, some care is required when we eliminate one observation. First, the jackknife method is implemented at the record-level. When we say "leave-one-out" we mean leaving out a record level observation, not a cell from the summary table. Second, we must exclude one record-level observation from the original data set (Table 10.1) and, equivalently, two record-level observations from the weighted data (Table 10.3). The jackknife procedure operates in sequence on each observation, i, in the dataset from 1, . . ., N as follows:

1. Eliminate the ith observation from the data. This implies both replicates of the ith observation in the weighted data in Table 10.3
2. Estimate the bias-adjusted OR from the weighted table and save it as OR^i

These two steps are iterated over each observation in the original data. In our example, we will eventually have N = 1144 estimates of OR^i, one for each observation we have excluded from the dataset. The jackknife standard error (SE_{jack}) is then computed from these estimates as:

$$SE_{jack} = \sqrt{\frac{n-1}{n} \sum \left(\ln(OR^i) - \overline{\ln(OR)} \right)^2}$$

where the OR's are first ln-transformed and $\overline{\ln(OR)}$ is the average of the OR^i. This standard error can be used with Wald-style confidence interval formula to produce a 95% confidence interval:

$$e^{\left(\ln(OR) \pm 1.96 \ SE_{jack} \right)}$$

where $\ln(OR)$ is estimated from the full weighted dataset in Table 10.3. Applying this procedure to the data in Table 10.3, we find an OR = 1.16 and $SE_{jack} = 0.076$ resulting in 95% a confidence interval of (1.00, 1.35).

At this point we have only considered random error in the outcome (as we typically do in conventional statistics) and not uncertainty in the bias parameter. Lyles and Lin propose an approach to incorporate this source of error that is similar to the approach discussed in Chapter 8. First, a user of this method must specify a probability distribution for the bias parameters as discussed in Chapters 7 and 8. Second, we take a large number, call it N, of draws from these bias distributions. For each of these N bias parameters, we estimate the bias-adjusted effect estimate and jackknife standard error. Next, we specify a normal distribution for each of the N estimates and their standard errors (on the ln-scale for relative measures of effect) and draw M random samples from each of the N distributions, resulting in $N \times M$ random samples. The 2.5 and 97.5th percentiles of this distribution constitute our simulation interval and the median is our point estimate.

For example, continuing with the antibiotic use and sudden infant death syndrome example, the validation dataset in Table 10.2 are used to specify beta distributions for the predictive values:

$$PPV_1 \sim Beta(29, 22)$$

$$PPV_0 \sim Beta(21, 12)$$

$$NPV_1 \sim Beta(143, 17)$$

$$NPV_0 \sim Beta(168, 16)$$

We set $N = 2000$, in accordance with Lyles and Lin, and sample 2000 predictive values from each distribution. For each of these 2000 sets of values, we generate a new Table 10.3. Each table has different predictive values. We estimate the bias-adjusted ln-OR by fitting a weighted logistic regression to each of the 2000 data tables. We use the jackknife procedure described above to estimate 2000 standard errors for the ln-OR. Normal distributions are specified with a mean equal to the ln-OR and standard error equal to the jackknife standard error, for each of the 2000 estimates. We sample $M = 500$ new ln-OR from each of these normal distributions,

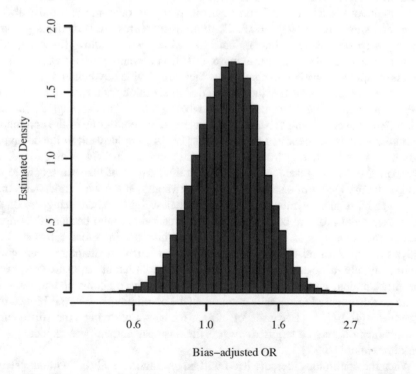

Figure 10.1 Histogram of simulation using the bias-adjustment method of Lyles and Lin to bias-adjust for misclassification and incorporate uncertainty about the bias parameters and random sampling error. Observed data from Greenland, 1988 [5].

resulting in $2000 \times 500 = 10^6$ bias-adjusted ln-OR estimates. The resulting distribution, shown in Figure 10.1, encompasses the bias-adjustment as well as the uncertainty in the bias parameters and the conventional sampling error. From these simulations, we estimate a median bias-adjusted OR = 1.17 and 95% simulation interval of (0.75, 1.81). The point estimate is identical to the estimate obtained using the approach of Marshall above. The simulation interval is extremely close (0.75, 1.81), with any minor discrepancy in other examples potentially due to either the explicit specification of a beta distribution for the predictive values or the use of jackknife (versus asymptotic) standard errors.

Classical and Berkson Measurement Error for Continuous Exposures

Like bias due to misclassification, bias due to measurement error is often described only qualitatively—despite the fact that methods to quantify the direction, magnitude and uncertainty of the bias have been long known—and often even these qualitative descriptions are in error [10]. For exposures and covariates measured on a continuous scale, the relation to a dichotomous outcome is almost always estimated using a regression model. Continuous measures are sometimes grouped into N categories using boundary cut-offs, which would allow for categorical analyses. For example, body mass index (BMI) is a continuous measure of body habitus computed as weight (a continuous measure) divided by height squared (also a continuous measure). BMI is therefore a derived continuous measure, and is often categorized using boundary cutoffs with labels such as "underweight," "normal weight," "overweight" and "obese" [11]. Mismeasurement of one of the continuous measures, such as self-reported body weight [12–14], might result in the computed body mass index deviating from the true body mass index. When categorized, the misreported body weight might then result in assignment of the data record to the wrong category [15]. For example, a truly overweight person who understates their body weight might be misclassified as normal weight. When categorized, the mismeasurement problem becomes a misclassification problem because the impact of mismeasurement is to classify data records into the incorrect category. Bias analysis methods for misclassification problems presented in Chapter 6 then apply. Importantly, non-differential mismeasurement of a continuous exposure (i.e., exposure measurement errors not expected to depend on the outcome status) does not assure non-differential misclassification when the continuous measure is binned into categories [16, 17]. Because of this, methods allowing for differential misclassification may be required, even if the mismeasurement was expected to be non-differential [16, 17].

When the continuous measure itself will be used in the analysis, without categorization, the analytic method will usually require regression analyses. Accordingly, any bias analysis method for continuous measures should naturally incorporate regression modeling. Although we recognize that continuous measures may also be analyzed using methods that are not strictly speaking regression methods, such as t-tests or F-tests, these can almost always be recast as regression analyses. Furthermore, the parameters of the bias models that we have used for misclassification problems (sensitivity, specificity, and positive and negative predictive value) will not be used for measurement error problems. These rely on categorization, so cannot be used for measurement error problems. Instead, in this section, we will rely on linear relations between two continuously measured variables: the imperfectly measured variable that is available for all (or almost all) members of the study population and the gold standard measure of that variable, which is available for only a subset of the study population. In some cases, the linear relation will be measured in a second, external population. Measurement error models are thoroughly discussed by Carroll et al. in their textbook [18].

To start, we will assume a general model in which a risk ratio relates change in a continuously measured exposure to risk of some dichotomous outcome. We will represent this as an exponential function of the change in the continuous variable x:

$$RR(x) = e^{\beta x}$$

where e^{β} represents the proportional change in risk per unit change in x [19]. Assume x is the gold standard measure of the continuous variable with mean μ_x and variance σ_x^2, and that m is an imperfect surrogate measure of x. This surrogate variable is available for all participants and will be used in the implementation of the regression model. For example, x might be true BMI, which could be derived from clinical measurement of height and weight, and m might be mismeasured BMI derived from self-reported height and weight. The mismeasured variable is a function of the true value and the error (e) due to relying on the imperfect measure:

$$m = x + e$$

This type of measurement error, in which the extent of error does not depend on the outcome, is referred to as "classical measurement error" in the measurement error literature. This classical measurement error is a type of non-differential error, which we discussed in Chapter 6. In classical measurement error, the only reason for the discrepancy between the true and surrogate measure is random noise. In fact, the surrogate is an unbiased estimate of x, in purely statistical terms. The regression model above can be implemented with the surrogate, rather than true exposure, allowing an estimate of the relative risk associated with a unit change in m:

$$RR(m) = e^{\beta' m}$$

In expectation, $|\beta'| < |\beta|$, so $RR(m)$ will be nearer to the null value of 1.0 than $RR(x)$. As in Chapter 6, nondifferential measurement error leads to a bias toward the null in expectation. And as in Chapter 6, it is crucial to remember that this result only holds in expectation. Random noise could render a given study result in any direction, and large deviations from expectation are less likely with large study sizes [20].

If we assume that the error in measurement of m is normally distributed with variance σ_e^2, then the expectation for β' is a function of β, σ_x^2, and σ_e^2:

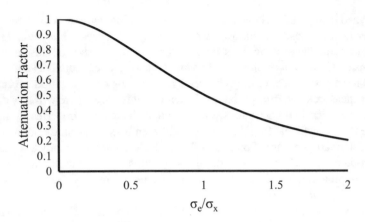

Figure 10.2 Strength of the attenuation factor (see Equation 10.7) as a function of the ratio of the standard error of the measurement error (σ_e) to the standard error of the gold standard exposure measure (σ_x).

$$B' = \frac{\sigma_x^2}{\sigma_x^2 + \sigma_e^2} \beta \tag{10.7}$$

The attenuation factor $\left(\frac{\sigma_x^2}{\sigma_x^2 + \sigma_e^2}\right)$ is a proportion known as the "reliability" of m as a measure of x and equals the square of the correlation between x and m in the source population [19].

Figure 10.2 depicts the strength of attenuation as a function of the ratio of the standard deviation of the error (σ_e) to the standard deviation of the gold standard measure (σ_x). When the standard deviation of the error is 10% of the standard deviation of the gold standard measure, one would expect $\beta' = 0.99\beta$. It is not until the standard deviation of the error is one-third the standard deviation of the gold standard measure that $\beta' = 0.90\beta$, and the two standard deviations would have to be equal before $\beta' = 0.50\beta$. To put these ratios in context, consider the example of human height, which has a standard deviation of about 3 inches (7.62 cm) in the United States. If self-report of height is made in error, the standard deviation of that error would have to be 1 inch (2.54 cm) before ratio estimates of association would be expected to be attenuated by 10%. For many continuous measures, the standard deviation of the error could plausibly be expected to be a small fraction of the standard deviation of the gold standard measure based on substantive knowledge.

If there is concern about measurement error, x would not be available for all members of the source population or study population, otherwise it would be used in the regression instead of m. To estimate the attenuation factor, one must therefore estimate the correlation between x and m in a representative subset of the source population or in a subset that can be reweighted to represent the source population. When no estimate of the reliability is available from internal or external sources, one can implement a sensitivity analysis of the bias analysis using selected values from a reasonable range, such as the range illustrated in Figure 10.2.

Rearrangement of Equation 10.7 suggests a straightforward bias-adjustment to address the error in measuring x [19]:

$$\hat{\beta} = \hat{\beta}' \Big/ \frac{\sigma_x^2}{\sigma_x^2 + \sigma_e^2} \tag{10.8}$$

where $\hat{\beta}'$ is the effect estimate from the conventional regression using the surrogate variable m and $\hat{\beta}$ is the bias-adjusted estimate one would expect from a regression using the gold standard variable x, conditional on the accuracy of the bias model. Interval limits for $\hat{\beta}$ can be computed using methods described below in the description of regression calibration.

Many continuously measured exposure variables always take on positive values and are often right skewed. Typical examples include body weight, pollutant exposures, and micronutrient intakes. In the case of right skewed variables, it is common practice to substitute $ln(x)$ and $\ln(m)$ for x and m above. When comparing a regression model that includes only an intercept and the term $\ln(x)$ versus a regression model that includes an intercept and $\ln(m)$, the attenuation described previously for non-transformed data will still hold. However, note that the ln transformation implies the relative risk for a 1-unit change in x depends on the reference level of x:

$$RR(x) = e^{\beta\,[\ln(x+1)-\ln(x)]} = \left(\frac{x+1}{x}\right)^{\beta}$$

The implication is that when analysts examine dose response relations for log transformed predictors with measurement error, the shape of the relationship could change [19]. For example, if there is truly a perfectly linear relation between $\ln(x)$ and the log-risk of the outcome, an analyst could find a quadratic relation between ln (m) and the log-risk of the outcome.

In some exposure settings, the error structure is of the form:

$$x = m + e$$

where the notation is as above. Under this structure, which is called the Berkson error model, the true value of the exposure (x) is a function of the measured value (m) and error (e) [21]. For example, consider a case-control study by Lambert et al. estimating the association between elongate mineral particles and mesothelioma among taconite miners [22]. Individual measures of the amount of elongate mineral particles (the gold standard exposure, x) for each worker were not available. Instead, the researchers obtained work records and estimated the amount of time each worker spent in various jobs at the taconite mine. The average concentration of elongate mineral particles was known for each job and workers were assigned time-weighted job-specific average exposure levels (the surrogate exposure, m). This is a Berkson model of measurement error because the true exposure, x, is a function of the surrogate, m, rather than the reverse. The true value of the exposure (x) would no doubt vary due to variations in time spent at locations within the workplace. These would result in deviations (e) from the time-weighted average concentration (m). If e is symmetrically distributed with mean of 0, then averaged over the workplace, the true mean exposure $(\bar{x})$ would be expected to equal m. With measurement error of this structure and the risk ratio again as an exponential function of the continuous variable x, we expect:

$$\beta' = \beta$$

and therefore:

$$\widehat{\beta'} = \widehat{\beta}$$

That is, with this error structure, there is no attenuation of the estimate expected to result from the measurement error, although there is additional uncertainty, which ought to be incorporated into interval estimates using methods described below [19, 23]. It is possible for both Berkson error and classical error to apply [24]. For example, if the time-weighted average concentration in the workplace is measured with error, then the error structure for x would have both Berkson and classical components.

Regression Calibration

Regression calibration [25, 26] is a general strategy for bias-adjustment of a mismeasured continuous variable in regression analyses. Regression calibration can be implemented in a variety of ways (see Carroll *et al* for a complete description) [18]. However, the general idea of regression calibration is to replace the imperfectly measured variable (m) with an estimate of the gold standard variable (x) and implement the regression with this new estimate of x. In practice, this calibration can be implemented with the following, relatively simple, procedure. First, in a validation study, fit a regression with x as the dependent variable and m and other relevant covariates as independent variables in the model. Second, predict values of x, $\widehat{x}$, from this calibration model. Third, use $\widehat{x}$ in the main regression model in place of m. Fourth, adjust standard errors for the calibration. We first present a very simple strategy to implement this approach in a setting with no additional covariates; that is, we only have an outcome, D, the mismeasured exposure, m, and the gold-standard exposure, x. In this general strategy, one first estimates (for example):

$$\text{logit}[\text{Pr}(D = 1|m)] = \beta_0 + \beta' m \qquad (10.9)$$

Since D is a binary indicator for disease status, this equation estimates the natural logarithm of the case-control odds ratio or of the disease-odds ratio for cohort data, both of which estimate the risk ratio when the disease is rare. The case-control odds ratio will also estimate the risk ratio when the disease is common, so long as controls are selected by case-cohort sampling, or the rate ratio, so long as controls are selected by incidence-density sampling [27]. Following the derivation in Rosner et al. and Spiegelman et al. [25, 26], we can estimate the bias-adjusted regression calibration coefficient and its variance using

$$\hat{\beta} = \frac{\hat{\beta}'}{\hat{\gamma}}$$

and   (10.10)

$$Var(\hat{\beta}) = \frac{Var(\hat{\beta}')}{\hat{\gamma}^2} + \frac{\hat{\beta}'^2}{\hat{\gamma}^4} Var(\hat{\gamma})$$

where $\hat{\beta}'$, the estimate of the ln(OR) for the mismeasured variable, is obtained by fitting the regression model in Equation 10.9 to the complete data and $\hat{\gamma}$ as well as its variance, $Var(\hat{\gamma})$, are obtained by fitting a simple linear regression model to the validation data with measures of both x and m:

$$E(x|m) = \alpha + \gamma m$$

and   (10.11)

$$Var(x|m) = \sigma^2$$

The bias-adjusted point estimate is then $e^{\hat{\beta}}$ and its interval limits are obtained from $e^{\hat{\beta} \pm z_{\alpha/2} \sqrt{Var(\hat{\beta})}}$. The same bias-adjustment can be applied to estimates of the rate ratio obtained from proportional hazards regression [23], assuming that m is independent of the disease rates given x and that losses to follow-up are independent of m [26]. The first of these is the conditional independence assumption required by most measurement error bias-adjustments and can be subject to inspection with sufficient validation data. The second of these is not verifiable and is equivalent to the independent censoring assumption routinely invoked for perfectly measured variables (or assumed to be perfectly measured variables) in most survival regression models. The method for bias-adjustment by regression calibration outlined above and applied to either logistic or proportional hazards models assumes a rare disease, modest relative effect, and modest measurement error [25, 26]. The first two of these correspond to the usual rare disease assumption required for disease odds ratios or rate ratios to provide valid estimates of risk ratios. A final assumption is that the measurement error is independent of disease status [25], which we would label "nondifferential" (see Chapter 6). Ordinarily, this assumption is best justified by recording m and x before the disease occurs (i.e., prospectively), although there may be other design settings in which the assumption can be defended. When differential measurement error is expected by design, bias-adjustment by nonparametric [28] or semi-parametric [29] methods ordinarily yield estimates with better interval coverage properties [30].

The ability of regression calibration to obtain relatively unbiased estimates of effect with nominal coverage probability relies on the valid and precise estimation of γ, the coefficient relating m to x. In the simplest setting, m is a single measure of x, and participants in the validation substudy are those with measures of both m and x. In the case of Berkson measurement error, $E(\gamma) = 1$ and regression calibration only serves to properly inflate $\mathrm{Var}\left(\widehat{\beta}\right)$ (see Equation 10.10 and the discussion above, where the true mean exposure $(\bar{x})$ was expected to equal m). More often, m is an error-prone measure of x, so $E(\gamma) \neq 1$ and we will not have Berkson measurement error. The estimate of γ is then often obtained from an internal validation substudy, in which a representative set of participants from the main study (in which m is obtained from all participants) provide a measure of the gold standard x. Occasionally, the validation data are obtained from a study that includes no participants in the main study, which is an external validation study. Designs and strategies for resource allocation between main and validation studies have been reviewed elsewhere [31, 32], and adaptive designs for longitudinal collection of validation data with simultaneous monitoring and design revisions have been recently proposed [33]. See also Chapters 3 and 6.

As described in Chapter 3, obtaining a representative sample of the main study participants may be difficult because validation substudies often require an additional respondent burden. As an alternative, one might also estimate γ from an external validation sample, one that includes no participants who were in the main study. An external validation design assumes that the estimate of γ is transportable to the main study participants, which may require reweighting the data by relevant characteristics [34]. When an internal validation sample is large and includes substantial information from persons with the outcome, an extension to the general method for mismeasurement of an independent variable [35] provides an optimized estimate that recognizes that the gold-standard measure is available for a substantial proportion of the main study population. For participants with the gold standard measurement, no adjustment for measurement error is needed, so applying the regression calibration only to those without the gold standard measure is statistically efficient. Effect estimates can be computed separately for those for whom regression calibration was and was not used and then combined using inverse variance weighting. When the internal validation sample is small, or the information from those with the outcome is sparse, there is little added benefit from the optimized extension [35]. Strategies to optimize the allocation of study resources to the main study versus the validation substudy have been described [31]. Design of the optimal allocation requires initial specification of values for α, γ, and σ in Equation 10.11, which are the very parameters meant to be estimated by the validation design. Adaptive or sequential solutions have been proposed [31, 33], but seldom implemented. In many cases, x cannot be well-measured in the validation study, which instead yields an alloyed gold standard x' [36]. For example, x might be total hours of vigorous physical activity over the course of 5 years, m might be a self-reported estimate of total hours of vigorous activity over the course of 5 years, and x' might be total hours of vigorous activity measured by actigraphy over 1 week's time multiplied by 260.7 (the number of weeks in 5 years). Regression calibration methods are largely robust to the imperfect correlation between x and x' [37]. The bias-adjusted estimate usually lies between the estimate without regression calibration ($e^{\widehat{\beta'}}$) and the estimate that would have been obtained from regression calibration were x itself measured ($e^{\widehat{\beta}}$) in the validation study. The expected attenuation can be estimated as a function of the correlations of x' with x and m with x, the ratio of the variances of x' and m, and the correlation between the errors in x' and m [37]. If the errors between x' and m are uncorrelated, then no attenuation is expected. While this lack of correlation may be difficult to verify, a third measure of x may sometimes be available, which will allow an estimate of the correlation that can be brought forward to the regression calibration [37]. Wong et al. describe validation study designs and associated analyses to address more complex scenarios such as these [38].

If v is a continuously measured confounder, and w is a non-differentially mismeasured (with respect to the dichotomous outcome) surrogate for v (such that $w = v + e'$), then control for confounding by v by including w in a regression model would result in incomplete control for v as a confounder. The principles are the same as for incomplete control of a categorized confounder because of its nondifferential misclassification (see Chapter 6). The estimate of effect would suffer from residual confounding, with an expectation that the estimate would lie between the estimate that would have been obtained had v been included in the regression model instead of w and the estimate that would have been obtained had neither v nor w been included in the regression model (*i.e.*, the estimate without adjustment for any measure of the confounder) [19].

The regression calibration models above can be extended to this multivariable setting [19]. The extension uses covariance matrices in the solution to account for the multiple variables (at a minimum, a continuously measured exposure and a continuously measured confounder) in place of the variances for the single mismeasured exposure variable used above [39]. These methods assume that the measurement error of all covariates are independent of true values for all covariates [19]. If both v and x are measured with error, then the conventional estimate of effect ($\widehat{\beta'}$) would lie between the estimate that would have been obtained had x and v been included in the regression model instead of m and w ($\beta_{(x|v)}$) and the estimate that would have been obtained had neither v nor w been included in the regression model (β_x) [19].

$$\widehat{\beta'} = \left(\frac{\sigma^2_{(x|v)}}{\sigma^2_{(m|w)}} \right) \left(\frac{\sigma^2_{(v|x)}}{\sigma^2_{(w|m)}} \right) \beta_{(x|v)} + \left(1 - \frac{\sigma^2_{(v|x)}}{\sigma^2_{(w|m)}} \right) \beta_x \qquad (10.12)$$

Rearranging to solve for the bias-adjusted estimate:

$$\widehat{\beta}_{(x|v)} = \left[\widehat{\beta}' - (1 - \frac{\sigma^2_{(v|x)}}{\sigma^2_{(w|m)}})\beta_x\right] \Bigg/ \left[\left(\frac{\sigma^2_{(x|v)}}{\sigma^2_{(m|w)}}\right)\left(\frac{\sigma^2_{(v|x)}}{\sigma^2_{(w|m)}}\right)\right] \qquad (10.13)$$

If the confounder is measured without error—so that $v = w$, $\frac{\sigma^2_{(v|x)}}{\sigma^2_{(w|m)}} = 1$, and $\left(1 - \frac{\sigma^2_{(v|x)}}{\sigma^2_{(w|m)}}\right)\beta_x = 0$—then Equation 10.13 simplifies to the form of Equation 10.8 above, but now conditional on control of the confounder. If the exposure is measured without error—so that $x = m$ and $\frac{\sigma^2_{(x|v)}}{\sigma^2_{(m|w)}} = 1$—then Equations 10.12 and 10.13 simplify to a form dependent only on the reliability of the measure of the confounder:

$$\widehat{\beta}' = \left(\frac{\sigma^2_{(v|x)}}{\sigma^2_{(w|m)}}\right)\beta_{(x|v)} + \left(1 - \frac{\sigma^2_{(v|x)}}{\sigma^2_{(w|m)}}\right)\beta_x$$

and

$$\widehat{\beta}_{(x|v)} = \left[\widehat{\beta}' - \left(1 - \frac{\sigma^2_{(v|x)}}{\sigma^2_{(w|m)}}\right)\beta_x\right] \Bigg/ \left(\frac{\sigma^2_{(v|x)}}{\sigma^2_{(w|m)}}\right)$$

These and other specific cases are described more completely in Armstrong et al. (1989) [40].

Extending regression calibration to a more general multivariable setting using most common epidemiologic regression tools is relatively straightforward. Rosner et al., 1990 and Carrol and Stefanski [39, 41] offer closed form solutions for standard error calculations, however they are somewhat involved for practical implementation and most modern regression calibration users tend to use bootstrapping to estimate standard errors. Here, we lay out the general strategy for implementing regression calibration in the common case of multiple confounders and predictors of the mismeasured variable. As above, let x be a gold standard measure of an exposure that is measured in the study of interest by the imperfect surrogate m. Let v be a vector of j covariates measured without error. The effects of m and v on a dichotomous outcome (D) can be estimated, as before, by a general linear model with a specified distribution, D, and a link function g():

$$g([\Pr(D = 1|m, v)]) = \beta_0 + \beta m + \omega'v \qquad (10.14)$$

where β is the regression coefficient relating the effect of the mismeasured variable to the outcome and ω' is a vector of regression coefficients relating the effects of the variables measured without error to the outcome, conditional on adjustment for the mismeasured variables. Note that this general specification of the regression model allows the most common epidemiologic regression models such as linear, logistic, Poisson, and log-linear regression. The parameter β is a biased estimate of effect. The approach to regression calibration requires the availability of validation data to estimate a linear regression model of the form:

$$x = \alpha + \gamma m + \rho' v + e \qquad (10.15)$$

where γ is the coefficient relating the mismeasured surrogate to the gold standard measures of the variables measured by imperfect surrogates, ρ is a vector of coefficients relating the variables measured without error to the gold standard measures of the variables measured by imperfect surrogates, α is an intercept, and e is an error term with normal distribution. The importance of Equation 10.15 is as a prediction model. That is, after fitting model 10.15, we use the estimated regression coefficients to estimate $\widehat{x}$, the predicted gold standard value for the exposure variable. These predicted values are retained in the dataset and used in place of the observed mismeasured variable m in Equation 10.14:

$$g([\Pr(D = 1|\widehat{x}, v)]) = \theta_0 + \theta_1\widehat{x} + \omega'_{rc}v \qquad (10.16)$$

The parameter θ_1 is the regression calibration-adjusted estimate of effect. If we had chosen a binomial family for the distribution of the outcome and $g() = \text{logit link}$, $\widehat{\theta}_1$ would be the bias-adjusted $\ln(\text{OR})$. Conventional standard errors would be incorrect and generally underestimate uncertainty. Instead, bootstrap standard errors are often implemented in regression calibration approaches. Stata has a regression calibration procedure to allow easy implementation (rcal) that employs bootstrap or robust standard errors. In the absence of a prewritten function for running a regression calibration, a user could program their own bootstrap function by resampling unique observations from the original data and repeatedly fitting 10.15 and 10.16, retaining $\widehat{\theta}_1$ after each iteration.

We have several comments about this general approach to regression calibration and the text by Carroll et al. is a very useful reference (18). This procedure appears generally robust to choice of standard epidemiologic regression models. Less standard models (for instance, non-linear toxicology models) should be approached with more caution. When used with logistic regression, there is some evidence it does not work well for large ORs (e.g., larger than OR = 3.0). When internal validation data are used, efficiency can be improved by fitting two models, as we described above: first, a model for the outcome that uses the available gold standard data; second, a regression calibration model for the remaining data for which the gold standard was not available. The two estimates can then be combined using inverse variance weights. Finally, the regression calibration approach, generally speaking, is extremely similar to the imputation methods for missing data, which we will discuss next in this chapter.

However, before discussing missing data methods, consider the influence of measurement error when a continuous <u>outcome</u> y is non-differentially mismeasured by a surrogate variable s, such that:

$$s = y + \epsilon \tag{10.17}$$

Using linear regression, we wish to estimate the effect of x on y using:

$$y = \alpha + \beta x + e \tag{10.18}$$

As above, both e and ϵ have an expectation of 0. Substituting Equation 10.18 into Equation 10.17, we have:

$$s = \alpha + \beta x + e + \epsilon$$

We see that classical mismeasurement of the continuous outcome yields an unbiased estimate of β, albeit with less precision because $\sigma_e^2 + \sigma_\epsilon^2 > \sigma_e^2$. If one can estimate σ_ϵ^2 from validation data, then the intervals can be narrowed accordingly.

When the continuous outcome is measured differentially (with respect to the exposure), we modify Equation 10.17 to account for the influence of the exposure x on the measurement error:

$$s = y + \delta x + \epsilon \tag{10.19}$$

Substituting Equation 10.18 into Equation 10.19, we have:

$$s = \alpha + (\beta + \delta)x + e + \epsilon$$

The estimate of the effect of the exposure on the outcome yielded by the regression is now a combination of the unbiased effect (β) and the bias due to differential measurement error (δ). Validation data comparing the gold-standard measure of the outcome (y) with the mismeasured surrogate (s) within categories of the exposure (x) would be required to estimate δ and ϵ, and the bias-adjustment proceeds by subtraction [42]. See Shu and Yi (2019) for more flexible outcome mismeasurement model forms and inverse probability weighting bias-adjustments [43] and Shaw et al. (2018) for regression calibration methods when mismeasurement of the outcome is correlated with mismeasurement of the exposure [44].

Missing Data Methods

All sources of bias can be viewed as emanating from a missing data problem [45, 46]. The general curse of causal inference is that we do not know the risk of the counterfactual outcome under both treatment regimes. If we assume exchange-ability between the index and reference groups within levels of measured and controlled covariates, standard counterfactual inference proceeds by imputing miss-ing data. For instance, if there were perfect exchangeability (perhaps within strata of other confounders) we can impute the missing counterfactual outcome of what would have happened to the exposed group, had they been unexposed, by examining what factually happened to the unexposed group. If, however, information is missing on one of the covariates in the sufficient set needed for control [47], then the uncontrolled confounding problem is a missing data problem. When bias emanates from information bias, the gold-standard classified or measured version of the variable is missing and must be imputed based on information about the outcome, measured exposure, and classification parameters such as predictive values. For selection bias, the data are missing for a subset of the source population (baseline selection bias, and not by design, such as in case-control sampling or survey sampling) or for a subsection of the source population's follow-up (loss to follow-up). The exposure or outcome information is missing for those not included. In what follows, we summarize missing data problems and methods to address them, and then illustrate application of missing data methods to bias analysis problems that might also be addressed by methods presented earlier in this text. Missing data problems and methods to address them have been carefully and thoroughly described elsewhere [48–52]; we present only an overview herein.

Overview of Missing Data Structures

Missing data can occur in numerous ways, and the ways in which it occurs for any particular variable in a dataset (*i.e.,* the determinants of missingness), as well as how the missing data are treated analytically, will determine the direction and magnitude of any bias resulting from the missingness. For example, missing data can occur through non-response to questions on a baseline survey and the analyst could conduct descriptive analyses on the observations for which data were collected. In some cases, the missingness does not depend on the actual value of the variable being measured or any other variable in the data. That is, the proportion of missingness does not differ within levels of the actual value or any other variable. In other cases, the missingness will be determined in part by another variable in the dataset or the actual value of the missing variable itself. For example, a question about having ever been diagnosed with a sexually transmitted infection might be more likely to go unanswered by men than by women, or by those who truly have had a previous infection than by those who have not. In this case, the extent of missingness differs by a measured variable (gender) or by the unknown actual value of the missing variable itself. We will discuss the implications of these mechanisms for missing data below. Regardless of the mechanism by which missing data arise, an initial consideration will always be whether to include only participants with no missing data in the analysis ("complete case analysis") or to impute guesses for the missing data so that participants with missing data are included in the analysis ("imputation"). We will describe these analytic choices in greater detail below.

In recognition of these different mechanisms, and their importance for differences in how they might affect bias, a standard lexicon has evolved to characterize patterns of missing data. In the simplest case, missing data are described as "missing completely at random" (MCAR). This pattern arises when the missing data are a random subset of the full data. The probability of missingness does not depend on the actual value of the variables with missing values or on the observed value of any other variable, whether observed or unobserved. When missingness is MCAR, estimates yielded by complete case analysis are unbiased, although they may be less precise than estimates that could be obtained with imputation of missing values. While the implications for MCAR are quite rosy from an analytic perspective, MCAR missingness is probably rare in typical epidemiologic studies. It might occur due to a random error in a data collection instrument. For example, imagine that in a 96-well plate, biospecimens are randomly assigned to wells, and the analytic instrument fails to collect a measurement for the 46th well on every plate. The missing measures of biospecimen collection for subjects assigned to the 46th well would be MCAR.

Data can also be "missing at random" (MAR), which differs from "missing completely at random" in that the probability of data being missing depends on measured covariates (but not on unmeasured covariates). For example, in a highly detailed study estimating physical activity among youth, younger children may be more likely to leave answers blank as they grow weary from the lengthy survey. When exposure or confounder data are MAR, biased effect estimates will generally be obtained from a complete case analysis. When the outcome is MAR, effect estimates will be unbiased as long as the variables predicting bias are included in the regression model. However, because missingness can be explained by the measured values, when data are MAR we can impute the missing data based on the observed data and obtain unbiased estimates. One would never know for certain that the mechanisms that generated missing data can be fully predicted by measured data, so an assumption that data are MAR will require prior subject matter knowledge and specific knowledge of the mechanisms that gave rise to missing data in the particular dataset. This assumption cannot be verified empirically.

Finally, data can be "missing not at random" (MNAR), in which case the missing data depend on the value of the missing information and the missingness cannot be explained by measured covariates. For example, in a study that ascertains the extent of teen substance use, those teens who use substances may be more likely to leave a question on substance use blank. The measured variables do not provide enough information to be able to impute the missing data and use of multiple imputation will not remove the bias, unless valid external information is available. When data are purely MNAR, the observed data may offer no help in controlling for biases induced by the missingness. Although external information, perhaps from a clever validation study, could be used to adjust for MNAR, this is quite rare. Further, it will generally be the case that multiple types of missingness are intertwined. A variable might be missing based on the actual value of that data (MNAR) but also missing dependent on an observed variable (MAR). We are unaware of studies indicating whether it would be better to impute or not in such a setting. If the imputation only uses information from observed data in the study, there will be residual bias after imputation, which would depend on how well the measured variables predict the missing data relative to the amount that unmeasured variables predict missingness. This evaluation depends on subject matter knowledge, an understanding of the mechanisms that gave rise to missing data in the dataset, and an assumption of how well measured variables would predict the MAR component of the missing data and how much of the original missingness was due to MNAR assumption. In some cases, quantitative bias analysis methods such as those described in earlier chapters may be preferred over the assumptions about the missingness needed to impute validly.

Imputation Methods

Imputation methods, and their application, have become common in recent years [53], with some considering it essential for valid inference when missing data are included in a dataset. Many statistical computing packages have integrated algorithms for missing data imputation. However, like with any method (including quantitative bias analysis), if done poorly it could lead to worse inferences than simply ignoring the bias and using a complete case analysis. What is clear is that ignoring the bias is never preferable to carefully considering whether implementation of either a quantitative bias analysis or missing data method will lead to more valid estimates and as such, both should be considered when substantial or impactful missing data exist within a dataset. It is critical that analysts understand the subject matter and missing data mechanism as part of the process for planning an analytic solution. For example, one might notice that age at first birth is missing for 15% of adult women in a data set. Age at first birth could easily be imputed from observed variables, such as marital status, age at marriage, age in years, birth cohort, socio-economic position, sexual orientation, and years of education (for example). However, it is very possible that most of the 15% of women with no age at first birth are nulliparous—a simple cross check of parity against age at first birth should reveal this association. The statistical computing program will readily impute an age at first birth for the nulliparous women using measured variables like those listed above. It is up to the analyst to understand the subject matter and data set well enough to avoid a mistake.

Imputation replaces a missing value with a guess of the value it would have had, had it been measured. The imputed value is guessed using a model that links the missing data with the observed data. We have already seen one type of missing data imputation: regression calibration. The amount of missingness can provide important information on whether we should be using imputation methods. If a tiny fraction of the data are missing, imputation may be relatively unimportant. When exposure status is missing for some non-trivial fraction of records in the dataset, but present for most, we would impute exposure information for those in whom it was missing. Numerous statistical methods have been proposed to account for missing data in an analysis using both frequentist and Bayesian frameworks [52]. However, we recommend a more recent method, with a distinctly Bayesian flavor, that is both flexible and relatively easy to implement: Multiple Imputation by Chained Equations (MICE) [51, 54–56]. Despite its flexibility, MICE is not without limitations. In addition to needing to correctly specify missing data mechanisms (which we give examples of below), analysts have raised mild theoretical concerns about MICE. As we describe below, MICE imputes missing data through a series of univariate regression models. It is possible to specify a series of these regression models such that they do not guarantee stable behavior of the imputation algorithm; that is, in some situations there is no theoretical reason why we should expect to obtain the correct answer [57–59]. That said, it appears that MICE works very well in practice and that when these theoretical difficulties cause practical difficulties they are likely to be easily spotted by examining diagnostics [58].

MICE is implemented in most major statistical packages including R, SAS and Stata. The general idea behind MICE is to iteratively use a model to predict a value for the missing covariate given other covariate information. This predicted value would itself be computed with random error; for example, if the missing variable is dichotomous, we could use a Bernoulli trial to guess the exposure status (based on explanatory variables in the data record). To capture the uncertainly in this guess, multiple imputation techniques repeat the imputation and average over repeated guesses [52]. To explain in more detail, consider an example in which an analyst intends to estimate the effect of body mass index (BMI) on risk of 5-year mortality, adjusting for biological sex. While mortality status is ascertained on all subjects, BMI and sex both have missing values. If we assume that the missingness for BMI and sex are both due strictly to measured covariates (BMI, sex, and mortality), we know that a complete case analysis could cause bias. To implement MICE to adjust for the missingness, we first need to specify a regression model for the variables that have missing data. In this case, there are natural choices. BMI, a continuous variable can be modelled by a linear regression while biological sex, typically a dichotomous variable, can be modeled by logistic regression. The MICE algorithm proceeds through the following steps.

1. To start the imputation, we must choose a starting imputed value for the missing values of BMI and sex. Statistical software will do this automatically. It may be a simple random draw from non-missing values.
2. Fit the linear regression model for BMI, conditional on all other variables (sex and mortality) and using imputed values of sex.

3. Sample a predicted value of BMI for each missing BMI, conditional on sex and mortality.
4. Fit the logistic regression model for sex, conditional on BMI and mortality using previously imputed values of BMI.
5. Sample a predicted value of sex, conditional on BMI and mortality.
6. Repeat steps 2–5 for a given number of iterations, K.

At the end of step 6, the current imputed values of BMI and sex are retained in an imputed dataset. This process (steps 1–6) is repeated to produce a number, M, of imputed datasets. Note that at step 2 the linear regression model includes the most recent imputed value of sex and in step 3 the logistic model includes the most recent imputed value of BMI. At each step of the algorithm, the regression is run on all of the data (both imputed and non-imputed). We also note that the prediction steps are slightly more complicated than simply taking a prediction. They also include first resampling the parameter estimates prior to obtaining a prediction. This incorporates random error in the imputation process and will be automatically accomplished by the statistical software.

After M datasets have been produced, the statistic of interest, for example $\ln(RR)$, and its variance, $V(\ln(RR))$, are estimated in each dataset. The M effect estimates are averaged to produce an overall multiple imputation estimate [60]:

$$\ln\left(\widehat{RR}\right)_{mi} = \frac{1}{M}\sum_{i=1}^{M}\ln\left(\widehat{RR_i}\right) \tag{10.20}$$

Where $\ln\left(\widehat{RR}\right)_{mi}$ is the overall estimate obtained from MICE and $\ln\left(\widehat{RR_i}\right)$ are the imputed dataset-specific estimates of effect. The variance is obtained by combining the average of the imputation dataset-specific variances and the between imputation variance [60]:

$$V[\ln\widehat{RR}]_{mi} = \frac{1}{M}\sum_{i=1}^{M}V[\ln\left(\widehat{RR_i}\right)] + \left(\frac{M+1}{M(M-1)}\right) \cdot \sum_{i=1}^{M}[\ln\left(\widehat{RR_i}\right) - \ln\left(\widehat{RR}\right)_{mi}]^2 \tag{10.21}$$

The first term in the summation is the average of the within-imputation variances and the second term is the between imputation variance. To put this in the context of probabilistic bias analysis we outlined in Chapter 8, the within imputation variance is study random error. The between imputation variance is the additional error introduced due to adjusting for missing data.

We have not mentioned how large the number of imputed datasets (M) should be or how many iterations (K) the MICE algorithm should run before creating an imputed dataset. Rubin gave a value of M = 5 for the number of imputed datasets. However, more simulations will yield better estimates and White et al. offer guidance on the number of total imputed datasets, generally much larger than M = 5

[51]. Similarly, there is no hard and fast rule for the total number of iterations that the MICE algorithm should run, but many authors seem to choose K>20 (up to 100) to obtain a stable sampling distribution. With modern computers, increasing the number of M and K costs little time. If there is a high proportion of missing data, it is important for authors to choose relatively high values for M and K. In fact, repeating analyses with even higher numbers for M and K, to test sensitivity of final results, is an excellent idea.

A special case merits attention: if the missing data arise from longitudinal data with repeated measures, analysts often have substantially more information to impute data. While it may be tempting to use simple imputation techniques such as 'last value carried forward,' these methods are generally known to perform quite poorly for reasons that are not difficult to imagine [61]. If a person reported they did not smoke consistently for 5 waves of a study and then failed to report their smoking status at wave 6, there may well be a MNAR mechanism at play with the participant failing to report because they have started smoking. Carry forward imputation implies that nothing has changed from one wave to another, a heroic assumption whose believability is belied by the fact the analyst is collecting these data at multiple waves. It is generally better to use data reported in the current wave and historical data to impute the missing longitudinal data [62].

When missing data are MCAR, complete case data (records with no missing data) can be viewed as a random sample of all data (including records with missing data). Estimates obtained by complete case analysis, which is the default for many statistical computing packages, would be unbiased in expectation, as noted also above. When missing data are MAR or MNAR, ignoring the missing data by using complete case analysis will generally yield biased estimates of the magnitude, uncertainty, and potentially even the direction of the effect [48, 52]. Simple approaches to missing data, like using an average value for all missing participants or using a missing indicator [48], can also lead to substantial bias and should, in most cases, be avoided.

For most missing data problems, multiple imputation is likely to provide more valid bias-adjusted estimates (provided proper model specification), with intervals that have better coverage properties, than complete case analysis, last observation brought forward, mean imputation, or use of missing data indicators [48, 52]. As with quantitative bias analysis methods described earlier in this text, the goal of missing data imputation is to obtain the estimate one would have had, had perfect and complete data been available. Missing data imputation seeks to identify the exposure, outcome and covariate information for records that would be lost in a complete case analysis, by attempting to predict (and then impute) the missing values in the data to make each record excluded from complete case analysis into a complete case. Multiple imputation uses information from the record with missing data and from the rest of the records to complete this imputation. This approach differs from the other bias analysis methods we have described, which use information external to the data to complete the bias-adjustments. In what follows, we use missing data methods to address example problems that have also been addressed earlier by these other quantitative bias analysis methods.

Table 10.4 Data on the association between male circumcision (E_1 = circumcised, E_0 = uncircumcised) and HIV status (D_1 = HIV+, D_0 = HIV−) stratified by being Muslim (C_1 = Muslim, C_0 = other religion), with observed data from Tyndall et al., 1996 [63].

	Total		C_1		C_0	
	E_1	E_0	E_1	E_0	E_1	E_0
D_1	105	85	75	3	30	82
D_0	528	93	431	6	97	87
	633	178	506	9	127	169

Table 10.5 Hypothetical percent missing data within categories of male circumcision (E_1 = circumcised, E_0 = uncircumcised) and HIV status (D_1 = HIV+, D_0 = HIV−) stratified by being Muslim (C_1 = Muslim, C_0 = other religion), with observed data from Tyndall et al., 1996 [63].

	Total		C_1		C_0	
	E_1	E_0	E_1	E_0	E_1	E_0
D_1	105	85	20%	30%	20%	30%
D_0	527	93	30%	20%	30%	20%
	632	178				

Uncontrolled Confounding Example

The overlap between missing data methods and bias analysis methods has already been noted. Regression calibration is a method for replacing mismeasured exposure data based on an imputation model. We can draw further parallels between bias analysis methods and missing data by considering an uncontrolled confounding problem in which the confounder was measured, but many records have missing data for the confounder. The first analytic option is complete case analysis, which would only be unbiased if the data were MCAR. If the data were MAR or MNAR, a second analytic option would be to use the quantitative bias analysis methods for an unmeasured confounder from Chapter 5. The method we outlined in Chapter 5 would make no use of the measured data, but it could be modified, in a straightforward manner, to do so. We could use the available data to inform estimates for the values of the bias parameters needed for quantitative bias analysis, and to combine this information with other external sources of information to assign distributions to the values of the bias parameters. However, this use of available study data is potentially biased since some of that data are, by definition, missing. A better option, which we will demonstrate by example, is to use missing data imputation to impute values for the uncontrolled confounder for the records with missing data.

Imputation would likely be best for this bias analysis when data were collected from multiple sites and all sites but one (or a small number) collected the confounder information. In such a case, imputation of the missing confounder data can be accomplished by using the measured information on other variables. As an example, we return to the uncontrolled confounder problem described in Chapter 5, in which we were interested in the relation between HIV and circumcision and were

Table 10.6 Hypothetical complete case data on the association between male circumcision (E_1 = circumcised, E_0 = uncircumcised) and HIV status (D_1 = HIV+, D_0 = HIV−) stratified by being Muslim (C_1 = Muslim, C_0 = other religion), with observed data from Tyndall et al., 1996 [63].

	Total		C_1		C_0	
	E_1	E_0	E_1	E_0	E_1	E_0
D_1	84	59	60	2	24	57
D_0	370	75	302	5	68	70
	454	134	362	7	92	127

Table 10.7 Summary level data structure used for missing data imputation, combining information in Tables 10.4 and 10.6 on the association between male circumcision (E_1 = circumcised, E_0 = uncircumcised) and HIV status (D_1 = HIV+, D_0 = HIV−) stratified by being Muslim (C_1 = Muslim, C_0 = other religion, . = missing), with observed data from Tyndall et al., 1996 [63].

Frequency	D	E	C
60	1	1	1
24	1	1	0
21	1	1	.
2	1	0	1
57	1	0	0
26	1	0	.
302	0	1	1
68	0	1	0
157	0	1	.
5	0	0	1
70	0	0	0
18	0	0	.

concerned about confounding by religion. Imagine that the true underlying data are as in Table 10.4. The data are slightly modified from Chapter 5 (Table 5.5) and rounded to the nearest integer to yield a crude risk ratio of 0.35 with an adjusted risk ratio (by standardization) of 0.46. For this example, we will assume that these data yield an unbiased estimate of the risk ratio, so no other confounding variables need to be controlled and there is no measurement error or selection bias.

We will now introduce missing data, with religious affiliation available in 73% of participants, and with the proportion missing within categories of the disease, exposure, and confounder as shown in Table 10.5. Note that the proportion with missing data depends on the exposure and outcome, but not on the actual value of the confounder. This is a MAR mechanism. The observed data that would be obtained from this missing data process, in expectation, is shown in Table 10.6 (fractions are rounded to the nearest whole number). If we were to analyze the data using a complete case analysis, we would obtain a crude risk ratio of 0.42 and an adjusted risk ratio of 0.58 from a log-binomial regression (biased towards the null compared with the true adjusted risk ratio of 0.46). The complete case analysis is biased because the missing data mechanism is MAR.

Table 10.8 Six imputed data sets with missing values of C imputed. Data on the association between male circumcision (E_1 = circumcised, E_0 = uncircumcised) and HIV status (D_1 = HIV+, D_0 = HIV−) stratified by being Muslim (C_1 = Muslim, C_0 = other religion), with observed data from Tyndall et al., 1996 [63]. RR = stratum specific risk ratio and $RR_{adj,\ i}$ equals regression adjusted risk ratio in the ith set of imputed data. Value in parentheses is the variance of $\ln(RR_{adj,\ i})$.

	C_1		C_0			C_1		C_0			C_0		C_0	
Set 1	E_1	E_0	E_1	E_0	Set 2	E_1	E_0	E_1	E_0	Set 3	E_1	E_0	E_1	E_0
D_1	76	5	29	80	D_1	76	5	29	80	D_1	74	4	31	81
D_0	431	6	96	87	D_0	431	6	96	87	D_0	420	7	107	86
Total	507	11	125	167	Total	507	11	125	167	Total	494	11	138	167
RR	0.33		0.48		RR	0.33		0.48		RR	0.41		0.46	
$RR_{adj,\ 1}$	0.50 (0.027)				$RR_{adj,\ 2}$	0.45 (0.029)				$RR_{adj,\ 3}$	0.46 (0.027)			

Set 4	C_1		C_0		Set 5	C_1		C_0		Set 6	C_1		C_0	
	E_1	E_0	E_1	E_0		E_1	E_0	E_1	E_0		E_1	E_0	E_1	E_0
D_1	75	3	30	82	D_1	74	2	31	83	D_1	74	2	31	83
D_0	432	6	95	87	D_0	428	7	99	86	D_0	420	6	107	87
Total	507	9	125	169	Total	502	9	130	169	Total	494	8	138	170
RR	0.44		0.49		RR	0.66		0.49		RR	0.60		0.46	
$RR_{adj,\ 4}$	0.49 (0.028)				$RR_{adj,\ 5}$	0.50 (0.027)				$RR_{adj,\ 6}$	0.47 (0.028)			

To address the missing data, we imputed missing values of the confounder using the MICE package in R. The structure of the data used for the imputation is shown in Table 10.7, which combines information in Tables 10.4 and 10.6. The MICE package in R specifies an imputation model (analogous to the calibration model in regression calibration). In this case, the model used to impute religious affiliation was a logistic model with the HIV status and religious category as the predictors in the model.

As an example, we imputed six datasets, which are shown in Table 10.8. However, our main analysis included more than six imputed datasets, which we present at the end of this section. In each dataset, we fit a log-binomial regression to estimate the association between circumcision and HIV infection, adjusted for imputed religious affiliation (Table 10.8). We also saved the variance of the ln(RR) from each regression model. Following the equations in 10.20 and 10.21, we estimate the multiple imputation adjusted effect as: $RR_{MI} = 0.48$ and $var(\ln(RR_{MI})) = 0.03$, resulting in 95% CI = (0.34, 0.67). By properly imputing for the missing data, we have removed the bias associated with the MAR mechanism. The main point is that the unmeasured confounder can be imputed for the records where it is missing and the imputed data sets, on average, yield a less biased estimate of association than the complete case analysis. As mentioned above, it is generally preferable to increase the number of iterations and imputed datasets. When we increased M from 6 to 100 and set the number of MICE iterations to K=200, we obtained RR = 0.48 (95%CI: 0.34, 0.69), a small change from the original implementation. Both estimates are very similar to the RR=0.46 from the analysis of the complete data.

Misclassification Example

The conceptual link between missing data methods and misclassification methods has already been emphasized with respect to regression calibration and can be further seen in a misclassification problem in which a variable was measured with error among all participants, and a gold standard measurement was made in a subset. If the subset is a simple random sample of the whole, then the missing data on the gold standard are expected to be MCAR. A complete case analysis using only the gold standard subset would be unbiased. However, the subset for whom the gold standard was measured is often a small proportion of the whole, because collecting data on the gold standard is often expensive compared with collecting data by the mismeasured method. It is not statistically efficient, therefore, to limit the data analysis to the subset with complete data, and imputation would provide data on the missing values of the gold standard measure that would be informed by the subset. This analysis will often yield a more precise estimate than limiting the analysis to the subset by complete case analysis.

As described in Chapter 3, a simple random sample is often not a statistically efficient design for a validation substudy. When the subset with the gold standard measure is selected based on other data, such as their values on the outcome and a mismeasured exposure variable, the resources expended on validation will yield a more statistically efficient design. With a validation study of this design, the data are MAR by design. That is, the validation study design introduces an expectation that the subset with complete data is a representative sample within strata used to inform selection into the validation substudy. The probability that missing data is encountered differs for those with and without the exposure variable (and possibly the outcome). Missing data imputation would generally need to respect the study design and impute data conditional, for instance, on the outcome variable.

In some instances, the subset with validation data is a convenience sample. This circumstance can arise because there is a subgroup with both the mismeasured and gold standard versions of the variable available for reasons that were not envisioned by design. It can also arise when participation in the validation substudy requires additional participant burden or requires additional participant consent. In these circumstances, the validation subset is not necessarily representative of the whole within strata of the measured variables, so the data may be MNAR. The analyst must evaluate the potential for data to be MAR or MNAR and, if MNAR, whether imputation is still a reasonable analytic approach. Note that in either situation, complete case analysis would also be a biased analytic solution. These considerations correspond to topics discussed in Chapter 3 and elsewhere in the context of applying information from validation substudies to the complete study population. The descriptions here use the terminology of missing data methods, but the underlying concepts are the same.

As an example of use of missing data imputation for a misclassification problem, we return to the example presented above (Table 10.1), in which we were interested in the relation between antibiotic use during pregnancy and the occurrence of Sudden Infant Death Syndrome in the offspring. Self-reported use of antibiotics

Table 10.9 Data structure used for missing data imputation, combining information in Tables 10.1 and 10.2 on the association between antibiotic use during pregnancy (S_1 = self-reported use, S_0 = no self-reported use) and Sudden Infant Death in the offspring (D_1 = case, D_0 = control). Gold standard measure of antibiotic use from medical record review (G_1 = antibiotic use during pregnancy documented in medical record, G_0 = no documented antibiotic use during pregnancy in medical record, G. = missing), with observed data from Greenland, 1988 [5].

Frequency	S	G	D
29	1	1	1
22	1	0	1
122	1	.	1
143	0	0	1
17	0	1	1
442	0	.	1
21	1	1	0
12	1	0	0
101	1	.	0
168	0	0	0
16	0	1	0
479	0	.	0

was available for all participants and was considered susceptible to misclassification. The conventional case-control odds ratio using the self-reported exposure information was 1.31. Medical records were used to validate antibiotic use in a subset, and bias-adjustment using the validation data yielded a bias-adjusted estimate of the case-control odds ratio of 1.16 using the predictive value methods above. Complete case analysis of only the validation sub-study [1, 4] yielded a case-control odds ratio of $\frac{46/37}{165/180} = 1.36$. The complete case analysis yields an estimate different from the bias-adjusted estimate, suggesting that either the validation data are not a representative subset, even within strata of the exposure and outcome, or that the small size of the validation study resulted in substantial random error. To address the missing data, we imputed missing values of the gold standard using the MICE package in R. The structure of the data used for the imputation is shown in Table 10.9, which combines information in Tables 10.1 and 10.2.

Paying careful attention to the imputation model is critical. In this example, we are imputing the gold standard exposure conditional on observed exposure and case-control status. Because the gold standard is dichotomous, our imputation will be based on Pr(Gold standard | observed exposure), which is simply a predictive value. In the example using these data above, we assumed that positive predictive value and negative predictive value could vary by case/control status and we must ensure that our imputation regression allows this as well. A simple way to do that is to have our imputation regression model include an interaction between observed exposure and case-control status:

Table 10.10 Summary multiple imputation OR and 95% confidence intervals estimated using different number of imputed datasets (M) and chain lengths (k). Estimates for the association between antibiotic use during pregnancy and Sudden Infant Death in the offspring, with observed data from Greenland, 1988 [5].

Imputed datasets	Iterations		
	$K = 25$	$K = 200$	$K = 500$
$M = 5$	1.399 (0.937, 2.088)	1.324 (0.728, 2.408)	1.352 (0.870, 2.100)
$M = 25$	1.223 (0.785, 1.906)	1.130 (0.749, 1.705)	1.160 (0.768, 1.752)
$M = 50$	1.185 (0.732, 1.917)	1.256 (0.822, 1.920)	1.230 (0.800, 1.892)
$M = 100$	1.194 (0.771, 1.850)	1.254 (0.793, 1.983)	1.192 (0.791, 1.796)
$M = 500$	1.214 (0.785, 1.877)	1.208 (0.770, 1.894)	1.218 (0.783, 1.896)

$$\text{logit}[\Pr(G = 1)] = \beta_0 + \beta_1 S + \beta_2 D + \beta_3 S D$$

where G is the gold standard value and S is self-report of exposure. To demonstrate the potential importance of imputing a large number of datasets (M) using a large number of iterations (K), we repeat the MICE algorithm using M = 5, 25, 50, 100 and 500 with all combinations of $K = 25$, 200 and 500. Results are shown in Table 10.10. We note that imputed estimates of effect vary over the number of datasets and iterations. Results with M = 5 are somewhat misleading, which is not unexpected. The standard advice to choose M = 5 is often in regard to datasets with a small fraction of missing. In this example, however, 73% of records are missing the gold standard exposure and estimates benefit from a larger number of imputed datasets calculated using a larger number of iterations. With M = 500 datasets with K = 500 iterations, the imputed OR = 1.22 with $se(\ln(OR)) = 0.226$ and 95% CI: 0.78, 1.90. This result is similar to what we observed in the example above using the method of Lyles, where we found OR = 1.17, 95%CI: 0.75, 1.81. However, notice that we have combined the validation data and the data without a gold standard using this missing data approach. When implementing the approach of Lyles above, we ignored the outcome data in the validation dataset when estimating the overall OR. We can combine the misclassification adjusted OR in Lyles with the observed OR (using the gold standard) to obtain an estimate more readily compara-ble to the missing data approach. In the validation data, we found OR = 1.36, SE = 0.25, 95% CI: 0.84, 2.20. To pool the two estimates, we weight the ln(OR) by the respective variances, as in a fixed-effects meta-analysis. First we estimate the se (OR) from the Lyles method by transforming the 95% confidence limits to a standard error:

$$\frac{\ln(1.81) - \ln(0.75)}{2 \cdot 1.96} = 0.225$$

which we can then use to pool results:

$$\ln\left(OR_{pool}\right) = \frac{\left(\frac{\ln(1.16)}{0.226^2} + \frac{\ln(1.36)}{0.25^2}\right)}{\left(\frac{1}{0.226^2} + \frac{1}{0.25^2}\right)} = 0.22$$

$$V(\ln(OR_{pool})) = \frac{1}{\left(\frac{1}{0.225^2} + \frac{1}{0.25^2}\right)} = 0.028$$

These pooled estimates can be converted to pooled ORs and 95%CI: 1.25 (0.90, 1.73). These pooled results from Lyle's predictive method are relatively close to the results that we get from multiple imputation. Our main point is that the gold standard measure can be imputed for the records where it is missing and the imputed data sets, on average, yield a less biased estimate of association than the complete case analysis.

Selection Bias Example

In Chapter 4, we showed how inverse probability of participation weights could be used to address selection bias from differential baseline participation and inverse probability of attrition weights could be used to address selection bias from differential loss to follow-up. In these approaches, each subject in the dataset (or the summary cells of a contingency table) are weighted by the inverse of the probability of being selected. We here show how imputation of missing information can be used to address the same types of selection bias problems.

The example in Chapter 4 pertained to loss to follow-up of women diagnosed with breast cancer. Table 10.11 shows the 449 breast cancer patients stratified by receipt of less than guideline therapy or guideline therapy (the exposure) and hospital of diagnosis (diagnosed at one of the two hospitals where a tumor registry operated during only part of the study enrollment period or at one of the hospitals

Table 10.11 Characteristics of 449 Rhode Island breast cancer patients followed over 5 years after diagnosis for breast cancer mortality, with observed data from Lash et al., 2000 [64].

Hospital group	Less than guideline therapy		Guideline therapy	
	Complete registry	Incomplete registry	Complete registry	Incomplete registry
N died from breast cancer	29	3	35	2
N did not die from breast cancer	67	5	238	11
Observed person-years	356.2	32.0	1241.3	62.1
N missing follow-up	3	10	11	35
N total patients	99	18	284	48
Breast cancer mortality (N/100PY)	8.1	9.4	2.8	3.2
IPAW	1.03	2.25	1.04	3.70

IPAW inverse probability of attrition weight.

where a tumor registry operated for the entire study enrollment period). Diagnosis at one of the two hospitals where a tumor registry operated during only part of the study enrollment was a strong predictor of loss to follow-up, because these women could not be reidentified and linked with the US National Death Index to ascertain the outcome (death from breast cancer). Baseline characteristics (receipt of definitive therapy or not, age at diagnosis, breast cancer stage at diagnosis, and hospital of diagnosis) were known for all breast cancer patients, regardless of whether they could be linked to the National Death Index. The crude rate ratio associating less than guideline therapy (relative to guideline therapy) with breast cancer mortality collapsed over both types of hospitals is $((29 + 3)/(356.2 + 32))/((35 + 2)/ (1241.3 + 62.1)) = 2.90$, which comes from the complete case analysis. The corresponding crude hazard ratio is 2.96 (95% CI 1.84, 4.75) and the hazard ratio adjusted for age category, stage, and hospital group is 2.00 (95% CI 1.16, 3.44).

Within both therapy categories, the complete case breast cancer mortality rate was higher in the patients diagnosed at hospitals with incomplete reporting to the registry than in the patients diagnosed at hospitals with complete reporting to the registry. Furthermore, because the patients diagnosed at hospitals with incomplete reporting to the registry were much more likely to be lost to follow-up, they also had much higher inverse probability of attrition weights. The prevalence of diagnosis at hospitals with incomplete reporting to the registry was only 15%, so the reweighting had little influence on the estimated rate ratio. The inverse probability of attrition weighted rate ratio is 2.897, as shown in Equation 10.22. Note that in this chapter,

Table 10.12 Summary results from six data sets with missing values of breast cancer (BC) deaths and person-years of follow-up imputed. Data on the association between receipt of less than guideline breast cancer therapy, versus guideline therapy, and death from breast cancer in the first 5 years of follow-up, with observed data from Lash et al., 2000 [64]. RR = rate ratio and HR = hazards ratio.

Set 1	<guideline	guideline	Set 2	<guideline	guideline	Set 3	<guideline	guideline
BC deaths	39	44	BC deaths	42	51	BC deaths	39	47
Person-years	435.3	1507.5	Person-years	416.5	1486.0	Person-years	434.5	1504.7
Crude RR	3.07		Crude RR	2.94		Crude RR	2.87	
Adjusted HR (95% CI)	2.07 (1.27, 3.38)		Adjusted HR (95% CI)	2.21 (1.38, 3.54)		Adjusted HR (95% CI)	2.11 (1.30, 3.44)	
Set 4	<guideline	Guideline	Set 5	<guideline	Guideline	Set 6	<guideline	Guideline
BC deaths	37	49	BC deaths	39	48	BC deaths	35	42
Person-years	439.7	1493.9.6	Person-years	434.5	1519.2	Person-years	441.0	1507.1
Crude RR	2.57		Crude RR	2.84		Crude RR	2.85	
Adjusted HR (95% CI)	1.80 (1.11, 2.92)		Adjusted HR (95% CI)	2.24 (1.38, 3.63)		Adjusted HR (95% CI)	1.82 (1.10, 3.03)	

follow-up was terminated at 5 years, whereas in Chapter 4 patients were followed for as long as 12 years.

$$RR_{IPAW} = \frac{29\frac{1}{96/99} + 3\frac{1}{8/18}}{35\frac{1}{273/284} + 2\frac{1}{13/48}} \Bigg/ \frac{356.2\frac{1}{96/99} + 32.0\frac{1}{8/18}}{1241.3\frac{1}{273/284} + 62.1\frac{1}{13/48}} = 2.897 \qquad (10.22)$$

As an alternative approach to address the missing data, we imputed missing values of breast cancer mortality status and person-time of follow-up using the MICE package in R. We imputed six datasets, with summary results shown in Table 10.12. The mean of the imputed datasets yielded a crude rate ratio of 2.57, which is different than the complete case and IPAW-adjusted rate ratios of 2.90. This difference suggests that other factors in the data set, beyond hospital group, also influence missingness. The pooled adjusted hazard ratio equals 1.84 (95% CI 1.08, 3.15), which is less than the adjusted hazard ratio of 2.00 (95% CI 1.16, 3.44) from the complete case analysis. This result suggests that the missing values due to loss to follow-up may have biased the complete case analysis away from the null.

When to Impute, When to Use Bias Analysis

Multiple imputation methods work best when the data are MCAR or MAR with likely predictors well-measured. Evaluation of these considerations will require subject matter expertise and familiarity with the dataset. In addition, because multiple imputation uses the observed data to estimate associations between parameters and impute missing values, it is usually not very helpful if the amount of missingness is quite large (such that reliable associations cannot be estimated). As the number of missing data points increases, the likelihood of the imputation creating bias rather than removing it may increase. Missing data and imputation diagnostics are available to inform judgments about particular imputation problems.

When the validity of missing data imputation is questionable or when only external information (external validation studies or subject matter knowledge) is available, quantitative bias analysis methods such as those described in earlier chapters may provide a better alternative. Using these methods effectively substitutes one set of assumptions (about the bias model and values assigned to the bias parameters) for another set of assumptions (about the structure of the missing data and the quality of information available to inform the imputation). These assumption sets may overlap, depending on the mechanisms generating the biases. For some problems, both methods might be implemented as a sensitivity analysis. If results of both methods agree, that might provide some assurance that the bias has been addressed, although of course this agreement can occur in cases where both approaches are invalid, and the agreement arises by chance or because of shared modeling structures in the bias analysis and the imputation. As we have emphasized throughout, there is no perfect bias analysis; but no bias analysis is most often a worse alternative to some bias analysis, even if only to reduce overconfidence in the conventional result.

References

1. Greenland S. Variance estimation for epidemiologic effect estimates under misclassification. Stat Med. 1988 Jul;7:745–57.
2. Fink AK, Lash TL. A null association between smoking during pregnancy and breast cancer using Massachusetts registry data (United States). Cancer Causes Control. 2003 Jun;14:497–503.
3. Piper JM, Mitchel EF Jr, Snowden M, Hall C, Adams M, Taylor P. Validation of 1989 Tennessee birth certificates using maternal and newborn hospital records. Am J Epidemiol. 1993;137:758–68.
4. Marshall RJ. Validation study methods for estimating exposure proportions and odds ratios with misclassified data. J Clin Epidemiol. 1990;43:941–7.
5. Greenland S. Statistical Uncertainty Due to Misclassification - Implications for Validation Substudies. J Clin Epidemiol. 1988;41:1167–74.
6. Lyles RH, Lin J. Sensitivity analysis for misclassification in logistic regression via likelihood methods and predictive value weighting. Stat Med. 2010 Sep 30;29:2297–309.
7. Lyles RH, Zhang F, Drews-Botsch C. Combining Internal and External Validation Data to Correct for Exposure Misclassification: A Case Study. Epidemiology. 2007;18:321–8.
8. Magder LS, Hughes JP. Logistic Regression When the Outcome Is Measured with Uncertainty. Am J Epidemiol. 1997;146:195–203.
9. Lyles RH, Tang L, Superak HM, King CC, Celentano DD, Lo Y, et al. Validation Data-Based Adjustments for Outcome Misclassification in Logistic Regression: An Illustration. Epidemiol Camb Mass. 2011;22:589–97.
10. Shaw PA, Deffner V, Keogh RH, Tooze JA, Dodd KW, Küchenhoff H, et al. Epidemiologic analyses with error-prone exposures: review of current practice and recommendations. Ann Epidemiol. 2018;28:821–8.
11. Weir CB, Jan A. BMI Classification Percentile And Cut Off Points. In: StatPearls [Internet]. Treasure Island (FL): StatPearls Publishing; 2020 [cited 2020 Nov 18]. Available from: http://www.ncbi.nlm.nih.gov/books/NBK541070/
12. Connor Gorber S, Tremblay M, Moher D, Gorber B. A comparison of direct vs. self-report measures for assessing height, weight and body mass index: a systematic review. Obes Rev Off J Int Assoc Study Obes. 2007;8:307–26.
13. Keith SW, Fontaine KR, Pajewski NM, Mehta T, Allison DB. Use of self-reported height and weight biases the body mass index–mortality association. Int J Obes. 2011;35:401–8.
14. Bodnar LM, Abrams B, Bertolet M, Gernand AD, Parisi SM, Himes KP, et al. Validity of Birth Certificate-Derived Maternal Weight Data. Paediatr Perinat Epidemiol. 2014;28:203–12.
15. Stommel M, Schoenborn CA. Accuracy and usefulness of BMI measures based on self-reported weight and height: findings from the NHANES & NHIS 2001-2006. BMC Public Health. 2009;9:421.
16. Flegal KM, Keyl PM, Nieto FJ. Differential misclassification arising from nondifferential errors in exposure measurement. Am J Epidemiol. 1991;134:1233–44.
17. Wacholder S, Dosemeci M, Lubin JH. Blind assignment of exposure does not always prevent differential misclassification. Am J Epidemiol. 1991;134:433–7.
18. Carroll RJ, Ruppert D, Stefanski LA, Crainiceanu CM. Measurement error in nonlinear models: a modern perspective. CRC press; 2006.

19. Armstrong BG. The effects of measurement errors on relative risk regressions. Am J Epidemiol. 1990;132:1176–84.
20. Loken E, Gelman A. Measurement error and the replication crisis. Science. 2017;355:584–5.
21. Berkson J. Are there two regressions? J Am Stat Assoc. 1950;45:164–80.
22. Lambert CS, Alexander BH, Ramachandran G, MacLehose RF, Nelson HH, Ryan AD, et al. A case–control study of mesothelioma in Minnesota iron ore (taconite) miners. Occup Environ Med. 2016;73:103–9.
23. Prentice RL. Covariate Measurement Errors and Parameter Estimation in a Failure Time Regression Model. Biometrika. 1982;69:331–42.
24. Heid IM, Küchenhoff H, Miles J, Kreienbrock L, Wichmann HE. Two dimensions of measurement error: classical and Berkson error in residential radon exposure assessment. J Expo Anal Environ Epidemiol. 2004;14:365–77.
25. Rosner B, Willett WC, Spiegelman D. Correction of logistic regression relative risk estimates and confidence intervals for systematic within-person measurement error. Stat Med. 1989;8: 1051–69.
26. Spiegelman D, McDermott A, Rosner B. Regression calibration method for correcting measurement-error bias in nutritional epidemiology. Am J Clin Nutr. 1997;65:1179S-1186S.
27. Pearce N. What Does the Odds Ratio Estimate in a Case-Control Study? Int J Epidemiol. 1993;22:1189–92.
28. Pepe MS, Fleming TR. A Nonparametric Method for Dealing With Mismeasured Covariate Data. J Am Stat Assoc. 1991;86:108–13.
29. Carroll RJ, Wand MP. Semiparametric Estimation in Logistic Measurement Error Models. J R Stat Soc Ser B Methodol. 1991;53:573–85.
30. Sturmer T, Thurigen D, Spiegelman D, Blettner M, Brenner H. The performance of methods for correcting measurement error in case-control studies. Epidemiology. 2002;13:507–16.
31. Spiegelman D, Gray R. Cost-Efficient Study Designs for Binary Response Data with Gaussian Covariate Measurement Error. Biometrics. 1991;47:851–69.
32. Holford TR, Stack C. Study design for epidemiologic studies with measurement error. Stat Methods Med Res. 1995;4:339–58.
33. Collin LJ, MacLehose RF, Ahern TP, Nash R, Getahun D, Roblin D, et al. Adaptive Validation Design. Epidemiol Camb Mass. 2020;31:509–16.
34. Webster-Clark M, Lund JL, Stürmer T, Poole C, Simpson RJ, Edwards JK. Reweighting Oranges to Apples: Transported RE-LY Trial Versus Nonexperimental Effect Estimates of Anticoagulation in Atrial Fibrillation. Epidemiology. 2020;31:605–13.
35. Spiegelman D, Carroll RJ, Kipnis V. Efficient regression calibration for logistic regression in main study/internal validation study designs with an imperfect reference instrument. Stat Med. 2001;20:139–60.
36. Wacholder S, Armstrong B, Hartge P. Validation studies using an alloyed gold standard. Am J Epidemiol. 1993;137:1251–8.
37. Spiegelman D, Schneeweiss S, McDermott A. Measurement error correction for logistic regression models with an "alloyed gold standard." Am J Epidemiol. 1997;145:184–96.
38. Wong MY, Day NE, Bashir SA, Duffy SW. Measurement error in epidemiology: the design of validation studies I: univariate situation. Stat Med. 1999;18:2815–29.
39. Rosner B, Spiegelman D, Willett WC. Correction of logistic regression relative risk estimates and confidence intervals for measurement error: the case of multiple covariates measured with error. Am J Epidemiol. 1990;132:734–45.
40. Armstrong BG, Whittemore AS, Howe GR. Analysis of case-control data with covariate measurement error: Application to diet and colon cancer. Stat Med. 1989;8:1151–63.
41. Carroll RJ, Stefanski LA. Approximate quasi-likelihood estimation in models with surrogate predictors. J Am Stat Assoc. 1990;85:652–63.
42. VanderWeele TJ, Li Y. Simple Sensitivity Analysis for Differential Measurement Error. Am J Epidemiol. 2019;188:1823–9.

43. Shu D, Yi GY. Causal inference with measurement error in outcomes: Bias analysis and estimation methods. Stat Methods Med Res. 2019;28:2049–68.
44. Shaw P, He J, Shepherd B. Regression calibration to correct correlated errors in outcome and exposure. Statistics in Medicine. 2021;40:271–286.
45. Edwards JK, Cole SR, Westreich D. All your data are always missing: incorporating bias due to measurement error into the potential outcomes framework. Int J Epidemiol. 2015;44:1452–9.
46. Howe CJ, Cain LE, Hogan JW. Are all biases missing data problems? Curr Epidemiol Rep. 2015;2:162–71.
47. Greenland S, Pearl J, Robins JM. Causal diagrams for epidemiologic research. Epidemiology. 1999;10:37–48.
48. Greenland S, Finkle WD. A Critical Look at Methods for Handling Missing Covariates in Epidemiologic Regression Analyses. Am J Epidemiol. 1995;142:1255–64.
49. Donders ART, van der Heijden GJMG, Stijnen T, Moons KGM. Review: a gentle introduction to imputation of missing values. J Clin Epidemiol. 2006;59:1087–91.
50. White IR, Carlin JB. Bias and efficiency of multiple imputation compared with complete-case analysis for missing covariate values. Stat Med. 2010;29:2920–31.
51. White IR, Royston P, Wood AM. Multiple imputation using chained equations: Issues and guidance for practice. Stat Med. 2011;30:377–99.
52. Little RJ, Rubin D. Statistical analysis with missing data, 2nd ed. New York: Wiley; 2002.
53. Hayati Rezvan P, Lee KJ, Simpson JA. The rise of multiple imputation: a review of the reporting and implementation of the method in medical research. BMC Med Res Methodol. 2015;15:1–14.
54. Van Buuren S. Flexible imputation of missing data. CRC press; 2018.
55. Stuart EA, Azur M, Frangakis C, Leaf P. Multiple imputation with large data sets: a case study of the Children's Mental Health Initiative. Am J Epidemiol. 2009;169:1133–9.
56. Raghunathan TE, Lepkowski JM, Van Hoewyk J, Solenberger P. A multivariate technique for multiply imputing missing values using a sequence of regression models. Surv Methodol. 2001;27:85–96.
57. Van Buuren S, Brand JP, Groothuis-Oudshoorn CG, Rubin DB. Fully conditional specification in multivariate imputation. J Stat Comput Simul. 2006;76:1049–64.
58. Arnold BC, Castillo E, Sarabia JM. Conditionally Specified Distributions: An Introduction (with comments and a rejoinder by the authors). Stat Sci. 2001;16:249–74.
59. Van Buuren S. Multiple imputation of discrete and continuous data by fully conditional specification. Stat Methods Med Res. 2007;16:219–42.
60. Rubin DB. Multiple imputation for nonresponse in surveys. Vol. 81. John Wiley & Sons; 2004.
61. Lachin JM. Fallacies of last observation carried forward analyses. Clin Trials. 2016;13:161–8.
62. Biering K, Hjollund NH, Frydenberg M. Using multiple imputation to deal with missing data and attrition in longitudinal studies with repeated measures of patient-reported outcomes. Clin Epidemiol. 2015;7:91.
63. Tyndall MW, Ronald AR, Agoki E, Malisa W, Bwayo JJ, Ndinya-Achola JO, et al. Increased risk of infection with human immunodeficiency virus type 1 among uncircumcised men presenting with genital ulcer disease in Kenya. Clin Infect Dis. 1996;23:449–53.
64. Lash TL, Silliman RA, Guadagnoli E, Mor V. The effect of less than definitive care on breast carcinoma recurrence and mortality. Cancer. 2000;89:1739–47.

Chapter 11
Bias Analysis Using Bayesian Methods

Introduction

Chapter 2 introduced Bayesian inference and alluded to the close relationship between probabilistic bias analysis and Bayesian methods. The probabilistic bias analysis methods that were discussed in Chapters 7–9 can be considered approximately Bayesian methods. That is, inferential results obtained from a probabilistic bias analysis that adjusts for a bias would typically be very similar, or even indistinguishable, from inferential results obtained from a corresponding Bayesian analysis [1, 2]. If results are typically so similar, it is natural to wonder when analysts might want to use formal Bayesian inference for quantitative bias analysis. In many cases the results from both approaches will be similar. In some settings, however, a formal Bayesian inferential approach may result in bias-adjusted estimates that are quite different from the probabilistic bias analysis approaches that we have discussed so far. Although there has not been a formal theoretical examination of exactly when the two approaches differ to an appreciable extent, there is some literature we can use to identify when this might occur. We highlight three common situations when one might want to use Bayesian inference below.

As we have described in Chapter 6, when bias-adjusting for misclassification using simple quantitative bias analysis by relating the observed data to the bias-adjusted data using sensitivity and specificity, it is possible to obtain negative bias-adjusted cell counts. Recall that the bias-adjusted estimate of the number of exposed cases (A) can be obtained from this equation, which we presented in Chapter 6:

$$A = \frac{a - D_{1,total}(1 - Sp_1)}{Se_1 - (1 - Sp_1)} \tag{11.1}$$

where a is the number of observed exposed cases, $D_{1,\,total}$ is the total number of cases, Sp_1 is the specificity of exposure classification among cases, and Se_1 is the sensitivity of exposure classification among cases. Both the numerator and the

M. P. Fox et al., *Applying Quantitative Bias Analysis to Epidemiologic Data*, Statistics for Biology and Health, https://doi.org/10.1007/978-3-030-82673-4_11

denominator of this equation could yield a negative value, and if either one is negative (but not both), the bias-adjusted estimate of the number of exposed cases will be negative, which is an impossible result. Indeed, we give bounds on the sensitivity and specificity values in Chapter 6 and note that values assigned to these bias parameters outside of those bounds will result in a negative bias-adjusted cell count. These negative cell counts are an indication of an incompatibility between the observed data and some portion of the bias distribution, and the analyst should endeavor to understand that incompatibility, as we discussed in Chapter 8. When specifying distributions for sensitivity and specificity for use in probabilistic bias analysis, if there is only a slight discrepancy between the specified distributions for the bias parameters and the observed data then the analysis can proceed. On the other hand if, for example, values of specificity less than 90% are impossible given the data but we have defined our distribution to include values for specificity as low as 70%, this may indicate a flaw in either our data or in our prior knowledge, and both are worth exploring before proceeding. In other words, the incompatibility of the data with the prior belief about the bias parameters is providing useful modeling information. It may also be an indication that conventional probabilistic bias analysis techniques may not work well in these settings, and Bayesian methods may be more useful because they allow updating of the distributions assigned to the bias parameters to resolve the incompatibilities.

Turning to Bayesian methods as a panacea for incompatibility between values assigned to bias parameters and data is poor practice if it is implemented in lieu of understanding the underlying incompatibility. However, in more complicated bias analyses with many confounders and possibly multiple types of bias, it may be difficult for an analyst to know if they are nearing a situation in which probabilistic bias analysis would not be a good approximation for a formal Bayesian analysis. For example, in a crude model, Equation 11.1 could result in an overall negative cell count. However, if we have multiple confounders then, roughly speaking, Equation 11.1 is applied to many strata resulting in many chances for negative cells. In instances where negative cell counts are an issue, Bayesian methods are a more suitable approach than probabilistic bias analysis as the former formally adapts the prior distribution in light of the observed data.

Further, Bayesian analysis allows users to more completely and formally incorporate substantive prior information into their analysis. Quantitative bias analysis already incorporates prior information on bias parameters. Bayesian inference can also incorporate prior information on the occurrence of the outcome, the effect of the exposure on the outcome, and the effect of measured confounders. There is a large literature on the potential benefit of incorporating this prior information relative to typical frequentist inference [3–6]. Incorporating credible prior information can lead to estimates with lower mean squared error, relative to estimates generated in the absence of prior information, and provides a final estimate of effect that incorporates not only the current study but relevant prior information as well [3, 7]. Analysts should be careful, however, when incorporating prior information in situations where there is likely bias. If studies used to inform the prior distribution are subject to bias as well, then a Bayesian analysis that incorporates that distribution could lead to even more biased estimates than an analysis that ignores the prior information. In practice, if an analyst is worried that the prior literature is also biased (perhaps due to publication bias or routine exposure misclassification in the prior literature), they would be wise to also perform a systematic review and bias analysis as part of a meta-analysis of the prior literature before incorporating information from it in any Bayesian analysis.

Perhaps the most useful aspect of Bayesian inference in probabilistic bias analysis, as alluded to above, is the ability to accommodate more complicated models. As described below, implementing Bayesian analyses will often rely on Markov chain Monte Carlo (MCMC) techniques. While MCMC techniques may be somewhat more difficult to implement than a probabilistic bias analysis (which typically uses Monte Carlo techniques, but not a Markov chain), they also offer the ability to fit data to more complicated model structures that might otherwise be extraordinarily complicated [8, 9]. For example, Lian et al. (2019) used a Bayesian modeling approach to simultaneously model a meta-analysis of the association between a misclassified exposure and its gold standard as well as a meta-analysis of the misclassified exposure and disease, while also accounting for heterogeneity between studies [10].

Regardless of the rationale for implementing a formal Bayesian analysis, the practical implementation of Bayesian models requires tools that are usually unfamiliar to most epidemiologists. In this chapter, we present a brief introduction to the aspects of Bayesian inference that are most relevant to quantitative bias analysis and discuss Bayesian approaches to adjusting for measurement error, selection bias and uncontrolled confounding bias-adjustment. Our implementation of these models will require MCMC techniques and we provide a brief description of these techniques as well as software capable of performing them.

An Introduction to Bayesian Inference and Analysis

Chapter 2 included a brief introduction to Bayesian inference, but to implement a Bayesian quantitative bias analysis a more thorough introduction is required for foundational knowledge. Many excellent textbooks have been written on this topic and our purpose here is to provide a relatively basic introduction of concepts necessary for implementing bias analysis [7, 11–14]. A simple example will suffice to explain the fundamental aspects of Bayesian methods and inference. Chung et al. estimated the prevalence of depressive symptoms among a cohort of low-income pregnant and postpartum women [15]. The authors administered the Center for Epidemiologic Studies Depression (CES-D) scale to assess depressive symptoms to 774 women attending Philadelphia public health centers. Women who scored greater than or equal to 16 were categorized as having depressive symptoms. As the authors noted, the CES-D score is not perfectly predictive of a person actually having clinical depression. Indeed, the authors noted that the sensitivity was estimated to be 95% while the specificity was estimated to be 70%. Rather than estimate the prevalence of depressive symptoms, a misclassified surrogate of depression, it may be more useful to formally incorporate the sensitivity and specificity to estimate the prevalence of clinical depression in the population, which we can accomplish using Bayesian methods.

Bayesian analyses consist of three key components: the likelihood, the prior distribution and the posterior distribution. The likelihood is the combination of the observed data with the model representing the process by which the data are assumed to arise. It is also used in frequentist data analysis. It is a function of the data conditional on unknown parameter values. For a dichotomous variable like clinical depression, a binomial likelihood is the most common model for the data. The data themselves consist of the total number of participants (N=774) as well as the number who tested positive by CES-D score ≥ 16 (x=372) giving the following likelihood function:

$$L(N, x|p) = \text{binomial}(N, x|p) = \binom{N}{x} p^x (1 - p)^{N-x}$$

$$= \binom{774}{372} p^{372} (1 - p)^{402} \tag{11.2}$$

where p is the prevalence of the misclassified outcome, depressive symptoms measured by CES-D score and $\binom{N}{x}$ is the binomial coefficient, $\frac{N!}{x!(N-x)!}$. A typical approach to analyzing these data from a frequentist perspective is to choose the value of p that maximizes the likelihood function for these data, which occurs at $\widehat{p}$=372/ 774$\approx$0.48.

The likelihood function in Equation 11.2 will suffice to estimate the prevalence of the misclassified outcome, depressive symptoms; however, if the goal is to estimate the prevalence of clinical depression, it is useful to re-express Equation 11.2 in terms of the true prevalence we wish to estimate [16, 17]. The misclassified prevalence, p, can be expressed as a function of the true prevalence, π, using the following equation:

$$p = \pi Se + (1 - Sp)(1 - \pi) \tag{11.3}$$

where Se and Sp are the sensitivity and specificity of classification, respectively. This equation is a re-expression of bias models from Chapter 6, a=A(SE_{D1}) + B(1 − SP_{D1}), with both sides of the equation divided by the sample size, N. We can express the likelihood of the model in terms of the parameter we wish to estimate by substituting Equation 11.3 into Equation 11.2:

$$L(N, x|p) = \binom{774}{372} [\pi Se + (1 - Sp)(1 - \pi)]^{372} [1 - (\pi Se + (1 - Sp)(1 - \pi))]^{402}$$

$$\tag{11.4}$$

If we were not interested in Bayesian inference, and we assume a fixed sensitivity (Se=0.95) and specificity (Sp=0.70), we could use the likelihood function in Equation 11.4 to obtain a maximum likelihood estimate, $\hat{\pi}$, by finding the value of π that maximizes the function. For these data, the maximum likelihood estimate occurs at $\hat{\pi} \approx 0.28$.

However, suppose an analyst knows there exists prior knowledge on the true prevalence of depression and she believes that incorporating that knowledge could be beneficial. In Bayesian inference, this substantive knowledge is represented by the prior distribution. Indeed, in a Bayesian analysis all unknown parameters must have prior distributions assigned to them. In this example, the unknown parameters are π, Se, and Sp; however, we will initially assume that both sensitivity and specificity are known with certainty to simplify the explanation, leaving π as the only unknown parameter. Assigning prior distributions to unknown parameters requires careful consideration; if the distributions are poorly chosen (in the sense of being at odds with the true state of nature), the resulting Bayesian analysis could be worthless. A vast literature exists on how to specify distributions in Bayesian analyses [3, 4, 7, 18, 19]. For purposes of bias analysis, we believe subjective Bayesian inference has the best logical basis. In subjective Bayesian inference, prior distributions should represent the belief an analyst has in the magnitude of the unknown parameter as well as the uncertainty in that unknown parameter. The median of the prior distribution represents the parameter value (in this case, the true prevalence of disease) that the analyst believes is most likely. The spread of the distribution represents the uncertainty of that belief. For example, the analyst might specify they are 95% certain the true value lies between the 2.5th and 97.5th percentiles of their prior distribution. The believability and utility of a Bayesian analysis will rest, to a large extent, on the plausibility of these prior distributions. Subjective specification of prior distributions should not be conflated with arbitrary specification or specification based on wishful thinking. Analysts should use the literature to inform prior distributions before seeing their study results. Assigning a prior distribution in a Bayesian analysis is exactly analogous to the discussion of assignment of bias parameter distributions in Chapters 7–10. The information to inform the prior distribution can come from expert opinion or, usually better, from internal sources (such as an internal validation study) or external sources (such as previous literature), or from combinations of all of these. In this example, we will use different prior distributions to demonstrate their impact on study results. A beta distribution (discussed in Chapter 7) is a convenient choice for priors on probabilities since the distribution is very flexible, is naturally bounded by 0 and 1, can be empirically defined when data are available, and has appealing mathematical properties when combined with a binomial likelihood:

$$f(\pi) = \text{beta}(\pi|\alpha,\beta) = C(\alpha,\beta)\pi^{\alpha-1}(1-\pi)^{\beta-1} \qquad (11.5)$$

where $C(\alpha,\beta)$ is a constant (based on gamma functions), with respect to π, that ensures the density integrates to 1. For purely pedagogical purposes, we initially assume that we have no knowledge about the prevalence of clinical depression among low-income mothers and we believe every prevalence from 0 to 1 is equally likely. We emphasize that this belief is completely implausible, and we will alter this specification later. A convenient way to specify a uniform distribution is by specifying a beta($\alpha=1,\beta=1$), which simplifies Equation 11.5 to $f(\pi) = C(\alpha,\beta)=1$. That is, each value of π is equally likely, with constant density, which is identical to a uniform distribution, as we discussed in Chapter 7.

The prior distribution and likelihood are combined to produce the posterior distribution, which is a mathematical expression that represents subjective belief about the distribution of the parameter estimate when we combine the study data and modeling assumptions for that data (the likelihood) along with prior evidence and modeling assumptions for that prior evidence (the prior distribution).

The posterior distribution is calculated by use of Bayes' theorem. The version of Bayes' theorem for discrete parameters was presented in Chapter 6 and is commonly used in epidemiology to translate between sensitivities and specificities and positive and negative predictive values. For instance, if T is an imperfect dichotomous test of true disease status D and we wish to convert a sensitivity and specificity to a positive predictive value once we have seen a positive test, $D = 1$:

$$\Pr(D = 1|T = 1) = \frac{\Pr(T = 1|D = 1)\Pr(D = 1)}{\Pr(T = 1|D = 1)\Pr(D = 1) + \Pr(T = 1|D = 0)\Pr(D = 0)}$$

In Bayesian terms, $\Pr(D = 1)$, is the prior distribution (the probability of a patient having the disease prior to seeing any test results). The sensitivity and specificity are both part of the likelihood function. They determine the probability of seeing the data (the test result, T) given the unknown true parameter D (disease status). The positive predictive value is the posterior distribution, the probability of the unknown parameter (disease status) now that we have collected data (test result) and combined it with our prior knowledge of disease prevalence. We can rewrite Equation 11.6 slightly to show that the denominator is simply a summation of all the ways the data ($T=1$) could be obtained under every true parameter value ($D=1$ or $D=0$):

$$\Pr(D = 1|T = 1) = \frac{\Pr(T = 1|D = 1)\Pr(D = 1)}{\sum_d \Pr(T = 1|D = d)\Pr(D = d)}$$

In general, our parameters will be continuous rather than discrete. Indeed, in the depression screening example our parameter is π, the proportion of women who truly have clinical depression, which can take any value between 0 and 1. We cannot sum the denominator over an infinite number of values for π; however, we can modify Bayes' theorem by switching the summation in the denominator to an integration:

$$f(\pi|x) = \frac{f(x|\pi)f(\pi)}{\int f(x|\pi)f(\pi)d\pi} \tag{11.7}$$

where $f(\pi)$ is the prior distribution, $f(x|\pi)$ is the likelihood and $f(\pi|x)$ is the posterior distribution.

The goal is to find the posterior distribution of the prevalence of true clinical depression, π, given the data we have observed using an imperfect depression screening test. To use Equation 11.7, we need a likelihood and a prior. The likelihood of the data, given the screening test's sensitivity and specificity, $f(x|\pi)$, was derived in Equation 11.4. For a simple prior distribution, we will assume every value of π is equally likely, $f(\pi) = beta(1, 1) = C(\alpha, \beta)$. We can insert the likelihood and the prior distribution directly into Bayes' theorem in Equation 11.7 to obtain:

$$f(\pi|x) = \frac{f(x|\pi)f(\pi)}{\int f(x|\pi)f(\pi)d\pi}$$

$$= \frac{\binom{774}{372}[\pi Se + (1 - Sp)(1 - \pi)]^{372}[1 - (\pi Se + (1 - Sp)(1 - \pi))]^{402}C(\alpha, \beta)}{\int_0^1 \binom{774}{372}[\pi Se + (1 - Sp)(1 - \pi)]^{372}[1 - (\pi Se + (1 - Sp)(1 - \pi))]^{402}C(\alpha, \beta)\, d\pi}$$

$$= \frac{\binom{774}{372}[\pi Se + (1 - Sp)(1 - \pi)]^{372}[1 - (\pi Se + (1 - Sp)(1 - \pi))]^{402}}{\int_0^1 \binom{774}{372}[\pi Se + (1 - Sp)(1 - \pi)]^{372}[1 - (\pi Se + (1 - Sp)(1 - \pi))]^{402}\, d\pi}$$

$$(11.8)$$

Note that because the prior distribution is a constant, $C(\alpha, \beta)$, it cancels out of the numerator and denominator for this simple example. This cancellation will not generally occur. Calculating the posterior distribution in Equation 11.8 involves an integration in the denominator for which there is, outside of trivial examples, no closed-form solution. When the posterior cannot be computed in closed-form, a general solution is to turn to MCMC techniques to estimate the posterior distribution. MCMC techniques avoid the impossible integration that lurks in the denominator of Equation 11.7. Instead, MCMC algorithms are designed to randomly sample from the posterior distribution even if that distribution cannot be written down in closed form, a remarkable achievement of mathematical statistics! The series of random samples that is drawn is referred to as a chain. Many MCMC algorithms have been developed and technical details of these algorithms and requisite assumptions are available elsewhere and are beyond the scope of this book [7, 12]. Our purpose is to gain the familiarity necessary for epidemiologists to implement MCMC algorithms in quantitative bias analyses. We begin by implementing a simple MCMC algorithm, called the Metropolis algorithm, to provide necessary intuition for implementing more complicated MCMC algorithms in realistic bias analyses [20]. Readers may be more familiar with the name Metropolis-Hastings algorithm, which is a generaliza-tion of the original Metropolis algorithm [21].

The Metropolis algorithm was the first MCMC algorithm and remains the easiest to explain [20, 21]. The Metropolis algorithm is designed to draw random samples of the parameter from the space of all possible parameter values. In our example, that means the Metropolis algorithm samples the parameter π from values between 0 and 1 (the possible parameter space). The algorithm accomplishes this by taking a "random walk" through the parameter space. The 'steps' of this random walk are the random samples from the parameter space and produce the "Markov chain." The algorithm starts at any location in the parameter space and then randomly steps to another location in the parameter space (the first step). It then steps to a new location in the parameter space (the second step). It would do us no good, however, if the algorithm simply took completely random steps as this would simply return a uniform distribution. Instead, the Metropolis algorithm was created to take a directed random walk over the posterior parameter space, which makes taking a step toward areas of higher posterior probability more likely.

We will implement a Metropolis algorithm to draw samples from the posterior distribution in Equation 11.8. This random walk must start somewhere, and this initial value is chosen either by the analyst or by the software package. In theory, the initial value does not matter and MCMC algorithms are guaranteed to find the posterior distribution eventually, provided that the posterior distribution actually exists. In practice, for complicated models, it can take a very long time for the MCMC algorithm to find the posterior distribution; providing an initial value near the posterior can hasten the analysis substantially. We will start our random walk at a value of $\pi=0.999$, a value that will turn out to be quite far from the center of the posterior distribution. To begin the process of deciding where to take the first "step" in this random walk, we must draw a new parameter value. We will sample our new values from a normal distribution with some mean and standard deviation. Since this distribution is giving us proposed steps to take, it is referred to as the proposal distribution. The mean of the proposal distribution will always be the current prevalence (in our example, 0.999). We must also specify a standard deviation for this proposal distribution. We start with a standard deviation of 0.1 and offer more guidance on choosing a value for this (tuning) parameter later. That each sample is drawn conditional on the preceding sample is what makes this a Markov chain: a sequence has a Markovian property in statistics if it only depends on the previous value and is independent of all other preceding values in the chain. We now sample a proposal value from the proposal distribution, $N(mean = 0.999, sd = 0.1)$. If the sampled value lies outside the 0–1 interval, it is discarded. In this case, we sample a value of 0.943. At this point we have an initial value ($\pi=0.999$), from which we started, and a proposal ($\pi=0.943$). Our algorithm is not guaranteed to take a step to a new parameter value. Indeed, if the parameter value is more incompatible with the posterior distribution than the current parameter value, our algorithm could be unlikely to take a step in that direction. After the proposed value is drawn, we have to decide whether to take a step to the new value or not. The probability of accepting the proposed value and taking a step is computed as the ratio of the posterior distribution in Equation 11.8 under the proposed and current prevalence values. Note that the denominator of Equation 11.8 does not depend on the prevalence, since it has been integrated out. Thus, when computing the ratio under the proposed and current prevalence values, the denominator cancels out. Notice that this allows us to avoid the complexity we initially faced of how to compute the integral in the denominator of Equation 11.8 and the acceptance probability is simply the likelihood times the prior (given the proposed value in the numerator) and the likelihood times the prior (given the current value in the denominator):

$$pr_{acc} = \min \left(1, \frac{f(\pi^{proposed}|x)}{f(\pi^{current}|x)} \right)$$

$$= \min \left(1, \frac{f(x|\pi^{proposed})f(\pi^{proposed})}{f(x|\pi^{current})f(\pi^{current})} \right)$$

(11.9)

Note that the likelihood and prior are easily computed by any modern statistical software package, making this acceptance probability relatively simple to calculate. Further, we have initially specified that the prior distribution is uniform; that is $f(\pi)=$ beta($\alpha=1$, $\beta=1$)$=C(\alpha, \beta)$. This distribution is constant with respect to π, it is always equal to $C(\alpha, \beta)$, and therefore cancels out of the numerator and denominator in Equation 11.9. The acceptance probability in this case is simply the ratio of the likelihood under the proposed prevalence to the current prevalence:

$$pr_{acc} = \min \left(1, \frac{f(x|\pi^{proposed})}{f(x|\pi^{current})} \right)$$

The Metropolis algorithm takes a "step" to the proposed value with probability$=$ pr_{acc}. If the proposed value has higher posterior probability than the current value, we always take a step to that value. However, even if it has a smaller probability than the current value, we have a chance of taking a step in that direction. The steps to high probability proposals mean that we will take a directed walk toward the highest probability regions of the posterior distribution and the less likely steps to lower probability proposals guarantee that we will also explore the rest of the posterior distribution. This acceptance probability dictates the behavior of the Metropolis algorithm, with the MCMC chain of prevalence values moving, on average, toward more likely values. Although we start at a very unlikely posterior prevalence of 0.999, we rapidly move toward more likely values of prevalence. In this example, the first proposed prevalence in an example chain had a value of 0.943 and was accepted because it was vastly more compatible with the posterior distribution than a prevalence of 0.999. To see this, we use the likelihood in Equation 11.4 to estimate $f(x|\pi^{proposed}) = 1.9 \times 10^{-210}$ and $f(x|\pi^{current}) = 1.1 \times 10^{-298}$. Although both are very unlikely values, the proposed value has a much higher relative probability.

After the first step, the newly accepted value (0.943) became the current value in the example chain, resulting in a chain$=$(0.999, 0.943). We then sample a new proposed value from the proposal distribution, centered on the current value: N (mean$=$0.943, sd$=$0.1). In this case, the proposed value was 0.774, which again was far more compatible with the posterior distribution than the current value, resulting in $pr_{acc}=1$ and a step from 0.943 to 0.774. The example Markov chain now consisted of three values: (0.999, 0.943, 0.774). In the next step of the Metropolis algorithm, 0.774 was the current value and a proposal of 0.787 was drawn from the distribution N(mean$=$0.774, sd$=$0.1). Because this proposed value was more extreme and further from the posterior distribution, its acceptance probability was close to zero, so at the 3rd iteration of this algorithm, no step was taken, resulting in an example chain of (0.999, 0.943, 0.774, 0.774). Subsequent steps in the algorithm mimicked the first steps: a proposal value was drawn from a normal distribution centered on the current value and either accepted or rejected based on how well it corresponded to the posterior distribution.

We chose to implement the Metropolis algorithm for 10,000 iterations, which took approximately 0.2 seconds to run and generated an example chain too long to list. Analyses are performed in R and code is provided at the text's website. After all iterations have been completed, it is typical to examine a trace plot (plot of posterior samples generated in the random walk) of the Markov chain (Figure 11.1). It is clear from this graph that after the first few proposed values have been accepted, the Markov chain remains in the same part of the sample space (roughly between a prevalence of 0.20 and 0.35). This pattern is an excellent example of a chain converging to the posterior distribution. The chain quickly moves to a single region of the sample space and randomly explores that area with no noticeable trend over the subsequent iterations. The first few samples in the chain, where it has not yet converged on the posterior distribution, are referred to as the burn-in period. Because these are not samples from the posterior distribution, they are discarded. There is no appreciable penalty for erring on the side of caution, so a large burn-in period is typically chosen. In this case, discarding the first 1,000 iterations (or even the first 100) is more than sufficient.

Although the trace plot in Figure 11.1 provides evidence that the chain has converged to the posterior distribution, it is not a guarantee. In more complicated models, which might have posterior distributions that have multiple modes, it is possible for a Markov chain to get "stuck" near a local mode without ever finding the global mode with the highest density. For this reason, it is good practice to start multiple Markov chains from different initial values to see if they all converge to the same posterior distribution. When we have many variables in the model, it can be difficult to specify initial values in a way suitable to convince ourselves we have

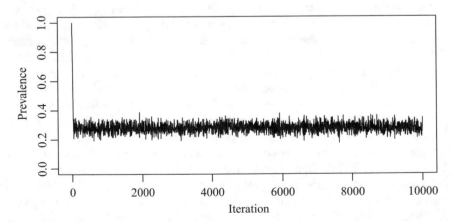

Figure 11.1 Trace plot of posterior values drawn from a Metropolis MCMC algorithm to explore the posterior density of the misclassification-adjusted prevalence of depression.

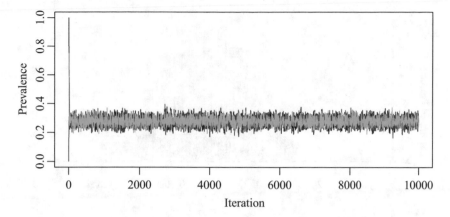

Figure 11.2 Trace plots for Metropolis algorithm estimating the prevalence of depression using four different starting points. Each Markov chain is a different color, with chains started from initial values of 0.999 (black), 0.001 (red), 0.25 (blue) and 0.75 (green).

explored the entire posterior distribution. In lower dimensional problems such as this one, however, it is relatively easy. Figure 11.2 shows trace plots of four separate Metropolis chains started from initial values of 0.999, 0.001, 0.25 and 0.75. The trace plots show that regardless of starting location, the chains quickly converge to the same posterior distribution. More information on diagnosing convergence can be found in Chapter 11 of Bayesian Data Analysis [12].

In this example of estimating the prevalence of clinical depression in pregnant women, we started four Markov chains from different locations and are comfortable that the chains have converged. Since all four chains are random samples from the same posterior distribution, we can combine the four chains, after excluding the burn-in period of 1000 samples from each chain, and treat all $9000 \cdot 4 = 36,000$ values as draws from the posterior distribution. A histogram of these 36,000 samples is shown in Figure 11.3 with a superimposed kernel density smoothed estimate. The histogram and kernel smoothed estimate show some jitteriness in the distribution, but probably not to an extent that would prove problematic in making inferences about the prevalence of depression in this population. MCMC methods are referred to as "exact" methods, in the sense that they can represent the exact posterior distribution of interest, if the chain is run for a long enough period of time. In this case, if the histogram is not smooth enough, or if we are concerned about whether the chain has converged, it would be straightforward to run the chain until the desired smoothness or convergence has been obtained.

Once we have completed the Metropolis algorithm and are convinced it has converged, our interest will be in reporting statistics based on the posterior distribution, typically the posterior median and posterior 95% credible interval. The Bayesian credible interval is a measure of precision, analogous to a frequentist confidence interval. A Bayesian credible interval will not, however, typically have the same properties as a frequentist confidence interval. For example, a series of Bayesian 95% credible intervals, produced from exactly replicated and perfectly

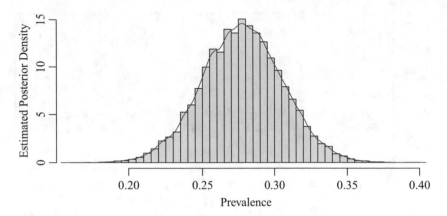

Figure 11.3 Posterior distribution of the prevalence of depression. A Metropolis algorithm was used to draw samples from the posterior distribution. Boxes are draws from the posterior that have been normalized to fit under the kernel density estimate of those draws (solid red line).

conducted randomized trials, will not necessarily contain the true parameter value 95% of the time, as would be expected from 95% frequentist confidence intervals computed from the same data. Rather, the 95% credible interval can be translated into a subjective statement regarding our uncertainty about the magnitude of the effect. How well our uncertainty aligns with the reality will depend on our prior belief. If our prior belief is accurate (in the sense that it is "close" to the truth), Bayesian 95% credible intervals perform quite well. On the other hand, if our prior belief is inaccurate (in the sense of being a bad guess at the truth) our Bayesian estimates could perform quite poorly, at least compared with a frequentist perspective. Thus, it is important to pay careful attention to specification of prior distributions.

In our example, now that we have completed the Metropolis algorithm, it is simple for us to characterize the posterior distribution. For the four Markov chains that we ran, we can compute the median prevalence estimate and 95% credible intervals (using the 2.5th and 97.5th percentiles of the distribution) as: $\pi = 0.28$; 95 % CrIn $(0.22, 0.33)$. We can compare this point estimate to the maximum likelihood estimate we found earlier, $\hat{\pi} = 0.28$, and we see that they are the same. This is not an unexpected result. We placed an uninformative prior on π and it is well known that, in relatively simple models, frequentist results and Bayesian results will be identical when the Bayesian model uses uninformative priors. A few technical and implementational notes are warranted.

First, note that the Bayesian parameter does not include a circumflex while the frequentist parameter does. In the frequentist world view the parameter, π, is a fixed (not random) quantity and the circumflex signifies that an estimator, $\hat{\pi}$, is the random quantity. In Bayesian inference, however, all parameters are random and there is no reason to adorn them with bonnets to help us remember they are random. Second, the interpretation of a Bayesian credible interval requires care (as does interpretation of a frequentist confidence interval). A subjective Bayesian interpretation of these results is that our best guess is the prevalence of clinical depression is 28% and we are 95% certain that the true prevalence lies between 22% and 33%. The validity of this statement rests, in part, on the assumption that the prior distribution truly represents the analyst's prior belief. If it does not, this subjective interpretation is meaningless. We placed a uniform, beta($\alpha=1$, $\beta=1$), prior on the prevalence of disease. This implies we believe every prevalence from 0 to 1 is equally likely, which we do not. As such, a subjective interpretation of this credible interval is meaningless. We chose this prior to ease our explanation of the Metropolis algorithm. We also note that the validity of the posterior distribution also relies on accurately specifying the likelihood (a binomial distribution) and that no other sources of error exist, as do frequentist analyses. Third, different analysts can and should use different prior distributions, resulting in different posterior distributions. We view this as a strength of the Bayesian approach as it allows natural and transparent sensitivity analyses based on scientists' beliefs. Fourth, the purpose of this chapter is the use of Bayesian methods in probabilistic bias analysis. It is commonly the case that prior distributions are placed on bias parameters (very similar to the bias parameter distributions discussed in Chapters 7–9). Often other parameters (such as the main effect of exposure on outcome) will have non-informative priors placed on them but there is no reason this has to be the case, an analyst could place a substantively informed prior distribution on the main effect. If the analyst did this, a subjective interpretation of the 95% credible intervals could be used; "We are 95% confident the truth lies between these two points, after incorporating what was known in the literature and adjusting for misclassification bias." However, most of the literature does not do this and our examples in this chapter follow common practice in not placing informative priors on main effects. There are clearly situations in which incorporating prior information on the main effect will be quite beneficial, however, and analysts should seriously consider this approach. For example, consider a situation in which an exposure increases risk of an outcome, albeit slightly, and the true RR=1.05. A study is conducted and the observed $RR_{obs} = 0.98$ and, furthermore, there is nondifferential exposure misclassification. An almost null protective effect was observed by random chance and when we adjust for nondifferential exposure misclassification, the bias-adjusted estimate will be made even more protective. Prior information on the RR, if it exists, could be extremely helpful in this scenario. If prior information existed that the effect of exposure increased risk, a Bayesian (non bias-adjusted) RR_{obs} could be greater than zero, allowing bias adjustment to estimate closer to the truth. For the reasons mentioned above, interpretation of 95% credible intervals should be very similar to the interpretation of simulation intervals from Chapters 8 and 9 and confidence intervals from Chapter 10: a range of parameters that are consistent with the data after we have adjusted for the bias, and conditional on all other modeling assumptions being correct.

In implementing Markov chains, the validity of the results rest on whether the chain has adequately sampled from the posterior distribution and this can be difficult to diagnose. It is not uncommon for Markov chains to exhibit high autocorrelation, which implies they are moving through the posterior parameter space very slowly. To demonstrate a scenario like this, we modify the Metropolis algorithm described

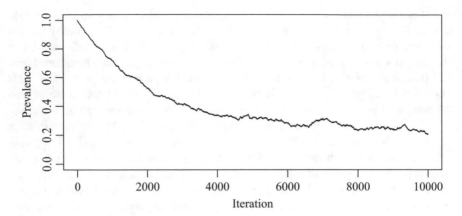

Figure 11.4 Trace plot of posterior values drawn from a Metropolis MCMC algorithm with very small proposal jumps, exploring the posterior density.

above to make much smaller steps, using a normal proposal density with a standard deviation of 0.001 (rather than 0.1). The result of this MCMC sampling algorithm is shown in Figure 11.4. The high autocorrelation is immediately obvious, as each parameter value is very close to the prior one. Since we already know that 95% of the posterior density lies between 0.22 and 0.33, we can see that this MCMC algorithm is, in fact, exploring that parameter space; however, it is doing so very, very slowly. We could explore the sample space adequately by running the algorithm even longer, but that could take an exorbitant amount of time. In this example, modifying the size of the Metropolis jumps (referred to as a tuning parameter) is an easy solution. In the following sections, we will not be programing our own MCMC algorithms. Instead, we will rely on standard software such as JAGS and STAN to implement the MCMC algorithms. These software packages attempt to find an optimal tuning parameter before beginning the Markov chain and will alert the user if they have trouble doing so.

    A principal benefit of Bayesian inference is the ability to coherently incorporate prior knowledge. Our first analysis of these data assumed that we had no a-priori belief that any value of π was more plausible than any other. In reality, there will typically be information to inform the distributions for these parameters and uninformative distributions will generally have little face validity. To illustrate how a more informative prior could change results, we specify a prior distribution of:

$$\pi \sim \text{beta}(\alpha = 10, \beta = 32)$$

which implies our prior belief is that the true prevalence of depression in this cohort is about 0.24 (10/(10+32)) with 95% prior intervals of: 0.12, 0.38. We repeat the Metropolis algorithm outlined above but note that the prior distribution is no longer constant and does not cancel out of the numerator and denominator in Equation 11.7. We programmed this Metropolis algorithm in R (posted at the text's website) and

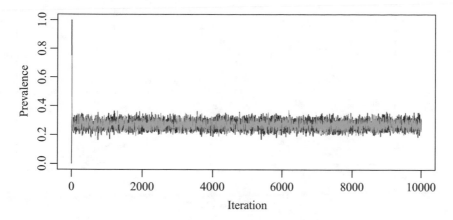

Figure 11.5 Trace plot from the four Markov chains implemented using a Metropolis algorithm. Each chain has a different color and is started from a different location, with chains started from initial values of 0.999 (black), 0.001 (red), 0.25 (blue) and 0.75 (green). This is an example of excellent convergence, with a random spread around a constant mean that does not increase or decrease over the iterations. All four chains appear to converge to exactly the same posterior distribution.

implemented four chains with different starting values for each chain, allowing each chain to run for 10,000 iterations. As before, we excluded the first 1,000 results as a burn-in. This model took 0.5 seconds to run. Convergence was determined by examining the traceplots (see Figure 11.5). There was no evidence that the traceplots were trending toward higher or lower values; rather, the Markov chain oscillated randomly around a mean. The posterior median prevalence was $\pi = 0.27$; 95% CrIn (0.22, 0.32). A smoothed density plot of the posterior samples with this second prior is shown in Figure 11.6, along with the smoothed density resulting from the uniform (uninformative) prior distribution in the first example. The posterior distribution derived from the informative prior has a smaller standard deviation (narrower credible interval) than the posterior distribution starting with the (uninformative) uniform prior. This result is to be expected, as the former result incorporates more prior information and bringing more information to the analysis should, intuitively, make posterior distributions more precise. We can also see that the posterior distribution is shifted slightly toward the mean of the prior distribution when we use an informative prior. This is a natural consequence of coherently incorporating prior knowledge via Bayesian inference: our posterior belief can be thought of as a weighted average of our prior belief and our observed data.

Convergence of a chain (or set of chains) to a posterior distribution is critically important; we can only trust the results of an analysis after we are satisfied that our chains have converged. Many diagnostics have been proposed to assess convergence. The first diagnostic we strongly suggest is running multiple chains and plotting traceplots as in Figures 11.2 and 11.5. These figures give examples of excellent convergence. Regardless of where the chain is started, they arrive at the same posterior distribution very quickly. An example of poor convergence is shown

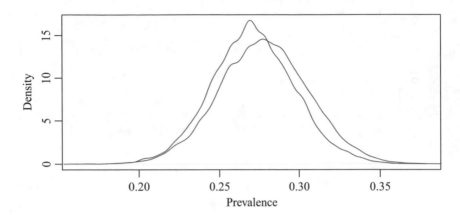

Figure 11.6 Smoothed posterior distribution of the prevalence of depression using a non-informative prior (blue line) and an informative prior (red line).

in Figure 11.4, where the chain seems to very slowly move to the posterior distribution. Examining traceplots can quickly highlight convergence difficulties. In addition, a number of statistics have been proposed to examine convergence. The Gelman-Rubin diagnostic monitors the parameter variability within and between each MCMC chain [12, 22, 23]. When this test statistic is greater than 1 it indicates that there is substantial variability between the chains and a larger number of iterations is necessary to achieve convergence. This statistic is easily computed in R. The effective sample size is a final useful diagnostic to assess convergence. The effective sample size takes the number of iterations in a Markov chain and divides it by a function of the autocorrelation within that chain to estimate the effective number of iterations that were independently sampled. For example, if 1,000 iterations were drawn and had perfect correlation (such that all samples had the same value), the effective sample size would be 1. If there were no autocorrelation, the effective sample size would be 1,000. If an analyst intended to draw 10,000 samples from the posterior distribution but the effective sample size indicates that only 1,000 effective independent samples were drawn, it would be beneficial to increase the total number of samples by a factor of 10 to obtain the intended number of samples [12].

Software for Bayesian Analyses

A wide variety of software packages are available for implementing Bayesian analyses for quantitative bias analyses. Standard epidemiologic software such as SAS and Stata have some ability to implement Bayesian analyses. However, in the opinion of the authors, JAGS (Just Another Gibbs Sampler) and Stan offer more flexible approaches to implementing MCMC methods [24, 25]. Each package implements a different type of MCMC. JAGS uses Gibbs sampling, which can be a relatively quick version of MCMC and resembles probabilistic bias analysis in that it samples parameter values in sequence, conditional on other parameters (that is, it samples a sensitivity and specificity and then samples an adjusted prevalence of exposure among the exposed, etc). Stan uses Hamiltonian Monte Carlo methods to update all parameters simultaneously. Either package can be called through R (as well as SAS or Stata). We opt to use R because of its excellent graphics, variety of packages that aid in the analysis of MCMC samples, and ability to hold multiple arrays in memory simultaneously. While either Stan or JAGS can be used for Bayesian bias analysis, we have found Stan to be preferable on many occasions. We have encountered several applied examples where JAGS took a considerable amount of time to converge to the posterior distribution, whereas Stan seemingly found the posterior distribution with little effort. Presumably, this is due to the fact that Stan updates all model parameters simultaneously, whereas JAGS updates them one at a time. The benefit of this speedy convergence is hard to overstate as one author of this text has spent far too many hours of his life toiling to find clever reparameterizations that would help models converge more quickly. For the examples in this chapter, we present results from Stan called through R. However, in the code posted on this text's website, we also include JAGS code. Indeed, since JAGS and Stan implement MCMC using different approaches, running models in both packages is an excellent sensitivity analysis.

Bayesian Adjustment for Exposure Misclassification

There exists a large literature on Bayesian approaches that adjust for misclassification. Interested readers are encouraged to read Gustafson (2004) for an extensive discussion of misclassification and measurement error [16]. Applied Bayesian approaches that adjust for misclassification in epidemiologic settings can be found in Chu et al. (2006), MacLehose et al. (2008) and Lesko et al. (2018) [17, 26, 27]. In this section we illustrate how to implement Bayesian adjustments for misclassification first using sensitivities and specificities and second using positive and negative predictive values. The applications are shown with regard to exposure misclassification and can be easily adapted to misclassification of an outcome or covariate.

Exposure Misclassification Using Sensitivities and Specificities

In Chapter 6, we described an approach to adjust for exposure misclassification in a case-control study examining the association between self-reported smoking during pregnancy, collected from birth certificate data, and breast cancer. In Chapter 8, we used probabilistic bias analysis to incorporate uncertainty about the bias parameters (sensitivities and specificities) into bias-adjusted effect estimates. We now demonstrate, using the same example, how to implement a Bayesian approach to exposure misclassification. It is important to note that there are many ways to implement these adjustments (see Chapter 8) that involve reparametrizing the model we present. However, we follow the general approach of Chu et al. because it is relatively easy to implement [17]. In the appendix at the end of this chapter we show that the misclassification approach from Chapter 8 is roughly analogous to a Bayesian approach.

To begin the Bayesian approach, we first specify the probability model by which the observed outcome data occur. Conceptually, case-control studies are efficient versions of cohort studies. Cases are information dense, so we usually include all of them. The persons and person-time giving rise to the cases is less information dense, so it is more efficient to sample controls from it than to include a census of it [28]. Statistically speaking, however, it is common to think of the exposure as the dependent variable in case-control studies. For misclassification models it is convenient to retain the statistical parameterization with the exposure as the dependent variable. Although there are various ways to model this data, including using the hypergeometric distribution, we represent this using a common statistical model:

$$a \sim \text{binomial}(N_1, P_1)$$
$$c \sim \text{binomial}(N_0, P_0)$$
(11.10)

where a and c are the number of cases and controls, respectively, who are classified as exposed, N_1 and N_0 are the total numbers of cases and controls, respectively and P_1 and P_0 are the proportions of cases and controls classified as exposed. The expressions in Equation 11.10 can be combined to form the likelihood for these data:

$$f(a, c | P_1, P_0) \propto \left[P_1{}^a (1 - P_1)^{N_1 - a} \right] \left[P_0{}^c (1 - P_0)^{N_0 - c} \right]$$
(11.11)

Note that we only write out the likelihood involving the random variables and that constants, such as the binomial coefficient in Equation 11.2, are omitted. As we saw with the Metropolis algorithm example, these constants generally play no part in calculations. The proportions P_1 and P_0 are easily estimated from the observed data and can be used to compute an estimated OR. However, since a and c are potentially misclassified, these estimates are biased. Our objective is to estimate π_1 and π_0, the proportions of cases and controls who are truly exposed. We did not directly observe π_1 and π_0, but we can relate them to quantities that we did observe, as in the previous section:

$$P_1 = \pi_1 Se_1 + (1 - Sp_1)(1 - \pi_1)$$
$$P_0 = \pi_0 Se_0 + (1 - Sp_0)(1 - \pi_0)$$
(11.12)

where Se_0 and Se_1 are the sensitivities among the controls and cases, respectively and Sp_0 and Sp_1 are the specificities among controls and cases, respectively. In this example, we assume non-differential misclassification such that $Se_0 = Se_1$ and

$Sp_0 = Sp_1$. Combining Equations 11.11 and 11.12 defines the likelihood in terms of the quantities of interest:

$$f(a, c | \pi_1, \pi_0, Se, Sp)$$
$$\propto \left[\{\pi_1 Se + (1 - Sp)(1 - \pi_1)\}^a (1 - \{\pi_1 Se + (1 - Sp)(1 - \pi_1)\})^{N_1 - a} \right] \quad (11.13)$$
$$\cdot \left[\{\pi_0 Se + (1 - Sp)(1 - \pi_0)\}^c (1 - \{\pi_0 Se + (1 - Sp)(1 - \pi_0)\})^{N_0 - c} \right]$$

Although this likelihood expresses the observed data in terms of the quantities of interest, these quantities are not formally identified from the data. That is, it is impossible to estimate π_1 and π_0 from the observed data without making additional assumptions. To estimate these quantities, we must place prior distributions on all unobserved quantities: π_1, π_0, Se, and Sp. Prior distributions on sensitivity and specificity have already been developed in Chapter 8 when describing probabilistic bias analysis. Indeed, bias parameter distributions are equivalent to prior distributions:

$$Se \sim \text{Beta}(\alpha = 50.6, \beta = 14.3)$$
$$Sp \sim \text{Beta}(\alpha = 70, \beta = 1) \quad (11.14)$$

Again, we only have one sensitivity distribution and one specificity distribution because we have assumed a nondifferential misclassification mechanism. If we had a differential misclassification mechanism, we would have had two sensitivity and two specificity distributions (and presumably a correlation between the respective distributions, see Chapter 7). We now need to place prior distributions on π_1 and π_0. If there is substantive information that can be used to inform these distributions, it may be worth doing so. However, to mirror common practice and the probabilistic bias analysis approach more closely, we will place uninformative priors on these parameters, as if we were ignorant about the true prevalence of exposure among cases and controls:

$$\pi_1 \sim \text{Beta}(\alpha = 1, \beta = 1)$$
$$\pi_0 \sim \text{Beta}(\alpha = 1, \beta = 1) \quad (11.15)$$

Expressions 11.13 to 11.15 complete the model specification for exposure misclassification. When discussing probabilistic bias analysis, we often presented results that included only conventional random error or only error due to uncertainty in the bias parameters or both. In Bayesian bias analysis both sources of error will be combined to produce credible intervals that account for both random error and uncertainty in the bias parameters. This combination of both sources of error stems from the fact that we are simultaneously combining them in one model. Both Equation 11.10 (conventional random error) and Equation 11.14 (uncertainty in the bias parameters) are combined in the same model. While it would be possible to manipulate the Bayesian model to produce results that include only conventional random error or only uncertainty in the bias parameters, we do not do so here. If this is of interest, an easy approximate approach to this would be as follows: to produce results that only account for random error, alter the beta distributions in Equation 11.14 so they have virtually no randomness by multiplying the alpha and beta parameters for each distribution by 10,000. If one is interested in only the systematic error and not conventional random error, the analyst should do the reverse: increase the sample size of the study to make it so large there is virtually no random error. Multiplying a, c, N_1 and N_0 by 10,000 would accomplish this.

As discussed above, there will generally be no closed form solution to Bayesian problems of this sort and we will need to rely on MCMC techniques. We implemented four MCMC chains from initial values chosen randomly by the Stan package. The MCMC technique implemented by Stan operates similarly to the Metropolis algorithm we discussed above. The algorithm randomly samples from the posterior distribution of each unknown parameter in the model (*se, sp*, π_1, π_0). Each subsequent step of the model is based on the previous step. After the MCMC algorithm has been run, we have a large number of parameters. To estimate the effect of interest, the misclassification-adjusted OR, we simply take each sampled π_1 and π_0 to compute $OR = \left(\frac{\pi_1}{1 - \pi_1}\right) / \left(\frac{\pi_0}{1 - \pi_0}\right)$ at each iteration. This final step is the "Monte Carlo" aspect of MCMC. We implemented this model using four MCMC chains that were run for 60,000 iterations and took approximately 55 seconds to complete. The first 10,000 samples of each chain were treated as the burn-in period and discarded. The remaining 50,000 iterations from each chain were retained for the analysis.

Ensuring that the MCMC algorithm has converged to the posterior distribution is important and easily overlooked. It is incumbent upon the analyst to examine trace plots and distributions for each of the parameters in the model. We present these plots in Figures 11.7, 11.8 and 11.9 for the parameters *Se, Sp*, π_1, π_0, and the main parameter of interest, the misclassification-adjusted OR. The parameter estimates presented here incorporate both conventional random error and systematic error. Each of the trace plots suggests excellent convergence. For each parameter's

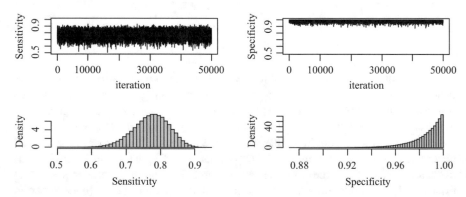

Figure 11.7 Trace plots and histograms of MCMC samples from the posterior distribution of sensitivity and specificity.

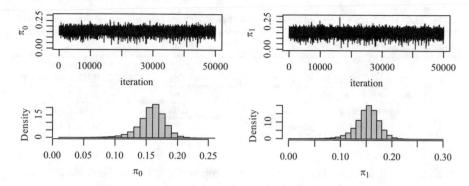

Figure 11.8 Trace plots and histograms of MCMC samples from the posterior distribution of the true prevalence of exposure among cases and controls, π_1 and π_0, respectively.

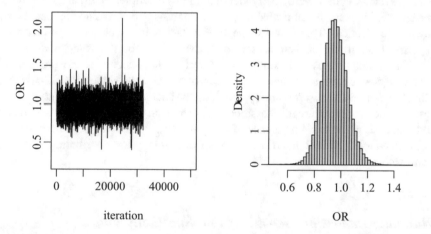

Figure 11.9 Trace plots and histograms of MCMC samples from the posterior distribution of the OR for the effect of smoking on breast cancer, bias-adjusted for misclassification.

traceplot, we only show 1 chain, for simplicity. Convergence was achieved rapidly in this example and nearly identical results are obtained if the MCMC algorithm was run for only 5000 iterations, instead of 50,000. However, we encourage users to run the algorithm much longer than they think is needed to help detect problems with convergence. Histograms can also be useful in detecting modeling problems. If the distribution does not look smooth, it may be an indication the user should run the algorithm longer. The Gelman-Rubin statistic for each parameter in our model was almost exactly 1.0 and the effective sample size for the bias-adjusted odd ratio was approximately 150,000. All of these diagnostics combine to suggest excellent convergence.

Final results (Table 11.1) of the Bayesian approach to bias-adjustments for misclassification are very similar to results obtained from the probabilistic bias

Table 11.1 Results of a Bayesian analysis, probabilistic bias analysis and conventional analysis of the relationship between smoking and breast cancer bias-adjusting for nondifferential misclassification of smoking.

Analysis	OR	95% Interval (2.5th, 97.5th percentile)
Conventional result, assuming no bias	0.95	(0.81, 1.13)
Probabilistic Bias Analysis	0.95	(0.78, 1.14)
Bayes Bias-Adjustment	0.95	(0.77, 1.15)

analysis approach that includes both random and systematic error (and, indeed, to the conventional approach). The point estimate is identical across all three approaches and the 95% credible intervals are slightly larger in the Bayes approach than in the probabilistic bias analysis approach; however, the practical distinction is negligible. Still, just because they were nearly identical in this example, we do not wish to give readers the impression that probabilistic bias analysis and Bayesian approaches will always be identical. There are two general instances in which the approaches will diverge: first, if the analyst uses informative priors for the true prevalence of exposure among cases and controls (or for the OR); second, if the probabilistic bias analysis returns a substantial number of impossible values (as discussed in Chapters 7–10). The first possibility, in which the analyst uses an informative prior is less common for exposure misclassification because an informative prior would need to be placed on the true prevalence of exposure (π_0, π_1), however without use of a gold standard it would be difficult to specify this prior distribution with any confidence.

Handling Impossible Sensitivity and Specificity Values in Bayesian Bias Analyses

To illustrate divergence between probabilistic bias analysis and Bayesian analysis, we slightly modify the example discussed above. Instead of choosing relatively precise beta distributions for the sensitivity and specificity, we specify a more diffuse beta distribution for the specificity parameter. In particular, we assume, as above, that $se \sim beta(\alpha=50.6, \beta=14.3)$ but now assume far less certainty about specificity, $sp \sim beta(\alpha=10, \beta=1)$. The beta distribution implies an expected specificity of $10/11=0.91$ with 95% prior interval (0.65, 1.00). As we showed in Chapter 6, there are bounds on what values of sensitivity and specificity are admissible given the observed data [1, 29]. In the present context of nondifferential misclassification, these bounds are:

$$\max\left[\Pr(E^* = 1|D = 1), Pr(E^* = 1|D = 0)\right] \leq Se \leq 1$$

$$1 - \min \left[\Pr(E^* = 1|D = 1), Pr(E^* = 1|D = 0) \right] \leq Sp \leq 1$$

or

$$0 \leq Se \leq \min \left[\Pr(E^* = 1|D = 1), Pr(E^* = 1|D = 0) \right]$$
$$0 \leq Sp \leq 1 - \max \left[\Pr(E^* = 1|D = 1), Pr(E^* = 1|D = 0) \right]$$

Note that if the prevalence of exposure is low, as in this example, the second set of bounding conditions may be unrealistic, requiring sensitivity and/or specificity to be implausibly low. Therefore, we focus on the first set of conditions. In this example, $\Pr(E^* = 1|D = 1) = 0.129$ and $\Pr(E^* = 1|D = 0) = 0.135$ which gives bounding conditions among the cases and controls, respectively, of:

$$0.135 \leq Se \leq 1 \text{ and } 0.871 \leq Sp \leq 1$$

The prior we propose for sensitivity has virtually its entire density within these bounds and, thus, the sensitivity parameter is unaffected by the bounds. The prior distribution for specificity, however, has $\int_0^{0.871} \text{Beta}(sp|\alpha = 10, \beta = 1) d\,sp = 0.25$ of its density outside the admissibility bound. The 25% of specificity values less than 0.871 are inadmissible. In a probabilistic bias analysis (Chapters 7–9), these specificity values would result in negative cell sizes.

A key distinction between probabilistic bias analysis and Bayesian analysis is how the two approaches deal with these inadmissible parameter values. There are three general approaches to dealing with inadmissible bias parameter values: 1) alter the prior distribution to prohibit most of the impossible values, 2) in a probabilistic bias analysis, exclude all iterations in which a negative cell count is obtained from an inadmissible parameter, and 3) in a Bayesian analysis, formally update the prior distribution in light of this bound. We encourage readers to revisit the section headed "Impossible Values for Bias Parameters and Model Diagnostic Plots" in Chapter 8 regarding the root causes of impossible values and how they relate to the choice of approach to deal with them.

We demonstrate the subtle distinction between the three approaches by examining the shape of the prior (bias) distribution of $beta(\alpha = 10, \beta = 1)$ and the distributions implied by the three approaches (Figure 11.10). We also compute the probability densities for the specificity parameter under each distribution at three unique specificity values: specificity=(0.84, 0.88, 0.92). Under the beta(α=10, β=1) prior, the density of these three specificities is 2.08, 3.16, and 4.72, respectively. Densities (vanishingly thin probability slices) are difficult to directly interpret; instead, we consider the density of each point relative to the specificity=0.88. Under the prior distribution this gives 2.08/3.16=0.66, 3.16/3.16=1, 4.72/3.16=1.49 for the three specificities, respectively. These numbers are directly interpretable and imply our prior belief is that a specificity of 0.84 is two-thirds as

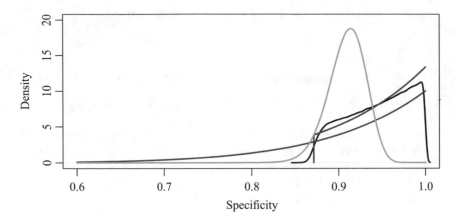

Figure 11.10 Density or estimated density of the distribution for the specificity parameter. The inadmissibility boundary occurs at specificity=0.871. The blue line is the density for the prior distribution, beta(α=10, β=1). The orange line is the density for the prior density modified to reduce inadmissible draws, beta(α=160, β=16). The red line is a prior distribution truncated at specificity=0.87. The black line is the estimated density of the posterior specificity distribution from MCMC sampling.

likely as a specificity of 0.88 and a specificity 0.92 is 1.49-times more likely than a specificity of 0.88.

In the first approach to dealing with inadmissible parameters, the analyst changes the prior distribution such that only a small amount of prior density lies in the inadmissible region. From a Bayesian perspective, this amounts to changing your prior belief after having seen the data. This is an inappropriate procedure that should not be used. To examine why this is inappropriate, consider an analyst who endeavors to keep the mean specificity=10/11 while specifying a new prior distribution that limits the inadmissible proportion of the prior distribution (that less that 0.871) to <5%. This would result in the specificity being assigned a far more precise distribution: $sp \sim beta(\alpha = 160, \beta = 16)$. As can be seen in Figure 11.10, this radically changes our subjective belief about the specificity of classification, with values near the mean becoming far more likely in this new distribution. The density of the three specificity values (0.84, 0.88, 0.92) under consideration are 0.31, 6.69, and 17.9. In relative terms, this new prior distribution implies a specificity of 0.84 is 5% as likely as a specificity of 0.88. This seems beneficial since probabilities of 0.84 are inadmissible and we want to give them little weight. However, the specificity of 0.92 becomes 2.7 times more likely than the specificity of 0.88. This is radically different than the prior distribution we had initially intended to specify. If our initial prior distribution was based on subject matter literature, this new prior would drastically misstate our belief about specificity values.

The second approach, used frequently in probabilistic bias analysis, serves to truncate the prior (or bias) distribution. Any value of the prior distribution that has an inadmissible value is excluded. Equivalently, we could set the density of the prior distribution at zero for all inadmissible parameter values. The shape of the remainder of the distribution (for admissible values) is unchanged; it is simply shifted up. The upshift in all density values occurs because the inadmissible probability mass to the left of specificity 0.871 (approximately 0.25) is uniformly redistributed to the admissible specificities (1). The density of the three unique specificities (0.84, 0.88, 0.92) under the truncated distribution is now 0, 4.2, 6.3, respectively with relative densities 0, 1 and 1.49. We can see that the inadmissible value has 0 density, as we desire, and the ratio of densities for specificity=0.92 to specificity=0.88 remains the same as it was in the prior specification. This distribution makes some intuitive sense; unfortunately, it is technically incorrect. The 25% inadmissible probability mass should not be uniformly distributed to all admissible points. Intuitively, that probability mass should be more heavily redistributed to admissible specificities that are closer to the bound of 0.871 than to specificities values that are further from the bound. This is exactly what the Bayesian approach does. It automatically redistributes the inadmissible probability mass closer to the admissibility bound. Examining Figure 11.10, we see that the Bayesian posterior distribution for specificity gives values close to the admissibility higher density than the probabilistic bias analysis approach. Examining the specificities (0.84, 0.88, 0.92) we calculate estimated densities 0, 4.5, 6.8, respectively. The Bayesian distribution gives higher density to admissible specificities near the admissibility bound. However, this will only be noticeable when there is a substantial amount of the prior distribution that is inadmissible, as in this example. To put it more succinctly, when the prior distribution overlaps with the inadmissibility bounds, we can expect the Bayes and probabilistic bias analysis approaches to differ with the former having a more solid theoretical foundation.

Returning to our example with a new $beta(\alpha = 10, \beta = 1)$ prior distribution for specificity, we repeat probabilistic bias analysis and Bayesian approaches using Excel spreadsheets and R code, respectively. Results of these analyses are included in Table 11.2. The probabilistic bias analysis result (OR=0.92; 95% SI: 0.48, 1.18) has a point estimate that is similar to the result obtained above with a more precise distribution for the specificity and a simulation interval that is substantially wider. The Bayesian result is much less precise (OR=0.93; 95% CrIn: 0.40, 1.41) than the probabilistic bias analysis approach. While the probabilistic bias analysis approach provides some limited suggestion of a decreased risk, the width of the credible intervals in the Bayesian approach make such a statement impossible. The reason for

Table 11.2 Results of a conventional, probabilistic bias, and Bayesian analysis of the relationship between smoking and breast cancer bias-adjusting for nondifferential misclassification of smoking with beta(10,1) prior on specificity.

Analysis	OR	95% Interval (2.5th, 97.5th percentile)
Conventional result, assuming no bias	0.95	(0.81, 1.13)
Probabilistic Bias Analysis	0.92	(0.48, 1.18)
Bayes Bias-Adjustment	0.93	(0.40, 1.41)

the difference in interval width between the two approaches is that the Bayesian approach places higher probability on specificity values near the admissibility bound. The values near the admissibility bound will generally introduce more variability in bias-adjusted point estimates, which intuitively makes sense as those imputed categorical tables are dominated by very small cell sizes.

Exposure Misclassification Using PPV and NPV

When data from an internal validation study have been collected, positive and negative predictive values may be available to the analyst instead of sensitivities and specificities. Switching the bias parameters necessitates a change to the structure of the Bayesian bias model used to bias-adjust for misclassification. As above, we note that there are many ways to structure a bias model and we opt for a straight-forward parameterization that is easily implemented in summary data. When internal validation data are available, model structures that treat misclassification as a missing data problem are likely to be more efficient (see Chapter 10). In particular, it is not necessary to impute exposure values if the gold standard was actually measured on a subject, as occurs with internal validation data.

As above, we specify the likelihood for the model using Equation 11.11. How-ever, rather than expressing the likelihood in terms of sensitivity and specificity, we retain the observed proportion of exposed (P_1) and unexposed (P_0) in the likelihood for ease of computation. Rather than specifying prior distributions on the true proportions, π_1 and π_0, we place prior distributions on the observed proportions P_0 and P_1. As above we choose uninformative priors:

$$P_0 \sim \text{Uniform}(0, 1)$$

$$P_1 \sim \text{Uniform}(0, 1)$$

We must also place a prior distribution on the positive and negative predictive values (just as we did with sensitivity and specificity, when those were the bias parameters). Recall that if the misclassification is nondifferential, and there is a non-null association, the PPV among cases (ppv_1) will not equal the PPV among controls (ppv_0), nor will the NPV among cases (npv_1) equal the NPV among controls (npv_0). We place beta distributions on these four parameters:

$$ppv_1 \sim \text{beta}(\alpha = a_1, \beta = b_1)$$

$$ppv_0 \sim \text{beta}(\alpha = a_0, \beta = b_0)$$

$$npv_1 \sim \text{beta}(\alpha = c_1, \beta = d_1)$$

$$npv_0 \sim \text{beta}(\alpha = c_0, \beta = d_0)$$

When we implement the MCMC algorithm, values of P_0, P_1, and all four predictive values are updated at each step of the algorithm. Bayesian software packages such as Stan and JAGS make it easy to include additional steps that transform P_0, P_1, and the four predictive values into the parameters of inferential interest, π_1 and π_0:

$$\pi_1 = P_1 \cdot ppv_1 + (1 - P_1)(1 - npv_1)$$
$$\pi_0 = P_0 \cdot ppv_0 + (1 - P_0)(1 - npv_0)$$

These parameters are then combined to form the misclassification-adjusted $OR = \left(\frac{\pi_1}{1 - \pi_1}\right) \Big/ \left(\frac{\pi_0}{1 - \pi_0}\right)$.

Before proceeding to the applied example, we note that this model and the way it will be sampled in an MCMC framework is nearly identical to the alternative version we proposed for probabilistic bias analysis using positive and negative predictive values in Chapter 8. In that chapter, the alternative approach sampled first from PPV and NPV distributions and then, second, from beta distributions for the observed probabilities. Finally, those samples were combined to estimate π_0, π_1, and OR. Indeed, because the distributions of PPV, NPV, P_1 and P_0 are independent, an MCMC algorithm to estimate the posterior distribution of π_0, π_1, and OR, based on the specification above can be implemented by sampling from the prior distributions for ppv_1, ppv_0, npv_1 and npv_0 and then sampling from the posterior distribution of the observed probabilities, P_1 and P_0. Because the data follow a binomial distribution and the prior follows a beta distribution, it can be shown that the posterior distributions for P_1 and P_0 are available in closed form as:

$$P_1 \sim beta(\alpha = a + 1, \beta = N_1 - a + 1)$$
$$P_0 \sim beta(\alpha = b + 1, \beta = N_0 - b + 1)$$

These distributions are nearly identical to the ones we sampled from in Chapter 8 with the only difference the addition of "+1" here. This will be unimportant in practice if we have moderate sample sizes. Other than this small difference, the alternative probabilistic bias analysis approach for misclassification based on PPV and NPV is the same as a Bayesian approach. Technically, the probabilistic bias analysis approach in Chapter 8 is a Bayesian approach in which an improper $\beta(\alpha = 0, \beta = 0)$ prior is placed on the parameters P_1 and P_0 to minimize the impact of prior knowledge to the greatest extent possible.

In Chapter 6, we implemented a bias analysis to bias-adjust for misclassification in a study of pre-pregnancy maternal BMI and preterm delivery, pretending it was a case-control study. Maternal BMI was determined via self-report and pre-pregnancy medical records were used as a gold standard to estimate predictive values of underweight (or normal weight) for a random sample of mothers who did and did not experience preterm delivery. Validation data for this study are shown in Table 11.3 and we use them to directly inform the parameters of our bias distributions. For example, the PPV among cases is the fraction of mothers whose infants were preterm who actually are underweight given they reported being underweight (50/(50+26)). Analyzing the validation data, we obtain a maximum likelihood

Table 11.3 Validation data comparing birth certificate (misclassified) BMI category to medical record review (gold standard) by early preterm birth. Observed data from Lash et al., 2014 [30].

	Term		Preterm	
	Medical Record			
Birth Certificate	Underweight	Normal Weight	Underweight	Normal Weight
Underweight	132	47	50	27
Normal weight	2	115	0	120

Table 11.4 Results of a conventional, probabilistic bias analysis, and Bayesian analysis of the relationship between maternal pre-pregnancy weight and preterm delivery.

Analysis	OR	95% Interval (2.5th, 97.5th percentile)
Conventional result, assuming no bias	1.51	(1.39, 1.65)
Probabilistic Bias Analysis	1.06	(0.67, 1.50)
Bayes Bias-Adjustment	1.07	(0.68, 1.52)

estimate of $ppv_1 \approx 66\%$ with exact 95% CI: 54%, 76%. We choose a prior distribution for the bias parameter ppv_1 that reflects these values by assuming $ppv_1 \sim beta$ ($\alpha = 50, \beta = 26$). This prior implies that our best guess for ppv_1 is 66% with a 95% interval of 55%, 76%. We use similar logic and data from Table 11.3 to inform the remaining parameters and note that we add 0.5 to the zero cell in the table, otherwise the distribution of npv_1 would be a point mass at 1, with no uncertainty:

$$ppv_1 \sim \text{beta}(\alpha = 50, \beta = 27)$$
$$ppv_0 \sim \text{beta}(\alpha = 132, \beta = 47)$$
$$npv_1 \sim \text{beta}(\alpha = 120, \beta = 0.5)$$
$$npv_0 \sim \text{beta}(\alpha = 115, \beta = 2)$$

The code to implement this model in Stan is included on the website for this text. The MCMC algorithm was run for 60,000 iterations with the first 10,000 discarded as a burn-in. We used four MCMC chains. The algorithm took approximately 25 seconds to complete. Table 11.4 gives results of this analysis. The conventional approach found that women who had reported being underweight had 1.5 times the odds of preterm birth as women who reported being normal weight prior to pregnancy (OR=1.51; 95%CI: 1.39, 1.65). In Chapter 6, we treated the predictive values as constant (at the mean of the bias distributions specified above) and found a bias-adjusted OR of 1.1. The Bayesian model adjusting for misclassification found a nearly identical effect estimate (OR=1.07; 95% CrIn: 0.68, 1.52). The Bayesian point estimate (OR=1.07) is nearly identical between the analyses because unform priors were placed on P_1 and P_0 and no external information was brought into the analysis. Had we incorporated prior information, the Bayesian estimate and probabilistic bias analysis estimate would be expected to differ. For comparison purposes, we also include results from a probabilistic bias analysis that used the prior distributions above as bias parameter distributions. The probabilistic bias analysis was run for 10,000 iterations. Results from the probabilistic bias analysis were nearly identical to the Bayesian results.

Selection Bias

As discussed in earlier chapters, quantitative bias analysis for selection bias is often the easiest bias for which to adjust mathematically and simply requires specification of weights that can be applied to the data. However, specifying those weights in practice is difficult because it requires knowledge that is often unavailable, exactly because of the selection bias. Modeling selection bias from a Bayesian standpoint can be accomplished in a variety of ways. If selection bias is due to enrolment in a study depending on the exposure and the outcome and information is available on how many people did not enroll, selection bias could be handled from a Bayesian approach by treating unenrolled participants as missing data (see Chapter 10). Because this is a less common scenario, we do not focus on it here.

Quantitative bias analysis to address selection bias in epidemiology is often addressed by reweighting participants so that they represent non-participants, such as inverse probability of participant weighting or inverse probability of attrition weighting [31–33]. As described in Chapters 4 and 8, if the observed data are weighted by the inverse of the probability of selection, bias-adjusted effects can be obtained. To provide a parallel approach, we present Bayesian models to bias-adjust for selection bias using weights. Unfortunately, Bayesian inference using weights is not thoroughly developed and not without controversy [34, 35]. To implement a fully Bayesian approach, an analyst should specify a model for the outcome of interest as well as a model for the weights. Accomplishing this in sensible ways, however, has proven problematic. A natural consequence of Bayesian inference in some models involving weights would lead to updating sample weights based on the remaining observed data. However, this is not always desirable and can lead to poorly performing estimators [34–37]. Robins et al. have argued that, in many instances, a fully Bayesian solution incorporating weights is unlikely to produce estimators with good characteristics because the parameters of interest and weights are independent *a priori* and any Bayesian solution must ignore the weights [34, 37]. Notwithstanding these difficulties, a number of interesting methods have been proposed that either function in two-steps to prevent updating weights from the observed data [38, 39] or estimate parameters via weighted likelihood bootstrap procedures [40, 41].

To parallel the probabilistic bias analysis approach, we present two methods to bias-adjust for selection bias in a Bayesian framework that mimic the two-step approaches mentioned earlier [38, 39]. In the first, we specify a distribution for the selection odds ratio and in the second we specify distributions for the individual selection probabilities that constitute the selection odds ratio. Some statisticians would not consider this 2-step approach to be fully Bayesian since it does not operate on the joint likelihood of all parameters but, instead, a Bayesian pursuit of estimators with certain frequentist properties [37]. We agree with this critique but do not find it particularly troubling. Whether these estimators are fully Bayesian or not is of little importance to practicing analysts and we think most analysts would prefer estimators with good frequentist operating characteristics (such as low mean squared error and unbiasedness) rather than dogmatic adherence to statistical approaches.

In Chapters 4, 8 and 9, we applied bias analysis methods to bias-adjust for selection bias in a case-control study of the association of mobile phone use and uveal melanoma. Selection bias was a concern because people may have agreed to participate in the study at different rates if they were a uveal melanoma case than if they were a control, and at different rates if they used a mobile phone than if they did not, so analyzing only those people who participated in the study could result in selection bias when estimating the association between uveal melanoma and mobile phone use. In Chapter 8, we implemented probabilistic bias analysis using selection probabilities and the selection odds ratio for the bias model. In implementing the probabilistic bias analysis approach, we used data on the number of people who were eligible for the study and did not enroll to inform our selection probabilities (Table 8.12 in Chapter 8). Our first Bayesian approach specifies a prior distribution for the selection probabilities, similar to the probabilistic bias analysis distribution. Because this study is a case-control study, specifying the likelihood proceeds similar to our last example. We posit that the number of exposed among the cases (a) and number of exposed among controls (c) follow binomial distributions:

$$a \sim \text{binomial}(N_1, P_1)$$
$$c \sim \text{binomial}(N_0, P_0)$$

(11.16)

where N_1 and N_0 are the total number of cases and controls, respectively, and P_1 and P_0 are the unknown proportion of cases and controls who are exposed. The estimated probabilities in these models are used to estimate an odds ratio and is not adjusted for selection bias, $OR = (\frac{P_1}{1-P_1})/(\frac{P_0}{1-P_0})$. This crude odds ratio is then bias-adjusted for selection bias using the sampled selection probabilities. All that remains is to specify prior distributions for the selection probabilities and exposure proportions (P_1 and P_0). As above, we specify uniform priors on the exposure probabilities among cases and controls, indicating our lack of knowledge of the true effect of mobile phone use on uveal melanoma. There are four selection probabilities, s_{DE}, which indicate the probability of participating in the study for people who were cases ($D=1$) or controls ($D=0$) and exposed ($E=1$) or unexposed ($E=0$). Recall from the example introduced in Chapter 4, and used also in Chapters 8 and 9, that some non-participants answered a brief survey about mobile phone use, and this information was used to estimate the prevalence of mobile phone use among non-participating cases and controls. This information then allows estimation of the selection proportions (see Table 4.2). We place beta prior distributions on each of the selection probabilities. The parameters of the beta distributions are the observed number of people in a category who enrolled (set equal to α) and the modeled number who did not enroll (set equal to β). For example, there were 139 exposed cases who enrolled and an estimated 5.1 exposed cases who did not enroll (as derived in Table 8.12). This leads to a prior distribution of $s_{11} \sim beta(\alpha = 139, \beta = 5.1)$ for the first selection probability. The remaining prior distributions are:

$$s_{11} \sim \text{beta}(\alpha = 139, \beta = 5.1)$$
$$s_{10} \sim \text{beta}(\alpha = 114, \beta = 11.9)$$
$$s_{01} \sim \text{beta}(\alpha = 369, \beta = 96.1)$$
$$s_{00} \sim \text{beta}(\alpha = 377, \beta = 282.9)$$

$$(11.17)$$

Care should be exercised in the specification of these distributions. A distribution that gives probability to values near zero could have a dramatic impact on the final result since cell counts are being divided by these weights. In practice, this approach almost perfectly mirrors the probabilistic bias analysis approach and incorporates conventional random error associated with the outcome (Equation 11.16) and uncertainty in the selection probabilities (Equation 11.17). In this MCMC algorithm, we sample selection probabilities (*via* the priors on the selection probabilities in Equation 11.17) and sample a crude odds ratio (via the likelihood specification in Equation 11.16 and priors on exposure probabilities). When we sample the selection probabilities, we combine them to compute a selection OR, $sOR = \frac{s_{11} \times s_{00}}{s_{10} \times s_{01}}$. When we sample from the posterior distribution for the exposure probabilities, we combine them to estimate the crude $OR_{crude} = \left(\frac{P_1}{1-P_1}\right) \Big/ \left(\frac{P_0}{1-P_0}\right)$. Finally, we combine these two sets of results to estimate the selection bias-adjusted odds ratio at each step of the MCMC algorithm: $OR_{adj} = \frac{OR_{crude}}{sOR}$. This specification of the model separates the prior information on the selection probabilities from the study data. It is similar in nature to the two-step process of McCandless (2009) [38]. This model was implemented in Stan and run for 60,000 iterations across four chains with the first 10,000 discarded as a burn in. Sampling took less than a second to complete. Results are shown in Table 11.5. Using the Bayes adjustment for selection bias we find OR of 1.62 (95% CrIn: 1.20, 2.20), which is similar to what we found in the probabilistic bias analysis approach in Chapter 8.

Table 11.5 Results of a Bayesian analysis, probabilistic bias analysis, and conventional analysis of the relationship between mobile phone use and uveal melanoma. Original data from Stang et al., 2009 [42].

Analysis	OR	95% Interval (2.5th, 97.5th percentile)
Conventional result, assuming no bias	0.71	(0.51, 0.97)
Probabilistic Bias Analysis	1.62	(1.20, 2.20)
Bayes Bias-Adjustment	1.62	(1.20, 2.20)

As with probabilistic bias analysis, we could implement Bayesian bias-adjustments for selection bias by placing a prior on the selection odds ratio itself rather than the selection probabilities. As mentioned in Chapter 8, placing a distribution on the selection odds ratio could prove problematic as information to directly inform this distribution may be unavailable. To specify a prior distribution for the selection OR in this example, we sample from the selection probabilities in Equation 11.17. We use these sampled selection probabilities to compute the log-selection odds ratio and estimate the mean and variance of that parameter. We use that estimated mean and variance to directly specify our selection OR on the log scale:

$$\ln(sOR) \sim \text{Normal}(\mu, \tau)$$

In this example, we specify $\mu = -0.265$ and the standard deviation, $\tau = 0.05$. This implies the selection odds ratio itself has a median 0.77 and prior 95% intervals of (0.70, 0.85). Implementing this prior distribution in Stan, we find identical results to the above with OR of 1.62 (95% CrIn: 1.20, 2.20). These two approaches are expected to return identical results and choice of which to use will depend on the source of bias parameter information (selection probabilities or selection OR).

Uncontrolled Confounding

In Chapter 5, bias analyses for uncontrolled confounding were implemented by specifying values for three bias parameters: the association between the uncontrolled confounder and the outcome, the prevalence of the confounder among the exposed, and the prevalence of the confounder among the unexposed. In a Bayesian approach to uncontrolled confounding, we specify values for these same bias parameters, albeit in a slightly different form. The approach we outline below is motivated by, and similar to, the approaches taken by multiple authors [2, 43–45]. While other Bayesian approaches exist, the approach we present is most in keeping with probabilistic bias analysis approaches already introduced in this book.

In Chapters 5 and 8, we introduced an example of uncontrolled confounding in a cross-sectional study that estimated the association between circumcision and HIV infection in Nairobi, Kenya. Although the original authors found a strong protective effect of circumcision on HIV infection, religion category—which is plausibly associated with both HIV risk and circumcision—was not controlled in their analyses. Specifying a model (whether using Bayesian techniques or probabilistic bias analysis) will often depend on the information available. It might seem natural to suspect that we should specify a model that follows the direction of causality: 1) HIV risk (y) conditional on circumcision (x) and religion category (u), $f(y_i|x_i, u_i)$; 2) circumcision risk conditional on religion category, $f(x_i|u_i)$; 3) the prevalence of Muslim religious category, $f(u_i)$. Combining these three components gives the full likelihood: $f(y_i, x_i, u_i) = f(y_i|x_i, u_i)f(x_i|u_i)f(u_i)$. After placing prior distributions on relevant parameters, this model is intuitive and relatively straightforward to implement. However, it is less commonly used in epidemiology. Dating at least back to Schlesselman (1978) [46], it has been more common for epidemiologists to reverse the second model above and, instead, specify the probability of religion category conditional on circumcision: $f(u_i|x_i)$. Presumably this is in part because this information is easier for analysts to obtain. In this parameterization of the model, the exposure is viewed as a fixed, rather than random, variable and the full likelihood is conditional on the exposure: $f(y_i, u_i|x_i) = f(y_i|x_i, u_i)f(u_i|x_i)$. This latter approach is explained in more detail in Lin (1998), McCandless (2007), and Gustafson (2010) [45, 47, 48]. Because this model has fewer parameters, it may be somewhat easier to fit. In addition, it is a natural Bayesian analog to the approaches presented in Chapters 5 and 10.

To implement this Bayesian approach adjusting for uncontrolled confounding, we need to specify models for the outcome, $f(y_i|x_i, u_i)$, and uncontrolled confounder, $f(u_i|x_i)$. The outcome model is relatively easy to specify. Because HIV infection is relatively common in this study, it is natural to estimate a risk ratio. A large literature exists on estimating risk ratios from common dichotomous data and many of the models specified in this literature can be adapted to this framework [49–53]. We specify a very simple log-linear model with the caveat that it can often fail to converge in frequentist model fitting algorithms or crash Bayesian algorithms because the model does not constrain probabilities to be less than 1. This model can be altered to impose constraints that circumvent this problem [52]. However, the ease of using this model is that it allows for very easy interpretation of model parameters as measures of effect.

$$f(y_i|x_i, u_i) = \text{Bernoulli}(p_i)$$

$$p_i = \exp{(\beta_0 + \beta_1 x_i + \beta_2 u_i)}$$

Specifying a model for $f(u_i|x_i)$ requires careful consideration and there are multiple acceptable choices. Because our uncontrolled confounder (religion category: Muslim or other) is dichotomous and we would prefer to bound the probability between 0 and 1, we choose a logistic specification:

$$f(u_i|x_i) = \text{Bernoulli}(q_i)$$

$$q_i = \text{expit}(\gamma_0 + \gamma_1 x_i)$$

Considering both of these equations together, we see that $\exp(\beta_1)$ is the risk ratio of interest: the association between circumcision and HIV infection, controlling for religion category (and assuming no interaction between religion category and circumcision). The parameter $\exp(\beta_2)$ is the risk ratio of the association between Muslim religion (versus other religions) and HIV infection (again assuming no interaction). The parameter $\text{expit}(\gamma_0)$ is the prevalence of Muslim religion among those who are uncircumcised and the parameter $\text{expit}(\gamma_0 + \gamma_1)$ is the prevalence of Muslim religion among those who are circumcised. These three parameters, $\exp(\beta_2)$, $\text{expit}(\gamma_0)$, and $\text{expit}(\gamma_0 + \gamma_1)$, are analogs of the bias parameters RR_{CD}, p_0, and p_1, respectively, from Chapter 5. Prior distributions—analogs of the distributions assigned to the bias parameters in Chapter 8—must be placed on the parameters of this model to estimate associations. We choose distributions consistent with those chosen in Chapter 8. Because γ_0 is the logit of the prevalence of the confounder among the unexposed and can be any real number, it is common to assign a normal distribution for its prior. In Chapter 8, a trapezoidal distribution was assigned to p_0, with a minimum of 0.03 and maximum of 0.10. To estimate the mean of the normal distribution, we calculate the logit of these values and then use their midpoint, $\frac{logit(0.10) + logit(0.03)}{2} = -2.84$, to compute the mean of the prior distribution. We also use these values to back-calculate a standard deviation: $\frac{logit(0.10) - logit(0.03)}{2 \cdot 1.96} = 0.326$, which completes the specification of the prior distribution for γ_0:

$$\gamma_0 \sim Normal(-2.84, 0.326)$$

To ensure that we have the prior specification we intend, we can draw samples from this distribution, take the expit of each random draw, and examine the 2.5th, 50th and 97.5th percentiles. In this case, taking 10^5 random draws gives values of 0.03, 0.06 and 0.10, respectively, for these percentiles. That is, our prior belief is that the prevalence of Muslim religion among those who are uncircumcised is 6% and we are 95% certain the prevalence is between 3% and 10%.

Using the same approach, we can specify the mean and standard deviation of the logit of the probability of Muslim religion among the circumcised as $\frac{logit(0.9) + logit(0.7)}{2} = 1.52$ and $\frac{logit(0.9) - logit(0.7)}{2*1.96} = 0.344$. However, this is the mean of standard deviation of the distribution of $\gamma_0 + \gamma_1$ and we only need the distribution of γ_1. To specify the mean of γ_1 we subtract the mean of γ_0 and compute $1.52 - (-2.84) = 4.36$. Assuming independence of prevalence estimates, we can do the same for variances, $0.344^2 - 0.326^2 = 0.012$, or a $sd = \sqrt{0.012} = 0.11$, and arrive at the following prior specification:

$$\gamma_1 \sim Normal(4.36, 0.11)$$

As above, it is important to check our prior specification. To do so, we sample from γ_0 and γ_1. We add the random samples together, take the expit of the sum and calculate the 2.5th, 50th and 97.5th percentiles. In this case, with 10^5 samples, we find the respective percentiles of 0.70, 0.82 and 0.90, as we wanted.

Finally, we need to specify a prior distribution for the log(RR) of the association between Muslim religion (compared with other religions) and HIV risk. In Chapter 8, the trapezoidal distribution had a range of 0.5 and 0.8. We translate this into a normal distribution treating the limits of the trapezoidal distribution as the lower and upper 95% confidence intervals, as above. The mean of the distribution is the midpoint of the log of the interval endpoints, $\frac{\log(0.8) + \log(0.5)}{2} = -0.46$, and the intervals are also used to compute the standard deviation, $\frac{\log(0.8) - \log(0.5)}{2*1.96} = 0.120$:

$$\beta_2 \sim \text{Normal}(-0.46, 0.120)$$

Checking this prior specification is straightforward, sampling from the distribution for β_2, exponentiating it and examining percentiles. Doing so with the distribution specified above gives percentiles of 0.50, 0.63, 0.80, as expected.

To complete the Bayesian specification of our model, we place relatively uninformative and independent N(mean=0, sd=10) priors on β_0 and β_1. The model above can be implemented in either Stan or JAGS and we provide code for both of these on the text's website. In JAGS, the algorithm iterates through two general steps: first, sampling from the posterior distribution for each parameter in the model conditional on the observed data and imputed confounders; and second, using those parameters and the observed data to sample from the posterior distribution of the uncontrolled confounder. JAGS iterates through these two steps many times to obtain samples from the joint posterior distribution, which can then be analyzed.

Stan requires an alternative approach since it does not have the capacity to sample categorical unobserved confounders. Stan requires the user to rewrite the likelihood, integrating out the unobserved confounder. [44] In theory, this marginal likelihood allows for a more efficient MCMC algorithm, because the unobserved confounders do not need to be updated. The text's website gives an example of the Stan code implementing a marginal likelihood. In practice the results should be similar whether working with the marginal likelihood in Stan or full likelihood in JAGS. Analysts may find the process of writing the marginal likelihood in Stan unduly complex, and may prefer JAGS for this reason.

We implemented the model in Stan, running the algorithm for 5,500 iterations and excluding the first 500 iterations as a burn-in period. Four chains were simultaneously run, resulting in 20,000 samples from the posterior distribution for final analysis. Convergence was assessed for each parameter and was deemed acceptable. We compare this with the conventional results that ignore uncontrolled confounding and results from the probabilistic bias analysis in Table 11.6. The Bayesian point estimate RR=0.49 was nearer to the null than the conventional RR=0.35, indicating that uncontrolled confounding by religion category may have had some impact on the results. The Bayesian analysis yielded a confounding-adjusted RR almost identical to the probabilistic bias analysis approach. The interval was slightly narrower in the Bayesian analysis than in the probabilistic bias analysis. The slight difference between the two approaches is likely due to the different shape of the bias and prior

Table 11.6 Results of a Bayesian analysis, probabilistic bias analysis and conventional analysis of the relationship between circumcision and HIV infection. Original data from Tyndall et al. 1996 [56].

Analysis	RR	95% Interval (2.5th, 97.5th percentile)
Conventional result, assuming no bias	0.35	(0.28, 0.50)
Probabilistic Bias Analysis	0.47	(0.29, 0.73)
Bayes Bias-Adjustment	0.49	(0.36, 0.65)

distributions for the same parameters. While trapezoidal distributions were chosen for the probabilistic bias analysis, we opted for normal distributions in the Bayesian approach. However, the probabilistic bias analysis had wider confidence intervals and McCandless and Gustafson point out that there are good reasons to expect that the results of the Bayesian and probabilistic bias analyses should be different [44]. In particular, there appears to be some information in the observed data to update the prior guess at the association between the unmeasured confounder and the outcome. That is, the prior distribution can be updated, even though there is no information to directly estimate this parameter (because we did not observe the unmeasured confounder). Rather, the information available to impute the unmeasured confounder appears to be sufficient to update the prior distribution. Because probabilistic bias analysis is essentially a Bayesian approach that only samples from the prior (without accounting for the observed data), this implies that this approach should differ from the Bayesian approach in some situations. While this topic has not been fully explored, McCandless and Gustafson (2017) offer some reassurance that the departures between the approaches might not be large, at least in the limited scenarios they examine [44]. Further, in response to their article, Greenland (2017) points out that when the prior distribution is misspecified (that is, the analyst guesses poorly in regard to the prior distribution), not updating the prior distribution could be beneficial in some cases [54]. Similarly, as discussed in Chapter 8, updating a prior that is incompatible with data may be counter-productive if the incompatibility arises from sparse data and not from poorly assigned values for the bias parameters. The potential discrepancy between Bayesian and probabilistic bias analysis approaches, however, with some exceptions, has not been thoroughly explored, and neither has the effect of mis-specifying prior distributions [55].

Conclusion

The Bayesian approach is natural for quantitative bias analysis. The bias distributions we discussed in Chapters 7, 8, and 9 are nothing more than prior distributions from a Bayesian perspective. Indeed, as we discussed above, the probabilistic bias analysis approach is simply an approximate Bayesian approach. In many cases the approximation is excellent. The natural question is: when will the two approaches differ? For misclassification, it appears the two approaches will be quite similar unless the prior or bias distribution overlaps with the admissibility bounds [1]. For selection bias, given our implementation, the Bayesian and probabilistic bias analysis approaches should be indistinguishable. Finally, it appears that while there is some difference in Bayesian and probabilistic bias analysis approaches to uncontrolled confounding they are often not large [44, 54]. However, if an analyst uses informative priors and incorporates substantive information on effect sizes, the Bayesian and probabilistic bias analysis approaches could be substantially different.

Even here the data augmentation approach of Greenland offers an easy way to incorporate prior knowledge in probabilistic bias analysis [3–5]. Thus, in many cases, the decision to implement a formal Bayesian approach vs a probabilistic one may come down to ease of implementation. It should be noted that this potential difference between probabilistic bias analysis results and a full Bayesian approach's results may emanate primarily from the prior placed on the estimate itself. Conventional frequentist statistical analyses place no prior on the estimate itself, whereas a full Bayes analysis would place a prior on the estimate itself. When this use of a prior causes a difference in the ultimate estimates, this difference would carry through to any bias analyses conducted as well. This difference does not arise from a difference in the bias analysis methods and results, but rather from a difference in the conventional analysis. Bayesian analysis requires special knowledge to diagnose convergence. The nature of the probabilistic bias analysis algorithm should ensure it has no issues with failing to converge, given sufficient number of iterations. Both approaches may require a user to learn new software. For summary level quantitative bias analysis, the spreadsheets provided with the text may make probabilistic bias analysis easier. However, when working on record level bias analysis, either approach is likely to require analyst programming (code we have provided may be adapted to the particular problem). We note that each approach above is quickly adapted to record level data by altering the binomial distributions to Bernoulli distributions. A Bayesian approach may be no more difficult for record level bias analysis due to the flexibility and automated nature of the programming languages.

Appendix: Justification for Probabilistic Bias Analysis for Exposure Misclassification in a Case-Control Study

The justification for the probabilistic bias analysis approach detailed in Chapter 8 is that it is an approximately Bayesian approach to inference. The probabilistic bias analysis algorithm we detailed is very similar to a Markov chain Monte Carlo algorithm laid out by Gustafson [16]. We first present Gustafson's model for exposure misclassification in a case-control study; second, we detail the algorithm Gustafson proposed to draw samples from the joint distribution of all variables and make inferences regarding the bias-adjusted OR; third, we show the similarity between Gustafson's algorithm and our probabilistic bias analysis.

The model proposed by Gustafson is similar to the one we presented in the Appendix in Chapter 8. Our interest is on estimating $OR = \frac{\pi_1}{1-\pi_1} / \frac{\pi_0}{1-\pi_0}$ where π_i is the prevalence of the true exposure among cases (i=1) and controls (i=0). Rather than observing the true exposure status (which we denote using capital letters in Table 11.7), we observe a misclassified exposure (lower case letters, Table 11.7) as a result of a classification process with some, possibly differential, sensitivities (se_i)

Table 11.7 Latent table with true exposure status and observed table with misclassified exposure status*.

	True Table				Observed Table		
	E=1	E=0	Total		E^*=1	E^*=0	Total
Y=1	A	B	N_1	Y=1	a	b	N_1
Y=0	C	D	N_0	Y=0	c	d	N_0

E is a persons true exposure status; E is their observed (misclassified) exposure; Y is case-control status

Table 11.8 Cross classification of true exposure by apparent exposure by case status*.

	Cases (Y=1)				Controls (Y=0)		
	E=1	E=0	Total		E=1	E=0	Total
E^*=1	W_1	X_1	a	E^*=1	W_0	X_0	c
E^*=0	Y_1	Z_1	b	E^*=0	Y_0	Z_0	d
Total	A	B	N_1	Total	C	D	N_0

E is a persons true exposure status; E is their observed (misclassified) exposure; Y is case-control status

and specificities (sp_i). We write the statistical model in terms of the latent true exposure by apparent exposure categories from Table 11.8:

$$\pi_i \sim \text{beta}(\epsilon_{a_i}, \epsilon_{b_i})$$

$$se_i \sim \text{beta}(\alpha_i, \beta_i)$$

$$sp_i \sim \text{beta}(\gamma_i, \delta_i)$$

$$[W_i, X_i, Y_i, Z_i] \sim \text{Multinomial}(N_i, [p_{1_i}, p_{2_i}, p_{3_i}, p_{4_i}])$$

where the distribution on π_i represents prior belief about this parameter or, as is generally the case, this might be specified as $\text{beta}(\epsilon_{a_i} = 1, \epsilon_{b_i} = 1)$ to represent a general lack of knowledge about the *a priori* magnitude of π_i. The beta distributions on sensitivity and specificity are equivalent to the prior used at the beginning of this chapter. All cases and controls in the dataset are distributed to one of four true-by-apparent exposure categories in Table 11.8 based on the parameters $p_{1_i}, p_{2_i}, p_{3_i}, p_{4_i}$ which are defined as:

$$p_{1_i} = se_i \pi_i$$

$$p_{2_i} = (1 - sp_i)(1 - \pi_i)$$

$$p_{3_i} = (1 - se_i)\pi_i$$

$$p_{4_i} = sp_i(1 - \pi_i)$$

That is, a person is both truly and apparently exposed with probability $p_{1_i} = se_i \pi_i = \Pr(E^* = 1|E = 1)\Pr(E = 1) = \Pr(E^* = 1 \& E = 1)$. A person is classified as both truly unexposed and apparently exposed with probability $p_{2_i} = (1 - sp_i)(1 - \pi_i) = \Pr(E^* = 1|E = 0)\Pr(E = 0) = \Pr(E^* = 1 \& E = 0)$, and so forth. By assuming that W_i, X_i, Y_i, and Z_i follow a multinomial distribution, we are ensuring that the sum of these random variables will always equal N_i. We note that, by examining Table 11.8, we can see the relation between the observed and latent variables:

$$a = W_1 + X_1$$
$$b = Y_1 + Z_1$$
$$c = W_0 + X_0$$
$$d = Y_0 + Z_0$$

Given the model described in the preceding paragraphs, we can now turn our attention to inference. Gustafson detailed a type of Markov chain Monte Carlo algorithm, known as a Gibbs sampler, to allow analysts to draw samples from the posterior distribution of $\theta_i = (\pi_i, W_i, X_i, Y_i, Z_i, se_i, sp_i)$. To derive this algorithm, we first write out the joint posterior distribution given the observed data implied by the statistical model and then factor this joint distribution into conditional distributions for each parameter in θ_i. It turns out that each conditional distribution is available in closed form by familiar distributions:

$$f(se_i|-) = \text{beta}(\alpha_i + W_i, \beta_i + Y_i)$$
$$f(sp_i|-) = \text{beta}(\gamma_i + Z_i, \delta_i + X_i)$$
$$f(\pi_i|-) = \text{beta}(W_i + Y_i + \epsilon_{a_i}, X_i + Z_i + \epsilon_{b_i})$$
$$f(W_i|-) = \text{binomial}(a, ppv_i)$$
$$f(Z_i|-) = \text{binomial}(N_i - a, npv_i)$$

Where $ppv_i = \frac{se_i\pi_i}{se_i\pi_i + (1-sp_i)(1-\pi_i)}$ and $npv_i = \frac{sp_i(1-\pi_i)}{sp_i(1-\pi_i) + (1-se_i)\pi_i}$. Note that, in the above algorithm, we do not sample from the random variables X_i or Y_i. As can be seen from Table 11.8, once we know the value (or distribution) of W_i we immediately know the value (or distribution) of $X_i = a - W_i$, since we have observed a. Similar logic holds for Y_i. The conditional distributions above are obtained by assuming differential misclassification. Under non-differential misclassification, there is only one sensitivity and one specificity parameter which have the following conditional posterior distributions:

$$f(se|-) = \text{beta}(\alpha + W_1 + W_0, \beta + Y_1 + Y_0)$$
$$f(sp|-) = \text{beta}(\gamma + Z_1 + Z_0, \delta + X_1 + X_0)$$

The remaining conditional distributions are unchanged if nondifferential misclassification is assumed.

In order to implement this Gibbs sampler algorithm for this example of nondifferential misclassification, an analyst proceeds through the following steps. First, the analyst must choose the values of parameters in the prior distributions (often called hyperparameters). In this case that would entail choosing values for $\epsilon_{a_i}, \epsilon_{b_i}, \alpha, \beta, \gamma, \delta$. Notice that in Chapter 8, we discussed how to choose values for α, β, γ, δ and set them equal to $\alpha = 50.6$, $\beta = 14.3$, $\gamma = 70$, $\delta = 1$. All that remains is to specify $\epsilon_{a_i}, \epsilon_{b_i}$ which we set to $\epsilon_{a_i} = 1, \epsilon_{b_i} = 1$ to represent our general ignorance in the prevalence of the exposure among cases and controls. Second, an analyst must choose initial values for θ_i. For example, suppose we chose initial values of $\theta_1 = (\pi_1 = 0.13, W_1 = 200, W_0 = 600, X_1 = 15, X_0 = 68, Y_1 = 449, Y_0 = 296,$ $Z_1 = 1000, Z_0 = 4000, se = 0.78, sp = 0.99)$. Note that the initial values were chosen with π_1 at the prevalence of apparent exposure, sensitivity and specificity at the mean of the bias parameter (prior) distribution and $W_1, X_1, Y_1,$ and Z_1 were chosen to sum N_1. Third, we draw random samples for each of the distributions above. For our example, that would mean drawing random samples from the following distributions (for the cases):

$$se \sim \text{beta}(50.6 + 200 + 600, 14.3 + 449 + 296)$$

$$sp \sim \text{beta}(70 + 1000 + 4000, 1 + 15 + 68)$$

$$\pi_1 \sim \text{beta}(200 + 449 + 1, 15 + 1000 + 1)$$

$$W_1 \sim \text{binomial}(215, ppv_1)$$

$$Z_1 \sim \text{binomial}(1449, npv_1)$$

We update the values of ppv_1 and npv_1 after we draw values of se, sp and π_1. For example, when sampling the sensitivity, specificity, and prevalence from the distributions above, we obtain values of $se_1 = 0.362$, $sp_1 = 0.990$, $\pi_1 = 0.395$. Inserting these values into formula for ppv_1 and npv_1 yields $ppv_1 = \frac{0.362 \cdot 0.395}{0.362 \cdot 0.395 + (1 - 0.990)(1 - 0.395)} = 0.959$ and $npv_i = \frac{0.990(1 - 0.395)}{0.990(1 - 0.395) + (1 - 0.362)0.395} = 0.704$. We then sample $W_1 \sim binomial(215, 0.959)$ which returns a value of $W_1 = 206$ and sample $Z_1 \sim binomial(1449, 0.704)$ which returns a value of $Z_1 = 1028$. We would need to repeat this process for the parameters among the controls.

At this point, we have completed the first iteration of the Gibb's algorithm and we store our results for future use. We can also compute a misclassification-adjusted OR by using the sample values of π_1 and π_0: $OR = \frac{\pi_1}{1-\pi_1} / \frac{\pi_0}{1-\pi_0}$. To draw a sample in a second pass through the Gibb's algorithm, we simply replace our intial values for θ_i with the values we sampled from the previous iteration: $\theta_1 = (\pi_1 = 0.395,$ $W_1 = 206, X_1 = 9, Y_1 = 421, Z_1 = 1028, se_1 = 0.362, sp_1 = 0.990)$. After we sample a new parameter, we replace the old parameter in this list and proceed to sample the next parameter. This procedure is repeated a very large number of times and, as described above we would account for the burn-in phase, during which we are attempting to locate the posterior distribution. At each step, we calculate the bias adjusted OR using the sampled prevalences, π_1 and π_0. After a large number of these iterations has been performed, we can examine the median of the distribution of the OR to get a summary of the bias-adjusted OR and the 2.5th and 97th percentiles for the Bayesian credible interval, which we shall see is quite similar to the probabilistic bias analysis simulation interval. In this example, running the Gibbs algorithm for cases and control, using the parameters described above, gives a result of OR=0.95 (0.77, 1.15).

Having developed a Bayesian solution to exposure misclassification that imputes the number of truly exposed and unexposed individuals, we are ready to compare the Gibb's sampling algorithm to the probabilistic bias analysis algorithm introduced in Chapter 8. Both algorithms proceed, generally, by sampling 1) se_i, sp_i, 2) π_i, and 3) W_i, X_i, Y_i, Z_i. We discuss similarities and differences of each step, in order.

The probabilistic bias analysis samples sensitivities and specificities without updating the prior (or bias parameter) distribution in light of the observed data. That is, our probabilistic bias analysis approach samples from $se_i \sim beta(\alpha_i, \beta_i)$ rather than $se_i \sim beta(\alpha_i + W_i, \beta_i + Y_i)$, which we would sample from in the Bayesian approach. The observed data generally have little influence on the prior distribution, as shown by MacLehose, so this difference is unlikely to have much impact on final results (1). To illustrate this, we compare the posterior distribution of the sensitivity and specificity from a proper Bayesian analysis with the prior distribution, which is sampled from in the probabilistic bias analysis approach in Chapter 8. We illustrate using the example data on smoking and breast cancer discussed in Chapter 8. Figure 11.11 shows the histogram (filled bars) of the specificity (left panel) and sensitivity (right panel) compared to the density of the beta prior distribution. For both parameters, we see the prior and the posterior are a near perfect match, suggesting the probabilistic bias analysis approach in this setting is an adequate approximation of a Bayesian approach.

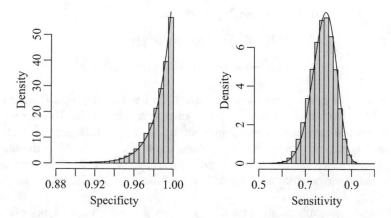

Figure 11.11 Comparison of posterior estimates of specificity (left panel) and sensitivity (right panel) (histograms) with their prior distribution (solid lines) used in probabilistic bias analysis.

In the next step, both algorithms sample the prevalence of exposure among cases and controls. In both the Bayesian approach and probabilistic bias analysis approach, the sampling is done using very similar distributions. In the Bayesian approach, we sample from $f(\pi_i|-) = beta(W_i + Y_i + \epsilon_{a_i}, X_i + Z_i + \epsilon_{b_i})$. While in the probabilistic bias analysis approach, we sample from $f(\pi_i|-) = beta(W_i + Y_i, X_i + Z_i)$. The only difference between the methods for this sampling step is the inclusion of prior information via ϵ_{a_i} and ϵ_{b_i}. In the probabilistic bias analysis, we assume both of these terms are 0, an improper prior designed to allow prior information to have as little impact as possible on the results. However, this also makes clear how one could alter the probabilistic bias analysis algorithm to incorporate prior knowledge.

In the final step, both algorithms sample the latent cell probabilities. In the Bayesian approach, we sample from $f(W_i|-) = binomial(a, ppv_i)$ and $f(Z_i|-) = binomial(N_i - a, npv_i)$. In the probabilistic bias analysis approach, we sampled from $f(W_i|-) = binomial(a, ppv_i)$ and $f(Y_i|-) = binomial(N_i - a, 1 - npv_i)$. Clearly these two sampling approaches will yield the same result since $N_i - a = Y_i + Z_i$.

At this point, the two algorithms are nearly identical. The primary difference between the two is that the Bayes algorithm updates parameter values sequentially, using previously sampled values. In contrast, the probabilistic bias analysis approach essentially resets itself and starts anew after each iteration of sampling π_i, W_i, X_i, Y_i, Z_i, se_i, sp_i. To be more specific, in the first pass of the Bayes algorithm, we sample π^1_i, W^1_i, X^1_i, Y^1_i, Z^1_i, se^1_i, sp^1_i, where the superscript indicates these parameters were sampled in the first iteration. In the second iteration of the Bayes algorithm, we use the previously drawn values. For example, we sample from:

$$f(se_i^2|-) = \text{beta}(\alpha_i + W_i^1, \beta_i + Y_i^1)$$

$$f(sp_i^2|-) = \text{beta}(\gamma_i + Z_i^1, \delta_i + X_i^1)$$

$$f(\pi_i^2|-) = \text{beta}(W_i^1 + Y_i^1 + \epsilon_{a_i}, X_i^1 + Z_i^1 + \epsilon_{b_i})$$

$$f(W_i^2|-) = \text{binomial}(a, ppv_i^2)$$

$$f(Z_i^2|-) = \text{binomial}(N_i - a, npv_i^2)$$

We note that ppv and npv have a 2 superscript because they are calculated from values of $W_i^2, X_i^2, Y_i^2, Z_i^2, se_i^2, sp_i^2$. Subsequent iterations of the Bayes algorithm would proceed similarly. This is in contrast to the probabilistic bias analysis approach where each set of steps is independent of the samples from the previous steps. Because there appears to be relatively little information available in the data to update the prior parameters, there is little information lost in failing to do this updating.

However, the probabilistic bias analysis approach we outlined takes one more step. After sampling π_i, W_i, X_i, Y_i, Z_i, se_i, sp_i at a given step of the algorithm, it resamples π_i conditional on the current parameters. From a Bayesian perspective, this final step can be viewed as taking a step into the next iteration of the Gibbs sampler. In simulation studies we have found that adding this one additional step moderately improves the coverage probability of estimates.

In the end, the probabilistic bias analysis approach appears to be a very good approximation to a Bayesian result and simulations (not shown) bear this out. Indeed, in the example from Chapter 8, we found a bias adjusted OR=0.95 (0.78, 1.14) while the Bayesian result returned a bias adjusted result of OR=0.95 (0.77, 1.15). The benefit of the Bayesian approach, of course, is that it is designed to sample from the exact posterior distribution of interest while the probabilistic bias analysis approach is sampling something that appears to be a reasonable approximation of that posterior distribution. The two main benefits of the probabilistic bias analysis approach are 1) it is far faster than the Bayesian approach since a vector of parameters can be drawn once at each step as opposed to the Bayes approach in which one must draw 1 parameter at a time in order to continually update the distribution from which we sample and 2) implementing a Bayesian approach requires careful consideration of burn-in periods and convergence, which is not required in the probabilistic bias analysis approach.

References

1. MacLehose RF, Gustafson P. Is probabilistic bias analysis approximately Bayesian? Epidemiology. 2012;23:151–8.
2. Greenland S. Multiple-bias modeling for analysis of observational data. J R Stat Soc Ser -Stat Soc. 2005;168:267–91.
3. Greenland S. Bayesian perspectives for epidemiological research: I. Foundations and basic methods. Int J Epidemiol. 2006;35:765–75.
4. Greenland S. Bayesian perspectives for epidemiological research. II. Regression analysis. Int J Epidemiol;36:195–202.
5. Greenland S. Bayesian perspectives for epidemiologic research: III. Bias analysis via missing-data methods. Int J Epidemiol. 2009;38:1662–73.
6. Dunson DB. Practical Advantages of Bayesian Analysis of Epidemiologic Data. Am J Epidemiol. 2001;153:1222–6.
7. Carlin BP, Louis TA. Bayesian Methods for Data Analysis. CRC Press; 2008.
8. MacLehose RF, Hamra GB. Applications of Bayesian methods to epidemiologic research. Curr Epidemiol Rep. 2014;1:103–9.
9. Hamra G, MacLehose R, Richardson D. Markov chain Monte Carlo: An introduction for epidemiologists. Int J Epidemiol. 2013;42:627–34.
10. Lian Q, Hodges JS, MacLehose R, Chu H. A Bayesian approach for correcting exposure misclassification in meta-analysis. Stat Med. 2019;38:115–30.
11. Congdon P. Bayesian Statistical Modelling. 704. John Wiley & Sons; 2007.
12. Gelman A, Carlin JB, Stern HS, Dunson DB, Vehtari A, Rubin DB. Bayesian Data Analysis. CRC Press; 2013.
13. Lunn D, Jackson C, Best N, Thomas A, Spiegelhalter D. The BUGS book: A Practical Introduction to Bayesian Analysis. CRC Press; 2012.
14. Berger JO. Statistical Decision Theory and Bayesian Analysis. Springer Science & Business Media; 2013.
15. Chung EK, McCollum KF, Elo IT, Lee HJ, Culhane JF. Maternal depressive symptoms and infant health practices among low-income women. Pediatrics. 2004;113:e523–9.

16. Gustafson P. Measurement Error and Misclassificaion in Statistics and Epidemiology: Impacts and Bayesian Adjustments. Boca Raton, Fla: Chapman & Hall/CRC; 2004.
17. Chu H, Wang Z, Cole SR, Greenland S. Sensitivity analysis of misclassification: a graphical and a Bayesian approach. Ann Epidemiol. 2006;16:834–41.
18. Press SJ. Subjective and Objective Bayesian Statistics: Principles, Models, and Applications. 590. John Wiley & Sons; 2009.
19. Kass RE, Wasserman L. The selection of prior distributions by formal rules. J Am Stat Assoc. 1996;91:1343–70.
20. Metropolis N, Rosenbluth AW, Rosenbluth MN, Teller AH, Teller E. Equation of state calculations by fast computing machines. J Chem Phys. 1953;21(6):1087–92.
21. Chib S, Greenberg E. Understanding the metropolis-hastings algorithm. Am Stat. 1995;49:327–35.
22. Gelman A, Rubin DB. Markov chain Monte Carlo methods in biostatistics. Stat Methods Med Res. 1996;5:339–55.
23. Gelman A, Rubin DB. Inference from iterative simulation using multiple sequences. Stat Sci. 1992;7:457–72.
24. Stan Development Team. Stan modeling language users guide and reference manual [Internet]. Available from: https://mc-stan.org
25. Plummer M. JAGS: A program for analysis of Bayesian graphical models using Gibbs sampling. In: Proceedings of the 3rd international workshop on distributed statistical computing. Vienna, Austria; 2003. p. 1–10.
26. MacLehose RF, Olshan AF, Herring AH, Honein MA, Shaw GM, Romitti PA. Bayesian methods for correcting misclassification an example from birth defects epidemiology. Epidemiol. 2009;20:27–35.
27. Lesko CR, Keil AP, Moore RD, Chander G, Fojo AT, Lau B. Measurement of current substance use in a cohort of HIV-infected persons in continuity HIV care, 2007–2015. Am J Epidemiol. 2018;187:1970–9.
28. Pearce N. What does the odds ratio estimate in a case-control study? Int J Epidemiol. 1993;22: 1189–92.
29. Gustafson P, Le ND, Saskin R. Case–control analysis with partial knowledge of exposure misclassification probabilities. Biometrics. 2001;57:598–609.
30. Lash TL, Abrams B, Bodnar LM. Comparison of bias analysis strategies applied to a large data set. Epidemiol. 2014;25:576–82.
31. Hernan MA, Hernandez-Diaz S, Robins JM. A structural approach to selection bias. Epidemiology. 2004;15:615–25.
32. Weuve J, Tchetgen Tchetgen EJ, Glymour MM, Beck TL, Aggarwal NT, Wilson RS, et al. Accounting for bias due to selective attrition: The example of smoking and cognitive decline. Epidemiol. 2012;23:119–28.
33. Bodnar LM, Tang G, Ness RB, Harger G, Roberts JM. Periconceptional multivitamin use reduces the risk of preeclampsia. Am J Epidemiol. 2006;164:470–7.
34. Robins JM, Ritov Y. Toward a curse of dimensionality appropriate (CODA) asymptotic theory for semi-parametric models. Stat Med. 1997;16:285–319.
35. Robins J, Wasserman L. Conditioning, likelihood, and coherence: A review of some foundational concepts. J Am Stat Assoc. 2000;95:1340–6.
36. McCandless LC, Gustafson P, Austin PC. Bayesian propensity score analysis for observational data. Stat Med. 2009;28:94–112.
37. Robins JM, Hernán MA, Wasserman L. On Bayesian estimation of marginal structural models. Biometrics. 2015;71:296.
38. McCandless LC, Douglas IJ, Evans SJ, Smeeth L. Cutting feedback in Bayesian regression adjustment for the propensity score. Int J Biostat. 2010;6.
39. Lunn D, Best N, Spiegelhalter D, Graham G, Neuenschwander B. Combining MCMC with 'sequential' PKPD modelling. J Pharmacokinet Pharmacodyn. 2009;36:19–38.

40. Saarela O, Stephens DA, Moodie EE, Klein MB. On Bayesian estimation of marginal structural models. Biometrics. 2015;71:279–88.
41. Capistrano ES, Moodie EE, Schmidt AM. Bayesian estimation of the average treatment effect on the treated using inverse weighting. Stat Med. 2019;38:2447–66.
42. Stang A, Schmidt-Pokrzywniak A, Lash TL, Lommatzsch PK, Taubert G, Bornfeld N, Jöckel KH. Mobile phone use and risk of uveal melanoma: Results of the risk factors for uveal melanoma case-control study. J Natl Cancer Inst. 2009;101:120–3.
43. Steenland K, Greenland S. Monte Carlo sensitivity analysis and Bayesian analysis of smoking as an unmeasured confounder in a study of silica and lung cancer. Am J Epidemiol. 2004;160: 384–92.
44. McCandless LC, Gustafson P. A comparison of Bayesian and Monte Carlo sensitivity analysis for unmeasured confounding. Stat Med. 2017;36:2887–901.
45. McCandless LC, Gustafson P, Levy A. Bayesian sensitivity analysis for unmeasured confounding in observational studies. Stat Med. 2007;26:2331–47.
46. Schlesselman JJ. Assessing effects of confounding variables. Am J Epidemiol. 1978;108:3–8.
47. Gustafson P, McCandless LC, Levy AR, Richardson S. Simplified Bayesian Sensitivity analysis for mismeasured and unobserved confounders. Biometrics. 2010;66:1129–37.
48. Lin DY, Psaty BM, Kronmal RA. Assessing the sensitivity of regression results to unmeasured confounders in observational studies. Biometrics. 1998;54:948–63.
49. Spiegelman D, Hertzmark E. Easy SAS calculations for risk or prevalence ratios and differences. Am J Epidemiol. 2005;162:199–200.
50. Muller CJ, MacLehose RF. Estimating predicted probabilities from logistic regression: different methods correspond to different target populations. Int J Epidemiol. 2014;43:962–70.
51. Richardson DB, Kinlaw AC, MacLehose RF, Cole SR. Standardized binomial models for risk or prevalence ratios and differences. Int J Epidemiol. 2015;44:1660–72.
52. Chu H, Cole SR. Estimation of risk ratios in cohort studies with common outcomes: A Bayesian Approach. Epidemiology. 2010;21:855–62.
53. Greenland S. Model-based Estimation of Relative Risks and Other Epidemiologic Measures in Studies of Common Outcomes and in Case-Control Studies. Am J Epidemiol. 2004;160:301–5.
54. Greenland S. A commentary on 'A comparison of Bayesian and Monte Carlo sensitivity analysis for unmeasured confounding.' Stat Med. 2017;36:3278–80.
55. Johnson CY, Howards PP, Strickland MJ, Waller DK, Flanders WD. Multiple bias analysis using logistic regression: an example from the National Birth Defects Prevention Study. Ann Epidemiol. 2018;28:510–4.
56. Tyndall MW, Ronald AR, Agoki E, Malisa W, Bwayo JJ, Ndinya-Achola JO, et al. Increased risk of infection with human immunodeficiency virus type 1 among uncircumcised men presenting with genital ulcer disease in Kenya. Clin Infect Dis. 1996;23:449–53.

Chapter 12
Multiple Bias Modeling

Introduction

Most nonrandomized epidemiologic studies, and even some randomized studies [1, 2], are susceptible to more than one threat to validity (i.e., multiple biases). Bias analysis applied to these studies requires a strategy to address each important threat to yield a reasonable estimate of the total bias affecting the study and the impact that it has on the magnitude and direction of the estimate of effect. The methods described in earlier chapters can be applied separately to adjust for one source of bias at a time; that is, a bias analysis would be conducted separately for misclassification and then for confounding and then for selection bias. Alternatively, they can be applied serially one after the other (with some important caveats on how this is done) to simultaneously quantify the biases and their associated uncertainties. Either type of adjustment, one at a time or serially, can be conducted using simple bias analysis techniques that do not produce simulation intervals or using probabilistic bias analysis techniques that do provide such intervals. In this chapter we will discuss serial adjustment using simple techniques and serial adjustment using probabilistic techniques.

The most easily accomplished approach to multiple bias analysis applies the simple bias analysis methods of Chapters 4, 5 and 6 separately for each source of bias. For each threat to validity, the analyst conducts one simple sensitivity analysis, without attempting to account for biases simultaneously. Although this method is the most straightforward, it has an important shortcoming beyond the general disadvantages of simple sensitivity analysis. Namely, the absence of an estimate of the joint or simultaneous effect of the biases. Each simple bias analysis will provide an estimate of the direction of the bias (i.e., toward or away from the null) and of the strength of the bias (i.e., how much does the estimate of association change with adjustment for the bias). The analyst can only guess that the joint effect will somehow average the effects of the individual biases, but, as we have demonstrated in previous chapters, it can often be difficult to intuit the direction and magnitude of the impact of a single source of bias; it is even more difficult to intuit the combined impact of multiple sources of bias.

© Springer Nature Switzerland AG 2021
M. P. Fox et al., *Applying Quantitative Bias Analysis to Epidemiologic Data*,
Statistics for Biology and Health, https://doi.org/10.1007/978-3-030-82673-4_12

For example, imagine a bias analysis that addresses the impact of independent nondifferential misclassification of a dichotomous exposure and the impact of an unmeasured confounder that is more prevalent in the exposed than the unexposed and increases the risk of the disease outcome. The simple bias analysis to address exposure misclassification will suggest that the original estimate of association was biased toward the null; the adjustment for misclassification will yield a bias-adjusted estimate further from the null. The simple bias analysis to address the unmeasured confounder will suggest that the original estimate of association was biased away from the null (presuming exposure increases risk of the outcome); the adjustment for the unmeasured confounder will yield an estimate of the association nearer to the null. The analyst may think that the joint effect will be a simple average of the two bias-adjustments, but that is not necessarily the case. This is because in some settings, one source of bias may dominate over the other, while in other settings there may be interactions between the impacts of the sources of bias. While these simple bias analyses may be a useful starting point to understand the general direction of each source of bias, they can easily yield an incorrect impression about the joint effects of more than one threat to validity, so ought to be avoided as a final analysis. Equally troubling, each bias analysis has its own simulation interval and there is no interval for the joint bias. The width of a final interval that incorporates all biases is even more difficult to intuit than the overall magnitude of the bias. At the least, separate bias analyses like this should be presented with the aforementioned caveat that the joint effects cannot be reliably estimated by averaging of the individually adjusted estimates. This warning applies particularly for bias adjustments that involve misclassification, which often do not yield a simple bias factor that can be included as a summed error term.

The second approach to multiple bias analysis is to conduct the first bias-adjustment and then, using the bias-adjusted data, perform a second bias adjustment, possibly within strata defined by the first bias adjustment (e.g., if the first bias-adjustment was for uncontrolled confounding). Finally, we would pool the results to obtain a summary estimate bias-adjusted for the various threats to validity. For example, to solve the multiple bias problem introduced in the preceding paragraph, one might first apply the simple bias analysis methods to address exposure misclassification, which were explained in Chapter 6. These misclassification-adjusted data could then be used in the bias-adjustment for the unmeasured confounder per the methods in Chapter 5. Finally, the stratum-specific associations calculated after rearrangement to address misclassification can then be pooled by conventional methods (e.g., standardization or information-weighted averaging such as Mantel-Haenszel methods).

Order of Bias Adjustments

The order in which bias adjustments ought to be made is an important consideration in any multiple bias analysis, including the one introduced in the preceding paragraph. Bias-adjustment does not generally reduce to independent multiplicative bias factors (with some exceptions [3]), so the order of bias-adjustments can affect the ultimate result. A general guiding principle for determining the order in which bias-adjustments should be made is that they should be done in the reverse of the order in which they occurred when the data were generated. For example, confounding is ordinarily perceived as a population-level phenomenon (i.e., correlation between factors occurs in the population regardless of whether a study is conducted), whereas misclassification often occurs as data are collected or tabulated. We conceive of confounding therefore as occurring before misclassification as the data are generated. Bias-adjustments should precede in the reverse order because we are seeking to adjust for each source of bias to get back to the data that would have occurred had the bias been absent. Doing this in reverse chronological order allows for adjustment on the data that would have existed (assuming the bias parameters are valid) at the time each bias occurred. The simple bias analysis to address misclassification should precede the simple bias analysis to address the unmeasured confounder. In this example, bias-adjustments for the unmeasured confounding should be done on the dataset that has already been bias-adjusted for misclassification because there would have been no misclassification at the time the confounding occurred.

Unfortunately, there is no constant rule regarding the order in which threats to validity arise. Furthermore, the order of bias-adjustment can be affected by the data source that informs the values assigned to bias parameters. For example, classification parameters might be measured in a population-based setting (i.e., negligible selection bias), but be applied to a data set where selection bias is a concern. In this setting, the analyst should bias-adjust for selection bias before bias-adjusting for misclassification, even if the selection bias preceded the misclassification in the data generation process. In this case, the data source requires application of the classification parameters to "unselected" data, even though the more general rule would suggest application of the bias analyses in the opposite order.

Although there is no general rule on the order in which the biases occur, some general principles should at least be considered in making the determination. As noted above, confounding occurs in the population before any study is done. Selection bias can occur at different times, but if it is related to selection into the study, before any data are collected, it is likely to occur after confounding but before any classification errors. If, on the other hand, selection bias is related to either selection out of the study (e.g., selective drop out from the study) or from analytic decisions (e.g., conditioning on a collider, or some analyses of subsets of the population) then the bias is likely to occur after classification errors. Even following these guidelines, the order can be difficult to determine. For example, if there are classification errors in both the exposure and outcome, selection bias could occur after mismeasurement of an exposure measured at study enrollment (e.g., in a prospective cohort study) but before mismeasurement of an outcome (loss to follow-up before study completion). In a retrospective case-control study, however, the selection bias may occur at enrollment into the study, before both the exposure and outcome are measured. The point is that the analyst must carefully grapple with the order of the biases as data generating mechanisms and bias-adjust in the reverse order in which they are believed to be generated in their specific study. If there is controversy around order in which the biases have occurred, a sensitivity analysis of the bias analysis can be conducted in which different orderings are used to determine if the assumptions lead to meaningful differences in the results.

It is also important to think through the timing of measurement errors in relation to when the data were collected. If data were collected prior to enrollment in the study, as might happen when collecting information from a medical record, the misclassification might precede selection bias. In other cases, several variables may be misclassified and the timing of the misclassification may be roughly at the same time (as exposures and confounders might be in a prospective study, or confounders, exposures and outcomes in a case-control study). In this case, if the measurement errors in each variable are independent of each other, the order of bias-adjustments in relation to other mismeasured variables is unlikely to matter. Finally, it is also worth noting that there can be multiple sources of measurement error for a single variable occurring at different times. For example, a medical record could contain error in the exposure that occurs prior to inclusion in the study, but then there may be additional misclassification that occurs through data entry errors at the time of entering the study. As noted above, it is therefore critical that the analyst pays careful attention to the timing and mechanisms of these data-generating events.

For each multiple bias analysis, the analyst must determine the order of application of the bias-adjustments, taking account of the order in which the threats to validity arise during the data generation process, the influence of the data source used to assign values to the bias parameters, and (for probabilistic bias analysis) the point in the analysis when random error should be included. The last topic will be discussed further in the section of this chapter on multiple bias analysis using probabilistic methods.

Multiple Bias Analysis Example

In this chapter, we will apply multiple bias analysis methods to a study of the association between use of antidepressant medications and the occurrence of breast cancer (observed data from Chien et al., 2006 [4]). We first introduced this study in an example in Chapter 7. This population-based case–control study used the US Surveillance Epidemiology and End Results cancer registry to enroll 975 women with primary breast cancer diagnosed in western Washington State between 1997 and 1999. The Centers for Medicare and Medicaid Services records were used to enroll 1007 controls. All women were 65–79 years old at enrollment. Participation rates were 80.6% among cases and 73.8% among controls.

In-person interviews were used to collect information on medication use in the 20 years preceding breast cancer diagnosis (cases) or a comparable index date (controls) and on other known and suspected breast cancer risk factors. Table 12.1 shows the frequency of ever and never use of antidepressants among cases and controls, as well as the crude and adjusted odds ratios associating antidepressant use with breast cancer occurrence. The crude and adjusted odds ratios are nearly identical, so we will use the crude frequencies to conduct the bias analyses.

Table 12.1 Crude and adjusted odds ratios associating 20-year history of ever use of antidepressants, versus never use of antidepressants, with incident breast cancer.

	Ever use of antidepressants	Never use of antidepressants
Cases	118	832
Controls	103	884
Crude odds ratio (95% CI)	1.22 (0.92, 1.61)	
Adjusted odds ratio (95% CI)	1.2 (0.9, 1.6)	

Observed data from Chien et al., 2006 [4].

The adjusted odds ratio controls for confounding by age and county of residence and was only reported to one significant digit in the manuscript. Adjustment for other measured potential confounders changed the odds ratio by less than 10%.

To illustrate the application of multiple bias analysis methods, we will consider three threats to the validity of the reported association. First, cases were more likely to participate than controls. If participation is also related to use of antidepressants, then the results are susceptible to selection bias. Second, although the analysts adjusted for many potential confounders, they did not adjust for physical activity. Physical activity may be related to use of antidepressants, and physical activity has been shown to reduce the risk of breast cancer in some studies. The study results may therefore be susceptible to bias from this unmeasured confounder. Last, medication history was self-reported, so the study results are susceptible to misclassification of the exposure. Bias analysis methods to address each of these threats to validity have been explained in the preceding chapters. The focus of this analysis, therefore, is to show how multiple bias analysis can be used to estimate an effect that is adjusted for all three threats to validity.

Serial Multiple Bias Analysis, Simple Methods

Due to the nature of the biases and the study design, we believe the threats to validity in this study were introduced in the following order: confounding, selection bias, misclassification. Confounding exists in the population as a relation between physical activity and breast cancer and an association between physical activity and use of antidepressant medications. While physical activity could directly increase or decrease the use of antidepressant medications (for instance, increased physical activity could reduce depressive symptoms reducing the need for medication), physical activity and antidepressant use also likely share common ancestors in a causal graph, which necessitates control of physical activity. Selection bias arises because the analysis can be conducted only among participants. It is clear from the enrollment proportions that cases were more likely to participate than controls. We do not know from the study data, however, whether use of antidepressant medications affected participation rates. It is reasonable to assume, however, that postmenopausal women who use antidepressants, or with a history of antidepressant use, might participate at different rates than women who never used antidepressants. If so, then the result is conditioned on a factor (participation) that is a descendant of both the exposure and the outcome, which would introduce a selection bias. Finally, among the participants, self-reported history of antidepressant use may be misclassified. Because cases were aware of their disease status at the time of their interview, they may have recalled or reported their history of medication use differently than did the controls. The analysts validated self-report of antidepressant medication use against pharmacy records in a subset of participants [5]. This internal validation study will inform the misclassification bias analysis. Given that the classification errors were introduced last in the data generation process, misclassification will be the first threat to validity examined in the bias analysis.

Simple Misclassification Bias Analysis

At the time of enrollment into the parent study [4], the investigators also sought consent from participants to compare their interview answers regarding medication use with pharmacy records [5]. Not every participant in the parent study agreed to this validation; 1.6% of cases refused to allow the validation and 7.3% of controls refused to allow the validation. Among those who agreed to allow the validation, pharmacy records were only available for those whose medical care was provided through an integrated health care system in western Washington and for those who said they always filled their prescriptions at one of two large retail pharmacies. Finally, for those who satisfied these criteria, not all had pharmacy records that could be located. After these restrictions, the validation study was conducted among 403 of the 1937 parent study participants. This raises the possibility of selection bias in the validation study; however, we set that concern aside for the moment and return to it in the probabilistic bias analysis section of this chapter.

The pharmacy records were available for only the preceding 2 years for all validation study participants, whereas the exposure classification in the parent study was based on a self-reported 20-year medication history. Comparing the self-report of medication use in the preceding 2 years with the pharmacy records yielded the cross-classification frequencies shown in Table 12.2.

These classification frequencies can be used to estimate the bias parameters. For cases, the sensitivity equals 24/(24 + 19) = 56% and the specificity equals 144/(144 + 2) = 99%. For controls, the sensitivity equals 18/(18 + 13) = 58% and the specificity equals 130/(130 + 4) = 97%. When we use these bias parameters with the simple sensitivity analysis spreadsheet used to address misclassification (Chapter 6), the odds ratio bias-adjusted for classification errors equals 1.63 (Table 12.3). Spreadsheets available at the website for this text can quickly perform this adjustment. We will use these bias-adjusted data to proceed with the multiple bias analysis by using them to adjust for selection bias.

Table 12.2 Cross-classification frequencies of interview responses and 2-year pharmacy records for cases and controls, as reported by Boudreau et al., 2004 [5].

Interview	2-year pharmacy record (cases)	
	Ever AD	Never AD
Ever AD	24	2
Never AD	19	144

Interview	2-year pharmacy record (controls)	
	Ever AD	Never AD
Ever AD	18	4
Never AD	13	130

Table 12.3 Bias-adjusted odds ratios associating 20-year history of ever use of antidepressants, versus never use of antidepressants, with incident breast cancer[a].

	Ever use of antidepressants	Never use of antidepressants
Cases	192.8	757.2
Controls	133.5	853.5
Bias-adjusted odds ratio	1.63	

Observed data from Chien et al., 2006 [4]. Validation data from Boudreau et al., 2004 [5].
[a]The bias-adjusted cell frequencies and odds ratio were computed using the methods of Chapter 6, the sensitivity and specificity values given in the text, and the observed crude data shown in Table 12.1. Frequencies are rounded, but exact bias-adjusted frequencies were used to compute the bias-adjusted odds ratio, which was then rounded.

Simple Selection Bias Analysis

As discussed above, the order biases likely occurred in this study was confounding then selection bias then misclassification. Working backward, now that we have misclassification adjusted data, we can address selection bias. As reported in the summary of the study, controls and cases enrolled in the study at different rates. The proportion of controls who agreed to participate equaled 73.8% and the proportion of cases who agreed to participate equaled 80.6%. It is not unusual for the participation rate in cases to exceed the participation rate in controls, most likely because cases are motivated by their disease to participate in research related to it. Controls do not have so salient a motivation.

Among those who did not participate, we do not know the proportion who used antidepressant medications in the preceding 20 years. Note that, because we have bias-adjusted for exposure misclassification, we are interested now in the gold-standard proportion, not in the proportion that would have self-reported such a history of antidepressant medication use in the preceding 20 years. Because the results have been bias-adjusted for classification errors, the gold-standard exposure classification is the classification concept to apply when conducting the simple bias analysis to address selection bias.

Not only is the proportion who used antidepressant medications unknown among nonparticipants, but it is also unknown whether this proportion is different for controls and cases. We have no data from the study to inform this selection bias analysis, so will have to assign values to the bias parameters from external data sources. To begin, we will assume that the proportion participating among controls or cases is a simple weighted average of the participating proportions among antidepressant users and nonusers. That is:

$$p_{observed} = (1 - w_{AD+})p_{AD-} + w_{AD+} \cdot p_{AD+}$$
$$= (1 - w_{AD+})p_{AD-} + R \cdot w_{AD+} \cdot p_{AD-}$$

where $p_{observed}$ is the observed participation proportion, w_{AD+} is the prevalence of antidepressant use, R is the participation risk ratio for antidepressant users against nonusers, and p_{AD-} is the participation proportion among women who did not use antidepressants in the 20 years preceding their interview. The participation proportion among women who did use antidepressants in the 20 years preceding their interview (p_{AD+}) equals Rp_{AD-}. The parameters p_{AD-} and p_{AD+} (for cases and controls) are the selection probabilities that we need to implement a bias analysis. We will calculate them by specifying values for w_{AD+} and R and solving Equation 12.1 for p_{AD-}:

$$p_{AD-} = \frac{p_{observed}}{(1 - w_{AD+}) + R \cdot w_{AD+}} \qquad (12.1)$$

and then calculating $p_{AD+} = p_{AD-} \cdot R$. We note that 12.1 will be estimated separately, with different parameters, for cases and controls.

Note that only the overall participation rate ($p_{observed}$) has been observed, among cases and controls; the values assigned to the other variables must be derived from external data sources or educated guesses. For the proportion of women using antidepressant medications among cases (w_{AD+}) we will use the prevalence of moderate or severe depression among women with benign breast disease (12%) [6]. Although this study was published in 1978, the assessment period corresponds well to the twenty-year look-back period of the Chien et al. study [4], which enrolled participants 1997–1999. We assume that this prevalence is the same in controls and cases, which effectively introduces a null-centered prior on the association we wish to measure through the bias parameters. This is not equivalent to placing a null-prior on the main effect of interest in a Bayesian context. In other applied settings, it may be reasonable not to assume this equality. Furthermore, we recognize that not all such women will have used antidepressant medications, and that not all women taking antidepressants will have been diagnosed with moderate or severe depression. Nonetheless, this exposure prevalence is consistent with the 20-year antidepressant exposure prevalence in the Chien et al. study [4] [10.4% in the observed data (Table 12.1) and 13.5% in the misclassification-adjusted data (Table 12.3)]. We assign R a value of 0.8 for controls and 0.9 for cases, indicating that women with a history of antidepressant use are less likely to participate than those without, but that this difference is likely to be less pronounced among cases than among controls.

Table 12.4 Misclassification and selection bias adjusted cell frequencies and odds ratio associating 20-year history of ever use of antidepressants, versus never use of antidepressants, with incident breast cancer.

	Frequency of ever use of antidepressants [modeled selection proportion]	Frequency of never use of antidepressants [modeled selection proportion]
Cases	262.6 [0.734]	928.1 [0.816]
Controls	220.7 [0.605]	1128.7 [0.756]
Bias-adjusted odds ratio	1.45	

Observed data from Chien et al., 2006 [4].

With these parameters specified, we can generate the selection probabilities we need for the bias analysis. We first calculate the probability of selection among unexposed controls as:

$$p_{AD-} = \frac{P_{observed}}{[(1 - w_{AD+}) + R \cdot w_{AD+}]} = \frac{0.738}{[(1 - 0.12) + 0.8 \cdot 0.12]} = 0.756$$

We can then calculate the probability of selection among exposed controls as $p_{AD+} = p_{AD-} \cdot R = 0.756 \cdot 0.8 = 0.605$. The selection probabilities among the cases are calculated in the same manner.

When we use these bias parameters with the simple bias analysis methods to address selection bias (Chapter 4), the odds ratio bias-adjusted for selection bias and misclassification equals 1.45, as shown in Table 12.4. We will use these bias-adjusted data to proceed with the multiple bias analysis.

The bias-adjusted cell frequencies and odds ratio were computed using the methods from Chapter 4. Selection proportions were computed using Equation 12.1 and the values assigned to the bias parameters as described in the text. The misclassification-adjusted data shown in Table 12.3 provided the input frequencies. For example, the bias-adjusted frequency of ever use of antidepressants among cases is computed as 192.8/0.734 ≈ 262.6. Frequencies are rounded, but exact bias-adjusted frequencies were used to compute the bias-adjusted odds ratio, which was then rounded. The cell counts in Table 12.4 and the bias adjusted OR = 1.45 have been adjusted for both misclassification and selection bias. We note that had we switched the order of operations and adjusted for selection bias prior to adjusting for misclassification bias, we would have obtained bias adjusted OR = 1.35. Bias adjustments, as we said earlier, are sensitive to the order in which they are performed. We retain the (misclassification and then selection) bias-adjusted cell counts and use them to conduct a final adjustment for an unmeasured confounder.

Simple Unmeasured Confounder Analysis

Chien et al. (2006) did not report using directed acyclic graphs to determine which variables to adjust for, but rather used a change in estimate of effect criteria to determine which variables to include in their models. They considered adjusting the association between antidepressant risk and breast cancer occurrence for confounding by race/ethnicity, income, marital status, education, time since last routine medical check-up, age at menarche, parity, age at first birth, type of menopause, age at menopause, duration of contraceptive use, use of menopausal hormone therapy, first-degree family history of breast cancer, cigarette smoking status, alcohol consumption, body mass index, and medical history of depression, hypertension, hypercholesterolemia, arthritis, diabetes mellitus, or thyroid problems. None of these potential confounders resulted in a change of the association by more than 10% upon adjustment, so ultimately the analysts adjusted only for age, county of residence, and reference year. Although some of these potential confounders may not satisfy the prerequisite conditions for a confounder of the association between antidepressant use and breast cancer occurrence, assuming these confounders were measured well, the fact that adjustment for such a comprehensive set of breast cancer risk factors (known and suspected) resulted in negligible change in the odds ratio suggests that the crude association is little affected by measured confounding (or, potentially, that the confounders suffered from misclassification). Nonetheless, for the purpose of illustration, we will proceed with a bias analysis to account for the potential confounding by physical activity. It is likely that physical activity would correlate with some of the confounders for which adjustment was made (e.g., cigarette smoking, alcohol consumption and body mass index) and for which there was negligible evidence of important confounding. The simple bias analysis does not account for the strong prior evidence that confounding by physical activity is likely to be negligible. Rather, the simple bias analysis treats confounding by physical activity as independent of the confounding by the measured confounders.

Recall from Chapter 5 that a simple bias analysis to address an unmeasured confounder requires information on the strength of association between the confounder and the outcome, as well as information on the prevalence of the confounder in the exposed and unexposed groups. For this simple bias analysis, we assigned a value of 0.80 to the association between any strenuous physical activity (versus no strenuous physical activity) and incident breast cancer. This is the value of the average association reported in Table 2 of a systematic review of physical activity and breast cancer studies [7]. For the prevalence of physical activity in postmenopausal antidepressant users and nonusers, we will substitute the prevalence observed in a contemporaneous community-based sample of perimenopausal women aged 45–54 [8]. In that cross-sectional survey, 43.6% of women with a Center for Epidemiologic Studies-Depression Scale (CES-D) score below 16 reported regular moderate or heavy exercise, whereas 29.9% of women with a CES-D score of 16 or greater (the standard cut-off for possible depression) reported regular moderate or heavy exercise. We recognize that this measure of depressive symptoms is not the

Table 12.5 Bias-adjusted odds ratios associating 20-year history of ever use of antidepressants, versus never use of antidepressants, with incident breast cancer.

	Regular exercise		No regular exercise	
	Ever use of antidepressant	Never use of antidepressant	Ever use of antidepressant	Never use of antidepressant
Cases	66.8	354.6	195.8	573.5
Controls	66.0	492.1	154.7	636.6
Bias-adjusted stratum odds ratio	1.40		1.40	
Bias-adjusted summary odds ratio	1.40			

Observed data from Chien et al., 2006 [4].

same as the measure of exposure to antidepressants. Some women with high CES-D scores would not have been taking antidepressants, and some women with low CES-D scores would have been taking antidepressants. Nonetheless, because we could not find a contemporaneous report of exercise habits in women taking antidepressant medication, and because these prevalences correspond with our expectation that exercise habits would be lower in depressed women (and hence in women taking antidepressants), they are a sound basis to start with the simple sensitivity analysis. When we use these bias parameters in the simple sensitivity analysis spreadsheet used to address an unmeasured confounder (Chapter 5), the odds ratio bias-adjusted for selection bias and misclassification equals 1.40, as shown in Table 12.5.

The bias-adjusted cell frequencies and odds ratio were computed using the methods of Chapter 5. Stratified frequencies were computed using the values assigned to bias parameters as described in the text. The misclassification- and selection-adjusted data shown in Table 12.4 provided the input frequencies. Frequencies are rounded, but exact bias-adjusted frequencies were used to compute the bias-adjusted odds ratios, which was then rounded.

The crude OR from Chien et al. was 1.22, indicating a slight increase in breast cancer risk among those with a history of antidepressant use. However, the presence of misclassification, selection, and confounding biases makes interpreting that association difficult. By serially adjusting for biases using simple quantitative bias techniques from Chapters 4, 5 and 6, we estimate a bias-adjusted effect (OR = 1.40) somewhat larger than the one that was reported. The largest change resulted from bias-adjustment for exposure misclassification (OR = 1.22 vs. 1.63). Both selection bias and uncontrolled confounding resulted in a reduction in the bias adjusted OR. That is, the selection and confounding bias were acting in a different direction than the misclassification bias. Accounting for both of these biases reduced the bias-adjusted estimate from 1.63 to 1.40, still higher than the original estimate of 1.22. It would be difficult to have any reliable intuition on the magnitude of effect (or even direction) that these three biases would have in conjunction. Based on the bias parameters, a clever analyst could postulate the direction that each bias would move the estimate; however, it would be almost impossible to judge the relative impact of each bias. Only a formal quantitative bias could provide such information.

Serial Multiple Bias Analysis, Multidimensional Methods

The preceding simple multiple bias analysis used one set of bias parameters to adjust serially the estimate of association for each threat to validity we identified (exposure misclassification, selection bias, and an unmeasured confounder). The multidimensional multiple bias analysis repeats these calculations using different values for the bias parameters, and different combinations of those assigned values. We begin by creating multiple bias parameter scenarios for each simple bias analysis, including for each a scenario that corresponds with no bias.

Misclassification Scenarios

A . None, all case and control sensitivity and specificity set to 100%
B . As reported by Boudreau et al. (2004) [5] and used in the simple misclassification bias analysis section; sensitivity in cases ≈56%, specificity in cases ≈99%, sensitivity in controls ≈58%, specificity in controls ≈97%
C . Nondifferential; sensitivity = 50%, specificity = 100%
D . Differential sensitivity; sensitivity in cases = 60%, sensitivity in controls = 50%; specificity in cases and controls = 100%
E . Differential; sensitivity in cases = 60%, sensitivity in controls = 50%; specificity in cases = 97%, specificity in controls = 100%

Selection Bias Scenarios

1. None; all selection proportions set to 100%
2. The scenario described in the simple selection bias analysis section
3. Proportion using antidepressants (w_{AD+}) = 20% in controls and 25% in cases, ratio participating (R for AD+ versus AD–) = 0.7 in controls and 0.8 in cases, resulting in participation proportions of 78.5% and 84.8% in AD– controls and cases, respectively, and 55.0% and 67.9% in AD+ controls and cases, respectively

Unmeasured Confounder Scenarios

α   None; no association between antidepressant use and physical activity, or no association between physical activity and breast cancer occurrence, or both
β   The scenario described in the simple unmeasured confounder bias analysis section
γ   Risk ratio associating strenuous physical activity, versus no strenuous physical activity, with incident breast cancer = 0.80; prevalence of strenuous physical activity among those taking antidepressants = 25%, and prevalence of strenuous physical activity among those not taking antidepressants = 40%.

Table 12.6 Estimates of the association between ever-use of antidepressants in the preceding 20 years, versus never use, adjusted for combinations of simple bias analyses to address misclassification (scenarios A–E), selection bias (scenarios 1–3), and an unmeasured confounder (scenarios α–γ).

	α			β			γ		
	1	2	3	1	2	3	1	2	3
A	1.22	1.08	1.07	1.20	1.07	1.05	1.18	1.05	1.03
B	1.63	1.45	1.42	1.61	1.43	1.41	1.58	1.40	1.38
C	1.25	1.11	1.10	1.24	1.10	1.08	1.21	1.08	1.06
D	0.99	0.88	0.87	0.98	0.87	0.86	0.96	0.85	0.84
E	0.75	0.67	0.66	0.74	0.66	0.65	0.73	0.65	0.64

Observed data from Chien et al. (2006) [4].

Table 12.6 presents the results of the multidimensional bias analysis that simultaneously incorporates all the preceding scenarios. The misclassification scenarios are in rows (labeled with their letters), the selection bias scenarios are in minor columns (labeled with their numbers), and the unmeasured confounder scenarios are in major columns (labeled with their Greek letters). To read the table, one selects a combination of scenarios that is of interest, and then finds the interior table cell at the intersection of those scenarios. For example, if one is interested in finding the conventional result, which corresponds to the combination of the three no-bias scenarios (A, 1, α), one can find it in the upper left interior cell. That odds ratio (1.22) corresponds with the conventional result reported by Chien et al., 2006 [4].

One can see from the table that, given these scenarios, misclassification of the exposure (comparisons across rows) and selection bias (comparisons across minor columns) have a larger influence than the unmeasured confounder (comparisons across major columns). The approximately nondifferential scenario informed by the validation substudy suggests that the classification errors created a bias toward the null (i.e., comparing row B with row A), although the size of the bias is largely influenced by assigning an imperfect specificity (i.e., comparing the change from row A to row B with the change from row A to row C). The differential scenarios that correspond with errors one might expect given the retrospective design [(i.e., recall bias, modeled by better sensitivity of exposure classification in cases than in controls (scenarios D and E) and some false-positive reports of antidepressant use in cases but not in controls (scenario E))] suggest that the classification errors created a bias away from the null.

The selection bias, given these scenarios, was away from the null (comparing minor columns 2 or 3 with column 1 in any major column), and the two choices of values assigned to the bias parameters had little effect on the estimated size of the bias (comparing minor column 2 with minor column 3 in any major column).

The unmeasured confounder, given these scenarios, created little bias of the estimate of association (comparing any cell in major columns β or γ with its corresponding cell in major column α).

Finally, one can see that the range of bias-adjusted odds ratios in the interior cells (0.64–1.63, for a width of $1.63/0.64 = 2.56$ (recall that we are rounding last)) is 44%

wider than the conventional 95% frequentist interval (0.9–1.6, for a width of 1.6/ 0.9 = 1.78), despite the fact that the range of values in the table incorporates no sampling error, indicating we have substantially more uncertainty about the magnitude of effect resulting from possible biases than from random error.

Serial Multiple Bias Analysis, Probabilistic Methods

To complete the example, we extend the analysis from a multidimensional simple multiple bias analysis to a probabilistic multiple bias analysis. One way to think about the multidimensional bias analysis is as a probabilistic bias analysis, with the probability density for the bias parameters evenly distributed between its assigned values. For example, in Table 12.6, the sensitivity of classification of antidepressant use among cases takes on four values: 100% (Scenario A), ~56% (Scenario B), 50% (Scenario C), and 60% (Scenarios D and E). Thus, one fifth of the probability density is assigned to each of the first three values, and two fifths of the probability density is assigned to the last value. In a conventional analysis, all of the probability density is implicitly assigned to a sensitivity of 100%.

A probabilistic multiple bias analysis assigns more conventional continuous probability density distributions to the values of the bias parameters, reflecting subject matter knowledge about uncertainty in those parameters. These density distributions are parameterized based on internal and external validation studies and evidence, combined with intuition and experience (i.e., educated guesses). This process was described in more detail in Chapters 3, 7, 8 and 9. Below we describe the process we applied to this example. As always, the results are conditional on the assigned distributions being approximately accurate. Other stakeholders may assign different and equally reasonable distributions, which would allow an analysis of the sensitivity of the bias analysis to the assigned distributions. We have provided the R code we used to conduct the probabilistic multiple bias analysis on the book's web site, which is good practice so that users can implement this code with different assigned density distributions to perform this sensitivity analysis of the bias analysis [9].

The bias analyses we describe here follow a familiar process, very similar to the one we described for simple bias analysis. First, we will conduct a misclassification probabilistic bias analysis. We will sample bias parameters and then bias-adjust the observed cell counts for misclassification using the algorithm in Chapter 8. At the end of that process we will retain the bias-adjusted cell counts and use them in the selection bias adjustment. If we implemented the exposure misclassification probabilistic bias analysis for 10^6 iterations, we would have 10^6 bias-adjusted tables at the end of our procedure. These will be used to estimate a misclassification-adjusted effect estimate, but we will use them for something else as well. In the second step, we will adjust for selection bias by taking the 10^6 saved bias-adjusted tables from the first step and applying a selection bias analysis to each of them. One slight change in approach is worth noting, however. Rather than taking a large number of samples from the bias distribution for each of the 10^6 tables, we will conduct a single bias-adjustment on each table. That is, we will draw one set of selection bias parameters and implement one iteration of a selection bias adjustment for each of the 10^6 tables. The reason for this modified procedure is computational time. If we were to implement a full bias analysis on each table, given that we have three sources of bias, we would need to draw a quintillion (10^{18}) samples. Needless to say, this would be time consuming. With one sample per bias-adjusted table, at the conclusion of this step, we will have 10^6 sets of misclassification and selection bias adjusted tables. This continues in the third step where we take those tables that were previously adjusted for misclassification and selection bias and adjust them for uncontrolled confounding. The final set of adjusted tables will be summarized to produce a point estimate and measure of uncertainty (variance) as in Chapter 8. As we mentioned in Chapters 8 and 9, there are other approaches to obtaining variance estimates (the bootstrap and jackknife) that could be used instead.

Probabilistic Misclassification Bias Analysis

Recall that the simple and multidimensional analyses of bias from misclassification addressed errors in self-reported use of antidepressant medication in the 20 years preceding the interview. An internal validation study showed ~56% sensitivity in cases, ~58% sensitivity in controls, ~99% specificity in cases, and ~97% specificity in controls. These internal validation data were collected from a subset of the study population (see the section in simple misclassification bias analysis). Although the internal validation data suggest that misclassification was approximately nondifferential, the study design suggests the very real potential for differential misclassification. Breast cancer cases interviewed shortly after their diagnosis are likely to recall and report antidepressant use differently than are controls without the memory stimulation of their diagnosis. One would expect, therefore, that the sensitivity of classification would be greater in cases than in controls (yielding fewer false-negatives in cases) and that the specificity of classification could be lower in cases than in controls (yielding more false-positives among cases).

The tension one faces in assigning bias parameters in this example, therefore, is to decide how strongly to be influenced by the internal validation study versus the inclination to favor the pattern expected for recall bias. The internal validation data, if accepted without scrutiny, would require that we assign exactly the classification parameters observed and reported by Boudreau et al. (2004) [5]. The probabilistic bias analysis would assign the entire probability density to the reported classification parameters, so would then become equivalent to the simple misclassification bias analysis shown earlier. In general, analysts should carefully scrutinize all validation data before applying it to a population in a quantitative bias analysis.

These classification parameters clearly do not pertain exactly as measured, however, since they are reports of the 2-year recall of antidepressant use compared with pharmacy records from the same period, whereas the exposure contrast is a measure of the 20-year history of antidepressant use. In addition, the reported classification parameters are proportions, so measured with binomial error. At the least, one would want to incorporate the random classification error into the probabilistic bias analysis. For example, we could use a beta distribution to translate the random error in the validation study into our uncertainty about the bias parameter values (see Chapter 7). Furthermore, an understanding of the study design's susceptibility to differential classification errors would support higher sensitivity in cases – but slightly higher sensitivity was actually observed in controls – and lower specificity in cases, but lower specificity was actually observed in controls. Recall that the validation study population was restricted to members of a single health maintenance organization and to study subjects who reported that they filled all prescriptions at one of two large retail pharmacies [5]. It is possible that the recall of antidepressant medication use in persons with these characteristics is less likely to be influenced by their disease state than the remainder of the study population and, as such, adjustments to the validation parameters may need to be made. We ultimately address the tension between the validation study findings and our prior belief of differential classification errors by conducting two probabilistic bias analyses. We will conduct a primary analysis that is based exclusively on the validation data in the study. A sensitivity analysis of this probabilistic multiple bias analysis will specify bias distributions for sensitivity and specificity that are more in line with our prior belief and will describe the sensitivity of the results and inferences to different assumption sets about misclassification of self-reported antidepressant use.

For the first probabilistic bias analysis, we use the validation data to directly assign values to beta distributions. Using the data in Table 12.2 [5], we assign the following distributions: sensitivity in cases follows a beta($\alpha = 24, \beta = 19$) distribution, sensitivity in controls follows a beta($\alpha = 18, \beta = 13$) distribution, specificity in cases follows a beta($\alpha = 144, \beta = 2$) distribution, and specificity in controls follows a beta($\alpha = 130, \beta = 4$) distribution. We induced a correlation (r = 0.8) between draws from the two sensitivity distributions and the same correlation between draws from the two specificity distributions (see Chapter 7).

For the second probabilistic bias analysis, we assign largely overlapping trapezoidal distributions to the case and control sensitivities and to the case and control specificities. We switch from beta distributions to trapezoidal distributions to give readers exposure to a wider variety of distributions. The location and width of these assigned distributions were informed by the validation data shown in Table 12.2, but the relative locations of the sensitivity and specificity distributions for cases and controls are informed more by the expectation under differential misclassification emanating from recall bias. For the case sensitivity, we assigned a minimum of 0.45, a lower mode of 0.5, an upper mode of 0.6, and a maximum of 0.65. This parameterization reflects our best representation, given the validation data and our educated guess about the range of sensitivity. We represent this parameterization by specifying that the case sensitivity is distributed as trapezoidal(min = 0.45, mode1 = 0.5, mode2 = 0.6, max = 0.65). We parameterized the control sensitivity as trapezoidal(min = 0.4, mode1 = 0.48, mode2 = 0.58, max = 0.63). Note that the distributions largely overlap and are centered near the results of the validation study, but the probability mass is centered somewhat lower for controls than for cases, which reflects our expectation of differential misclassification. We parameterized the case specificity as trapezoidal(min = 0.95, mode1 = 0.97, mode2 = 0.99, max = 1) and the control specificity as trapezoidal(min = 0.96, mode1 = 0.98, mode2 = 0.99, max = 1). Again, the two trapezoidal distributions largely overlap and are centered near the results of the validation study, but the probability mass is centered somewhat higher for controls than for cases, which again reflects our expectation of differential misclassification. As with the beta distributions above, we assume a correlation of (r = 0.8) between sensitivities and the same correlation between specificities. Figure 12.1 shows the probability density distributions for both bias analyses and their relations to one another.

Figure 12.2 is a scatter plot of the values selected from these distributions in the first 200 of 1,000,000 iterations. The plot shows the sensitivity in cases plotted against the sensitivity in controls, and of the specificity in cases plotted against the specificity in controls. The ascending diagonal line represents nondifferential scenarios, when the case sensitivity equals the control sensitivity, or the case specificity equals the control specificity. Red squares represent the draws from the beta distributions from the main analysis and blue dots represent the draws from the trapezoidal distributions from the sensitivity analysis. The draws from the specificity distributions are much more closely clustered than the draws from the sensitivity distributions, reflecting the much narrower distributions assigned to the specificities than to the sensitivities (Figure 12.1). For the draws from the beta distributions (red squares), most pairs drawn from the sensitivity distributions lie above the ascending diagonal, which is consistent with the parameterization that expects better sensitivity in controls than in cases. Most of the pairs drawn from the specificity distributions lie below the ascending diagonal, which is consistent with the parameterization that expects better specificity in cases than in controls. The opposite pattern applies to draws from the trapezoidal distributions (blue dots). For these draws, most of pairs drawn from the sensitivity distributions lie below the ascending diagonal, which is consistent with the parameterization that expects better sensitivity in cases than in controls. Most of the pairs drawn from the specificity distributions lie above the ascending diagonal, which is consistent with the parameterization that expects better specificity in controls than in cases.

We implemented a probabilistic bias analysis for misclassification under each of the two parameterizations above. Algorithms were run for 1,000,000 iterations. Under the primary bias model, which assigns beta distributions to the sensitivities and specificities directly from the validation data [5], the median odds ratio that also incorporated conventional sampling error equaled 1.63 [95% simulation interval (SI): 1.03, 3.04]. This result is depicted in panel B.1 of Figure 12.3. Under the second bias model, which assigns trapezoidal distributions to the sensitivities and specificities that are consistent with the expectations under recall bias, the median odds ratio in 1,000,000 iterations that also incorporated conventional sampling error equaled 1.15 (95% SI: 0.78, 1.68). This result is depicted in panel B.2 of

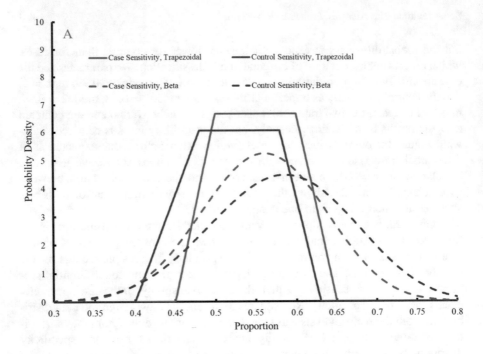

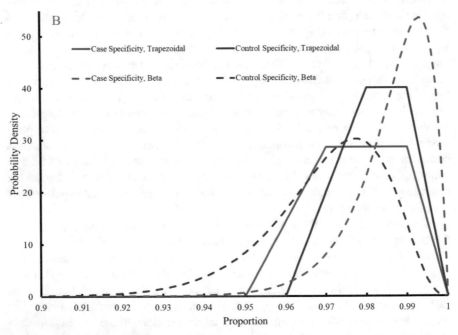

Figure 12.1 Density distributions for sensitivities (Panel A) and specificities (Panel B) of classification of self-reported twenty-year history of antidepressant use among cases and controls. Beta distributions parameterized directly from validation data of Boudreau et al., 2004 [5]. Trapezoidal distributions informed by that validation data and assigned as described in the text. Note that for sensitivities, the non-overlapping mass lies to the right for cases compared with controls for the beta distributions and to the left for the trapezoidal distributions. For specificities, the opposite pattern applies.

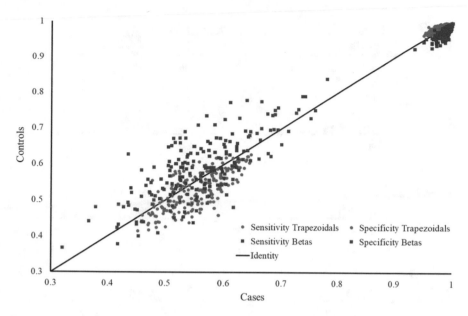

Figure 12.2 Sensitivity and specificity draws from the first 200 iterations of the probabilistic bias analysis. Note that the same set of correlated draws from the standard uniform ($r = 0.8$) was used for all four pairs of draws for cases and controls, following the generalized method for sampling from a distribution described in Chapter 7.

Figure 12.3. Recall that the conventional result, without any bias adjustment, was OR = 1.22, with 95% confidence interval (CI) 0.92, 1.61 (panel A of Figure 12.3). The primary bias model suggests that the conventional result was biased towards the null (bias-adjusted OR = 1.63 vs. crude OR = 1.22), whereas the second bias model suggests that the conventional result was biased away from the null (bias-adjusted OR = 1.15 vs. crude OR = 1.22). Both bias analyses convey that there is greater uncertainty than depicted by the conventional confidence interval (3.04/ 1.03 = 2.95 precision from the first model is 70% wider than conventional estimate precision 1.61/0.92 = 1.75 and 1.68/0.78 = 2.15 is 23% wider). With the trapezoidal distribution, no impossible values were found in the simulations. With the beta parameterization, we found 336 negative cell counts and 29 zero cell counts. These were dropped from all analyses. We carry the remaining bias-adjusted cell counts from both probabilistic bias analyses forward to the selection bias analysis.

Probabilistic Selection Bias Analysis

The simple and multidimensional analyses of selection bias addressed errors arising from the observed difference in participation between cases (80.6%) and controls

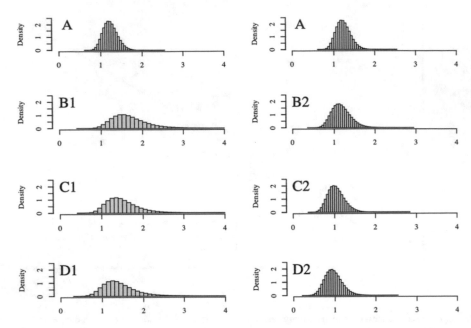

Figure 12.3 Histograms of bias analysis results obtained by bias modeling methods described in the text. X-axes are odds ratio estimates of the association between 20-year history of antidepressant use and breast cancer incidence, with original data from Chien et al. (2006) [4]. Panel A: conventional odds ratio with 1,000,000 resampling from conventional standard error. Panels B: addition to A of probabilistic bias analysis to address misclassification of twenty-year history of antidepressant use (B1 = beta distributions informed directly by Boudreau et al., 2004 validation study [5], B2 = trapezoidal distributions located and scaled with information from the validation study combined with expectation of recall bias). Panel C: addition to corresponding panel B of probabilistic bias analysis to address selection bias. Panel D: addition to corresponding panel C of probabilistic bias analysis to address uncontrolled confounding by physical activity.

(73.8%) and from the expectation that those with a history of antidepressant use in the preceding 20 years may have agreed to participate at different rates than never users of antidepressants. This latter expectation cannot be verified from the data because information on antidepressant use was gathered by interview, which of course could not be administered to nonparticipants. The tension one faces in assigning bias parameters in this case, therefore, is to decide what prevalence to assign to history of antidepressant use in cases and controls, what participation ratio to assign to those with a history of antidepressant use in the preceding 20 years versus those without such a history, and whether to allow this ratio to vary for cases and controls.

In our probabilistic selection bias analysis, we assign distributions to the bias parameters that are consistent with the simple bias analysis we performed earlier in this chapter. We assigned a beta distribution with $\alpha = 6$, $\beta = 44$ to the prevalence of antidepressant use in the preceding 20 years among cases (so median 11.5% and 95% density between 4.6% and 22.2%). We note the mean (12%) matches our previous specification perfectly. As above, we assigned a beta distribution with $\alpha = 6$, $\beta = 44$ to the prevalence of antidepressant use in the preceding 20 years among controls. Draws from the two distributions were correlated with $r = 0.8$. Note that these assignments effectively assume a null prior for the association between antidepressant use and breast cancer occurrence, which is the very association measured by the study. This prior information on exposure frequency in cases and controls does not inform the final parameter estimate in the same way that prior information would in a formal fully Bayesian analysis; it is being used only in the specification of prior parameters which might, only indirectly, influence the estimated bias-adjusted effect.

We also parameterized the participation ratio (R from above) among cases (ratio of participation given antidepressant use history relative to participation given never use history) as trapezoidal(min = 0.8, mode1 = 0.85, mode2 = 0.95, max = 1.0). The mean of this distribution (0.9) is the same as the specification for R among cases in the simple bias analysis. This parameterization reflects our belief that antidepressant users, or those with a history of antidepressant use, would be less likely to participate in the case–control study than never users of antidepressants. We parameterized the participation ratio among controls (antidepressant use history relative to never use history) as trapezoidal(min = 0.7, mode1 = 0.75, mode2 = 0.85, max = 0.9), which reflects the same belief. Draws from the two distributions were correlated with $r = 0.8$. The trapezoidal distribution for cases is located nearer the null to reflect our belief that the tendency of antidepressant users to refuse to participate in a research study would be attenuated by a diagnosis of breast cancer. All the selection bias models were constrained so that the overall participation rates among cases and controls equaled the observed participation rates.

Random samples were drawn from the four distributions specified in the previous two paragraphs. They were used in Equation 12.1 to produce selection probabilities among exposed and unexposed cases and controls. These specified distributions induced a prior probability of participation among exposed cases of 73.4%, 95% simulation interval (67.4%, 79.2%), among unexposed cases of 81.5% (80.7%, 83.1%), among exposed controls of 60.5% (54.8%, 66.0%), and among unexposed controls of 75.5% (74.4%, 77.7%). These point estimates are approximately the same as the participation probabilities from the simple bias analysis in Table 12.4.

We applied these participation probabilities to the misclassification-adjusted cell counts that were generated in the previous step. The process we implement is identical to those described for summary level data in Chapter 8. The only difference is that rather than using crude data, we used bias-adjusted data. As suggested in Chapter 8, we do not base our variance estimate on the selection bias adjusted data since it represents more people than were actually in the study. Rather, we use the misclassification-adjusted data to generate variance estimates for the misclassification and selection bias adjusted OR.

The median odds ratio in the 1,000,000 iterations applied in series to the bias adjustment for misclassification using beta distributions equaled 1.45 (95% SI: 0.91, 2.71). The median odds ratio in the 1,000,000 iterations applied in series to the bias adjustment for misclassification using trapezoidal distributions equaled 1.02

(95% SI: 0.69, 1.51). Figure 12.3 shows these results corresponding to panels C.1 and C.2, respectively. In both cases, adjustment for selection bias attenuated the estimated effect from OR = 1.63 to OR = 1.45 with the beta priors and from OR = 1.15 to OR = 1.02 with the trapezoidal. Simulation intervals indicated relatively little change in the precision of estimates after adjusting for selection bias.

Probabilistic Unmeasured Confounder Bias Analysis

The simple and multidimensional analyses of bias from an unmeasured confounder addressed uncertainty arising from the expectation that antidepressant medication use and regular exercise likely share common causal ancestors, that regular exercise may affect breast cancer occurrence, and that the original study did not adjust for confounding by regular exercise [4]. The challenge one faces in assigning bias parameters, therefore, is to decide what prevalence of regular exercise to assign to antidepressant users in the preceding 20 years and to never users of antidepressants in the preceding 20 years. One must also assign an estimate of the strength of association between regular exercise and breast cancer occurrence. Recall, however, that Chien et al. did examine the strength of confounding by many other candidate confounders, and that none of these substantially confounded the estimate of association. The combination of the assigned bias parameters should not, therefore, frequently yield estimates of confounding by exercise that are substantial, since such confounding is unlikely given the absence of other strong measured confounding.

In our probabilistic unmeasured confounder bias analysis, we assigned a beta distribution with $\alpha = 43$, $\beta = 55$ to the prevalence of regular exercisers among those without a history of antidepressant use in the preceding 20 years (so median 43.8% and 95% density between 34.2% and 53.7%). This prevalence is approximately centered on the prevalence of physical activity described in the simple bias analysis (43.6%) to address an unmeasured confounder. To place a prior on the prevalence of regular exercise among those with a history of antidepressant use, we could directly specify it in the same way with another beta distribution. These two prevalences, however, are naturally correlated and while we could specify a correlation between the two prevalences as we have with other bias parameters, a more natural approach that avoids needing to specify a correlation might be to specify the relative risk associating the exposure and confounder. Once we place a distribution on that relative risk, we can sample from it and the prevalence among the unexposed. Multiplying the two will give us the distribution of the prevalence of regular activity among those with a history of antidepressant use. This will naturally induce a correlation between the two prevalences. In the simple bias analysis earlier in the chapter, we assumed the prevalence among the exposed was 29.9% and among the unexposed was 43.6% giving $RR_{ec} = \frac{0.299}{0.436} = 0.69$. We specify a log-normal distribution for the exposure confounder relative risk, $\ln(RR_{ec}) \sim N(\ln(0.69), 0.09)$. This distribution gives a median RR_{ec} of 0.69 and 95% density between 57.8% and 81.8%. Multiplying draws from this distribution by draws from the prevalence among the unexposed gives a prevalence among the exposed of 30.2% (95% density between 22.4% and 39.7%). This approach gives a correlation of 78% between the prevalence of regular activity among those with and without depression. Finally, we parameterized the association between regular exercise and breast cancer occurrence as trapezoidal (min = 0.2, mode1 = 0.58, mode2 = 1.01, max = 1.24) based on a review of the cumulative evidence on this association in a period overlapping the study period [7]. Table 2 of that review provides 17 results associating leisure time physical activity with breast cancer occurrence in cohort studies. The minimum relative risk equaled 0.2 and the maximum relative risk equaled 1.24; we assigned these as the limits of the trapezoidal distribution. The third from lowest relative risk equaled 0.58 and the third from highest relative risk equaled 1.01; we assigned these as the lower and upper modes of the trapezoidal distribution.

The algorithm to bias-adjust proceeds using techniques presented for summary-level adjustment for uncontrolled confounding, with two slight differences. First, rather than using the crude data we instead use the misclassification and selection bias adjusted data. Confounding adjustments are performed directly on these bias-adjusted data. Second, the variance of the Mantel-Haenszel bias adjusted odds ratio is not calculated based on the final misclassification, selection bias, and confounding adjusted data. As mentioned earlier, because we adjust for selection bias, the final bias adjusted data will have a larger sample size than the original data. This presents a problem because an analyst could simply say their study was a sample from an enormous distribution in order to make all random error disappear. We calculate the bias-adjusted sample size at the end of this third step for each of the 1,000,000 bias-adjusted exposure by disease by confounder contingency tables. The sample size of the crude data is divided by the bias-adjusted sample size to create a scaling factor. Finally, the bias adjusted data are multiplied by this scaling factor resulting in a contingency table that is adjusted for misclassification, selection bias, and uncontrolled confounding but has the same size as the original data. The Mantel-Haenszel variance formula is computed for these data. This ad hoc method is easy to implement and prevents the problem of underestimating the random error. Alternative methods, such as bootstrap and jackknife standard errors, could also be explored.

Figure 12.3 shows the results of this probabilistic bias model, adding the model to bias-adjust for the uncontrolled confounder to the preceding bias models to address exposure misclassification and selection bias. The median odds ratio in the 1,000,000 iterations applied in series to the bias adjustment for the primary bias analysis using beta distributions equaled 1.35 (95% SI: 0.81, 2.60). The median odds ratio in the 1,000,000 iterations applied in series to the second bias adjustment using trapezoidal distributions equaled 0.94 (0.59, 1.43). Figure 12.3 shows these results corresponding to panels D.1 and D.2, respectively. Bias adjustment for this uncontrolled confounder attenuated the point estimates even further toward the null and, in the case of the trapezoidal prior, flipped the direction of median estimate of effect from causal to preventive. The ratio of the median estimate from the analysis adjusted for misclassification and selection bias to the median estimate from additional unmeasured confounder bias analysis equaled 1.35/1.45 = 0.93 and 0.94/1.02 = 0.92, respectively. By the criterion used to select confounders by Chien et al. (2006) [4], adjustment for exercise habits would not have been necessary. This result is consistent, therefore, with their observation that no substantial confounders were identified. In the trapezoidal analysis, no simulations were dropped due to impossible or zero values. In the beta distribution analysis, however, 484 confounding adjustments resulted in negative cells and 11 resulted in zero cells. In total 860 (<0.1%) simulations were dropped from the beta bias analysis.

Interpretation of Probabilistic Multiple Bias Analysis

The probabilistic multiple bias analysis for this example allows evaluation of the importance of three sources of uncertainty corresponding to the main threats to validity, conditional on the accuracy of the bias model. Using the conventional result (OR = 1.22, 95% CI 0.92, 1.61) as a starting point, bias modeling of exposure misclassification increased the point estimate 1.63/1.22 = 1.34-fold when the sensitivities and specificities were assigned beta distributions directly from the validation substudy [5]. The interval width increased 1.69-fold. These were the most important combined changes to the point estimates and interval widths associated with the bias models as implemented. Adding modeling for selection bias decreased the point estimate an additional 0.89-fold and only mildly increased the simulation intervals. Adding modeling for uncontrolled confounding decreased the OR an additional 0.93-fold and increased the simulation interval width 1.08-fold. Under this specification, although misclassification resulted in a large change from the crude estimate (crude OR = 1.22 vs misclassification adjusted OR = 1.63), both selection bias and uncontrolled confounding introduced bias in the opposite direction, offsetting the impact of misclassification. The estimate adjusted for all three sources of bias was closer to the original crude result (crude OR = 1.22 vs. all bias-adjusted OR = 1.35). Given the assumptions of this model, there is some modest evidence of effect (OR = 1.35, 95% simulation intervals: 0.81, 2.60); however, the effect is quite imprecise and efforts should be made to study the misclassification mechanism as much of the uncertainty around the final effect estimate stems from the uncertainty in the sensitivity and specificity parameters.

The sensitivity analysis using the trapezoidal distribution and different direction of differential misclassification resulted in a very different result. Under these assumptions, the bias-adjusted estimate was adjusted in a downward direction by each of the three biases, eventually becoming nearly null (all bias-adjusted OR = 0.94, 95% simulation interval: 0.59, 1.43). Again, these results are mildly imprecise, but there is little evidence of an effect from this analysis. An analyst viewing these results in conjunction with the main results above should recognize that their inference depends almost entirely on how the sensitivity and specificity parameters are specified. Slight changes in the prior distributions lead to substantially different results in this case.

Uncontrolled confounding by physical activity changed the point estimates very little, although did add some width to the simulation intervals. Given the small change to the point estimates, and the authors' report that adjustment for a large set of potential confounders changed the point estimates very little, one would be inclined to forego bias modeling for uncontrolled confounding by physical activity. Conditional on the accuracy of the bias models, therefore, it appears that misclassification of the 20-year history of antidepressant use is the most important source of systematic error in this study.

There is no information in the study, or externally, to choose between the two bias models for exposure misclassification; both are reasonable and consistent with the retrospective design of the case-control study [4] and the validation substudy [5]. This sensitivity analysis of the bias analysis suggests that the most productive avenue for further research would be to conduct a prospective study of the association between antidepressant use and breast cancer hazard, ideally with a better measure of antidepressant use than self-report. Conducting another retrospective case-control study (a replication effort) would add very little value to the accumulation of evidence, it is better to replicate and advance the topic area by designing a study that is less susceptible to the most influential sources of uncertainty [10]. Using bias analysis results to guide the most productive use of research resources for the next study is one of its most important utilities [11], although often underappreciated. Studies of prospective design published since this example have largely reported a null association [12–14].

There are also sources of uncertainty left unaddressed, such as errors in model specification, missing data bias, and data corruption by miscoding, programming errors, interviewer errors, or data processing errors. None of these sources of uncertainty have been incorporated, although they are often presumed to be small compared with the sources of error we have addressed quantitatively. This assumption, particularly as it pertains to data collection and management, rests on the foundation that the analysts are well-trained in research design and methods, that they have designed the data collection procedures to reduce the potential for errors to an acceptably low threshold, and that they have hired and trained staff to implement these procedures.

References

1. Fox MP, Lash TL, Hamer DH. A sensitivity analysis of a randomized controlled trial of zinc in treatment of falciparum malaria in children. Contemp Clin Trials. 2005;26:281–9.
2. Hazir T, Fox LM, Nisar YB, Fox MP, Ashraf YP, MacLeod WB, et al. Ambulatory short-course high-dose oral amoxicillin for treatment of severe pneumonia in children: a randomised equivalency trial. Lancet. 2008;371:49–56.

3. Lash TL, Schmidt M, Jensen AØ, Engebjerg MC. Methods to apply probabilistic bias analysis to summary estimates of association. Pharmacoepidemiol Drug Saf. 2010;19:638–44.
4. Chien C, Li CI, Heckbert SR, Malone KE, Boudreau DM, Daling JR. Antidepressant use and breast cancer risk. Breast Cancer Res Treat. 2006;95:131–40.
5. Boudreau DM, Daling JR, Malone KE, Gardner JS, Blough DK, Heckbert SR. A validation study of patient interview data and pharmacy records for antihypertensive, statin, and antidepressant medication use among older women. Am J Epidemiol. 2004;159:308–17.
6. Maguire GP, Lee EG, Bevington DJ, Kuchemann CS, Crabtree RJ, Cornell CE. Psychiatric problems in the first year after mastectomy. Br Med J. 1978;1:963–5.
7. Monninkhof EM, Elias SG, Vlems FA, Van Der Tweel I, Schuit AJ, Voskuil DW, et al. Physical activity and breast cancer: a systematic review. Epidemiology. 2007;18:137–57.
8. Gallicchio L, Schilling C, Miller SR, Zacur H, Flaws JA. Correlates of depressive symptoms among women undergoing the menopausal transition. J Psychosom Res. 2007;63:263–8.
9. Lash TL, Fox MP, MacLehose RF, Maldonado G, McCandless LC, Greenland S. Good practices for quantitative bias analysis. Int J Epidemiol. 2014;43:1969–85.
10. Lash TL. Advancing research through replication. Paediatr Perinat Epidemiol. 2015;29:82–3.
11. Lash TL, Ahern TP. Bias analysis to guide new data collection. Int J Biostat. 2012; 8(2):/j/ijb.2012.8.issue-2/1557-4679.1345/1557-4679.1345.xml
12. Brown SB, Hankinson SE, Arcaro KF, Qian J, Reeves KW. Depression, antidepressant use, and postmenopausal breast cancer risk. Cancer Epidemiol Prev Biomark. 2016;25:158–64.
13. Reeves KW, Okereke OI, Qian J, Tamimi RM, Eliassen AH, Hankinson SE. Depression, antidepressant use, and breast cancer risk in pre- and postmenopausal women: A prospective cohort study. Cancer Epidemiol Biomark Prev. 2018;27:306–14.
14. Haukka J, Sankila R, Klaukka T, Lonnqvist J, Niskanen L, Tanskanen A, et al. Incidence of cancer and antidepressant medication: record linkage study. Int J Cancer. 2010;126:285–96.

Chapter 13
Best Practices for Quantitative Bias Analysis

Introduction

Despite the extensive literature on Quantitative Bias Analysis methods, and despite the fact that the foundational methods for most of the material in this textbook were worked out in the 1950s and 1960s, there is still limited guidance on best practices for performing bias analysis [1]. As we have noted in earlier chapters, good practices for quantitative bias analysis follow good practices for study design and data analysis and therefore should not be new to the reader. For example, good research practices and good bias analysis practices both include: (a) development of a protocol to guide the work, (b) documentation of revisions to the protocol that are made once the work is underway, along with reasons for these revisions, (c) detailed description of the data used, (d) a complete description of all analytic methods used and their results, along with reasons for emphasizing particular results for presentation, and (e) discussion of underlying assumptions and limitations of the methods used. Good practices in presentation include (c)–(e) along with (f) description of possible explanations for the results, and (g) prudent and circumspect inference beyond the study, integrated with prior knowledge on the topic at hand.

We have described in other chapters best practices for deciding which biases to address, what method to use to model the biases, and assigning values to the parameters. In this chapter, we focus on best practices for when bias analysis is practical and productive, which biases to prioritize, how to select a method to model biases, and best practices for how to present and interpret a bias analysis. To date, no guidelines or checklists exist (like the CONSORT [2] or STROBE [3] guidelines) for reporting bias analyses, and this chapter is not meant to provide strict guidelines either. We rather focus on underlying principles for deciding on when to conduct a bias analysis and how to report one.

M. P. Fox et al., *Applying Quantitative Bias Analysis to Epidemiologic Data*,
Statistics for Biology and Health, https://doi.org/10.1007/978-3-030-82673-4_13

When Is Bias Analysis Practical and Productive?

In Chapters 1 and 2, we lay out some scenarios where bias analysis may be justified. We here lay out a more detailed rationale for when bias analysis is essential, when it may be justified, and when it can likely be forgone.

Cases in Which Bias Analysis Is Not Essential

We do not consider bias analysis to be essential in cases where a research report strictly limits itself to description of its motivation, conduct, and data, and stops short of discussing causality or other inferences beyond the observations. Although such purely descriptive reports are unusual and are even discouraged by many, they have been recommended as preferable to opposite extremes in which single studies attempt to argue for or against causality without regard to studies of the same topic or other relevant research [4].

This is not to say that bias analysis is not useful for descriptive epidemiology. Although the focus of this book has been on quantitative bias analysis for etiologic research, descriptive and predictive epidemiology can also be harmed by various sources of bias that can be explored through bias analysis methods [5, 6]. For example, estimates of prevalence of a disease can be subject to measurement error and selection bias that could be accounted for analytically using the methods we describe in this book [7]. While implementation of these methods is outside the scope of this textbook, we do not intend for this to imply that the methods should not be considered when conducting descriptive epidemiology.

Bias analysis may be helpful, but not necessary, when a report stops short of drawing inferences about causality or other targets beyond the observations, and instead offers alternative explanations for observations. This sort of report is among the most cautious seen in the literature, focusing on data limitations and needs for further research but refraining from substantive conclusions.

Bias analysis may be unnecessary when ordinary statistical analyses, encompassing only random error, show the study is incapable of discriminating among the alternative hypotheses under consideration within the broader topic community. Common examples include studies in which the precision of the effect estimate is so poor that the confidence interval includes all associations that are taken seriously by the topic community as possible effect sizes. This situation commonly arises when modest point estimates (e.g., relative risks around 2) are reported, the null value is included in the conventional frequentist confidence interval, and no one seriously argues that the effect, if any, could be large (e.g., relative risks above 5). Attempts to argue for or against causality in such cases would be ill-advised even if bias were absent, and discussion may be more appropriately focused by considering effects at both ends of the interval estimate with equal weight and as being broadly compatible with the observed data [8]. In the above situations, however, bias analysis becomes necessary if a reader attempts to draw substantive conclusions beyond those of the original study report, such as in public health policy, legal, and regulatory settings [9].

Bias analysis may also be unnecessary when the observed associations are dramatic, consistent across studies, and coherent to the point that bias claims appear unreasonable or motivated by obfuscation goals. Classic examples include associations between smoking and lung cancer, occupational exposure to vinyl chloride and angiosarcoma, estrogen replacement therapy and endometrial cancer, and outbreaks from infectious or toxic sources. In these situations, bias analysis may still be helpful to improve accuracy of uncertainty assessment. It may also be helpful for policy-makers seeking to incorporate the size of an effect estimate, and its total uncertainty, into hazard prioritization and regulation. Finally, bias analysis may be useful in this setting to demonstrate the unreasonableness of denialist claims, as did Cornfield et al. in response to claims that the smoking-lung cancer association could be attributed to a genetic factor affecting both tendency to smoke and cancer risk [10]. As a historical footnote, this paper is often cited as the first sensitivity analysis, although Berkson is an earlier example of quantitative bias analysis [11].

Cases in Which Bias Analysis Is Advisable

Bias analysis is advisable when a report of an association that is not dramatic goes beyond description and possible alternative explanations for results and attempts to draw inferences about causality or other targets beyond the immediate observations. In these cases, the inferences drawn from conventional statistics may not hold up to the scrutiny afforded by bias analysis, especially when conventional statistical analyses make it appear that the study is capable of discriminating among importantly different alternatives, or there is any attempt to interpret the study as if it does so. In public health policy, legal, and regulatory settings involving hazards, this situation frequently arises when the lower relative-risk confidence limit is above 1 or the upper limit is below 2 [12].

When conventional statistics appear decisive to some in the topic community, discussion needs to be appropriately focused by considering the potential impact of bias. Simple bias analyses will often suffice to adequately demonstrate robustness or sensitivity of inferences to specific biases. The aforementioned Cornfield et al. paper [10] is an example that addressed an extreme and unreasonable bias explanation (complete genetic confounding) for an extreme and consistent association (which was being promoted as calling for policy change) between smoking and lung cancer. The analysis by Cornfield et al. demonstrated the extremity of the bias explanation relative to what was known at the time (and has since been borne out by genetic and twin studies). It is uncertain whether they would have gone through this exercise had not a highly influential scientist raised this challenge [13], but the paper established the notion that one could not explain away the association between smoking and lung cancer as confounding alone without invoking associations and effects at least as large as the one in question.

Another case where bias analysis may be useful is in study design and grant writing [14]. We have noted previously that the assumptions that are necessary for sample size and power calculations provide enough information to create an expected summary 2×2 table. Such a table can then be used to conduct summary level bias analyses *before collecting any data* in order to see what biases are most likely to impact study results and any inferences drawn. We have shown in previous chapters of this book that biases that might seem like they will have a strong impact (*e.g.* variables measured with poor sensitivity) sometimes have little impact on point estimates (though they are likely to widen intervals about those point estimates) whereas in other cases, biases that may seem like they will have minimal impact (e.g. small amounts of selection bias) can have large impacts on point estimates. Determining which anticipated sources of bias are most likely to have the biggest impact can help investigators rationalize the use of limited resources in grant applications by limiting the impact of the most important sources of bias in the design of the study (*e.g.* through use of gold standard measures, or designing an internal validation study if the impact of misclassification is expected to be great) and putting less resources towards sources of bias that are likely to have less impact and leaving them for bias analysis methods in the analysis phase of the study. This can help grant applications be successful in highly competitive environments by demonstrating that the investigators understand the key sources of bias and have planned appropriately for mitigating their impacts.

Cases in Which Bias Analysis Is Arguably Essential

Bias analysis becomes essential when a report makes action or policy recommendations or has been developed specifically as a research synthesis for decision-making, and the decisions (as opposed to the statistical estimates) are sensitive to biases considered reasonable by the topic community. As with Cornfield et al. [10], simple bias analyses might suffice to demonstrate robustness or sensitivity of inferences. Nonetheless, multiple-bias analysis (see Chapter 12) might be necessary in direct policy or decision settings, and that in turn usually requires probabilistic inputs to deal with the large number of bias parameters.

As an example, by the early 2000s, over a dozen studies exhibited relatively consistent but weak (relative-risk estimates dispersed around 1.7) associations of elevated residential electromagnetic fields (EMFs) and childhood leukemia. Conventional meta-analyses gave relative-risk interval estimates in the range of 1.3 to 2.3 (P = 0.0001) [15, 16]. Consequently, there were calls by some groups for costly remediation (e.g., relocation of power lines). Probabilistic bias analysis found that more credible interval estimates could easily include the null [17], as well as very large effects that were inconsistent with surveillance data [18]. Thus, bias analysis showed that the evidence provided by the conventional meta-analysis should be down-weighted when considering remediation. In settings where immediate policy action is not needed, bias analysis results can provide a rationale for continued collection of better evidence, and can even provide a guide for further research [19].

In summary, simple bias analyses seldom strain resources so are often worthwhile. They are, however, not necessary until research reports contemplate alternative hypotheses and draw inferences. At this point, and certainly once policy decisions are contemplated, bias quantification by simple bias modeling becomes essential and more complex modeling may also be needed.

Which Sources of Bias to Model?

Once the sources of bias have been identified (as we describe in previous chapters), one must prioritize which biases to include in the analysis. We recommend prioritizing biases likely to have the greatest influence on study results. Judging this often requires relatively quick, simple bias analyses based on review of the literature and expert subject knowledge. Each of the sources of bias may be evaluated tentatively using simple bias analyses. Such an approach will often require a fair amount of labor, but is essential to informing the main part of the bias analysis and any conclusions that follow from it.

As an example, if little or no association has been observed, priority might be given to analyzing single biases or combinations of biases that are likely to be toward the null (e.g., independent nondifferential misclassification) and thus might explain the observation. In this regard, signed DAGs [20, 21] (see Chapter 5) can sometimes indicate the direction of bias and thus help to identify explanatory biases. A danger, however, is that by selecting biases to analyze based on expected direction, one will analyze a biased set of biases and thus reach biased conclusions. We thus advise that any bias that may be of substantively important magnitude be included in the final analyses, without regard to its likely direction.

Analysts may think that a source of bias is present, but that the magnitude of the bias is unimportant relative to the other errors present. For example, if the literature indicates that the association between an uncontrolled confounder and the exposure or outcome is small (e.g., as with socioeconomic status and childhood leukemia), then the amount of uncontrolled bias from this confounder is also likely to be small [10, 22]. In Chapter 5, we discussed methods to estimate bounds on the magnitude of bias due to uncontrolled confounding based on bounds for the component associations [23–29], which allow the analyst to judge whether that bias is likely to be important in their application.

Soliciting expert opinion (see Chapter 3) about possible bias sources can be a useful complement to, but no substitute for, the process described above in conjunction with a full literature review. Experts in the field may be aware of sources of bias that are not commonly mentioned in the literature. It is unlikely, however, that one will be able to obtain a random sample of expert opinions, a concern of special importance in controversial topic areas where experts may disagree vehemently.

Further, expert opinions are subject to all the cognitive biases mentioned in Chapter 1 and, despite being necessary and useful, should still be viewed with caution.

How Does One Select a Method to Model Biases?

As we have noted throughout this text, quantitative bias analysis encompasses an array of methods ranging from the relatively simple to the very complex. Bias analysts consider such factors as computational intensity and the sophistication needed to implement the method when selecting from among the methodologic options. The method's computational intensity is dictated, in part, by how the bias parameters are specified.

Simple bias analysis approaches can be computationally straightforward and require no detailed specification of a distribution for the bias parameters. Once the bias model and its initial values have been coded in a spreadsheet, for example, it is usually a small matter to change the values assigned to the bias parameters to generate a multidimensional analysis. However, as we have covered in extensive detail in previous chapters, such analyses do not explicitly incorporate uncertainty about the bias parameters in interval estimates or tests of the target parameter. While an analyst may wish to begin with simple and multidimensional methods, we recommend formal sensitivity analysis in cases where very small changes in values of bias parameters result in drastic changes in the bias-adjusted estimate, as often occurs in exposure-misclassification problems [30, 31] or when more complete depictions of uncertainty are desired [32].

Probabilistic bias analyses and Bayesian analyses are more computationally intensive but can provide a much more realistic assessment of the impact of sources of error on total study error. When multiple sources of bias are of concern, effect estimates can be bias-adjusted for each source simultaneously using multiple bias modeling (see Chapter 12). In these situations there are usually far too many bias parameters to carry out simple fixed-value bias analysis, and probabilistic bias analysis or Bayesian bias analysis becomes essential. This type of analysis is more realistic since it can incorporate all biases that are of serious concern, but there is little distributed software to do it.

While needless complexity should be avoided, there are areas in which too much simplification should also be avoided. Any realistic model of bias sources is likely to be complex. As with conventional epidemiologic data analysis, tradeoffs must be made between realistic modeling and practicality [33]. Simplifying assumptions are always required, and it is important that these assumptions are made explicit. For example, if the decision is made to omit some biases (perhaps because they are not viewed as being overly influential), the omissions and their rationales should be reported.

We encourage analysts using complex models to also examine simpler approximations to these models to both check coding and gain intuition about the more complex model. For instance, multiple bias models can provide realistic estimates of total study error but may obscure the impacts of distinct bias sources. We thus advise analysts implementing a multiple-bias model to examine each source of bias individually, which helps identify bias-adjustments with the greatest impact on results. One can also compare estimates obtained from probabilistic analysis to the estimate obtained when the bias parameters are fixed at the modes (or medians or means) of their prior distributions. In the event that the results of the simpler and more complex analyses do not align, the author should provide an explanation as to why.

Implications Regarding Transparency and Credibility

Transparency and credibility are integral to any quantitative bias analysis [34]. Unfortunately, increasing model complexity can lead to less transparency and, hence, reduce the credibility of an analysis. Analysts should take steps to increase the transparency of the methods they use. As with all analyses, analysts should avoid using models that they do not fully understand. Giving a full explanation of why the model specification produced the given results can increase transparency. We also encourage authors to make the data and code from their bias analyses publicly available. With the advent of electronic appendices in most major journals, providing bias analysis code as web appendices poses little problem. Online resources like GitHub allow for publishing code and data that allow for replication of the analysis. Published code will aid future analysts who need to implement bias analyses. Further, quantitative bias modeling can be complex, so public dissemination of code can help to identify and correct algorithmic errors.

Using Available Resources Versus Writing a New Model

Numerous resources are available to help analysts implement quantitative bias analysis. Many sources we cite contain detailed examples that illustrate the analyses. Several have provided code so future analysts could implement their analyses as well. See this text's website for our code, spreadsheet examples, and links to other resources we have vetted. When possible, we encourage authors to adopt code that has been previously developed, because it should help to identify and reduce coding errors. Existing resources may be difficult to adapt to new situations, however, particularly for multiple bias models. In that case, analysts have to write their own programs.

Implications of Poor Bias Analyses

We advocate for an increase in the use of bias analysis methods to aid in better inference, and we note that poor bias analysis can also cause harm [34]. As noted previously in this text, bias analysis that seeks a desired result (such as by focusing only on biases likely to bias toward the null or by choosing values for bias parameters that will create a larger effect size) can lead to even more confidence in biased results. We suspect that intentional manipulation of results is rare. Nonetheless, misunderstanding of bias analysis methods and good practices could lead to unintentionally drawing poor conclusions. This concern is true also for conventional analyses [33]. We have noted several instances of poorly implemented bias analyses [34] that either chose non-sensical values for bias parameters (e.g., noting a role for selection bias, then centering a selection odds ratio on 1, which would indicate no bias, and concluding selection bias was not likely an important source of bias [35]) or did not pay attention to the sensitivity of results to the choice of values for the bias parameters [36] (see also Chapter 12). Such analyses can lead to conclusions that are different from the conclusions that would have been drawn had different, more realistic parameters been chosen. Like with any analysis, careful attention to the implications of the underlying assumptions is key to drawing valid and defensible conclusions. This requires those who implement these methods to have a thorough understanding of the approaches and requires consumers to have the ability to identify potential problems in the approaches used. A notable strength of quantitative bias analysis is the transparency of the assumptions that are being made (if the analysis is well presented). If, for example, some analysts disagree with the bias analysis in question, it is relatively easy to focus on the source of the dispute (such as a particular bias parameter distribution) and the analyst can re-run the analysis with a new distribution to determine the sensitivity of results.

Who Should Be Calling for Bias Analysis?

Although there has been a considerable increase in the number of analysts that have used quantitative bias analysis since the first edition of this book, bias analyses are still rare relative to the size of the published epidemiologic literature. This likely has many reasons, but overall reflects the lack of incentive for analysts to implement the methods and lack of demand from reviewers, editors, stakeholders and grantmaking agencies [37]. This lack of demand will need to change if we are to get more accurate depictions of total study error, which will in turn allow us to make better inferences.

Responsibility for bias analysis starts with the analysts. Analysts have access to the primary data and should be invested in providing an honest, realistic picture of the findings and their uncertainty. Still, it is often clear that the incentives to conduct bias analyses are not aligned with the analysts' incentives to publish, and to publish in what are considered higher impact journals. While bias analyses can be manipulated in theory by focusing only on biases that will move point estimates further away from the null (aligning with publishers' desires to publish non-null results over null results), most bias analyses will lead to less certainty in study results and therefore to a more realistic representation of the uncertainty we should have given the biases inherent in the work. If analysts shy away from estimating the impacts that such biases have, then it is unlikely to have the desired impact of improving public health. Further, bias analyses often require additional time and resources to implement, through additional analyses and data collection to identify the values for bias parameters. Analysts are not incentivized to pass this additional hurdle. Thus, we cannot rely on analysts alone to conduct bias analyses as a routine aspect of data analysis.

Peer review is the time that the demand for bias analysis most needs to be strengthened. Reviewers are the ones who are most likely to identify sources of bias in a study and should want to know the impact that these biases have on the study results. Given the study write up and a search of the literature, there is often sufficient information for peer reviewers to be able to conduct a summary level quantitative bias analysis at the time of review and identify which source of biases are likely to have the biggest impact. This would have the benefit of allowing reviewers to be able to identify results and analyses where their intuitions about the impact of the bias are incorrect and focus on the sources where the greatest uncertainty lies. In addition, peer reviewers do not have the incentive to think only about biases that might "strengthen" the findings (even at the expense of certainty in those findings). However, we acknowledge that peer review is uncompensated work that busy analysts do in service to the field and it is unreasonable to ask them to conduct their own bias analyses as part of the peer review process.

The solution to this problem is for peer reviewers, as part of their peer review responses to the analysts, to ask that bias analyses be done in the revision stage if they feel the conditions we describe above for when bias analysis is most helpful are met. Asking the authors to conduct the bias analysis puts the onus back on the analysts and does not waste peer reviewers' time. It also puts the responsibility for the work on those who have access to the primary, record-level data. The resulting bias analyses can be shared with the reviewers (and eventually included with the manuscript) to allow for more rational evaluations of the work. As we have shown throughout this text, intuitions about the impacts of biases are not always correct and results that have far more uncertainty than they demonstrate with conventional methods can lead to poor inferences. At the same time, rejection of important findings because of speculation that bias may be strongly impacting the results, when in fact the impact is inconsequential, means we may miss out on important information that could benefit public health. Making bias analysis a more routine part of the peer review process has the potential to improve the overall quality of research [37].

Journal editors also have an important role to play in making bias analysis more commonplace. Editors are ultimately the ones who make decisions about what is accepted for publication and what is rejected. Supporting peer reviewers (or even calling for bias analysis themselves as part of the peer review process) in calling for more transparent assessments of the impact of systematic errors could go a long way towards improving the quality of epidemiologic research. This would require that editors are familiar with the methods and understand their importance. As part of this process, it will be important for editors to not penalize those who are more honest about the impact of bias on their results while prioritizing papers that do not make the effort but appear more certain about their findings.

Finally, we note getting bias analyses to be more routinely utilized will require a shift in the educational process, where courses on bias analysis become a standard part of epidemiologic training. We know of only a handful of courses employing these methods at the current time. Making bias analysis a part of doctoral level training, at a minimum, would provide analysts with the tools necessary to account for sources of systematic error in their results and give them the understanding of the importance of these methods in furthering the goals of etiologic research.

Conclusions

We advocate transparency in description of the methods by which biases were identified for analysis, models were developed, and values were assigned to the bias model's parameters. We also encourage bias analysts to make their data and computer code available for use by others, so that the results can be challenged by modifications to the model or by different choices for the values assigned to the parameters. When data cannot be made freely available, bias analysts at a minimum should offer to incorporate credible modeling modifications and changes to the values assigned to the model parameters, when these are suggested by other interested parties, and to report completely the results of these revised analyses. Bias models cannot be immediately verified as empirically correct, nor are values assigned to the model parameters identifiable in the given study. It is, therefore, crucial that credible alternatives be given thorough examination. And in fact, even if data cannot be made available, a representation of the full bias analysis can be approximated by creating a summary level version of the approach that only uses the summary contingency table and avoids the need for full record level data sharing.

Bias analysis is not a panacea. It cannot resolve fundamental problems with poor epidemiologic research design or reporting, although it can account for uncertainties arising from design limitations. If there is investigator bias that introduces fraud into the data collection or analysis [38], or incompletely represents the data collection and analysis process [39], then no analysis can be expected to correct the resulting bias. Because the bias analyses we have discussed are designed for unselected analyses of individual studies, they cannot resolve inferential errors arising from selective reporting of research results, whether that is due to selective reporting of "significant" associations or suppression of undesired associations [40–42]. Methods of publication-bias analysis [40, 41] and forensic statistics [43, 44] can help to investigate these problems.

References

1. Lash TL, Fox MP, MacLehose RF, Maldonado G, McCandless LC, Greenland S. Good practices for quantitative bias analysis. Int J Epidemiol. 2014;43:1969–85.
2. Schulz KF, Altman DG, Moher D. CONSORT 2010 Statement: Updated Guidelines for Reporting Parallel Group Randomized Trials. Ann Intern Med. 2010;152:726–32.
3. Vandenbroucke JP, von Elm E, Altman DG, Gøtzsche PC, Mulrow CD, Pocock SJ, et al. Strengthening the reporting of observational studies in epidemiology (STROBE): explanation and elaboration. Epidemiology. 2007;18:805–35.
4. Greenland S, Gago-Dominguez M, Castelao JE. The value of risk-factor ("black-box") epidemiology. Epidemiology. 2004 Sep;15:529–35.
5. Jurek AM, Maldonado G, Greenland S, Church TR. Uncertainty analysis: an example of its application to estimating a survey proportion. J Epidemiol Community Health. 2007;61:650–4.
6. Labgold K, Hamid S, Shah S, Gandhi NR, Chamberlain A, Khan F, et al. Estimating the unknown: Greater racial and ethnic disparities in COVID-19 burden after accounting for missing race and ethnicity data. Epidemiology. 2021;32:157–61.
7. Burstyn I, Goldstein ND, Gustafson P. Towards reduction in bias in epidemic curves due to outcome misclassification through Bayesian analysis of time-series of laboratory test results: Case study of COVID 19 in Alberta, Canada and Philadelphia, USA. BMC Med Res Methodol. 2020;20:146.
8. Poole C. Low P-values or narrow confidence intervals: which are more durable? Epidemiology. 2001;12:291–4.
9. Lash TL, Fox MP, Cooney D, Lu Y, Forshee RA. Quantitative bias analysis in regulatory settings. Am J Public Health. 2016;106:1227–30.
10. Cornfield J, Haenszel W, Hammond E, Lilienfeld A, Shimkin MB, Wydner E. Smoking and lung cancer: recent evidence and a discussion of some questions. J Natl Cancer Inst. 1959;22:173–203.
11. Berkson J. Limitations of the application of fourfold table analysis to hospital data. Biometrics Bulletin. 1946;2:47–53.
12. Greenland S, Robins JM. Epidemiology, justice, and the probability of causation. Jurimetrics. 2000;40:321–40.
13. Stolley PD. When genius errs: R. A. Fisher and the lung cancer controversy. Am J Epidemiol.;133:416–25.
14. Fox MP, Lash TL. Quantitative bias analysis for study and grant planning. Ann Epidemiol. 2020;43:32–6.
15. Ahlbom A, Day N, Feychting M, Roman E, Skinner J, Dockerty J, et al. A pooled analysis of magnetic fields and childhood leukaemia. Br J Cancer. 2000;83:692–8.
16. Greenland S, Sheppard AR, Kaune WT, Poole C, Kelsh MA. A pooled analysis of magnetic fields, wire codes, and childhood leukemia. Childhood Leukemia-EMF Study Group. Epidemiology. 2000;11:624–34.
17. Greenland S. Multiple bias modeling for analysis of observational data. J R Stat Soc A. 2005;168:1–25.
18. Greenland S. Bayesian perspectives for epidemiological research: I. Foundations and basic methods. Int J Epidemiol. 2006;35:765–75.
19. Lash TL, Ahern TP. Bias analysis to guide new data collection. Int J Biostat. 2012;8:/j/ijb.2012.8.issue-2/1557-4679.1345/1557-4679.1345.xml.

20. VanderWeele TJ, Robins JM. Signed directed acyclic graphs for causal inference. J R Stat Soc Ser B Stat Methodol. 2010;72:111–27.
21. VanderWeele TJ, Hernán MA. Results on differential and dependent measurement error of the exposure and the outcome using signed directed acyclic graphs. Am J Epidemiol. 2012;175: 1303–10.
22. Bross ID. Pertinency of an extraneous variable. J Chronic Dis. 1967;20:487–95.
23. Flanders WD, Khoury MJ. Indirect assessment of confounding: graphic description and limits on effect of adjusting for covariates. Epidemiology. 1990;1:239–46.
24. Yanagawa T. Case-control studies: Assessing the effect of a confouding factor. Biometrika. 1984;71:191–4.
25. Arah OA, Chiba Y, Greenland S. Bias formulas for external adjustment and sensitivity analysis of unmeasured confounders. Ann Epidemiol. 2008;18:637–46.
26. VanderWeele TJ, Arah OA. Bias formulas for sensitivity analysis of unmeasured confounding for general outcomes, treatments, and confounders. Epidemiology. 2011;22:42–52.
27. Ding P, VanderWeele TJ. Sensitivity analysis without assumptions. Epidemiology. 2016;27: 368–77.
28. VanderWeele TJ, Ding P. Sensitivity analysis in observational research: Introducing the E-Value. Ann Intern Med. 2017;167::268–74.
29. MacLehose RF, Ahern TP, Lash TL, Poole C, Greenland S. The importance of making assumptions in bias analysis. Epidemiology. 2021;32:617–24.
30. Copeland KT, Checkoway H, McMichael AJ, Holbrook RH. Bias due to misclassification in the estimation of relative risk. Am J Epidemiol. 1977;105:488–95.
31. Gustafson P, Le ND, Saskin R. Case–control analysis with partial knowledge of exposure misclassification probabilities. Biometrics. 2004;57:598–609.
32. Fox MP, Lash TL, Greenland S. A method to automate probabilistic sensitivity analyses of misclassified binary variables. Int J Epidemiol. 2005;34:1370–6.
33. Greenland S. Invited commentary: Dealing with the inevitable deficiencies of bias analysis–and all analyses. Am J Epidemiol. 2021;190:1617–1621.
34. Lash TL, Ahern TP, Collin LJ, Fox MP, MacLehose RF. Bias analysis gone bad. Am J Epidemiol. 2021:190:1604–1612
35. Di Forti M, Quattrone D, Freeman TP, Tripoli G, Gayer-Anderson C, Quigley H, et al. The contribution of cannabis use to variation in the incidence of psychotic disorder across Europe (EU-GEI): a multicentre case-control study. Lancet Psychiatry. 2019;6:427-436.
36. Chien C, Li CI, Heckbert SR, Malone KE, Boudreau DM, Daling JR. Antidepressant use and breast cancer risk. Breast Cancer Res Treat. 2006;95:131–40.
37. Fox MP, Lash TL. On the need for quantitative bias analysis in the peer-review process. Am J Epidemiol. 2017;185:865–8.
38. Greenland S. Transparency and disclosure, neutrality and balance: shared values or just shared words? J Epidemiol Community Health. 2012;66:967–70.
39. Phillips CV. Publication bias in situ. BMC Med Res Methodol. 2004;4:20.
40. Henmi M, Copas JB, Eguchi S. Confidence intervals and P-values for meta-analysis with publication bias. Biometrics. 2007;63:475–82.
41. Copas J, Dwan K, Kirkham J, Williamson P. A model-based correction for outcome reporting bias in meta-analysis. Biostatistics. 2014;15:370–83.
42. Lash TL. The harm done to reproducibility by the culture of null hypothesis significance testing. Am J Epidemiol. 2017;186:627–35.
43. Al-Marzouki S, Evans S, Marshall T, Roberts I. Are these data real? Statistical methods for the detection of data fabrication in clinical trials. BMJ. 2005;331:267–70.
44. Pogue JM, Devereaux PJ, Thorlund K, Yusuf S. Central statistical monitoring: detecting fraud in clinical trials. Clin Trials. 2013;10:225–35.

Chapter 14
Presentation and Inference

Introduction

Throughout the text we have illustrated methods to present the results of bias analysis and explained the inferences that might derive from those results. This chapter will briefly describe the overarching considerations we recommend for presentation and inference, and the reader should refer to specific examples throughout the text for the detailed implementations of these principles.

Presentation

Methods

Despite substantial progress since the first edition of this book came into print, the methods and terminology of quantitative bias analysis applied to epidemiologic data are still not readily familiar to most interested parties in epidemiologic research (by interested parties, we mean research colleagues, editors, reviewers, grant making institutions, readers, policymakers, and the interested public). In addition, despite the guidelines we have provided in the previous chapter, there are no established standards for good practice to which interested parties can compare the methods of a particular bias analysis for a preliminary assessment of its quality nor accepted guidelines to follow for those wishing to implement a bias analysis to ensure complete reporting of the methods used. What is clear is that many readers of bias analyses contained within a manuscript presenting scientific findings (*i.e.,* not as a standalone manuscript or as part of a methods paper) will not have previously encountered the approach and, as such, may find the terminology typically employed to describe such work difficult to follow. The first principle of presentation,

© Springer Nature Switzerland AG 2021
M. P. Fox et al., *Applying Quantitative Bias Analysis to Epidemiologic Data*,
Statistics for Biology and Health, https://doi.org/10.1007/978-3-030-82673-4_14

therefore, must be complete and accurate description of the bias analysis methods with a focus on language that can be easily understood.

How to present a bias analysis is something that should be thought about in the planning stages of the bias analysis so that the analyst can pay careful attention to how to make the presentation as clear and transparent as possible and generate the results necessary for this. In fact, we have argued there is a role for quantitative bias analysis in grant writing and study design [1] and have frequently included detailed description of future bias analyses in our own grants. As such, steps can be taken to ensure clear presentation of bias analyses even before data collection has begun by anticipating potential sources of bias and developing a plan to account for the impact of these biases analytically. It is this plan that can then be used to guide a clear presentation of what was done, what was found, and how to interpret the results.

This description of the bias analysis should begin with a clear statement of the bias analysis objectives. The description should include the nature of the biases to be evaluated (*e.g.,* selection bias, uncontrolled confounding, or information bias), and this description should relate to parts of the methods section pertaining to the conventional data collection or data analysis. In general, the objectives will parse into two related types, though other objectives may exist. The first type of objective is to estimate a bias-adjusted association and its uncertainty as well as to assess the direction and magnitude of the error introduced by the bias and any change in uncertainty associated with accounting for the bias. The second type of objective is to assess whether all or most of an observed association could be explained by the bias, sometimes referred to as nullification analysis.

The description should continue with a mathematical depiction of the model used to generate bias-adjusted results. For example, if adjusting for exposure misclassification in a simple bias analysis, the analyst should give the equation used to estimate the bias-adjusted cell counts or the bias-adjusted effect, depending on the chosen method. Given the space limitations sometimes imposed by journals, the full description of the model might be reserved for an appendix or online supplement. In many cases, this depiction will be most easily accomplished with mathematical equations, such as the ones we have presented in each of the chapters on specific biases, although citations to the methods literature will suffice for common equations such as those used to implement the simple bias analyses described in earlier chapters. Even when the equations used for the bias analysis seem simple, presentation is always better than citation, in that it allows those new to quantitative bias analysis to easily understand what was done and replicate the findings (when summary data are used) and it obviates the need for readers to seek other sources to comprehend the current work.

With this depiction of the model, an interested party should be able to relate the study's conventional results (*e.g.,* the frequencies of exposed and unexposed cases and controls) to the bias analysis results (*e.g.,* the frequencies of exposed and unexposed cases and controls after adjustment for exposure classification errors) through the set of bias parameters that were used for the analysis (*e.g.,* the sensitivities and specificities of exposure classification in cases and controls). If the equations that establish this relation are not presented, then the reader should be able to find these equations in the cited literature.

The description should then explain the values (and their rationale) assigned to the bias parameters in the case of a simple bias analysis or the distribution assigned to the bias parameter in the case of probabilistic bias analysis. If the assignments were based on a validation substudy, then the methods by which substudy participants were selected and validation data collected should be clearly explained so that a reader can draw conclusions about the validity of the bias parameters being used, and if the participants were not a representative sample, then the implications of the selection should be considered. If the assignments were based on external validity studies, then those studies should be cited and the transportability of the results to the present study should be considered. In both cases, how random error affected the results should also be presented so that the reader can assess the uncertainty in the bias parameters. If the assignments were based on educated guesses, then the reasoning behind the educated guess should be well-described and how the uncertainty in those educated guesses was determined. When there are multiple sources of data to inform the assignments, the description should include the rationale for how these multiple sources were used and/or integrated in the analysis. For example, if the analyst chose a simple bias analysis, then the description should explain why one set of bias parameters was preferred over all others and the implications for such a decision. Better still, the bias analysis should use multidimensional, probabilistic, or Bayesian methods to incorporate all of the sources of information, and the description should explain how these sources of data were used to inform the bias analysis. When distributions are used for the bias parameters, the set of distributions should be presented, the rationale for those distributions and any limitations put on those distributions (*e.g.* truncation, correlation of distributions) whether by the analyst or the data (*e.g.* if the distributions were truncated due to impossible values) so that the reader could replicate the results if needed or desired. For example, the analyst might report that all estimates of sensitivity of exposure classification from previous studies have been used in a multidimensional bias analysis (and should cite those previous studies, as well). Or the analyst might report that those previous studies were used to influence the choice of probability density and the values assigned to its parameters. In the latter case, the analyst should cite all studies that were used and be clear about *how* they were used to specify parameters (such as by using a meta-analysis). Graphical depictions of the probability density functions portray the analyst's understanding of the range of values assigned to bias parameters and relative preferences for certain values within the allowed range and make modeling choices clearer to interested parties with a less mathematical background. Any data sources that might have informed the values assigned to the bias parameters, but were rejected by the analyst, should be listed and the reasons for rejection should be explained.

Finally, if the bias analysis involves multiple biases, the description should be completed for each bias and the rationale for the order of bias-adjustments should be explained. In such cases, the order of bias-adjustments should also be reported for clarity.

Results

Guidelines for presentation of bias analysis results parallel good practices for presenting the results of conventional epidemiologic data analyses. If the bias analysis yields bias-adjusted frequencies in a 2×2 table, then it is useful to the reader to see these bias-adjusted frequencies just as presentation of descriptive frequencies is ordinarily recommended for conventional results. This presentation allows the reader to check their intuitions about the direction and magnitude of the bias, identify any errors or inconsistencies, and replicate the findings. For multidimensional or probabilistic bias analyses, the presentation can include representative bias-adjusted frequencies derived from one specified set of values assigned to the bias parameters, typically by choosing the median values of the bias-adjusted cell frequencies and conducting a simple bias analysis. This presentation provides the reader with intuition on the expected results and provides them with information to understand why any deviation from that expectation may have occurred. This representative presentation also allows stakeholders to verify calculations and to see the relation between the conventional results (cell frequencies) and the bias analysis results (bias-adjusted cell frequencies), which are connected by the bias model and the values assigned to the bias parameters. It would not, however, be a productive use of space or resources to present bias-adjusted cell frequencies for all combinations in most multidimensional bias analyses or for every iteration of a probabilistic bias analysis. And as noted previously, in the interest of space, some content should be put into an appendix or supplemental online file.

The presentation should continue with a description of the central tendency of the bias analysis result compared with the conventional result (*i.e.*, presentation of the direction and magnitude of the bias). For a simple bias analysis, this presentation amounts to a comparison of the single bias-adjusted result with the conventional result. In addition to presenting both results, the analyst may wish to say whether the bias analysis result is nearer to the null or further from the null than the conventional result (or potentially past the null), which indicates whether the original bias was away or toward the null, respectively (conditional, as always, on the accuracy of the bias model and the values assigned to the bias parameters). The direction and magnitude of the bias may ordinarily be conveyed as the ratio of the conventional result to the bias analysis result, which may be presented as the simple ratio or as a percentage. For bias-adjusted risk or rate differences, it is more informative to present differences of differences. For a multidimensional bias analysis, results favored because of the plausibility of the values assigned to the bias parameters might be highlighted, or certain combinations or subsets of the results may be selected for focus (e.g., the combinations of sensitivity and specificity that correspond to nondifferential misclassification).

For a probabilistic bias analysis, the central tendency of the results from the iterations should be substituted for the single result of a simple bias analysis. For a multidimensional bias analysis, one result may be selected as preferred over the others, perhaps because the values assigned to the bias parameters are deemed most plausible by the analyst, in which case this bias analysis result should be substituted for the single result of the simple bias analysis. If no single result from the multidimensional bias analysis is preferred over the others, then the range of values should be presented, along with a discussion of the direction(s) and magnitude(s) of the bias implied by that range.

Probabilistic bias analysis results can sometimes be best presented graphically. Density plots effectively portray the weight given to the values within the range determined by a particular bias parameter. Histograms of the bias parameter values used in the analysis can effectively portray the correspondence between draws from the probability density distributions and those actual density distributions, as shown in Chapter 7. Any lack of concordance between the two should be explained (*e.g.,* if sensitivity values were sampled but excluded because they resulted in negative adjusted cell counts). Histograms also portray the range and distribution of bias-adjusted results yielded by the bias analysis. As shown in Chapter 12, a series of these histograms readily compare the conventional result with different bias analyses, both in terms of the location of the central tendency and the width of the distribution of results. Such a series of histograms can be used to compare different assumptions used to conduct the bias analysis (*e.g.,* different density distributions assigned to the bias parameters, or different assumptions about the bias model, such as differential or nondifferential misclassification). These series can also show the change in location of the estimated association and the width of its simulation interval as the bias model incorporates ever more sources of uncertainty. For example, the series may begin with the conventional result (*i.e.,* depicting only random error), then report the uncertainty added from a bias-adjustment for classification error, then add uncertainty added from a bias adjustment for selection bias, and finally add uncertainty from a bias adjustment for an unmeasured confounder. Presenting each result alone is important, as with each addition, one can ascertain the change in location of the estimated association and the width of its simulation interval, and these changes will suggest which biases are most important with regard to both location and width, assuming the bias model and assigned values are accurate. It will also demonstrate how these sources of bias interact with each other to create the final adjusted result. Note that it is important to maintain constant scales on both the *y*-axis (frequency of result) and *x*-axis (measures of association). Of course, given limited space within the main text, some of these figures will need to be relegated to an appendix or online supplemental file, but presenting as many as possible will provide the reader with valuable information to replicate the analyst's findings.

If multiple bias analyses are conducted, an alternative to separate histograms of bias analysis results is a graph that includes overlapping histograms or density estimates from each bias analysis. A forest plot can also be used to show the results of each of the different analyses side by side in graphical form. For example, in Chapter 7 we used this graphical depiction to compare the results of a series of probabilistic bias analyses, each with a different probability density function assigned to the sensitivity or specificity of exposure classification. This method is somewhat simpler to prepare and allows stakeholders to visualize all of the relevant bias analyses in a single figure, rather than comparing histograms in different panels with one another. The disadvantage of this method is that the stakeholder loses the ability to see the "flattening out" of the histogram heights that occurs when the width of the distribution grows. This method is therefore best used when the distribution width changes little, as with the example in Chapter 7.

Availability of Data and Code

It is worth noting that much has changed since the early development of methods for quantitative bias analysis, including the expansion of online appendices and online repositories for data and code to replicate analyses. Quantitative bias analysis methods should be made as transparent as possible, including through publication of code that allows the replication of the analysis. Online repositories such as GitHub allow for posting of these materials, which in turn allows others to identify any errors in the code, to replicate the analyses, to see more clearly the assumptions made in the methodology, to make changes to the underlying assumptions as a sensitivity analysis of the bias analysis, and to learn from the approach used. While data cannot always be made available due to privacy concerns (though anonymized versions usually can), summary level bias analysis can be posted. Even when a record level dataset cannot be posted, a summary level version of the bias analysis that closely mimics any record level bias analysis can be implemented that would allow for nearly the same transparency as posting the full dataset. Thus, there is seldom reason that replication of the analysis is not possible or that code is not made available, and we strongly encourage use of these new tools to ensure transparency.

Inference

Inferential Framework

We have been asked whether these methods, particularly the probabilistic bias analysis methods, rest on a frequentist or Bayesian foundation. While some of the methods we presented (particularly in Chapter 10) are explicitly frequentist, the best answer is that most of the probabilistic bias analysis methods we have presented are semi-Bayesian, in that they place priors on some (i.e., the bias parameters), but not all, model parameters. Most important, with the exception of the Bayesian methods described in Chapter 11, the methods presented do not place a prior distribution on the parameter representing the effect of interest (the association between the exposure contrast and the outcome). Typical probabilistic bias analysis methods, then, assume the same noninformative prior on the effect of interest as is inherently assumed by frequentist statistical methods. This assumption allows a computationally simpler solution, traded-off against the ordinarily untenable belief that all possible values for the association between the exposure contrast and the outcome are equally likely prior to conducting the study. Furthermore, unlike the formal

Bayesian methods we describe in Chapter 11, the probabilistic bias analysis methods do not update the priors based on the data (save for the instance when negative cell frequencies result from misclassification corrections, although even then the method simply discards these iterations rather than formally updating the prior; see discussions in Chapters 8 and 11).

Although most of the methods presented in this text are not fully Bayesian, they could be altered to be fully Bayesian and incorporate prior information for the main effect. Data augmentation approaches could be applied to the probabilistic bias analysis methods presented here to implement a fully Bayesian procedure [2–4]. This would allow an analyst to adjust for biases while simultaneously incorporating prior knowledge about the magnitude of the effect of interest. As we stated in Chapter 1, however, we do not view Bayesian and frequentists statistics as adversarial, despite historical adversarial relationships between Bayesian and frequentists statisticians in the twentieth century. To highlight this important fact, we note that many of the probabilistic bias analysis methods developed in this text were developed without strict adherence to either school of statistical inference. It turned out that a theoretical justification was available for many of the procedures via a Bayesian framework. Further exploration of these procedures was conducted using a frequentist framework to determine their frequentist operating characteristics. For example, the coverage probability of probabilistic bias analysis adjusted estimates for confounding has been examined [5]. Similarly, the mean squared error and bias of exposure misclassification methods has been explored [6].

The values assigned to bias parameters may be the subject of disagreement among stakeholders, but it should often be the case that all stakeholders would agree that the values assumed by conventional methods, which assume there is no bias (or, equivalently, that all bias parameters are exactly equal to values that guarantee no bias), are very likely to be wrong. Simple bias analysis replaces these inherent and untenable fixed values assigned to bias parameters with more reasonable alternative values, though they too are fixed and likely unrealistic. Multidimensional bias analysis provides a sensitivity analysis of that bias analysis; it asks and answers the question of how the results and interpretation of those results would change as different sets of fixed values are assigned to the bias parameters. Probabilistic bias analysis and Bayesian bias analysis assigns credible and, hopefully more defensible, probability distributions to the values of the bias parameters (*e.g.,* distributions consistent with existing validation data). Thus, in addition to characterizing the main methods of probabilistic bias analysis we have presented as semi-Bayesian, one may also characterize bias analysis as an effort to achieve a more realistic estimate of the true association of some exposure on an outcome.

Caveats and Cautions

Given that most of the bias analysis methods described herein do not belong fully to either the frequentist or Bayesian schools of statistical methods and inference, one should be careful to avoid interpretations or inferences that rest on the assumption that they are solidly grounded in either school except when using the fully Bayesian approaches described in Chapter 11. For example, one should not characterize the simulation intervals described in Chapters 8 and 9 as the intervals within which the true parameter value is likely to lie with some stated probability (*e.g.*, 95%). Such an interpretation requires a fully Bayesian analysis. Likewise, one should not characterize the simulation interval as the interval which, with unlimited repetitions, will contain the true parameter value no less frequently than some stated probability (*e.g.*, 95%). While some of these intervals do have that property, it will depend on how well the analyst has specified the bias model.

The simulation interval is, instead, a reflection of the combined data and bias analysis assumptions. That is, estimation of the simulation interval begins with an assumption that the data were collected without intent to defraud or methodologic errors (*e.g.*, coding errors that reverse exposure categories). Second, estimation of the simulation interval requires a model of the relation between the observed data and systematic errors to be assessed (*e.g.*, that the observed data relate to classification errors through the equations provided in Chapter 6). Third, estimation of the simulation interval requires assumptions about the values to be assigned to the bias parameters used in the error model. These three together – observed data, error model, and values assigned to the model's bias parameters – fully define a simulation interval.

Comparison of the median of the simulations with the conventional point estimate of association provides some idea of the direction and magnitude of the systematic error acting on the conventional estimate of association, assuming that the model adequately depicts the influence of the systematic error and the values assigned to the bias parameters are near to the true values. Comparison of the limits of the simulation interval (*e.g.*, the 2.5th and 97.5th percentiles) with the limits of the conventional frequentist 95% confidence interval gives insight into the direction and magnitude of the systematic error acting on the conventional estimate of association. It also provides some insight into the understatement of total error reflected in the width of the conventional frequentist confidence interval, also assuming the model adequately depicts the influence of the systematic error and that the values assigned to the bias parameters are near to the true values. Simple bias analysis is just the point-density distribution simplification of the equivalent probabilistic bias analysis, and multidimensional bias analysis divides the probability density over several such point-density simplifications.

Utility

The major utility of bias analysis is as a safeguard against inferential errors resulting from overconfidence. When epidemiologic results appear precise and valid, interested parties may be tempted to action by unwarranted confidence in the accuracy and stability of the results. Bias analysis provides methods to test the susceptibility of the results to alternative assumptions about the strength of systematic errors. These analyses may reveal that the original confidence was overstated and should slow a rush to action. As noted in Chapter 1, the very act of bias analysis, by virtue of requiring alternative explanations for observed associations, should appropriately reduce confidence in research findings and provide a more accurate representation of the uncertainty we should have in our research findings.

A second utility of bias analysis is as an aid to identifying points of departure in interpretation among interested parties. For example, one party may believe that a result is unlikely to be susceptible to important classification error. A second party may believe that the same result is likely to be entirely attributable to errors in classification. Bias analysis allows these parties to compare how the association changes with different assumptions about the classification errors. In some cases, they may both learn that the association is not susceptible to substantial classification error, given the error model and reasonable choices for the values assigned to the bias parameters. In other cases, they may see that the assumptions about classification errors really do account for the difference in interpretation about the study results. Perhaps they disagree about the values that should be assigned to the bias parameters, or perhaps they disagree about the applicability of validation study results to the study setting. These disagreements provide guidance for further research [7], as reliable validation data should be a research priority to reconcile their two views. It also provides guidance on how to make tradeoffs in how to allocate resources for the next study. If, for example, measurement of a variable like the exposure or outcome is poor (e.g. low sensitivity) but does not lead to substantial bias in the results, whereas an unmeasured confounder leads to a shift in the point estimate as well as much additional uncertainty, future studies of the exposure outcome pair can prioritize collecting better data on the confounder over using a better measure of the misclassified exposure. Of course, care needs to be taken to ensure that the results of the bias analysis are generalizable to future studies in different populations, but still, the guidance bias analyses can provide in designing future studies can help rationalize the use of limited resources.

Bias analysis may therefore move the debate among stakeholders from the realm of qualitative criticism, which is often heavily influenced by politics and polemics, into the realm of quantitative analysis. Disagreements can move from dismissing research simply because a flaw can be identified, to a clear assessment of how much the resulting bias is likely to matter. If disagreements in the bias parameters' values leads to meaningful differences in the resulting estimates, then care needs to be taken to sort this issue out before strong inferences can be generated. But bias analysis can help prevent discarding potentially useful information simply because flaws existed. This approach to examining differences in interpretation of study results is far more consistent with the scientific enterprise, an inherently quantitative undertaking, than simple categorization of study results as "valid" or "invalid," since all studies are susceptible to imperfections [8].

Ultimately, bias analysis is a useful aid to inference. For example, when a bias analysis shows that systematic errors likely biased the measured association very little, inference from the research results can proceed with more confidence. Of course, one must always temper that confidence with the caveat that it is appropriate only so long as the assumptions about the error model and values assigned to the bias parameters are approximately accurate.

References

1. Fox MP, Lash TL. On the need for quantitative bias analysis in the peer-review process. Am J Epidemiol. 2017;185:865–8.
2. Greenland S. Bayesian perspectives for epidemiological research: I. Foundations and basic methods. Int J Epidemiol. 2006;35:765–75.
3. Greenland S. Bayesian perspectives for epidemiological research. II. Regression analysis. Int J Epidemiol. 2007;36:195–202.
4. Greenland S. Bayesian perspectives for epidemiologic research: III. Bias analysis via missing-data methods. Int J Epidemiol. 2009;38:1662–73.
5. McCandless LC, Gustafson P. A comparison of Bayesian and Monte Carlo sensitivity analysis for unmeasured confounding. Stat Med. 2017;36:2887–901.
6. MacLehose RF, Gustafson P. Is probabilistic bias analysis approximately Bayesian? Epidemiology. 2012;23:151–8.
7. Lash TL, Ahern TP. Bias analysis to guide new data collection. Int J Biostat. 2012;8:1–23.
8. Maldonado G. Adjusting a relative-risk estimate for study imperfections. J Epidemiol Community Health. 2008;62:655–63.

Index